Modeling of Dynamic Systems with Engineering Applications

This book provides cutting-edge insight into systems dynamics for both students and practicing engineers. Updated throughout for the second edition, this book serves as a firm foundation to develop expertise in design, prototyping, control, instrumentation, experimentation, and performance analysis.

Providing a clear discussion of system dynamics, this book enables students and professionals to both understand and subsequently model mechanical, thermal, fluid, electrical, and multi-domain (or multi-physics) systems in a systematic, unified, and integrated manner. Concepts of through- and across-variables are introduced and applied alongside tools of modeling and model representation in linear graphs. This book uses innovative worked examples and case studies alongside problems and exercises based on practical situations.

This book is a crucial companion to undergraduate and postgraduate engineering students alongside professionals in the engineering field. Complete solutions to end-of-chapter problems are provided in a solutions manual, which is available to instructors.

Modeling of Dynamic Systems with Engineering Applications

Second Edition

Clarence W. de Silva

CRC Press
Taylor & Francis Group
Boca Raton London New York

CRC Press is an imprint of the
Taylor & Francis Group, an **informa** business

Second edition published 2023
by CRC Press
6000 Broken Sound Parkway NW, Suite 300, Boca Raton, FL 33487-2742

and by CRC Press
4 Park Square, Milton Park, Abingdon, Oxon, OX14 4RN

CRC Press is an imprint of Taylor & Francis Group, LLC

First edition published by CRC Press 2023

ISBN: 9780367644215 (hbk)
ISBN: 9780367644284 (pbk)
ISBN: 9781003124474 (ebk)

DOI: 10.1201/9781003124474

Typeset in Times
by codeMantra

Access the Support Material: https://www.routledge.com/9780367644215

To all my teachers, some of whom are unknown to me.

Contents

Preface

This is the second edition of the book *Modeling of Dynamic Systems with Engineering Applications*. It concerns the analytical modeling of multi-physics systems (dynamic systems that consist of multiple physical domains). This is indeed considered the "mechatronic" approach to modeling, which is characterized by the integrated, unified, and systematic nature, leading a unique (i.e., single) model. This book systematically covers methodologies of understanding and analytical representation, or analytical modeling, of the dynamics of a physical engineering or mechatronic system, using proper principles of science. However, the presented concepts and approaches are applicable in non-engineering processes such as biological, economic, and social systems as well.

This book has all the features of a course textbook and is primarily intended for a course at the undergraduate level (typically, third or fourth year) or at the early graduate level. In many engineering curricula, a course in modeling is a prerequisite for the first course in control systems. In some other curricula, modeling is taught as a foundation course or as the first part of a control systems course. Also, it is an indispensable component of a curriculum in mechatronics. Since this book contains a wealth of practical information on the subject and illustrative examples of applying the developed methodologies in practical engineering systems, it is a valuable and practical reference source as well, primarily for researchers and practicing professionals. The primary emphasis of this book is on the engineering problem of model development rather than response analysis and simulation once a model is available. This book distinguishes itself from the existing books on dynamic system modeling in view of the following primary features:

- It concerns modeling of multi-physics (or multi-domain) systems consisting particularly of the mechanical electrical, fluid, and thermal domains.
- The presented modeling methodologies are centered on the use of integrated, unified, and systematic procedures that lead to a unique model. Integrated means all physical domains are considered simultaneously (or concurrently). Unified means analogous (or similar) methods are used for modeling different physical domains. Unique means only a single analytical model results when applying the procedures.
- The underlying approach modeling removes any doubt or confusion concerning what modeling method should be used in a given problem and the validity and the value of the end result of modeling.
- The presented "unified" and "integrated" approach to the modeling of multi-physics systems is rapidly becoming the standard in the modeling of mechatronic engineering systems, and of any engineering system, for that matter.
- The presented methodologies of modeling are properly justified.
- Modeling approaches that are commonly and effectively used in electrical engineering are extended to other physical domains, particularly mechanical, fluid, and thermal domains, so that those methods can also be applied to multi-domain (e.g., mechatronic or electromechanical or mixed) systems, in a unified manner.
- Equivalence or approximate equivalence (to the actual physical system or to another type of model) is considered as the primary basis in developing "equivalent models" and in "model reduction" using various criteria of equivalence, as presented in this book.

BACKGROUND

In the late 1970s, I taught a mandatory undergraduate course in dynamic system modeling at Carnegie Mellon University. The popular textbook *Introduction to System Dynamics* by Shearer, Murphy, and Richardson (Addison-Wesley, 1971) was used in this course. This excellent classic has not been revised to date (The late Professor Arthur Murphy had contacted me to undertake the revision, but that project did not materialize due to my other commitments and the untimely demise of Professor Murphy). After I moved to the University of British Columbia (UBC) in the late 1980s, I continued to teach the course. During this process, I had developed a wealth of material (including new approaches, extensions to existing approaches, new examples, problems, and projects). Subsequently, at UBC, there was a demand for an introductory graduate-level course as well in the subject, which materialized about 25 years ago.

At UBC, for teaching the course on modeling in the beginning, I used my own notes and a booklet on *Control System Modeling*, which I had prepared for the company Measurements & Data Corporation (Pittsburgh, PA) and serialized in their magazine (*Measurements and Control*) as a professional course. Subsequently, by incorporating as well the material that I had developed while teaching undergraduate and graduate courses in control systems, I published the book *Modeling and Control of Engineering Systems* (Taylor & Francis/CRC Press, 2009). Yet, I constantly felt the need for a single textbook on the subject of modeling of dynamic engineering systems that carries the features I have listed above. In 2018, this objective was materialized. This book is the new edition of *Modeling and Control of Engineering Systems*, where new material (procedure details, worked examples, and end-of-chapter exercises, in particular) has been incorporated and some supplementary material has been removed.

SCOPE OF THIS BOOK

Through this book, the student will learn to understand and model mechanical, thermal, fluid, electrical, and multi-physics (or multi-domain or mixed) systems in a systematic, unified, and integrated manner, leading to a unique (single) analytical model. For example, in this book, I explore the identification of lumped elements such as generalized sources (or input elements, both across-variable and through-variable sources), generalized capacitors (across-type energy storage elements), generalized inductors (through-type energy storage elements), and generalized resistors (energy dissipation elements) in different types of physical systems. I study analogies among the four main types of systems: mechanical, thermal, fluid, and electrical, in terms of these basic lumped elements and in terms of the system variables, leading to "unified" methodologies for all physical domains. I introduce and apply concepts of through- and across-variables. I study multi-physics (or multi-domain or mixed) systems, which consist of two or more of the basic system types (or physical domains such as mechanical, electrical, fluid, and thermal), as well.

The key procedures presented in this book are primarily applicable to linear systems. Since practical engineering systems are nonlinear to varying degrees, I treat the linearization of nonlinear systems in sufficient detail. Primarily, I present analytical and experimental/graphical methods of linearization. To supplement this knowledge, I present several other applicable approaches of linearization (e.g., energy equivalence and describing function methods) as well.

Even though the emphasis of this book is on lumped-parameter models, I address the treatment of distributed-parameter systems and their approximation using lumped-parameter models.

I present the graphical tool of modeling and model representation, linear graphs, which can be provided as a MATLAB® tool (see Appendix B). A main focus of this book is to develop a unique state-space model for a given system (*Note*: Generally, the state-space representation is not unique; many different state-space models may be presented for the same system) by following a unified, integrated, and systematic approach. I examine important considerations of input, output, causality, and system order.

The linear graph model representation, in the time domain, is extended to the frequency domain. I study Thevenin and Norton equivalent circuits and their application in nonelectrical (mechanical, fluid, and thermal) systems using linear graphs (through the concepts of transfer function linear graphs). I study the conversion of a mixed-domain (multi-physics) system model into an equivalent model in a single domain (typically the domain of the output segment of the system) and illustrate its application in practical systems using examples.

MAIN FEATURES

In the present context, modeling concerns understanding and analytical representation of the dynamics of a physical engineering system, using sound principles of science and through an integrated, unified, and systematic approach. The developed model must be suitable for meeting the subsequent purposes and tasks. For example, identification and selection of system components, system analysis and computer simulation, conceptual design, detailed design, prototyping, instrumentation, control, tuning (adjusting system parameters to obtain the required performance), testing, performance evaluation, and product qualification all are important tasks in engineering practice, and modeling plays a crucial and primarily role in all these tasks.

The main objective of this book is to provide a convenient, useful, and affordable textbook in the subject of *Modeling of Dynamic Systems with Engineering Applications*. The material presented in this book serves as a firm foundation for the subsequent building up of expertise in various aspects of engineering such as design, prototyping, control, instrumentation, experimentation, and performance analysis.

This book consists of seven chapters and two appendices. To maintain clarity and focus and to maximize the usefulness, this book presents its material in a manner that will be useful to anyone with a basic engineering background, be it civil, electrical, mechanical, manufacturing, material, mechatronic, mining, aerospace, or biomechanical. Complete solutions to the end-of-chapter problems are provided in a solutions manual, which is available to instructors who adopt this book.

In addition to presenting standard material on the modeling of dynamic engineering systems in a student-friendly and interest-arousing manner, this book somewhat deviates from other books on the subject in the following ways:

- This book presents systematic approaches of modeling, in both time domain and frequency domain, which lead to unique models (thereby removing the doubts on what method should be used in a given problem and the validity of the end result of modeling).
- This book provides modeling approaches that are equally applicable to engineering systems that contain one or more physical domains (electrical, mechanical, fluid, and thermal, in particular). Problems of multiple physical domains are also called multi-physics systems (or mixed systems).
- Since all the physical domains in the system are treated together, simultaneously or concurrently, while taking into account any dynamic coupling (dynamic interactions) among the domains, the presented modeling approach is "integrated." Since similar (or analogous) approaches are used for modeling different physical domains, the modeling approach is "unified." Since clear steps of modeling are used, the approach is "systematic." Since one analytical model is realized by following the approach, the approach is "unique."
- Popular modeling approaches that are commonly and effectively used in electrical engineering are extended to other domains, particularly mechanical, fluid, and thermal domains, so that they can also be applied to multi-physics (e.g., mechatronic or electromechanical) systems.
- I present linear graph (LG) approaches as well of model development, in view of the significant advantages of linear graphs. A MATLAB® toolbox of linear graphs is presented as well (in Appendix B), with examples.
- I present the "generalized" use of linear graphs, as graph-tree concepts, in Appendix A.
- I present physical principles and analytical methods using mathematics that is quite familiar to engineering students.
- I provide a large number of solved examples, analytical examples, numerical examples, and end-of-chapter problems (with solutions throughout this book) and relate them to real-life situations and practical engineering applications.

- I summarize the key issues presented in this book in point form at various places in each chapter, for easy reference, recollection, and presentation in PowerPoint form.
- I indicate the topics covered in each chapter at the beginning of the chapter. I provide the key material, formulas, and results in each chapter in a summary sheet at the end of the chapter.
- This book is concise, avoiding unnecessarily lengthy and uninteresting discussions, for easy reference and comprehension.
- There is adequate material in this book for two 12-week courses, one at the undergraduate level and the other at the introductory graduate level.
- In view of the practical considerations and techniques, tools, design issues, and engineering information presented throughout this book, and in view of the simplified and snapshot style presentation, including more advanced theory and techniques, this book serves as a useful reference source for engineers, technicians, project managers, and other practicing professionals in industry and research institutions.

Clarence W. de Silva
Vancouver, Canada

Acknowledgments

Many individuals have assisted in the preparation of this book, but it is not practical to acknowledge all such assistance here. First, I wish to recognize the contributions, both direct and direct, of my graduate students, research associates, and technical staff. I am particularly grateful to Nicola Sharpe, Editor, Mechanical Engineering, at CRC Press/Taylor & Francis Group, for her great enthusiasm and support throughout the project of preparing the second edition of this book. Other editors and support personnel of CRC Press also should be acknowledged here for their contribution in the production of this book. Finally, I wish to acknowledge here the unwavering love and support of my late wife and children.

Author

Dr. Clarence W. de Silva, PEng, Fellow ASME, Fellow IEEE, Fellow Canadian Academy of Engineering, Fellow Royal Society of Canada, is a Professor of Mechanical Engineering at the University of British Columbia, Vancouver, Canada since 1988. He has occupied the following chair professorships:

- Senior Canada Research Chair Professorship in mechatronics and industrial automation
- NSERC-BC Packers Research Chair in industrial automation
- Mobil Endowed Chair Professorship

He has served as a faculty member at Carnegie Mellon University (1978–1987) and as a Fulbright Visiting Professor at the University of Cambridge (1987–1988).

He has earned PhD from Massachusetts Institute of Technology (1978) and the University of Cambridge, England (1998), the Higher Doctorate, ScD from the University of Cambridge (2020), and an honorary DEng degree from University of Waterloo (2008).

- **Other Fellowships:** Lilly Fellow at Carnegie Mellon University; NASA-ASEE Fellow; Senior Fulbright Fellow at Cambridge University; Fellow of the Advanced Systems Institute of British Columbia; Killam Fellow; Erskine Fellow at the University of Canterbury, New Zealand; Professorial Fellow at University of Melbourne; and Peter Wall Scholar at the University of British Columbia.
- **Awards:** Paynter Outstanding Investigator Award and Takahashi Education Award, ASME Dynamic Systems & Control Division; Killam Research Prize; Outstanding Engineering Educator Award, IEEE Canada; Lifetime Achievement Award, World Automation Congress; IEEE Third Millennium Medal; Meritorious Achievement Award, Association of Professional Engineers of BC; and Outstanding Contribution Award, IEEE Systems, Man, and Cybernetics Society. Also, he has made 49 keynote addresses at international conferences.
- **Editorial Duties:** Served on 14 journals including *IEEE Transactions on Control System Technology, Journal of Dynamic Systems, Measurement & Control, Transactions of the ASME* and *IEEE-ASME Transactions on Mechatronics*; editor-in-chief, *International Journal of Control and Intelligent Systems*; editor-in-chief, *International Journal of Knowledge-Based Intelligent Engineering Systems*; senior technical editor, *Measurements and Control*; and regional editor, North America, *Engineering Applications of Artificial Intelligence—IFAC International Journal.*
- **Publications:** 25 technical books, 19 edited books, 51 book chapters, over 300 journal articles, and a similar number of conference papers.
- **Recent Books:** *Sensor Systems* (Taylor & Francis/CRC, 2017); *Sensors and Actuators: Engineering System Instrumentation* (2nd Ed., Taylor & Francis/CRC, 2016); *Mechanics of Materials* (Taylor & Francis/CRC, 2014), *Mechatronics: A Foundation Course* (Taylor & Francis/ CRC, 2010); *Modeling and Control of Engineering Systems* (Taylor & Francis/ CRC, 2009); *Vibration: Fundamentals and Practice,* (2nd Ed., Taylor & Francis/CRC, 2007); *Mechatronics; An Integrated Approach* (Taylor & Francis/CRC, 2005); *Soft Computing and Intelligent Systems Design: Theory, Tools, and Applications* (with F. Karray, Addison Wesley, 2004); and *Force and Position Control of Mechatronic Systems—Design and Applications in Medical Devices*, Springer, (with T.H. Lee, W. Liang, and K. Tan, Springer, 2021).

1 Introduction to Modeling

HIGHLIGHTS

- Objectives of the Chapter
- Importance and Applications of Modeling
- Modeling in Control
- Modeling in Design
- Dynamic Systems and Models
- Model Complexity
- Model Types
- Analytical Models
- Mechatronic Systems
- Steps of Analytical Model Development

1.1 OBJECTIVES

This book concerns the modeling of engineering dynamic systems. A model is a representation of an actual system. In this context, first we need to explore what a *dynamic system* (an engineering system in this case) is; what is specifically meant by *modeling*; what types of modeling are possible, and how to model a dynamic system. We will address all four topics in detail throughout this book. In brief, a dynamic system is a system where the "rates of changes" of its response variables (outputs) cannot be neglected. There are many types of engineering dynamic systems and many types of models, as we will learn. This book primarily concerns analytical modeling. We will primarily focus on "lumped-parameter" models, which depend on the independent variable "time," the more general "distributed-parameter" models, which have both time and space as the independent variables, are given some attention. We will learn a way to develop an analytical model that has the four characteristics: integrated, unified, unique, and systematic, for an engineering dynamic system. We will explore these four characteristics in detail, in this book.

The main learning objectives of this book are the following:

- Understand the formal meanings of a dynamic system, control system, mechatronic system, and multi-physics (or, multi-domain or mixed) system.
- Recognize different types of models (e.g., physical, analytical, computer, experimental) and their importance, usage, comparative advantages and disadvantages.
- Under analytical models, recognize the general and specific pairs of model categories.
- Learn the concepts of input (excitation), output (response), causality (cause–effect nature, what are inputs, and what are outputs in the system), and order (dynamic size) in the context of a dynamic system (or dynamic model).
- Understand the concepts of through-variables and across-variables, their physical significance, and relationship to state variables.
- Recognize similarities or analogies among the four physical domains: mechanical, electrical, fluid, and thermal (this is the basis of the "unified" approach to modeling).
- In each physical domain, recognize the lumped elements that store energy and that dissipate energy, based on the analogy among different physical domains.
- In each physical domain, recognize different types of source (input) elements, which possess independent input variables and are able to apply them to other components of a system, based on the analogy among different physical domains.

DOI: 10.1201/9781003124474-1

- Understand the "mechatronic" approach (i.e., the "integrated" or "concurrent" approach) to modeling a multi-physics (or multi-domain or mixed) system, which consists of two or more basic physical domains. Integrated means, all domains are modeled (and designed) simultaneously.
- Understand the "unified" approach to modeling a multi-domain system. Unified means, similar (i.e., analogous) methods are used to model the different physical domains in the system.
- Understand the meaning of state variables and the selection of them in a "unique" manner to generate a unique state-space model.
- Learn to apply the unified and integrated approach of modeling, in a systematic way, to develop a "unique" state-space model. Systematic means, the modeling steps are clear and there is no uncertainty associated with it. Unique means, a single model is obtained at the end.
- Understand the key steps of development of a unified, integrated, systematic, and unique approach for modeling an engineering dynamic system. Learn to develop state-space models using that approach, while using physically meaningful state variables that lead to a "unique" state-space model. Also, understand the physical meaning of "system order" or the dynamic size.
- Learn how to convert a state-space model into an input–output model, in the time domain.
- Learn to obtain a linear model of a nonlinear dynamic system, both analytically and experimentally. In the analytical context, learn different approaches to linearize a nonlinear system or model, particularly the slope-based local linearization and the energy-based global linearization.
- Understand and apply a graphical approach that uses linear graphs, to develop a state-space model.
- Understand the frequency-domain concepts of modeling; particularly, the concepts of "generalized" impedance, equivalent circuits, and circuit reduction of electrical systems (Thevenin and Norton concepts of equivalent circuits) and transfer function linear graphs (TFLGs) and apply them to mechanical, fluid, thermal, and multi-physics systems.
- Learn a systematic way to convert a model of a multi-physics system to an equivalent model of a single physical domain, which is preferably the output domain of the system. For this purpose, learn to use the concepts of energy transfer (or coupling) through generalized transformers and generalized gyrators.
- Gain the ability to relate the learned concepts of modeling to model a mechatronic system. For this purpose, understand the value of a more generalized definition of a mechatronic system.

Design, development, modification, implementation, operation, control, and performance monitoring and evaluation of an engineering system require a sufficient understanding of the system and a suitable "representation" of the system. In other words, a "model" of the system is required for these practical activities. A model is a convenient representation of the actual system. Properties established and results derived in various "model-based approaches" are associated with the model rather than the actual system, whereas it is to the actual system that the excitations (inputs) are applied to and the responses (outputs) are observed or measured from. This distinction is very important, particularly in the context of the treatment in this book. However, as customary, the terms the model and the system are often used interchangeably to refer to the model. This fact can be easily recognized depending on the specific context that is addressed and is usually not confusing.

An engineering system may consist of several different physical types of components, belonging to such physical "domains" as mechanical, electrical, fluid, and thermal. It is termed a *multi-physics* (or *multi-domain* or *mixed*) *system*. Furthermore, it may contain *multifunctional components*; for example, a piezoelectric component, which can interchangeably function as both a sensor and an actuator, is a multifunctional device. It is desirable to use analogous procedures in the modeling of multi-physics and multifunctional components. Then the individual *physical-domain models* or *functional models* can be developed using "unified" or "analogous" methodologies across the physical domains while considering all the physical domains simultaneously (i.e., "concurrent" or "integrated" manner), systematically, to obtain a "unique" (i.e., the "best" single) overall model.

Analytical models may be developed for mechanical, electrical, fluid, and thermal systems in a rather analogous manner, using the mentioned "unified" approach because clear analogies exist among these four types of systems and in their variables. This is an important focus of this book. In view of the existing *analogy*, then, a unified (analogous), integrated (concurrent), and systematic (having clear steps) approach may be adopted in the modeling, analysis, design, control, and evaluation of an engineering system. This integrated and unified approach is indeed the "mechatronic" approach to modeling. The "unified" approach goes beyond the conventional mechatronic procedures (which are "integrated" yet may not be unified) and exploits the similarities (analogies) of different physical domains of the system. In summary then, the studies and developments of this book target a modeling approach that has the following characteristics:

- Integrated (concurrent or simultaneous; considers all physical domains of the system simultaneously, while including "coupling" or "dynamic interactions" or "energy conversion" that exist among them)
- Unified (exploits analogies or similarities among different physical domains and uses similar/analogous procedures to model the dynamics in those physical domains)
- Systematic (follows a clearly indicated sequence of modeling steps, without any confusion as to the approach)
- Realization of a "unique" model (the modeling procedure leads to a single "best" model). This implicitly implies that some form of "optimization" is associated with the used procedures
- Physically meaningful (e.g., the system variables, particularly the state variables, are not chosen arbitrarily, and have physical meaning, and furthermore, it leads to a clear understanding of the dynamic size or "order" of the system).

1.1.1 Model Error of Science Error

A model is a representation of the actual system, with sufficient accuracy. Hence, there bound to be errors in the model, when compared to the real system, but often they can be neglected depending on the purpose of the model. Also, a model represents the physical phenomena in the actual system (again to an acceptable level of accuracy). Such phenomena are never in error, but our understanding and representation/formulation of these phenomena may not be accurate or may be evolving. So, the science itself can have errors, which are corrected from time to time.

1.2 IMPORTANCE AND APPLICATIONS OF MODELING

A dynamic model may be indispensable in a variety of engineering applications. The types of uses of a dynamic model include the following:

- Analysis of a dynamic system (particularly using mathematical methods and tools), even when the actual system is not available or developed yet
- Computer simulation, which can incorporate various types of models including mathematical (analytical) dynamic models and even some physical hardware (i.e., hardware-in-the-loop or SIL simulation)
- Determination of the required design of a dynamic system, prior to building the system (in fact it may assist in making the decision whether to build or not)

- Determination of the required modification of a dynamic system (or its model or the design), prior to the actual task of physical modification of the system
- Instrumentation (i.e., the exercise of "instrumenting") of a dynamic system. Specifically, instruments (such as sensors, actuators, and signal conditioning and component interconnecting hardware) needed for the operation and/or performance improvement of a dynamic system may be established (i.e., selected or sized) and analyzed through modeling and simulation
- Control or assistance in the physical operation of a dynamic system (e.g., for model-based control and for generating control signals and performance specifications)
- Testing of a dynamic system (where a test regiment is developed and evaluated through analytical and computational means) and in product qualification (where an available good-quality product is further tested and evaluated to determine whether it is suitable for a specialized application (e.g., seismic qualification of the components of a nuclear power plant; qualification of computer hardware for shipment))
- Performance evaluation (including online monitoring) of a system to detect deviations and diagnose malfunctions and faults (using a model as the reference for good performance).

Dynamic modeling is applicable in all branches of engineering (aerospace, biomechanical/medical, chemical, civil, electrical and computer, manufacturing, material, mechanical, mechatronic, mining, etc.) and even in non-engineering systems (e.g., social, economic, administrative, environmental). Analytical models are quite useful in predicting the dynamic behavior (response) of a system for various types of excitations (inputs). For example, vibration is a dynamic phenomenon and its analysis, practical utilization, and effective control require a good understanding (model) of the vibrating system. Computer-based studies (e.g., computer simulation) may be carried out using analytical models (sometimes incorporating some physical hardware as well—hardware in the loop or SIL simulation) while using suitable values for the system parameters (mass, stiffness, damping, capacitance, inductance, resistance, fluid inductance, or inertance and so on). A model may be employed when designing an engineering system for proper performance. Then, the system is first developed (designed) using a model, which is much easier (quick, flexible, inexpensive) to modify than a physical system or prototype. In the context of *product testing*, for example, analytical models are commonly used to develop test specifications and the input signals that are applied by the exciter in the test procedure. Dynamic effects and interactions in the test object, the excitation system, and their interfaces may be studied in this manner. Product qualification is a particular situation of this, which is the procedure that is used to establish the capability of a good-quality product to withstand a specified set of operating conditions, in a specialized application. In product qualification by testing, the operating conditions are generated and applied to the test object by an exciter (e.g., shaker). In product qualification by analysis, a suitable analytical model of the product replaces the physical test specimen that is used in product qualification by testing. In the area of automatic control, models are used in a variety of ways, as discussed next. In machine health monitoring, a bank of models that represent a non-faulty machine and that has different types of faults or malfunctions are used to assess the performance of the actual machine and determine whether it is sound or faulty. Such procedures are able to diagnose a fault as well.

1.2.1 The Use of Models in Control

The ways models are used in automatic control include the following:

- An analytical model of the control system is needed for representation, mathematical analysis, and computer simulation of the system
- A model of the system to be controlled (i.e., the plant or the process) may be used to develop the performance specifications, based on which a controller is developed for

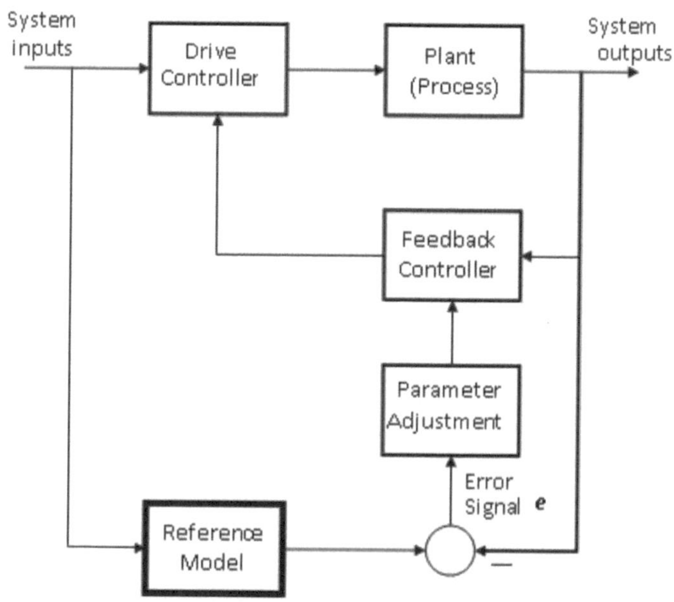

FIGURE 1.1 The scheme of model-referenced adaptive control.

the system. For example, in *model-referenced adaptive control*, a "reference model" dictates the desired behavior that is expected under control (see Figure 1.1). This is an implicit way of using a model to represent a set of required performance specifications. Then the controller seeks (by changing or adapting the controller parameters) to drive the actual behavior of the plant toward the desired behavior as dictated by the reference model

- In *model-based control*, a dynamic model of the actual process is employed to develop the necessary control schemes and control actions
- In the early stages of design of a control system, some (or all) parts of the desired system may not exist. In this context, a model of the anticipated system, particularly an analytical model or a computer model, can be very useful, economical, and time-efficient in the design of the control system. In view of the complexity of a design process, particularly when striving for an optimal design, it is useful to incorporate system modeling as a tool for design iteration.

In general, modeling is an "optimistic" approach where we attempt to accurately represent the system or the required system behavior by a model, possibly optimize the model, and then use it to design, implement (including instrument or modify the system), and evaluate the system while expecting (optimistically) that the implemented system will operate faithfully as represented by the model. On the other hand, control is a "pessimistic" approach where we suspect that the system will not behave according to the requirements (possibly as specified by a model) and then use sensing/monitoring and feedback control to make sure that the system performance meets the expectations.

1.2.2 The Use of Models in Design

Modeling and design can go hand in hand, in an iterative manner, both in the conceptual design and the detailed design. Of course, in the beginning of a design process, the desired system does not exist. Then, a model of the anticipated system can be very useful. In view of the complexity of a design process, particularly when striving for an optimal design, it is useful to incorporate system modeling as a tool for design iteration particularly because prototyping can become very costly and time-consuming. Initially, by knowing some information about the system (e.g., intended functions, performance specifications, past experience, and knowledge of related or similar systems) and using the design objectives, it is possible to develop a model of sufficient (low to moderate) detail and complexity. Through mathematical analysis and by carrying out computer simulations of the model, it will be possible to generate useful information that will guide the design process (e.g., generation of a conceptual or preliminary design). In this manner, iteratively, design decisions can be made and the model can be refined using the available (improved) design. This iterative link between modeling and design is schematically shown in Figure 1.2a.

It is expected that an integrated (i.e., considering all physical domains concurrently or simultaneously, while giving proper accounting for dynamic interactions and energy transfer between them) and unified (i.e., using analogous methodologies for different physical domains) approach (the mechatronic approach) for modeling and design will result in many benefits. It can be verified that the integrated and unified approach will result in cost-effective and high-quality products and services with optimally (or at least better than average) matched components, improved performance and controllability, and increased reliability, efficiency, and product life. Such products approach some form of optimality and are more compatible and easier to interact/interface with other similarly designed systems. From the design perspective alone, this will enable the development and production of multi-physics (or, multi-domain or mixed, e.g., electromechanical) systems efficiently, rapidly, and economically.

When performing a mechatronic design of a multi-physics system, the concepts of energy or power present a unifying thread. The reasons are clear. First, in a multi-physics system, ports of power transfer or energy transfer exist, which link the dynamics of various physical domains (mechanical, electrical, fluid, and thermal). Hence, modeling, analysis, and optimization of a multi-physics system can be carried out by using a hybrid-system formulation (a model) that integrates various aspects of different physical domains of the system in a unified manner. Second, an optimal design may aim for minimal energy dissipation and maximum energy efficiency. There are related implications; for example, greater dissipation of energy will mean reduced overall efficiency and increased thermal problems, noise, vibration, malfunctions, wear and tear, and increased environmental impact. Again, a hybrid model that presents an accurate picture of the energy/power flow within the system will present an appropriate framework for the multi-physics design. We can use linear graph models in particular, as discussed in Chapter 5, for this purpose, in an effective manner.

A design may use excessive safety factors and worst-case specifications (e.g., for mechanical loads and electrical loads). This will not provide an optimal design and may not lead to the most efficient performance. Designing for optimal performance may not necessarily lead to the most economical (least costly) design, however. When arriving at a truly optimal design, an objective function that takes into account all important factors (performance, quality, cost, speed, ease of operation, safety, environmental impact, etc.) has to be optimized. This is indeed the approach of *mechatronic design quotient* (MDQ). A complete design process for an engineering system should generate the necessary details for production or assembly of the system.

Note: Historically, an "integrated" approach that treats all physical domains concurrently was termed a "mechatronic" approach. An established definition of mechatronics is "synergistic application of mechanics, electronics, control engineering, and computer science in the development of electromechanical products and systems, through integrated design." However, in this book, we will generalize this definition by including all possible physical domains (not just electrical and mechanical) and call a modeling approach that is both integrated and unified (i.e., one that exploits domain similarities/analogies) while following systematic steps, leading to a single (unique) model as mechatronic modeling.

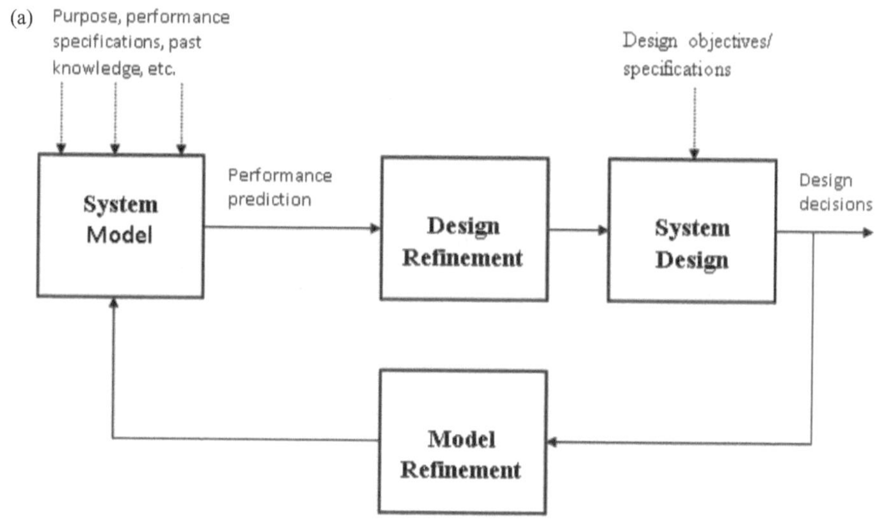

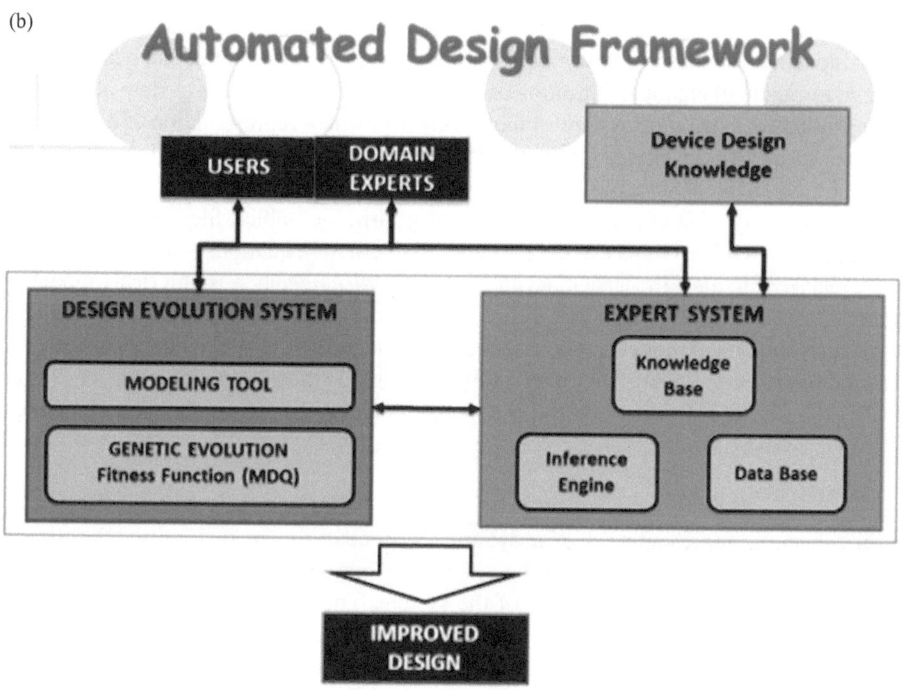

FIGURE 1.2 (a) Link between modeling and design; (b) An integrated framework with modeling, intelligent design, and optimization.

A more general framework that integrates modeling, intelligent design, and system optimization is shown in Figure 1.2b. Here, the task of design is carried out iteratively by an expert system, which incorporates the design knowledge of experts in the domain. In this manner, a design is "evolved," as more information is known (through available improvements to the design knowledge base, monitoring of the physical system or a prototype, if available, for design weaknesses, etc.). The overall design is optimized as well, using an evolutionary computing approach. Such schemes have been implemented and tested, but that subject is outside the scope of this book.

A model of the system will be useful in this context as well. The steps of the design evolution for a system that exists are as follows:

1. Develop a model of the existing physical system
2. Establish (using a design expert system and possibly a machine health monitoring system) what aspects or segments of the original system (represented by its model) may be modified/improved. These will provide "modifiable sites" for the existing system/model
3. Formulate a performance function to represent the "goodness" of the design. This is the Mechatronic Design Quotient (MDQ)
4. Use an optimization method (e.g., Genetic Programming or GP) to evolve the model so as to maximize the performance function
5. Implement in the existing physical system the design changes represented by the evolved model.

1.3 DYNAMIC SYSTEMS AND MODELS

Each interacted component or element of an engineering system will possess an *input–output* (or *cause–effect*, or *causal*) relationship. A "*dynamic*" *system* is one whose response variables (which are functions of time) have non-negligible "rates" of change. Also, its present output depends not only on the present input but also on some historical information (e.g., previous input and hence, previous output). A more formal mathematical definition can be given, but it is adequate to state here that a typical engineering system is a dynamic system. A model is some form of representation of a practical system. In particular, an *analytical model* (or *mathematical model*) comprises equations (e.g., differential equations) or an equivalent set of information, which represents the system to a required degree of accuracy. Alternatively, a set of curves, digital data (e.g., arrays or tables, files) stored in a computer or in some medium, and other numerical data—rather than a set of equations—may be termed a model, strictly a *numerical model.* In particular, an *experimental model* is a (numerical) model obtained from a set of test data (input–output data) through physical experimentation with the actual system. A representative analytical model can be established or "identified" from a numerical model (or from experimental input–output data). The related subject is called "model identification" or "system identification" in the field of automatic control. A dynamic model is a representation of a dynamic system.

1.3.1 TERMINOLOGY

A general schematic representation of a dynamic system is shown in Figure 1.3. The system is demarcated by a *boundary*, which may be either real (physical) or imaginary (virtual). What is outside this boundary is the *environment* of the system. There are *inputs*, which enter the system

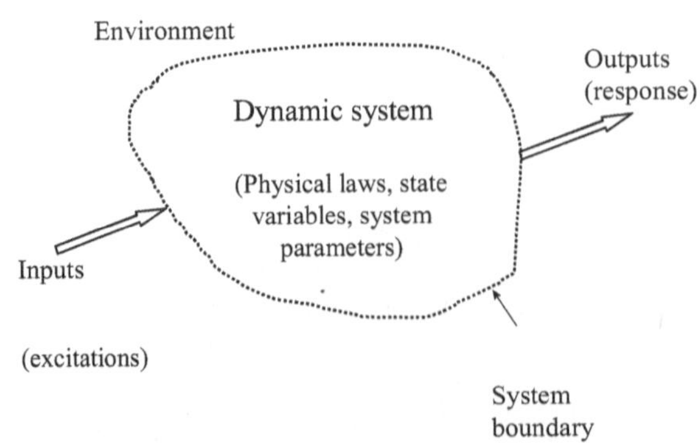

FIGURE 1.3 Nomenclature of a dynamic system.

from the environment, and there are *outputs*, which are generated by the system and provided to the environment. Some useful terms concerning dynamic systems are defined below.

- **System**: Collection of interacting components of interest, demarcated by a system boundary.
- **Dynamic System**: A system whose rates of changes of response/state variables cannot be neglected.
- **Plant or Process**: The system to be controlled.
- **Inputs**: Excitations (known or unknown; deliberate/desired or undesired) applied to the system.
- **Outputs**: Responses (desired or undesired) of the system.
- **State Variables**: A minimal set of variables that completely identify the "dynamic" state of a system. *Note*: If the state variables at one state in time and the inputs from that state up to a future state in time are known, the future state can be completely determined.
- **Control System**: A system that includes at least the plant and its controller. It may include other subsystems and components (e.g., sensors, signal conditioning, and modification devices).

Dynamic systems are not necessarily limited to engineering, physical, or man-made systems (e.g., social, economic, legal, management, political, and environmental, which may depend on engineering systems) even though the focus of this book is on engineering systems. Examples of dynamic systems with some inputs and outputs are given in Table 1.1. Exercise: add to this list some other inputs and outputs (intentional, unintentional, desirable, undesirable).

1.3.2 Model Complexity

It is unrealistic to attempt to develop a "universal model" that will incorporate all conceivable aspects of the system. For example, an automobile model that simultaneously represents ride quality, power, speed, energy consumption, emission, traction characteristics, handling, structural strength, capacity, load characteristics, cost, safety, and so on is not very practical and can be intractably complex. Furthermore, some aspects incorporated into the model can mask the behavior of other important aspects. The model should be as simple as possible and may address only a few specific

TABLE 1.1
Examples of Dynamic Systems

System	Examples of Inputs	Examples of Outputs
Human body	Neuro-electric pulses (due to external stimuli), desirable stimuli, distractions	Muscle contractions, body movements, cursing, unlawful activity
Company	Information that is used for operation, executive commands, weakening of stock market, price increase of raw material, drop in demand, or product market value (e.g., due to deteriorated product quality, alternative products, price)	Products, output information (on products that are physical or otherwise, etc.), decisions, wastage, product quality
Power plant	Fuel rate, environmental conditions (e.g., temperature), costs (personnel, material, etc.)	Electric power, pollution, reliability, noise
Automobile	Steering wheel movement, fuel, road disturbances, automobile accidents, weather, traffic, servicing	Front wheel turn, direction of heading, ride quality, noise, emissions, motion causing an accident, wear and tear
Robot	Voltage to joint motors, external force on end effector (e.g., from a collision), noise in input signal, obstacles	Robot motions (trajectory, pose or heading), accidents, malfunctions

Note: Categorize these inputs and outputs as: intentional, unintentional, known, unknown, desirable, and undesirable, as applicable.

aspects of interest in the particular study or application. Approximate modeling and model reduction are relevant topics in this context.

As an example, consider a hard-disk drive (HDD) unit, as shown in Figure 1.4a. If the objective is vibration analysis and control, a simplified model such as the one shown in Figure 1.4b or c may be used. In particular, Figure 1.4b provides a two-dimensional, multi-degree-of-freedom mechanical model where some of the inertia effects are modeled as lumped masses and the flexibility effects are represented by rectilinear springs. The mounting frame and the disk may be modeled as plates. Energy dissipation (damping) is not modeled. Figure 1.4c is a further simplified, one-dimensional, two-degree-of-freedom mechanical model. It is an entirely lumped-parameter model, and again energy dissipation is not modeled.

As another example, consider the innovative elevated-guideway transit system shown in Figure 1.5a, which has been extensively studied by me in a previous investigation. This is an automated transit system that is operated without drivers. The ride quality, which depends on the motion of the vehicle, may be analyzed using an appropriate model. Usually the dynamics (inertia, flexibility, and energy dissipation) of both the guideway and the vehicle, with their dynamic interactions, have to be incorporated into such a model. In a modeling exercise, first, the system is simplified by incorporating just one car of the train, as shown in Figure 1.5b. A further simplified model of it that may be used in a ride quality study is shown in Figure 1.5c. This is a lumped-parameter model where the distributed characteristics (e.g., mass) are approximated by lumped elements.

This model does not consider the dynamics of the guideway.

As the third example, consider the industrial machine shown in Figure 1.6a, which has been developed by us for the head removal of salmon. The conveyor, driven by an ac (induction) motor, indexes the fish in an intermittent manner. The image of each fish, obtained using a digital camera, is processed to determine the geometric features, which in turn establish the proper cutting location. A two-axis hydraulic drive unit positions the cutter accordingly, and the cutting blade is operated using a pneumatic

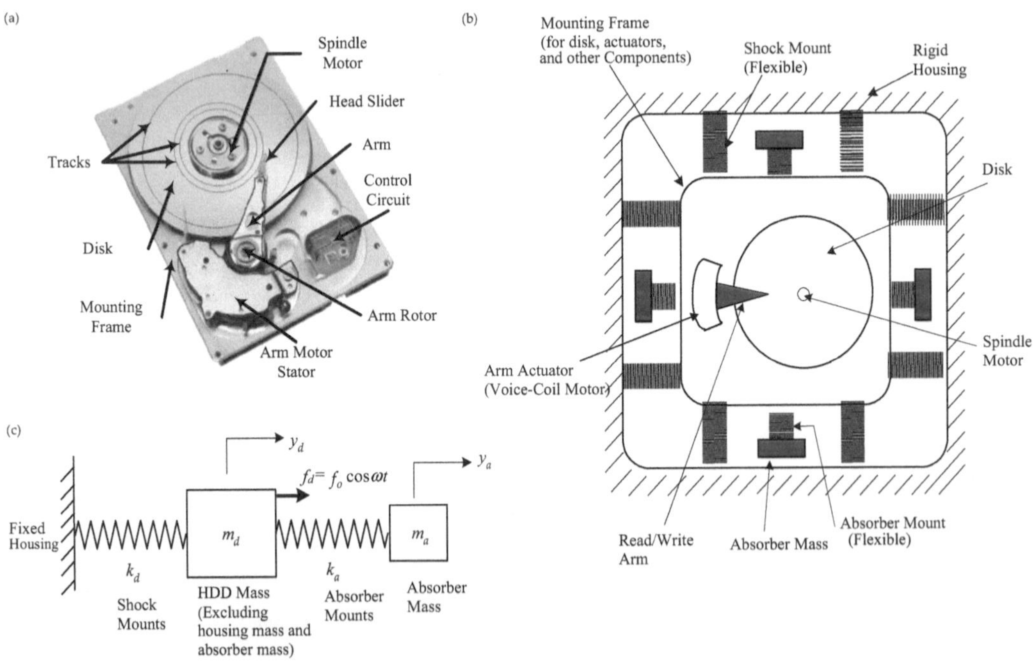

FIGURE 1.4 (a) A hard-disk drive (HDD) unit of a computer; (b) components for a simplified model; and (c) further simplified model for vibration analysis and control.

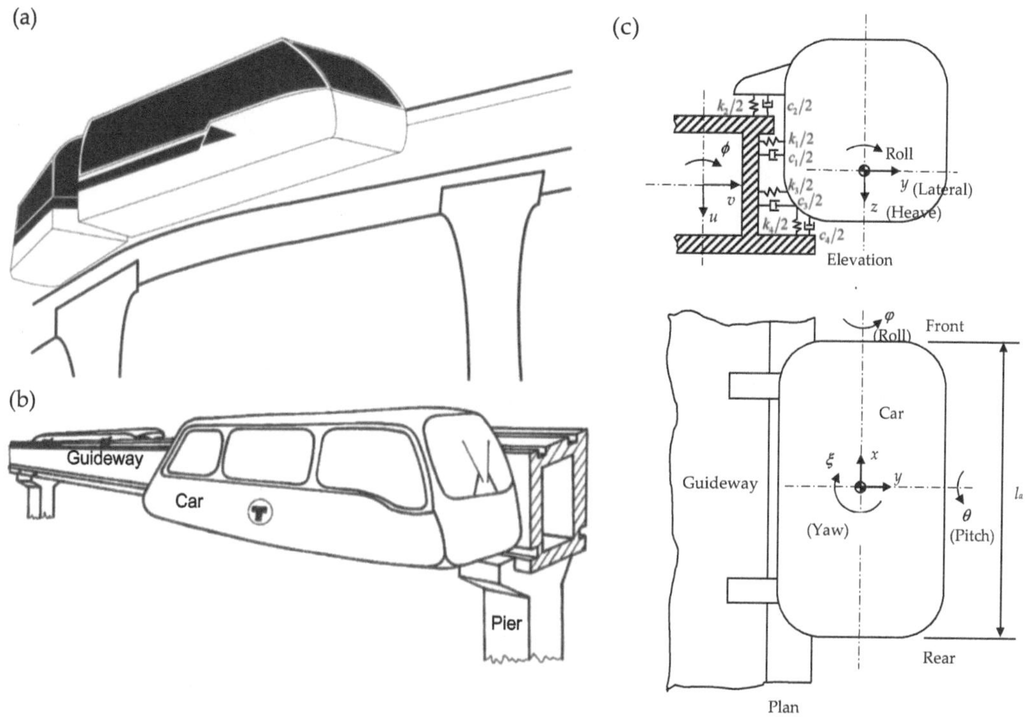

FIGURE 1.5 (a) An innovative elevated guideway transit system; (b) a simplified system that incorporates a single car; and (c) a further model for determining the ride quality of the mass transit system.

actuator. Position sensing of the hydraulic manipulator is done using linear magnetostrictive displacement transducers. A set of six gauge-pressure transducers are installed to measure the fluid pressure in the head and rod sides of each hydraulic cylinder and also in the supply lines. A high-level imaging system determines the product quality (cut fish), according to which adjustments may be made online, to the parameters of the control system so as to improve the machine performance. Clearly, this is a multi-physics system, which involves at least the mechanical, electrical, and fluid domains. A model that sufficiently represents the characteristics of these domains is shown in Figure 1.6b.

1.4 MODEL TYPES

One way to analyze a system is to apply excitations (inputs) to the system, measure the resulting responses (outputs) of the system, and fit the input–output data obtained in this manner to a suitable analytical model. This is known as *experimental modeling* or *model identification* or *system identification*. A model that is determined in this manner is called an *experimental model* or an *identified model*. Another way to analyze a system is by using an analytical model of the system, which originates from the physical (i.e., constitutive) equations of the constituent components or processes of the system. Analytical models include *state-space models* and *input–output models*. An important category of input–output models is the *transfer function models* (in the *Laplace domain* in general and in the *frequency domain* in particular).

Graphical techniques such as linear graphs, bond graphs, and block diagrams can assist in the development of analytical models. Since developing a physical model (or a prototype) of a system and testing it is often far less economical, flexible, quick, or practical than analyzing or computer-simulating an analytical model of the system, such analytical models are often used in the process of system design. In fact, analytical models are commonly used in practical applications, particularly

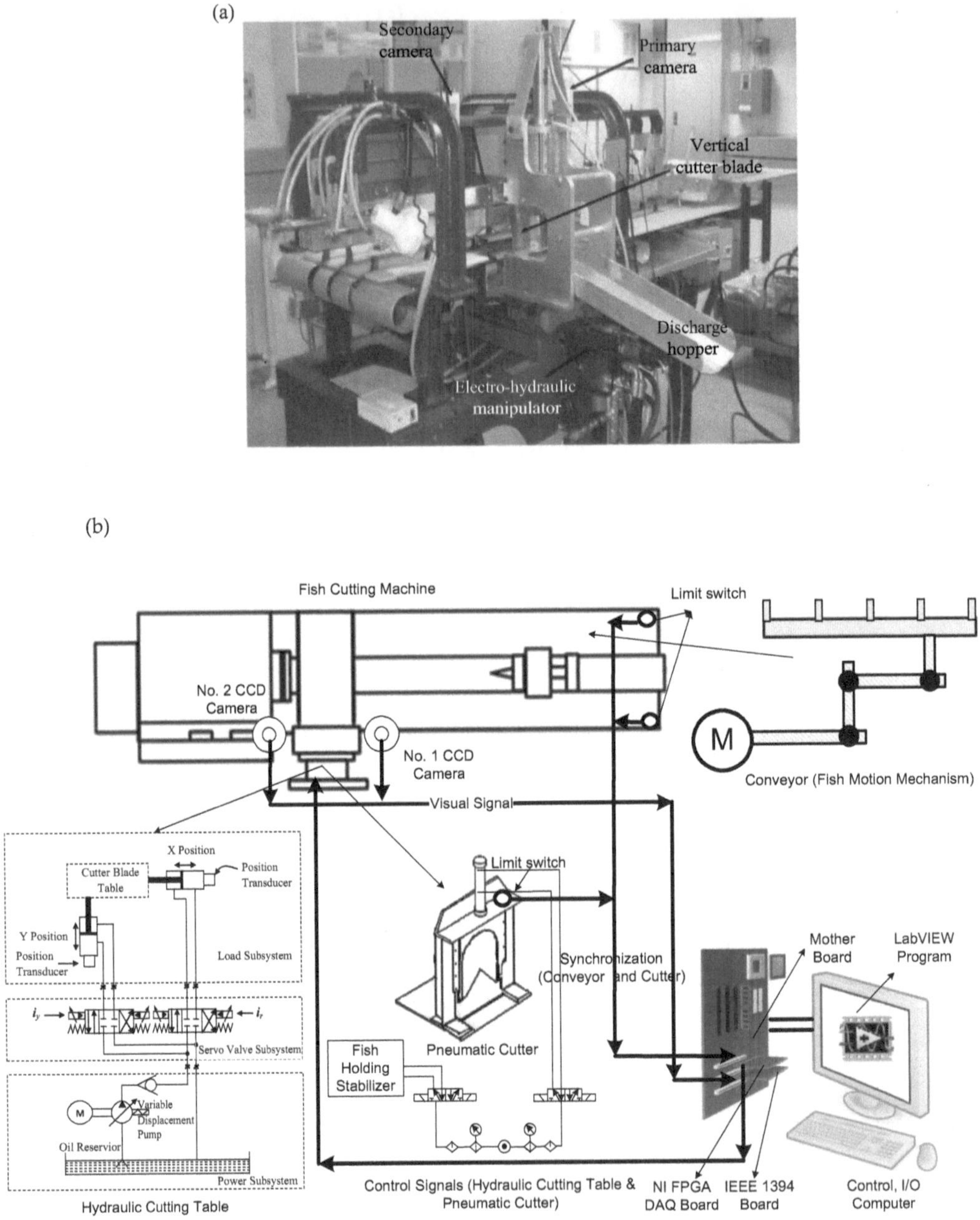

FIGURE 1.6 (a) An intelligent iron butcher and (b) a multi-physics model of the Iron Butcher.

during the pre-prototyping stage. Instrumentation (exciters, measuring devices, and analyzers) and computer systems for experimental modeling (e.g., modal testing and analysis systems) are commercially available. Experimental modeling is also done frequently, if less often than analytical modeling.

As discussed, in general, models may be grouped into the following categories:

- Physical models or prototypes
- Analytical models (mathematical models)

- Computer models or numerical models (computer programs, files, data tables and arrays, multidimensional graphs, etc.)
- Experimental models (where input–output experimental data are used for model "identification").

In this book, the main focus is on analytical models with a secondary focus on experimental and computer models, which have a direct impact from analytical models.

1.4.1 Advantages of Analytical Models

Normally, mathematical definitions for a dynamic system are given with reference to an analytical model of the system; for example, a state-space model or an input–output model. There are many advantages of analytical models over other types of models. The main advantages of analytical models (and computer models) over physical models include the following:

- Modern, high-capacity, high-speed computers can handle complex analytical models at high speed and low cost.
- Underlying physical principles are easily and directly included in analytical models.
- Analytical methods are more precise.
- Analytical models and computer models can be modified conveniently and at high speed and low cost.
- Analytical models provide much flexibility in making structural and parametric changes of a dynamic system/model. (*Note*: Structural changes are those that modify how the system components are interconnected—series, parallel, etc., the system structure. Parametric changes are changes to the parameter values of the system components.)
- Analytical models can be directly used in computer simulations.
- Analytical models can be easily integrated with computer/numerical/experimental models, to generate "hybrid" models (e.g., in hardware-in-the-loop computer simulations).
- Analytical modeling can be conveniently used in the system design and also in developing control strategies and control actions.
- Analytical modeling can be conveniently done well before a prototype is built. (In fact, this step can be crucial in deciding whether to prototype.)

1.4.2 Mechatronic Systems

The commonly accepted definition of Mechatronics is: synergistic application of mechanics, electronics, control engineering, and computer science in the development of electromechanical products and systems, through integrated design. However, the integrated, unified, unique, and systematic approach to modeling, as presented in this book, is similar to and even goes beyond what is formally known the "mechatronic" approach, which is systematic and integrated (concurrent) but may not be unified (uses analogous or similar approaches for modeling different physical domains) and unique (only a clear and single model is generated at the end). Mechatronic systems are multi-physics systems, but the initial focus of Mechatronics was on electromechanical (two-domain) systems. The modeling methodologies presented in this book go beyond this as well, by incorporating other physical domains (fluid and thermal in particular). Hence, this book is particularly suitable for studies in mechatronic systems. In this backdrop, it is useful to have some understanding of mechatronic systems, which had already been addressed earlier. Nevertheless, it should be emphasized that a "mechatronic" system is not just an electromechanical or even a multi-physics system but a multi-physics system that has been developed by using a mechatronic approach.

A typical mechatronic system consists of a mechanical skeleton, actuators, sensors, controllers, signal conditioning and modification devices, computer and digital hardware and software, interface devices, and power sources. Different types of sensing, information acquisition, and transfer are involved among all these various types of components. For example, a servomotor, which is a motor with the capability of sensory feedback for accurate generation of complex motions, consists of mechanical, electrical, and electronic components. The main mechanical components are the rotor, stator, and the bearings. The electrical components include the circuitry for the field windings and rotor windings (not in the case of permanent-magnet rotors), and circuitry for power transmission and commutation (if needed). Electronic components include those needed for sensing (e.g., optical encoder for displacement and speed sensing and tachometer for speed sensing). However, both thermal and fluid domains are also relevant in the design and operation of a servomotor, at least in dealing with its cooling through heat transfer and air flow (convection). The overall design of a servomotor can be improved by taking a mechatronic approach.

The humanoid robot shown in Figure 1.7a is a more complex and "intelligent" electromechanical system. It may involve many servomotors and a variety of mechatronic components, as is clear from the sketch in Figure 1.7b. As for a servomotor, a robot also incorporates fluid and thermal domains. A mechatronic approach can greatly benefit modeling, analysis, design, development, and control of a complex electromechanical (or in general, a multi-physics) system of this nature.

Technology issues of a multi-physics system are indicated in Figure 1.8. It is seen that they span the traditional fields of mechanical engineering, electrical and electronic engineering, control engineering, and computer engineering. Each aspect or issue within the system may take a multi-physics character. For example, as noted before, an actuator (e.g., dc servo motor) alone may represent a multi-physics device within a larger multi-physics system such as an automobile or a robot, all of which can benefit from the concepts of Mechatronics.

Study of mechatronic engineering should include all stages of modeling, design, development, integration, instrumentation, control, testing, operation, and maintenance of a mechatronic system.

1.4.3 Steps of Analytical Model Development

In a "systematic" approach to modeling, it is necessary to "clearly" indicate its key steps. As a preview of what yet to be presented, we give below the key steps of the systematic approach that is presented in this book.

1. Identify the system of interest (e.g., *purpose*, *components*, *system boundary*).
2. Identify or specify the *variables* of interest (e.g., *excitations* or *inputs*, *responses*, or *outputs*).
3. Approximate various segments (components, processes, phenomena) by *ideal elements*, suitably interconnected. Draw a *structural diagram* for the system (e.g., mechanical circuit, electrical circuit, linear graph), showing the structure (element or component interconnection) of the system.
4. Using the structural diagram:
    a. Write *constitutive equations* (physical laws) for elements (other than the "input elements" or "sources").
    b. Write *continuity equations* (or *conservation equations*) for through-variables (those variables that do not change through an element) at junctions (nodes), which connect two or more elements (e.g., equilibrium of forces at mechanical joints; current balance at circuit nodes).
    c. Write *compatibility equations* (or *loop equations*) for across-variables (potential variables, path variables), which are measured across an element, in closed paths (loops) (e.g., for velocities—geometric connectivity; for voltages— potential balance). Why is compatibility automatically satisfied (and hence, compatibility equations are redundant) for mechanical systems? (The answer is found elsewhere in this book).
    d. Eliminate *auxiliary variables* (i.e., unwanted variables).

(a)

(b)

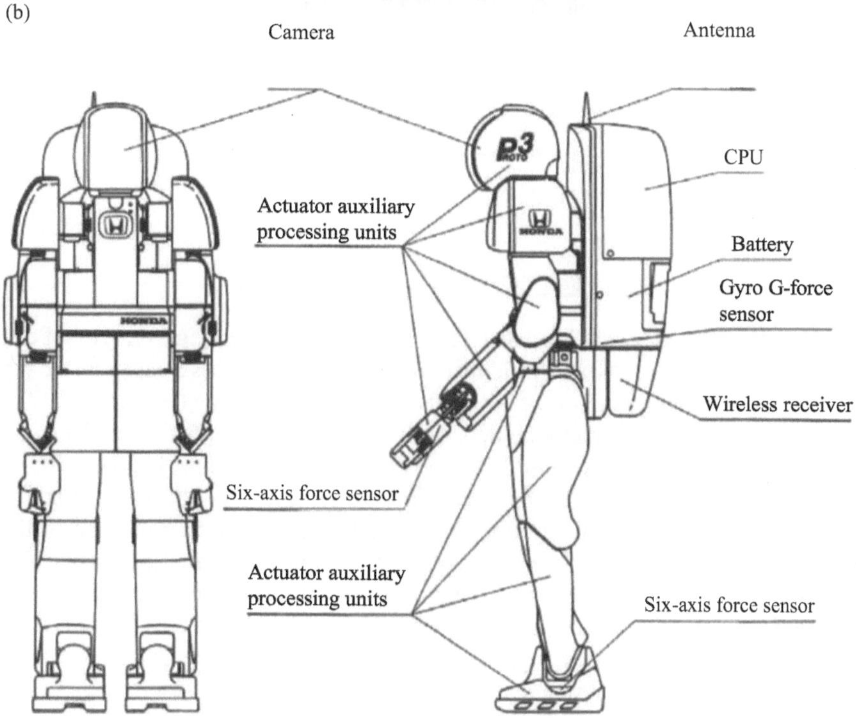

FIGURE 1.7 (a) A humanoid robot is a complex and "intelligent" multi-physics system and (b) components of a humanoid robot. (Courtesy of American Honda Motor Co. Inc., Torrance, CA.)

5. Express *boundary conditions* (only for distributed-parameter systems) and *initial conditions* using system variables.

Note: It should be clear that boundary conditions are needed only for distributed parameter systems (they are automatically satisfied in lumped-parameter models).

In this book, we will primarily focus on the following types of analytical models:

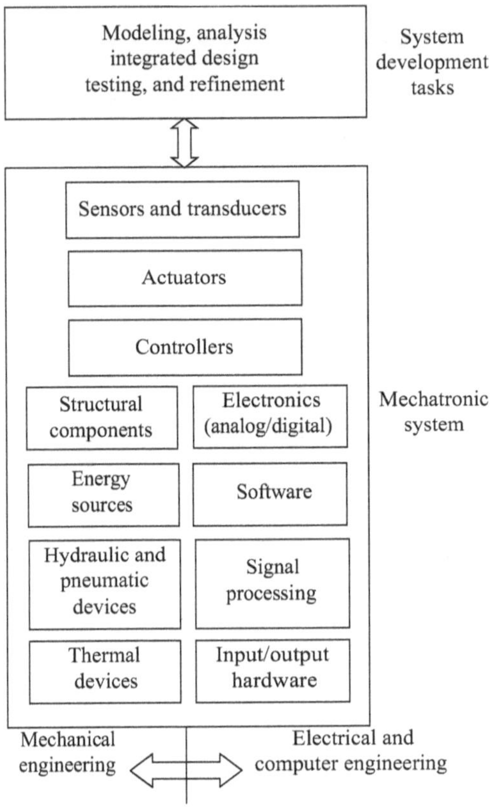

FIGURE 1.8 Concepts and technologies of a multi-physics system.

- **State Models (State-Space Models)**: They use state variables (e.g., velocity of a lumped mass, force in a spring, current through an inductor, voltage across a capacitor) to represent the state of the system, in terms of which the system response can be expressed. These are time-domain lumped-parameter models, with time t as the independent variable.
- **Transfer-function Models (Particularly Frequency-Domain Models)**: These are a type of *input–output models* expressed using transfer functions. A transfer function is the Laplace transform of the output variable divided by the Laplace transform of the input variable. Laplace variable s is the independent variable in a transfer function model. A frequency-domain model, expressed using frequency transfer functions, is a very practical transfer function model. It can be directly obtained from a Laplace transfer function simply by setting $s=j\omega$. In a frequency-domain model, the independent variable is the frequency ω at which the system operates.

In the development and application of these analytical models, we will particularly use the graphical representation known as *linear graphs*, of models. In a linear graph, a line represents a basic component (element) of the system, with one end as the *point of action* and the other end as the *point of reference*. The associated techniques are presented in this book (see Chapter 5 for time-domain linear graph techniques and Chapter 7 for frequency-domain linear graph techniques).

1.4.4 Modeling Criteria and Equivalent Models

This book pays particular attention to *equivalent models*. Notably, *approximate models* concern *approximate equivalence* of models. A model itself is "approximately equivalent" to the actual system. Indeed, various types of models (physical, analytical, computer, experimental, etc.) are approximately equivalent to the actual system. In the process of modeling, first, an equivalence is established based on such considerations as:

- The needs of the specific application (purpose of the model: design, analysis, control, etc., of what aspects?)
- Available resources for modeling: physical information, analytical methodologies, physical system, computer resources, accessibility to the physical system and the ability to acquire experimental data, ability to develop a physical prototype, past information, etc.
- The required accuracy and the constraints on model complexity.

The decision to develop an *analytical model* can be made based on such considerations.

In the development of an analytical model for a system, we have to first establish some *criteria of equivalence* or approximate equivalence of models. For example, we may consider the conversion of:

- A distributed-parameter system (model) into a lumped-parameter model
- A nonlinear system (model) into a linear model
- A time-domain system (model) into a transfer function (particularly, frequency-domain transfer function or FTF) model
- A component-based, detailed transfer-function model into a reduced TF model
- A complex and extensive subsystem of the system into a simple "equivalent model" such that: the dynamic interactions (of the original model and of the equivalent model) with the remaining "simple segment of interest" of the system remain the same.

In each case, we (explicitly or implicitly) use some criteria of equivalence; for example, energy equivalence, modal (natural-frequency) equivalence, analytical equivalence, computer-simulation (discrete-time or digital) equivalence, physical equivalence, or dynamic equivalence with respect to a "simple segment of interest" of the system. In this book, we will study various types of equivalent models.

1.5 ORGANIZATION OF THIS BOOK

This book, consisting of seven chapters and two appendices, is devoted to theory, techniques, and application of analytical modeling of engineering dynamic systems, with some focus on experimental and computer modeling. The chapters systematically cover the process of understanding and the analytical representation (modeling) of an engineering dynamic system, using proper principles of science. Common applications of analytical models are addressed as well.

Chapter 1 introduces the subject of modeling with a focus on multi-physics engineering dynamic systems. The importance of dynamic modeling in various applications is indicated. The use of modeling in the design and control of a dynamic system is highlighted. The terminology of dynamic systems and models is clarified. Common types of models and modeling techniques are introduced and their comparative advantages and disadvantages are indicated. Different types of dynamic systems and models are introduced. The subject of Mechatronics is introduced and expanded, and a pertinent approach for the dynamic modeling in Mechatronics is outlined. In this context, the importance of an integrated, unified, unique and systematic approach of analytical modeling of a multi-physics system and its relevance for mechatronic systems are highlighted. The main steps of developing an analytical model of a dynamic system are given. This introductory chapter sets the tone for the present study, which spans the remaining six chapters.

Chapter 2 identifies the basic elements of linear, lumped-parameter analytical models in the mechanical, electrical, fluid, and thermal domains. The across-variables and the through-variables that are used in the physical equations of these elements are discussed while identifying their analogies (similarities) across domains. The considered element categories are the sources (input elements), energy storage elements, and energy dissipation elements. The identification of proper and physically meaningful state variables is discussed in an analogous manner. Specifically, the independent capacitor-type energy storage elements (called *A*-type elements) use their across-variables as the state variable while the independent inductor-type energy storage elements (called *T*-type elements) use their through-variables as the state variables. This choice of state variables is the key reason for obtaining a "unique" state-space model, in the present approach. The resistor-type elements are energy dissipation elements (called *D*-type elements), and they do not determine state variables. Natural oscillations in a dynamic system are a manifestation of the presence of two types of energy storage elements (*A*-type and *T*-type).

Chapter 3 formally introduces analytical modeling of a dynamic system. First, several categories of analytical models are indicated, while comparing the general model form in each type with a more common special form. In particular, a comparison of continuous-time systems (models) with discrete-time systems (models) is given. As well, a comparison of distributed-parameter systems (models) with lumped-parameter systems (models) is presented. Also, the method of converting a general form of model into a more popular special is illustrated, while indicating the advantages and limitations of such conversions of models. Analytical models may be developed for mechanical, electrical, fluid, and thermal systems (and hence, for multi-physics systems) in a rather analogous manner because clear analogies exist among these four types of physical domains. This is the key basis of the "unified" approach to modeling that is presented in this book. The systematic development of state-space models of engineering systems in these four physical domains is studied with illustrative examples. Important properties of a state-space model are presented. A general approach for converting a state-space model into an input–output model is described, and illustrative examples are given.

Chapter 4 studies linearization of a nonlinear system/model. Specifically, it addresses the model category that concerns nonlinear model form and its conversion into a linear model form (see Chapter 3). In some methods, the conversion may be valid only in a restricted range of operation about an operating point, while in some other methods, the conversion may be valid globally. Real systems are nonlinear and they are represented by nonlinear analytical models consisting of nonlinear differential equations, in the time domain. Linear systems (models) are in fact idealized representations, and are represented by linear differential equations, in the time domain. First linearization of analytical models, particularly state-space models and input–output models, is treated. Then, linearization of experimental models (experimental data) is addressed. These methods are valid in a restricted range of operation. Finally, several methods of linearization that are globally valid are introduced.

Chapter 5 presents linear graphs—an important graphical tool for developing and representing a model of a dynamic system. State-space models of lumped-parameter dynamic systems, regardless of whether they are mechanical, electrical, fluid, thermal, or multi-domain (mixed) systems, and whether they are linear or nonlinear, can be conveniently developed by using *linear graphs*. The chapter systematically studies the use of linear graphs in the development of analytical models for mechanical, electrical, fluid, and thermal systems, and also of mixed-domain systems which consist of two or more physical domains. More advanced concepts of linear graphs, particularly the graph-tree concepts, are presented in Appendix A.

Chapter 6 treats transfer-function models and the frequency-domain analysis of dynamic systems. Both Laplace-domain concepts and frequency-domain concepts are presented on the basis of the Laplace transform and the Fourier transform. A linear, constant-coefficient (time-invariant) time-domain model can be converted into a transfer function, and vice versa, in a simple and straightforward manner. A unified approach is presented for the use of the transfer function approach in the modeling and analysis of multi-domain (e.g., electromechanical) systems. In this context, mechanical circuits are introduced (in a similar manner to electrical circuits), which make use of such mechanical transfer functions as mobility, mechanical impedance, force transmissibility, and motion transmissibility. Their practical applications such as vibration isolation are discussed. Maxwell's reciprocity principle is given for dynamic systems and is generalized for various physical domains (particularly mechanical and electrical domains).

Chapter 7 extends the concepts of equivalent circuits (Thevenin equivalence and Norton equivalence, which are common concepts in the electrical domain) to other physical domains such as mechanical and fluid domains. Frequency-domain linear graphs, called "transfer-function linear graphs," denoted by TFLG, are introduced. They are integrated with Thevenin and Norton equivalence, giving rise to linear graph reduction and equivalent linear graphs. Their application in various physical domains is illustrated. A mechatronic model of a multi-physics system may be simplified by converting all the physical domains into an equivalent single-domain system that is entirely in the output domain of the system. This approach of converting (transforming) physical domains is presented. In this manner, a multi-physics system may be represented by a linear graph model in a single physical domain (the output domain), which is far easier to formulate and analyze. For this purpose, the two-port linear graph elements, transformer, and gyrator (which are energy conversion elements) are generalized to two physical domains (where the input domain and the output domain of the element are not the same). This treatment generalizes transformer-coupled systems and gyrator-coupled systems. An illustrative example of a pressure-controlled hydraulic actuator system that operates a mechanical load is presented.

Appendix A presents the graph-tree approach, which broadens the application of linear graphs. Some issues concerning linear graphs that are not elaborated in Chapter 5 are discussed and illustrated in this appendix.

Appendix B provides a user guide for the MATLAB® Linear Graph Toolbox, which a toolbox developed in house and available to anyone.

1.6 SUMMARY SHEET

- **Book Objectives**: Learn the theory and application of a unified, integrated, systematic, and unique approach to analytical modeling (i.e., the "enhanced mechatronic" approach; learn modeling in both time domain and frequency domain.
- **Uses/Applications of Modeling**: Dynamic system analysis, computer simulation, design, modification/restructuring, instrumentation, control, testing, and performance evaluation.
- **Uses of a Model in Control**: (1) Mathematical analysis and computer simulation; (2) as a reference model for performance specification; (3) to develop a control scheme (in model-based control); (4) to design a control system.
- **Uses of a Model in Design**: (1) To represent the desired design of the entire system or part of it (particularly in the absence of a physical system or prototype); (2) to analyze and optimize a design; (3) to develop integrated and unified approaches of design (e.g., in mechatronic design); (4) to assist in instrumentation (which may be considered as an integral part of design) of a dynamic system.
- **System**: Interacting components of interest, demarcated by system boundary (real or virtual).
- **Dynamic System**: A system where rates of changes of response or state variables are not negligible.
- **Plant or Process**: The system to be controlled.
- **Inputs**: Excitations (known or unknown; deliberate or unintentional; desirable or undesirable) applied to the system.
- **Outputs**: Responses (desired or undesired; known or unknown) of the system.
- **State Variables**: A minimal set of variables that completely identify the "dynamic" state of the system. If the state variables at one state and the inputs from that state up to a future state are known, the future state can be completely and uniquely determined.

- **Control System**: Includes at least the plant and its controller. May include other subsystems and components (e.g., sensors or signal conditioning and modification devices, interfacing hardware).
- **Universal Model**: Incorporates all conceivable aspects of the system. Unrealistic. Too complex. Not practical.
- **Model Types**: Physical models (prototypes); analytical models (mathematical models); computer models or numerical models (computer programs, files, data tables and arrays, multidimensional graphs, etc.); experimental models (input–output experimental data are used for model "identification").
- **Advantages of Analytical Models**: Computers can handle complex analytical models at high speed and low cost; physical principles are easily and directly included in them; more precise; can be modified conveniently and at high speed and low cost; more flexible, when making structural and parametric changes (*Note*: Structural changes: modify component interconnection—series, parallel, etc. Parametric changes: change the parameter values—mass, capacitance, etc.); directly useful and applicable in computer simulation; can be easily integrated with computer/numerical/experimental models, and hardware (hardware-in-the loop models) → "hybrid" models; can be developed well before a prototype is built (can be crucial in deciding whether to prototype).

PROBLEMS

1.1 The modeling approach that is presented in this book is said to be: (1) integrated, (2) unified, (3) unique, and (4) systematic. Define these four terms in the present context.

1.2 A typical input variable is identified for each of the following examples of dynamic systems. Give at least one output variable for each system.

a. **Human Body**: neuro-electric pulsesb.
b. **Company**: informationc.
c. **Power Plant**: fuel rated.
d. **Automobile**: steering wheel movemente.
e. **Robot**: applied voltage to joint motorf.
f. **Highway Bridge**: vehicle force.

Also, indicate other possible inputs.

1.3 The use of solar energy is a sustainable way to generate electric power for houses. A schematic arrangement is shown in Figure P1.3a. Radiation from the sun is received at a solar panel, which consists of photovoltaic cells to convert solar energy to electric energy in the form of direct current (dc). Using an inverter, the dc power is converted into alternating current (ac) power of appropriate frequency (60 or 50 Hz) for household use. This electrical supply is connected through a two-way meter to the supply line of the house and to the main electricity grid (Figure P1.3b). In this manner, any excess power from the solar panels can be sold to the grid, and when the supply from the solar panel is not adequate (e.g., during cloudy days or at night), electricity can be purchased from the grid. The ac power is used for various household purposes such as operation of appliances, heating, and cooling.

a. Explain why this is a multi-physics (i.e., multi-domain or mixed) system.
b. Identify several key components of the system (*Note*: Some are shown in the figure). Discuss various processes within the components that may be categorized into the mechanical, electrical, fluid, and thermal domains. Indicate applicable modeling issues for the overall system.
c. Sketch the energy flow of the system, indicating relevant stages of energy conversion, which will be useful in the modeling process.

1.4 What is a system and what is a "dynamic" system?

1.5 Give four categories of uses of dynamic modeling. List advantages and disadvantages of experimental modeling over analytical modeling.

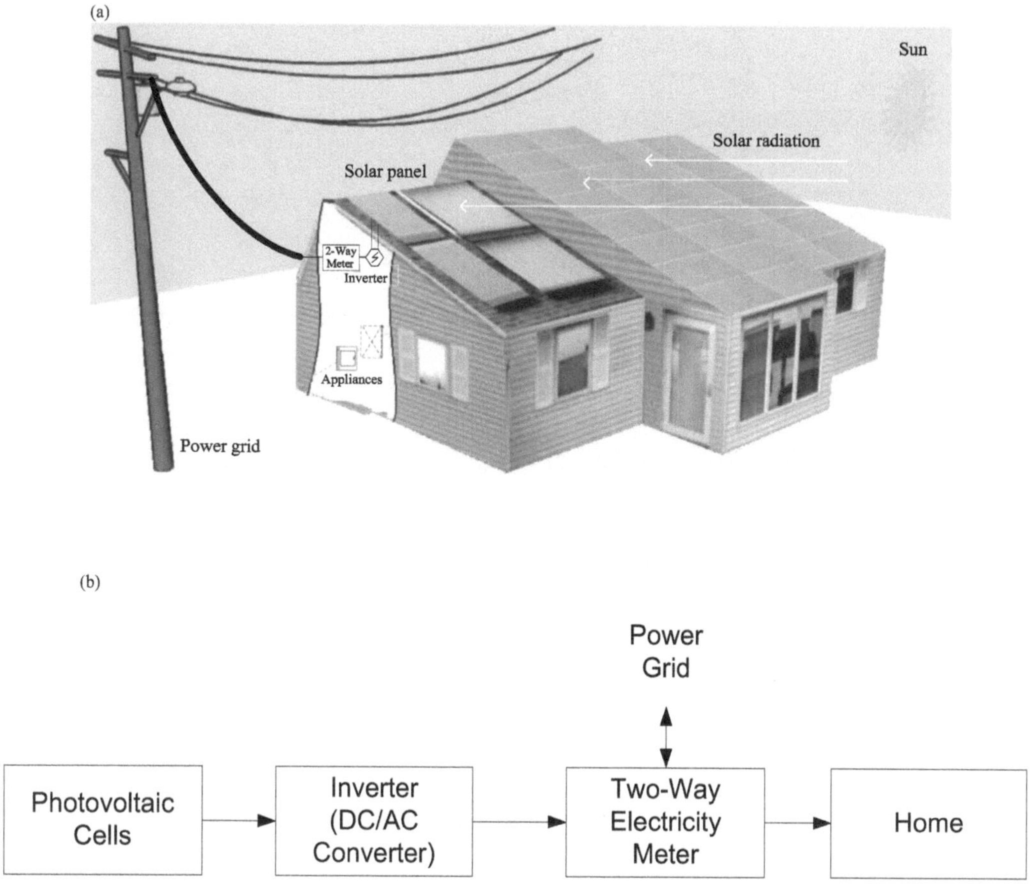

FIGURE P1.3 (a) A solar-powered house and (b) schematic diagram of the ac power supply.

1.6

a. Give logical steps of the analytical modeling process for a general physical system.
b. Once a dynamic model is derived, what other information would be needed for analyzing its time response (or for computer simulation)?
c. A system is divided into two subsystems, and models are developed for these two subsystems. What other information would be needed to obtain a model for the overall system?

1.7 Various possibilities of model development for a physical system are shown in Figure P1.7. For developing an approximate model, give advantages and disadvantages of the SM approach in comparison to a combined DM+MR approach.

1.8 Indicate several sources of error in a model-based approach. Your answer may be quite general even though specific application areas such as system analysis, computer simulation, and control may be considered.

1.9 What is the main consideration when deciding what aspects of the actual system should be included in its model. Consider the model of the HDD unit, as shown in Figure 1.4 (both b and c). What aspects can be analyzed using this model? Indicate, giving reasons, what other aspects may be modeled (in a more detailed model).

1.10 Consider the model of the vehicle in an elevated guideway system, as shown in Figure 1.5c. What aspects can be analyzed using this model? Indicate, giving reasons, what other aspects may be modeled (in a more detailed model).

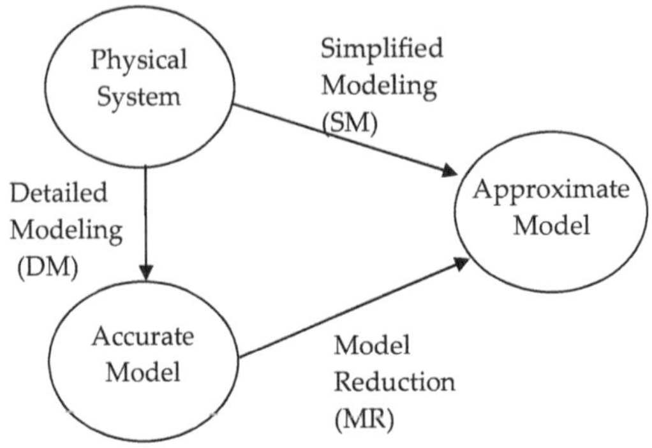

FIGURE P1.7 Approaches of model development.

1.11 Consider the representation of a dynamic system, as shown in Figure 1.3.
  a. What aspects are included in this model? What aspects are missing here, for a complete analysis or simulation of the system?
  b. In a model of a "dynamic" system, it is said the "rates of changes" of the outputs/state variables are not negligible. What is the analytical implication of this statement?
  c. What is causality?

1.12 Why is a "universal" model not appropriate for a given system? What is the best model for a system?

2 Basic Model Elements

HIGHLIGHTS

- Lumped Elements and Analogies
- Across-Variables and Through-Variables
- Energy Storage Elements
- Energy Dissipating Elements
- Input Elements (Sources)
- Mechanical Elements (Mass or Inertia Element, Spring or Stiffness Element, and Damper or Dissipation Element)
- Electrical Elements (Capacitor, Inductor, and Resistor)
- Fluid Elements (Capacitor or Accumulator, Inertor or Fluid Inductor, and Resistor)
- Fluid Capacitance due to Fluid Compressibility, Container Flexibility, and Gravity Head
- Thermal Elements (Capacitor and Resistor)
- Thermal Resistance due to: Conduction, Convection, and Radiation
- Natural Oscillations

2.1 INTRODUCTION

An engineering (physical) dynamic system typically consists of a mixture of different types of processes and components (e.g., those involving the four physical domains: mechanical, electrical, fluid, and thermal). A dynamic system of this type is called a *multi-physics system*, *multi-domain system*, or *mixed system*. In modeling such a system, an *integrated* approach is desirable where all physical domains are modeled together (i.e., a *concurrent* treatment), and also a *unified* approach is desirable, where similar approaches are used to model and analyze all physical domains (i.e., an *analogous* treatment). In this context, it is important to recognize the analogies that exist among various physical domains. In this chapter, we will study these analogies at the "element" level, which we will incorporate in model development in the subsequent chapters, at the "system" level. In particular, we will identify the basic elements of lumped-parameter analytical models in the mechanical, electrical, fluid, and thermal domains, in a systematic manner. The considered element categories are the sources (input elements), energy storage elements, and energy dissipation elements. For the energy storage elements, we will give particular consideration to their energy characteristics, input–output (causal) behavior, and governing variables (e.g., through- and across-variables; state variables), while identifying their analogies (similarities) across domains. In particular, the identification of proper and physically meaningful state variables is discussed in an analogous manner. Specifically, capacitor-type energy storage elements (called A-type elements) use their across-variables as the state variable while inductor-type energy storage elements (called T-type elements) use their through-variables as the state variables. Resistor-type elements are energy dissipation elements (called D-type elements) and they do not introduce state variables. The fundamental basis of generating a "unique" state-space model is this choice of state variables. Natural (i.e., free or unforced) oscillations in a dynamic system are a manifestation of the presence of two types of energy storage elements (A-type and T-type) that transfer energy between them, in that system.

DOI: 10.1201/9781003124474-2

2.1.1 Lumped Elements and Analogies

A system may possess various physical characteristics that arise from its physical domains; particularly, mechanical, electrical, thermal, and fluid processes. The procedure of model development will be facilitated by understanding the similarities (analogies) that exist among these domains and also among the basic physical elements of these domains.

The basic physical elements in an engineering system can be divided into two broad categories:

1. Energy storage elements
2. Energy dissipation elements

The dynamic "state" of a system is determined by its "independent" energy storage elements and the corresponding *state variables.* Depending on the type of the energy storage element, we can use either an across-variable or a through-variable as its state variable. As we will see, energy dissipation elements do not introduce state variables. The state of an energy dissipation element is determined by the state variables of the independent energy storage elements that are connected to it.

- **Independent Energy Storage Elements**: Two energy storage elements are independent if their dynamic state cannot be completely represented by a single equivalent energy storage element (hence by a single state variable).
- **Input Elements (Sources)**: In addition to the energy storage elements and energy dissipating elements, an engineering dynamic system may have input elements (or source elements), which perform the actuation (including control) functions in the system. An input element will have one "independent variable," which is the input variable that is applied to the connected system, and a "dependent variable" whose value will change according to the dynamics of the connected system. The value of the independent variable will not change in this manner (for an ideal source).

2.1.2 Across-Variables and Through-Variables

An across-variable is measured across an element, as the difference in the values at the two ends of the element. In other words, it is represented by the value at one end (point of action) of the element with respect to the value at the other end (point of reference). Velocity, voltage, pressure, and temperature are across-variables. If the across-variable of an element is an appropriate state variable for that element, it is termed an *A-type element.*

A through-variable represents a property that appears to pass through an element, unaltered. Force, current, fluid flow rate, and heat transfer rate are through-variables. If the through-variable of an element is an appropriate state variable for that element, it is termed a *T-type element.*

2.1.2.1 Energy Dissipating Element

An energy-dissipating element is called a *D*-type element. Unlike an independent energy storage element, it does not introduce a state variable. The variables (or the dynamic state) of a *D*-type element in an engineering system are completely determined by the independent energy storage elements (*A*-type and *T*-type) in the system that are connected to this element.

Analogies exist among mechanical, electrical, hydraulic, and thermal systems/processes. Next, we state the physical equations (i.e., *constitutive equations*) of the basic elements in these four domains, identify and justify the appropriate state variables for them and recognize analogies that exist across these domains.

2.2 MECHANICAL ELEMENTS

- **Energy Storage Elements**: For mechanical elements, we will show that the velocity (across-variable) of each independent inertia element (e.g., mass) and the force or torque (through-variable) of each independent flexibility element (e.g., spring) are the appropriate *state variables* (or response variables or output variables). Hence, mass is an A-type element and spring is a T-type element. These are energy storage elements that store either kinetic energy or potential energy. The corresponding constitutive equations form the "state-space shell" for an analytical model. These equations will directly lead to a unique *state-space model* of the system, as we will illustrate in subsequent chapters.
- **Energy Dissipating Element**: The energy-dissipating element in a mechanical system is the damper (a D-type element). It does not introduce a state variable. The dynamic state of a D-type element is completely established by the energy storage elements that are connected to it, in the system.
- **Input Elements**: There are two types of input elements (or source elements) for a mechanical system: A-type source and T-type source. A velocity source is an *A-type source* because its *independent variable* is an across-variable (velocity). The independent variable of a source is not affected by the dynamics of the system. For a velocity source, the associated force or torque variable—the *dependent variable*—will be affected by the dynamics of the system. A force source (or torque source) is a *T-type source* because its independent variable is a through-variable (force or torque), which is not affected by the dynamics of the system (while the associated velocity variable—the dependent variable—will be affected). These are *ideal sources* since in practice the source variable will be affected to some extent by the dynamics of the system (this effect is known as "loading") and is not entirely "independent."

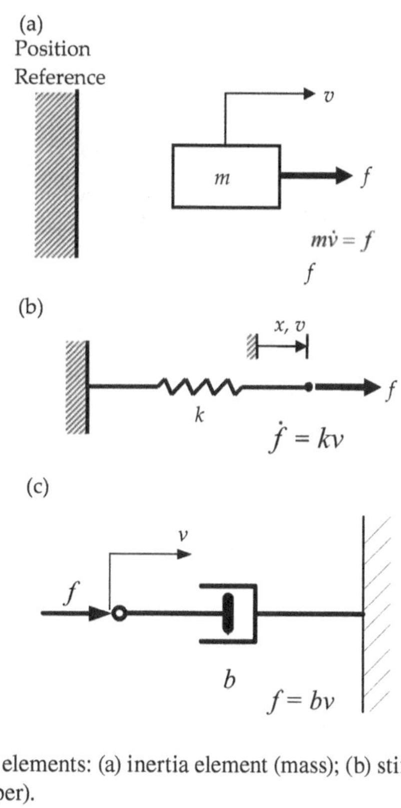

FIGURE 2.1 Basic mechanical elements: (a) inertia element (mass); (b) stiffness/flexibility element (spring); and (c) dissipation element (damper).

2.2.1 Inertia Element

Consider the mass element shown in Figure 2.1a. The constitutive equation (the physical law) of the element is given by Newton's second law:

$$m\frac{dv}{dt} = f \tag{2.1}$$

Here, v denotes the velocity of mass m, measured relative to an inertial reference (fixed on earth or moving at constant velocity), and f is the force applied "through" the mass. Since power=fv=rate of change of energy, the energy of the element may be determined by integrating it (after substituting equation 2.1):

$$E = \int fv\,dt = \int m\frac{dv}{dt}v\,dt = \int mv\,dv$$

or

$$\text{Energy } E = \frac{1}{2}mv^2 \tag{2.2}$$

This is the well-known *kinetic energy.*

Next, by integrating equation (2.1), we obtain

$$v(t) = v(0^-) + \frac{1}{m}\int f\,dt \tag{2.3}$$

By setting $t=0^+$ in equation (2.3), we see that if the applied force f is finite (which is a realistic assumption), the integral term becomes zero, and we have

$$v(0^+) = v(0^-) \tag{2.4}$$

Note: 0^- denotes the time instant just before $t=0$ and 0^+ denotes the time instant just after $t=0$.

In view of these observations, we may state the following facts:

1. An inertia is an energy storage element. It stores *kinetic energy.*
2. The velocity across an inertia element cannot change instantaneously unless an infinite force/torque is applied to it (not practical). In other words, a finite force/ torque input cannot cause an infinite acceleration (or step change in velocity) in an inertia element. Conversely, a finite instantaneous (step) change in velocity in an inertia element will need an infinite force/torque, which is not realistic. Hence, velocity v is a natural output (or response) variable for an inertia element, which can represent its dynamic state (i.e., state variable), and force/torque f is a natural input variable for an inertia element.
3. The fact that velocity is most appropriate to represent the "dynamic" state of an inertia element is further justified by several reasons: First justification is Item 2; second, from equation (2.3), the velocity of an inertia element at any time t can be completely determined with the knowledge of the initial velocity and the applied force/torque during the time interval 0 to t (*Note*: As we will see in Chapter 3, this is indeed the formal definition of state); and third, from equation (2.2), the energy of an inertia element can be represented by the variable v alone (kinetic energy).
4. Since its state variable (velocity) is an across-variable, an inertia is an A-type element.

   Note: In the present development, we used mass (in "translatory" motion) to represent the inertia element. The associate across-variable is the "rectilinear velocity" and the associated through-variable is the force. However, we could consider the moment of inertia (in "rotatory" motion) as the inertia element, leading to the same observations as before. The associate across-variable is the "angular velocity" and the associated through-variable is the torque. In general then, we can consider an inertia element having a "generalized force" (which includes both force and torque) and a "generalized velocity" (which includes both rectilinear velocity and angular velocity).

2.2.2 Spring (Stiffness or Flexibility) Element

Consider the spring element (linear) shown in Figure 2.1b. The constitutive equation (physical law) for a spring is given by Hooke's law:

$$\frac{df}{dt} = kv \tag{2.5}$$

Here, k is the stiffness (inverse of "flexibility") of the spring.

Note: We have differentiated the familiar force-deflection Hooke's law, in order to use variables that are consistent with the variable choice for an inertia element (i.e., velocity and force).

Now following the same steps as for the inertia element, the energy of a spring element may be expressed as follows:

$$E = \int fv\,dt = \int f\frac{1}{k}\frac{df}{dt}dt = \int \frac{1}{r} f\,df$$

or

$$\text{Energy } E = \frac{1}{2}\frac{f^2}{k} \tag{2.6}$$

This is the well-known (elastic) *potential energy*.

Also

$$f(t) = f(0^-) + k\int v\,dt \tag{2.7}$$

Furthermore, assuming that the applied velocity v is finite (which is a realistic assumption), we have

$$f(0^+) = f(0^-) \tag{2.8}$$

Through these results, we may state the following facts:

1. A spring is an energy storage element. It stores elastic *potential energy*.
2. The force through a spring element cannot change instantaneously unless an infinite velocity is applied to it (not practical). In other words, a finite velocity input cannot cause a step change in force in a spring element. Conversely, a finite instantaneous (step) change in force in a spring will need an infinite velocity, which is not realistic. Hence, force f is a natural output (or response) variable for a spring element, which can represent its dynamic state (i.e., state variable), and velocity v is a natural input variable for a spring element.
3. The fact that force is most appropriate to represent the "dynamic" state of a spring element is justified by several reasons: The first justification is Item 2; the second, from equation (2.7), the force of a spring element at any time t can be completely determined with the knowledge of the initial force and the applied velocity during the time interval 0 to t (*Note*: As we will see in Chapter 3, this is the formal definition of state); and the third, from equation (2.6), the energy of a spring element can be represented by the variable f alone (elastic potential energy).

4. Since its state variable (force) is a through-variable, a spring is a T-type element. *Note*: As stated under the inertia element, the term "force" in the context of a spring element represents a "generalized" force, which includes both force and torque. In other words, the same treatment as given above applies to a torsional (rotatory) spring whose variables are torque and angular velocity.

2.2.3 Damping (Dissipation) Element

Consider the mechanical damper (linear viscous damper or dashpot) shown in Figure 2.1c. It is a D-type element (energy-dissipating element). Its constitutive equation (physical law) is

$$f = bv \tag{2.9}$$

Here, b is the damping constant. Equation (2.9) is an algebraic equation. Hence, either f or v can serve as the natural output variable for a damper, and either one can determine its dynamic state. However, since the state variables v and f are established by an independent inertial element and an independent spring element, respectively, a damper will not introduce a new state variable. Its state is established by the states of the independent energy storage elements (inertia and spring elements) that are connected to it in the system.

In summary:

1. A mechanical damper is an energy-dissipating element (D-type element).
2. Either force f or velocity v may represent its dynamic state (it is established by the states of the independent energy storage elements that are connected to it in the system).
3. No new state variable is introduced by a damper.

   Note: Again, the term "force" in the context of a damper represents a "generalized" force, which includes both force and torque. In other words, the same treatment as given above applies to a torsional (rotatory) damper whose variables are torque and angular velocity.

2.3 ELECTRICAL ELEMENTS

In electrical systems, the *capacitor* is the A-type element, with voltage (across-variable) as its state variable; and the *inductor* is the T-type element, with current (through-variable) as its state variable. These are energy storage elements and their constitutive equations are differential equations. The *resistor* is the energy dissipater (D-type element) and as typical, with an algebraic constitutive equation; it does not introduce a new state variable. These three elements are discussed below. The input elements (or source elements) of an electrical system are the *voltage source*, where its voltage is the independent variable, which is not affected by the changes in the system (while the associated current—the dependent variable—will be affected); and the *current source*, where its current is the independent variable, which is not affected by the changes in the system (while the associated voltage—the dependent variable—will be affected). These are "ideal" sources since in practice the source variable will be affected to some extent by the dynamics of the system (this effect is known as "loading") and is not completely "independent."

2.3.1 Capacitor Element

Consider the capacitor element shown in Figure 2.2a. Its constitutive equation (the physical law) is given by the differential equation:

$$C\frac{dv}{dt} = i \tag{2.10}$$

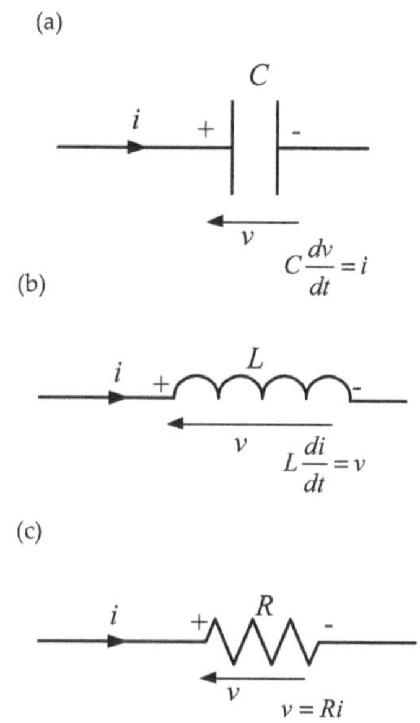

FIGURE 2.2 Basic electrical elements: (a) capacitor; (b) inductor; and (c) resistor.

Here, v denotes the voltage "across" the capacitor with *capacitance* C, and i is the current "through" the capacitor. Since power is given by the product iv, the energy in a capacitor may be obtained by substituting equation (2.10) in this product and integrating it, as

$$E = \int iv\,dt = \int C\frac{dv}{dt}v\,dt = \int Cv\,dv$$

or

$$\text{Energy } E = \frac{1}{2}Cv^2 \tag{2.11}$$

This is the familiar *electrostatic energy* of a capacitor. Also

$$v(t) = v(0^-) + \frac{1}{C}\int i\,dt \tag{2.12}$$

Hence, for a capacitor with a finite current, we have (as for a mechanical inertia element)

$$v(0^+) = v(0^-) \tag{2.13}$$

Using these results, we may state the following facts:

1. A capacitor is an energy storage element. It stores *electrostatic energy.*
2. The voltage across a capacitor element cannot change instantaneously unless an infinite current is applied to it (which is not practical). Specifically, a finite current cannot cause a step change in voltage in a capacitor element. Conversely, a finite instantaneous (step) change in voltage in a capacitor element will need an infinite current input (which is not realistic). Hence, voltage v is a natural output (or response) variable for a capacitor element, which can represent its dynamic state (i.e., voltage is a proper state variable for a capacitor), and current i is a natural input variable for a capacitor element.

3. The fact that voltage is most appropriate to represent the "dynamic" state of a capacitor element can be justified for several reasons: The first justification is Item 2; the second, from equation (2.12), the voltage of a capacitor element at any time t can be completely determined with the knowledge of the initial voltage and the applied current during the time interval 0 to t (*Note*: As we will see in Chapter 3, this is the formal definition of state); and the third, from equation (2.11), the energy of a capacitor element can be represented by the variable v alone (electrostatic energy).
4. Since its state variable (voltage) is an across-variable, a capacitor is an A-type element.

2.3.2 Inductor Element

Consider the inductor element shown in Figure 2.2b. Its constitutive equation (the physical law) is given by the differential equation:

$$L\frac{di}{dt} = v \tag{2.14}$$

Here, L is the *inductance* of the inductor. As before, it can be easily shown that energy in an inductor is given by

$$\text{Energy } E = \frac{1}{2}Li^2 \tag{2.15}$$

This is the well-known *electromagnetic energy* of an inductor. Also, by integrating equation (2.14), we obtain

$$i(t) = i(0^-) + \frac{1}{L}\int v\,dt \tag{2.16}$$

Hence, for an inductor with a finite voltage (realistic case), we have (as for a spring)

$$i(0^+) = i(0^-) \tag{2.17}$$

Through these results, we may state the following facts:

1. An inductor is an energy storage element. It stores *electromagnetic energy.*
2. The current through an inductor element cannot change instantaneously unless an infinite voltage is applied to it (not practical). In other words, a finite voltage cannot cause a step change in current in an inductor element. Conversely, a finite instantaneous (step) change in current in an inductor will need an infinite voltage input (which is not realistic). Hence, current i is a natural output (or response) variable for an inductor element, which can represent its dynamic state (i.e., state variable), and voltage v is a natural input variable for an inductor element.
3. The fact that current is most appropriate to represent the "dynamic" state of an inductor element may be justified by several reasons: The first justification is Item 2; the second, from equation (2.16), the current of an inductor element at any time t can be completely determined with the knowledge of the initial current and the applied voltage during the time interval 0 to t (*Note*: As we will see in Chapter 3, this is the formal definition of state); and the third, from equation (2.15), the energy of an inductor element can be represented by the variable i alone (electromagnetic energy).
4. Since its state variable (current) is a through-variable, an inductor is a T-type element.

2.3.3 Resistor (Dissipation) Element

Consider the resistor element shown in Figure 2.2c. It is a *D*-type element (energy-dissipating element). Its constitutive equation (physical law) is the well-known Ohm's law:

$$v = Ri \tag{2.18}$$

Here, R is the *resistance* of the resistor. Equation (2.18) is an algebraic equation. Hence, either v or i can serve as the natural output variable for a resistor, and either one can determine its dynamic state. However, since the state variables v and i are established by the independent capacitor elements and independent inductor elements of the system, a resistor will not introduce a new state variable.

In summary:

1. An electrical resistor is an energy dissipating element (*D*-type element).
2. Either current i or voltage v may represent its state.
3. No new state variable is introduced by a resistor element.

2.4 FLUID ELEMENTS

In a fluid component, pressure (P) is the across-variable and the volume flow rate (Q) is the through-variable. The three basic fluid elements are shown in Figure 2.3 and discussed below. Note the following:

1. The elements are usually distributed, but lumped-parameter approximations are used here.
2. The elements are usually nonlinear (particularly, the fluid resistor), but primarily linear models are used here.

The input elements (or source elements) of a fluid system are the *pressure source*, where its pressure is the independent variable, which is not affected by the changes in the system (while the associated flow rate variable—the dependent variable—will be affected); and the *flow source*, where its flow rate is the independent variable, which is not affected by the changes in the system (while the associated pressure variable—the dependent variable— will be affected).

2.4.1 Fluid Capacitor or Accumulator (*A*-Type Element)

Consider a rigid container with a single inlet through which fluid is pumped in at the volume flow rate Q, as shown in Figure 2.3a. The pressure inside the container with respect to the outside is P. Its linear constitutive equation is

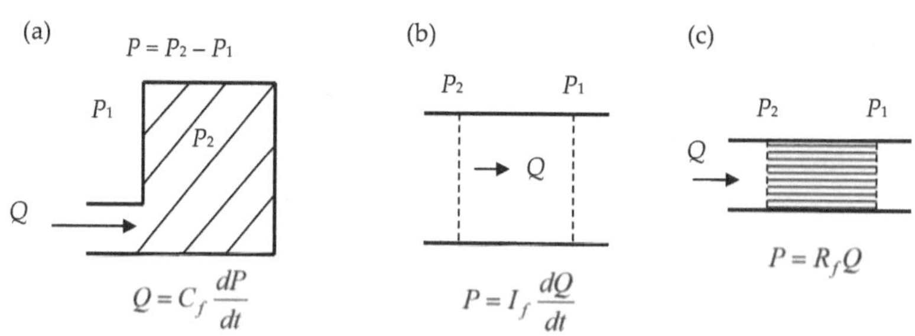

FIGURE 2.3 Basic fluid elements: (a) capacitor; (b) inertor; and (c) resistor.

$$C_f \frac{dP}{dt} = Q \tag{2.19}$$

Here, C_f=*fluid capacitance* (capacity). Several special cases of fluid capacitor will be discussed later.

A fluid capacitor stores potential energy, given by $\frac{1}{2}C_f P^2$. Hence, this element is like a fluid spring. However, it is an *A*-type element, and its appropriate state variable is the pressure difference (across-variable) *P*. Contrast here that the mechanical spring is a *T*-type element.

2.4.2 Fluid Inertor (*T*-Type Element)

A fluid inertor is sometimes called a fluid "inductor" in view of its analogy to an electric inductor. Consider a conduit with an accelerating flow of fluid, as shown in Figure 2.3b. The associated linear constitutive equation is

$$I_f \frac{dQ}{dt} = P \tag{2.20}$$

Here, I_f = fluid *inertance*. It should be clear that this parameter represents fluid inertia (not fluid spring) yet it is analogous to electrical inductance or mechanical spring. To explain further, a fluid inertor stores kinetic energy, given by $\frac{1}{2}I_f Q^2$. Hence this element is a fluid inertia. The appropriate state variable of it is the volume flow rate (through-variable) *Q*. Hence, fluid inertor is a *T*-type element, just like an electrical inductor or mechanical spring. Contrast here that mechanical inertia is an *A*-type element.

Energy exchange between a fluid capacitor and a fluid inertor leads to oscillations (e.g., water hammer of pipes in buildings) in fluid systems, analogous to oscillations in mechanical and electrical systems.

2.4.3 Fluid Resistor (*D*-Type Element)

Consider the flow of fluid through a narrow element such as a thin pipe, orifice, or valve. The associated flow will result in energy dissipation due to fluid friction. Its linear constitutive equation is (see Figure 2.3c)

$$P = R_f Q \tag{2.21}$$

Here, R_f = fluid resistance.

2.4.4 Derivation of Constitutive Equations

We now indicate the derivation of the constitutive equations for fluid elements.

2.4.4.1 Fluid Capacitor

The capacitance in a fluid element may originate from

a. Bulk modulus (compressibility) effect of liquids
b. Compressibility effect of gases
c. Flexibility of the fluid container
d. Gravity head of a fluid column.

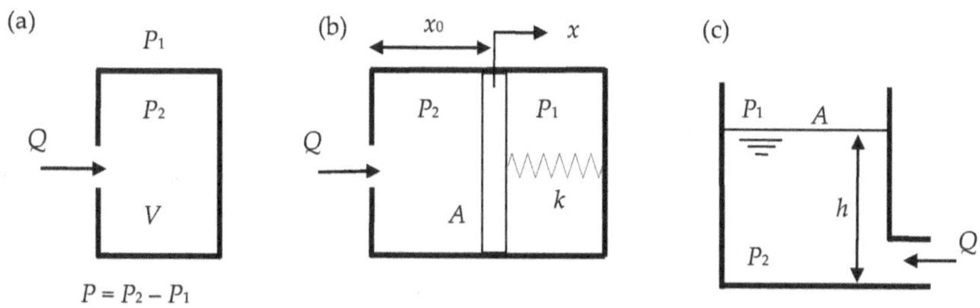

FIGURE 2.4 Three types of fluid capacitor: (a) compressible fluid (liquid bulk modulus or gas compressibility); (b) flexible container; and (c) liquid column with gravity head.

The derivation of the associated constitutive equations is outlined next.

2.4.4.1.1 Capacitance Due to Liquid Compressibility (Bulk Modulus)

Consider a rigid container. A liquid is pumped into it at the volume flow rate Q. An increase in the pressure in the container will compress the liquid volume in it, thereby letting in more liquid (see Figure 2.4a). From calculus, we have

$$\frac{dP}{dt} = \frac{\partial P}{\partial V}\frac{dV}{dt}$$

Here, V is the control volume of liquid. The volume flow rate (into the container) is given by

$$Q = -\frac{dV}{dt}$$

Note: The liquid volume V has to decrease (i.e., a negative increase) for a flow into the container (positive) to take place.

The bulk modulus of a liquid is defined by

$$\beta = -\lim_{\Delta V \to 0} \frac{\Delta P}{\Delta V / V} = -V\frac{\partial P}{\partial V} \tag{2.22}$$

Note: When the pressure increases, the liquid volume decreases. Hence, the negative sign in this expression ($\beta > 0$).

Or, from (2.22), $\beta = -V\frac{dP}{dt}\Big/\frac{dV}{dt} = V\frac{dP}{dt}\Big/Q$

Hence,

$$\frac{V}{\beta}\frac{dP}{dt} = Q \tag{2.23}$$

The associated fluid capacitance is

$$C_{\text{bulk}} = \frac{V}{\beta} \tag{2.24}$$

2.4.4.1.2 Capacitance due to Gas Compressibility

A perfect (ideal) gas is governed by the gas law:

$$PV = nRT \tag{2.25a}$$

Here

P=pressure (in units of pascals: 1 Pa=1 N/m^2)
V=volume (in units of m^3)
T=absolute temperature (in units of K or Kelvin)
n=[mass of gas (kg)]/molecular mass of the gas (kg)]
=number of "moles" in the gas in the volume
R=8.3145 J/mol/K=universal gas constant (same value for any ideal gas)

Note: 1 J=1 joule=1 Nm; 1 kJ=1000 J

Equation (2.25a) can also be expressed as follows:

$$PV = mR_{sp}T \tag{2.25b}$$

Here

m = mass of the volume of gas (kg)
R_{sp} = specific (or individual) gas constant (J/kg/K).

By comparing equations (2.25a) and (2.25b) and using the definition of n, it is easy to see that

$$R_{sp} = \frac{R}{\text{Molecular mass of the gas}}$$

For example, for oxygen, whose molecular mass is 32 g=32×10^{-3} kg, the specific gas constant is

$$R_{sp} = \frac{8.3145}{32\times10^{-3}} = 259.8 \text{ J/kg/°K}$$

Note: Clearly, the specific gas constant depends on the molecular mass of the gas and hence varies from gas to gas. The specific gas constants of some common gases are given in Table 2.1.

TABLE 2.1
Specific Gas Constants of Common Gases

Gas	Specific Gas Constant R_{sp} (J/kg/K)	Molecular Mass (kg)
Air	286.9	29×10^{-3}
Argon	208	40×10^{-3}
Ammonia	488	17×10^{-3}
Carbon dioxide	188.9	44×10^{-3}
Carbon monoxide	297	28×10^{-3}
Helium	2077	4×10^{-3}
Hydrogen	4154	2×10^{-3}
Methane	518.3	16×10^{-3}
Nitrogen	296.8	28×10^{-3}
Oxygen	259.8	32×10^{-3}
Propane	189	44×10^{-3}
Sulfur dioxide	130	64×10^{-3}
Water vapor	461.5	18×10^{-3}

- **Isothermal Case**: Consider a slow flow of gas into a rigid container (see Figure 2.4a) so that the heat transfer is allowed to maintain a constant temperature (i.e., isothermal condition). Differentiate equation (2.25) keeping T constant (i.e., RHS is constant):

$$P\frac{dV}{dt}+V\frac{dP}{dt}=0$$

In the case of a liquid in a rigid container,

$$Q=-\frac{dV}{dt}$$

Substitute this and equation (2.25b) into the above equation. We obtain

$$\frac{V}{P}\frac{dP}{dt}=\frac{mR_{sp}T}{P^2}\frac{dP}{dt}=Q \tag{2.26}$$

Hence, the corresponding fluid capacitance is given by

$$C_{comp}=\frac{V}{P}=\frac{mR_{sp}T}{P^2} \tag{2.27}$$

- **Adiabatic Case**: Consider a fast flow of gas (see Figure 2.4a) into a rigid (and possibly thermally insulated) container so that there is no time (or possibility) for heat transfer (*Note*: Adiabatic ⇒ zero heat transfer). The associated gas law is known to be

$$PV^k=c \text{ with } k=c_P/c_V \tag{2.28}$$

Here

c_p = specific heat when the pressure is maintained constant
c_v = specific heat when the volume is maintained constant
c = constant
k = ratio of specific heats

Note: Specific heat=amount of heat transfer that results in a unity change in temperature in an object of unity mass.

Differentiate equation (2.28):

$$PkV^{k-1}\frac{dV}{dt}+V^k\frac{dP}{dt}=0$$

Divide by V^k

$$\frac{Pk}{V}\frac{dV}{dt}+\frac{dP}{dt}=0$$

Now, use $Q=-\frac{dV}{dt}$ as usual, and also substitute equation (2.25):

$$\frac{V}{kP}\frac{dP}{dt}=\frac{mR_{sp}T}{kP^2}\frac{dP}{dt}=Q \tag{2.29}$$

The corresponding fluid capacitance is

$$C_{comp}=\frac{V}{kP}=\frac{mR_{sp}T}{kP^2} \tag{2.30}$$

2.4.4.1.3 Effect of Flexible Container

Without loss of generality, consider a cylinder of cross-sectional area A with a spring-loaded wall (of stiffness k), as shown in Figure 2.4b. As a fluid (assumed incompressible, but this can be relaxed, as noted later) is pumped into the cylinder, the flexible wall will move through x.

$$\text{Conservation of flow}: Q = \frac{d\left(A(x_0 + x)\right)}{dt} = A\frac{dx}{dt} \tag{2.31}$$

$$\text{Equilibrium of spring}: A(P_2 - P_1) = kx \text{ or } x = \frac{A}{k}P \tag{2.32}$$

Note: P_2 = interior pressure (of the fluid); P_1 = exterior (ambient) pressure; and x_0 = length of the cylinder interior when $P_2 = P_1$

Substituting (2.32) in (2.31), we obtain

$$\frac{A^2}{k}\frac{dP}{dt} = Q$$

The corresponding capacitance is

$$C_{\text{flex}} = \frac{A^2}{k}$$

Note: For an elastic container and a liquid having bulk modulus, the combined capacitance will be additive:

$$C_{\text{eq}} = C_{\text{bulk}} + C_{\text{flex}}$$

This is true because the pressure difference P is the same (i.e., common) with respect to the change in the fluid and container volumes, while the volume flow rates due to the two effects are additive. In other words, the capacitors of the two cases are connected in parallel, and hence, their capacitances are additive.

A similar result holds for a compressible gas and a flexible container; that is,

$$C_{\text{eq}} = C_{\text{comp}} + C_{\text{flex}}$$

Example 2.1

A container that carries a liquid has flexible walls. The wall flexibility is not uniform and is represented as in Figure 2.5, where there are three segments with area A_i and stiffness $k_i, i = 1,2,3$.

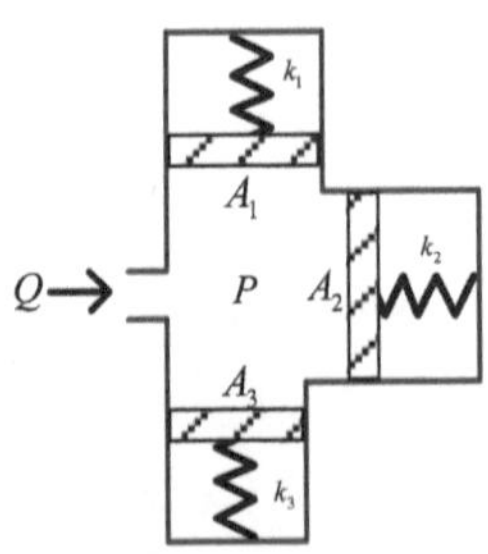

FIGURE 2.5 Model of a flexible container of liquid.

Liquid can enter the container through its opening at the volume flow rate Q. The internal pressure in the container is assumed uniform at P with respect to the ambient pressure.

i. Derive an expression for the fluid capacitance of the container (disregarding the compressibility of the liquid or assuming the liquid to be incompressible).
Note: Capacitance due to liquid compressibility can be incorporated separately (this corresponds to a parallel connection of the two capacitor elements, one for the container flexibility and the other for the liquid compressibility).
ii. Discuss why the present model (of three area segments) cannot be generalized to a case with a large number of area segments with different flexibilities.

Solution

i. Suppose that when the springs are not compressed (i.e., when the outside pressure=inside pressure, where $P = 0$), the volume of the liquid in the container is V_0. As more liquid enters the container, the inside pressure rises and the three walls deflect through x_1, x_2, and x_3.

The new volume of the liquid is

$$V = V_0 + A_1x_1 + A_2x_2 + A_3x_3 = V_0 + \sum_{i=1}^{3} A_i x_i \tag{i}$$

Also, force balance for each wall gives (this neglects the wall weight or assumes that the forces are horizontal)

$$PA_1 = k_1x_1; \quad PA_2 = k_2x_2; \quad PA_3 = k_3x_3$$

or

$$PA_i = k_i x_i, \quad i = 1,2,3 \tag{ii}$$

Differentiate (i) with respect to (wrt) time

$$\dot{V} = Q = A_1\dot{x}_1 + A_2\dot{x}_2 + A_3\dot{x}_3 = \sum_{i=1}^{3} A_i \dot{x}_i$$

Differentiate (ii) and substitute

$$Q = \frac{A_1^2}{k_1}\dot{P} + \frac{A_2^2}{k_2}\dot{P} + \frac{A_3^2}{k_3}\dot{P} = \left(\frac{A_1^2}{k_1} + \frac{A_2^2}{k_2} + \frac{A_3^2}{k_3}\right)\dot{P}$$

Hence, the equivalent fluid capacitance is

$$C_{\text{flex}} = \frac{A_1^2}{k_1} + \frac{A_2^2}{k_2} + \frac{A_3^2}{k_3} = \sum_{i=1}^{3} \frac{A_i^2}{k_i}$$

ii. This result "cannot" be generalized as $C_{\text{flex}} = \sum_{i=1}^{n} \frac{A_i^2}{k_i}$ for large n and continuous wall.

One reason is that the wall segments are interconnected and are not independent unlike the three-segment model in Figure 2.5. Hence, the stiffness values of the different segments are not independent. Also, the force balance equations for the different segments are not independent.

2.4.4.1.4 *Gravity Head of a Fluid Column*

Consider a uniform liquid column (uniform tank) having area of cross-section A, height h, and mass density ρ, as shown in Figure 2.4c. A liquid is pumped into the tank at the volume rate Q. As a result, the liquid level rises.

Relative pressure (wrt the ambience) at the foot of the column $P = P_2 - P_1 = \rho g h$ or $\frac{dP}{dt} = \rho g \frac{dh}{dt}$.

$$\text{Flow rate } Q = \frac{d(Ah)}{dt} = A\frac{dh}{dt}$$

Direct substitution gives

$$\frac{A}{\rho g}\frac{dP}{dt} = Q \tag{2.33}$$

The corresponding capacitance is

$$C_{\text{grav}} = \frac{A}{\rho g} \tag{2.34}$$

Note: The gravity capacitance cannot be simply added to the capacitances due to the fluid compressibility and the container flexibility, because the latter two assume a uniform pressure distribution of the fluid in the container, which is not the case in a liquid column (the pressure decreases from the bottom to the top). Hence, to combine these three capacitances, the liquid column can be divided into a series of elemental fluid layers (where the pressure is uniform), and the three capacitances added for each layer result has to be integrated over the height of the fluid column. Then, this elemental result has to be integrated, along the liquid column, to obtain the overall fluid capacitance.

2.4.4.2 Fluid Inertor

First, assume an "ideal" flow of fluid in a conduit, with a uniform velocity distribution across it. Along a "small" element of length Δx of fluid, as shown in Figure 2.6, the pressure difference is $P_2 - P_1 = P$, and the volume flow rate is Q (neglect the change ΔQ, over small distance Δx).

Mass density of the fluid = ρ
Area of cross-section of the tank = A
Mass of the fluid element = $\rho A \Delta x$
Net force in the direction of flow = PA
Velocity of flow (assumed uniform) = Q/A

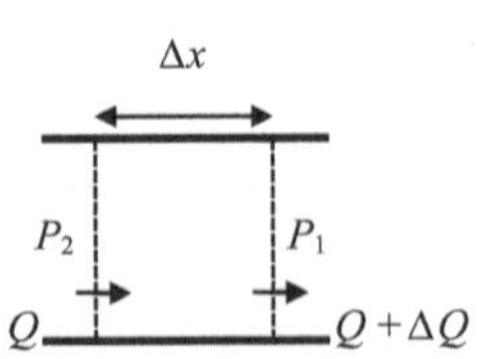

FIGURE 2.6 A fluid flow element.

Note: Since the fluid element (small), over its length in the conduit is uniform → A is constant.

$$\text{Fluid acceleration}: \frac{1}{A}\frac{dQ}{dt}$$

$$\text{From Newton's second law}: PA = (\rho A \Delta x)\frac{1}{A}\frac{dQ}{dt}$$

or

$$\frac{\rho \Delta x}{A}\frac{dQ}{dt} = P \tag{2.35}$$

Hence,

$$\text{Fluid } \textit{inertance } I_f = \frac{\rho \Delta x}{A} \tag{2.36a}$$

Using this result (applicable to a small element), for a nonuniform conduit with the area of cross-section $A = A(x)$ and length L, we can express the inertance as follows:

$$I_f = \int_0^L \frac{\rho}{A(x)}dx \tag{2.36b}$$

In a general (nonideal) flow, the velocity profile in a cross-section is not uniform, and a boundary layer may be present. The fluid velocity will be zero at the wall of the conduit and will increase over the boundary layer. Then, we can modify the result (equation 2.36a) as follows:

$$I_f = \alpha \frac{\rho \Delta x}{A} \tag{2.36}$$

Here, α is a correction factor.

Note: It should be clear that $\alpha > 1$ in general. To verify this, note that the flow rate Q that is used in the constitutive equation of an inertor corresponds to the average velocity (as given by Q/A) which is equal to the actual velocity only if the velocity profile (across the conduit) is uniform. The actual inertia force is governed by the local velocity (strictly, local acceleration) which is not uniform. In fact, in the core of the fluid flow, it is quite higher than the average value. Hence, the I_f that assumes uniform velocity should be corrected up.

It can be shown that for a conduit of circular cross-section and a parabolic velocity profile (a case of *laminar flow*, the opposite of turbulent flow), the inertance is given by

$$I_f = \frac{2\rho \Delta x}{A} \tag{2.36c}$$

2.4.4.2.1 Laminar Flow

This corresponds to low-velocity, high-viscosity flow where the flowing fluid moves in smooth and continuous layers that slide over one another. In fact, this is the case of low Reynolds number (Re), which is given by

$$\text{Re} = \frac{vL}{\upsilon} = \frac{\rho v L}{\mu} \tag{2.37}$$

Here

L = length of the pipe segment
ρ = absolute viscosity of the fluid (dynamic viscosity)
$\upsilon = \mu/\rho$ = kinematic viscosity
υ = fluid velocity along the pipe.

Note: Fluid stress $= \mu \frac{dv}{dy}$, where $\frac{du}{dy}$ is the velocity gradient across the pipe, and it governs the viscous force in the fluid.

2.4.4.3 Fluid Resistor

2.4.4.3.1 Laminar Flow

For the ideal case of viscous, laminar flow, we have the linear relationship for a fluid resistor (Figure 2.3c)

$$P = R_f Q \tag{2.38}$$

The fluid resistance is given by

$$R_f = 128 \frac{\mu L}{\pi d^4} \text{ for a circular pipe of diameter } d$$

$$R_f = 12 \frac{\mu L}{wb^3}, \text{ with } b \ll w \text{ for a pipe of rectangular cross-section of width } w \text{ and height } b.$$

Here

L = length of the pipe segment
μ = absolute (dynamic) viscosity of the fluid.

2.4.4.3.2 Turbulent Flow

For turbulent flow (i.e., high Reynolds number flow), the resistance equation will be nonlinear and is given by

$$P = K_R Q^n \tag{2.39}$$

2.5 THERMAL ELEMENTS

For a thermal element, temperature (T) is the across-variable, as it is measured with respect to some reference (e.g., ambient temperature) or as a temperature difference across an element, and heat transfer (flow) rate (Q) is the through-variable. Heat source and temperature source are the two types of source elements (input elements) in a thermal system. However, heat source is much more common than temperature source. A temperature source may correspond to a large reservoir whose temperature is hardly affected by heat transfer into or out of it.

There is only one type of energy (thermal energy) in a thermal system. The corresponding energy storage element is a *thermal capacitor* with the associated state variable, temperature. Hence, this is an *A*-type element. There is no *T*-type element in a thermal system (i.e., there are no thermal inductors). As a direct consequence of the absence of two different types of energy storage elements (unlike the case of mechanical, electrical, and fluid systems), a pure thermal system cannot exhibit natural (and free) oscillations. It can exhibit "forced" oscillations, however, when excited by an oscillatory input source or if it is coupled with a system in another domain (e.g., a fluid system resulting in a thermo-fluid system) that can oscillate.

2.5.1 Constitutive Equations

The constitutive equations in a thermal system are the physical equations for thermal capacitors (*A*-type elements) and thermal resistors (*D*-type elements). There are no *T*-type thermal elements. The constitutive equations of the basic thermal elements are discussed now.

2.5.2 Thermal Capacitor

A thermal capacitor has the "capacity" to store thermal energy. Consider a control volume of an object, with various heat transfer processes Q_i taking place at the boundary of the object (See Figure 2.7). The amount of thermal energy in the object$=\rho VcT$, where

T = temperature of the object (assumed uniform)
V = volume of the object
ρ = mass density of the object
c = specific heat of the object.

Note: Specific heat = amount of heat transfer into an object of unity mass that results in a unity change in temperature in the object.

Since the net heat inflow (transfer) is equal to the rate of change (increase) of thermal energy, the associated constitutive relation is

$$\sum Q_i = \rho Vc\frac{dT}{dt} \tag{2.40}$$

It is assumed that ρVc is constant. We write the result (2.40) as

$$C_h\frac{dT}{dt} = Q \tag{2.41}$$

Here, $C_h = \rho Vc = mc$ = thermal capacitance; with $m = \rho V$ = mass of the element.

2.5.3 Thermal Resistor

A thermal resistor provides resistance to heat transfer in a body or a medium. There are three general types of thermal resistance:

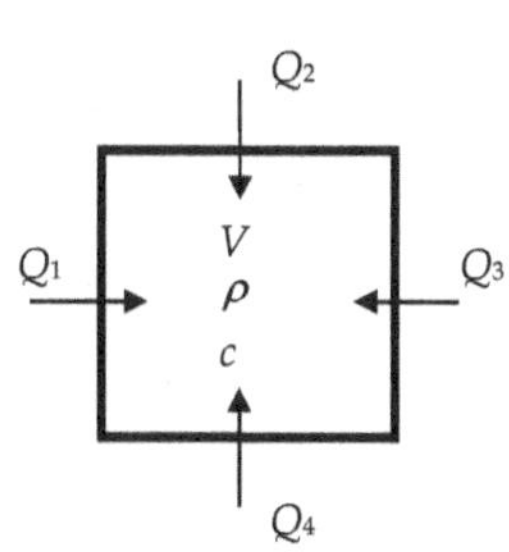

FIGURE 2.7 A control volume representing a thermal capacitor.

1. Conduction
2. Convection
3. Radiation

We now give the constitutive relation for each of these three types of thermal resistors.

2.5.3.1 Conduction

In the conduction heat transfer through a medium, the molecules of the medium itself do not move to carry the thermal energy (heat) from one place to another. Instead, heat transfer takes place due to the collision of adjoining microscopic particles and the associated energy transfer (e.g., change in the spinning speed of electrons). Conduction heat transfer takes place from a point of higher temperature to one of lower temperature. Specifically, heat conduction rate is proportional to the negative temperature gradient and is given by the *Fourier equation*:

$$Q = -kA\frac{\partial T}{\partial x} \tag{2.42a}$$

Here,

x = direction of heat transfer
A = area of cross-section of the path (element) of heat transfer
k = thermal conductivity.

The (Fourier) equation (2.42a) is a "local" equation. If we consider a finite object of length Δx and area of cross-section A, with temperatures T_2 and T_1 at the two ends, as shown in Figure 2.8a, the one-dimensional heat transfer rate Q can be written according to equation (2.42a) as

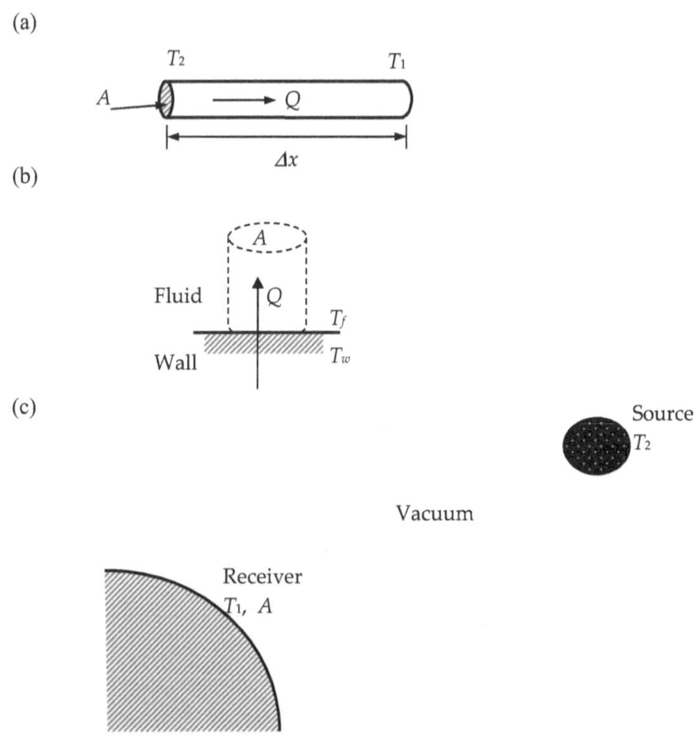

FIGURE 2.8 Three types of thermal resistance: (a) An element of 1-D heat conduction; (b) a control volume for heat transfer by convection: (c) heat transfer by radiation.

$$Q = kA\frac{(T_2 - T_1)}{\Delta x} = \frac{kA}{\Delta x}T \tag{2.42b}$$

Here, $T = T_2 - T_1$=temperature drop along the element.
or

$$T = R_k Q \tag{2.42}$$

with

$$R_k = \frac{\Delta x}{kA} = \text{conductive thermal resistance} \tag{2.43}$$

Nominal thermal conductivities of some useful materials are given in Table 2.2.

2.5.2.1.1 Three-Dimensional Conduction

In general, the conditions (and parameters) of heat transfer will change from location to location. Then, we need to introduce one or more space variables as independent variables, in addition to time t. In other words, we will need a *distributed-parameter model*, which is represented by partial differential equations (rather than ordinary differential equations, for lumped-parameter systems).

TABLE 2.2
Thermal Conductivities of Some Material (at 25°C)

Material	Thermal Conductivity (W/m/°C)
Acrylic	0.2
Air	0.025
Aluminum	210
Asphalt	0.75
Bitumen	0.17
Brass, bronze	110
Brick	0.75
Cast Iron	58
Cement, mortar	1.7
Chalk	0.09
Coal	0.2
Cobalt	70
Concrete	1.0
Copper	401
Cork	0.07
Glass, pyrex	1.0
Gold	310
Granite, marble	2.5
Paper	0.05
Rubber	0.07
Silver	430
Steel, structural	43
Steel, stainless	16
Water	0.6
Wood	0.15

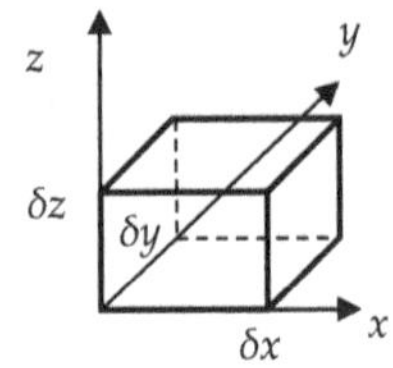

FIGURE 2.9 A 3-D heat conduction element.

Equation (2.42a) represents a one-dimensional (1-D) distributed parameter model, with the location x as an independent variable. This is further confirmed by the use of the partial derivative $\frac{\partial T}{\partial x}$, giving a "local" condition wrt x. Extending this, conduction heat transfer in a continuous 3-D medium is represented by a distributed-parameter model, with partial derivatives wrt all three dimensions. Then, the Fourier equation (2.42a) is applicable in each of the three orthogonal directions (x, y, z). To obtain the corresponding model, we use the thermal capacitance equation (2.40) as well.

Consider the small 3-D model element of sides δx, δy, and δz, in a conduction medium, as shown in Figure 2.9. Heat transfer into the element through the bottom surface (area $\delta x \times \delta y$) in the z direction, according to (2.42a), is $-k\,\delta x\,\delta y\frac{\partial T}{\partial z}$.

Since the temperature gradient at the top ($dx \times dy$) surface is $\frac{\partial T}{\partial z}+\frac{\partial^2 T}{\partial z^2}dz$ (which includes the temperature "increment" over the distance δz; from calculus), the heat transfer out of this surface is $k\,\delta x\,\delta y\left(\frac{\partial T}{\partial z}+\frac{\partial^2 T}{\partial z^2}\delta z\right)$. Hence, the net heat transfer into the element in the z-direction is $k\,\delta x\,\delta y\frac{\partial^2 T}{\partial z^2}\delta z$ or $k\,\delta x\,\delta y\,\delta z\frac{\partial^2 T}{\partial z^2}$. Similarly, the net heat transfer in the x and y directions are $k\,\delta x\,\delta y\,\delta z\frac{\partial^2 T}{\partial x^2}$ and $k\,\delta x\,\delta y\,\delta z\frac{\partial^2 T}{\partial y^2}$, respectively.

The thermal energy of the element is $\rho\,dx\,dy\,dz\,cT$, where $\rho\,\delta x\,\delta y\,\delta z$ is the mass of the element and c is the specific heat (at constant pressure). Hence, the capacitance equation (2.40) gives

$$k\,\delta x\,\delta y\,\delta z\left(\frac{\partial^2 T}{\partial x^2}+\frac{\partial^2 T}{\partial y^2}+\frac{\partial^2 T}{\partial z^2}\right)=\rho\,\delta x\,\delta y\,\delta z\,c\frac{\partial T}{\partial t}$$

or

$$\frac{\partial^2 T}{\partial x^2}+\frac{\partial^2 T}{\partial y^2}+\frac{\partial^2 T}{\partial z^2}=\frac{1}{\alpha}\frac{\partial T}{\partial t} \tag{2.44}$$

where $\alpha=\frac{k}{\rho c}$ = thermal diffusivity.

Equation (2.44) is called the *Laplace equation.*

As noted before, partial derivatives are applicable here because T is a function of many independent variables (space x, y, z, and time t); and derivatives with respect to all of them would be needed in the model equations. In summary, distributed-parameter models in 3-D have spatial variables (x, y, z) as well as the temporal variable (t) as independent variables and are represented by partial differential equations, not ordinary differential equations.

2.5.2.2 Convection

In convection, the heat transfer takes place by the physical movement of the heat-carrying molecules in the medium (i.e., through mass transfer). An example is the case of fluid flowing against a wall, as shown in Figure 2.8b. The constitutive equation is

$$Q = h_c A(T_w - T_f) \tag{2.45a}$$

Here,

T_w = wall temperature
T_f = fluid temperature at the wall interface
A = area of cross-section of the fluid control volume across which the heat transfer Q takes place through mass transfer
h_c = convection heat transfer coefficient.

In practice, h_c may depend on the temperature itself, and hence, equation (2.45a) is *nonlinear* in general. But, by approximating to a linear constitutive equation, we can write

$$(T_w - T_f) = R_c Q \tag{2.45b}$$

or

$$T = R_c Q \tag{2.45}$$

Here, $T = T_w - T_f$ and

$$R_c = \frac{1}{h_c A} = \text{convective thermal resistance} \tag{2.46}$$

In *natural convection*, the particles in the heat transfer medium move naturally (e.g., due to change in their density due to the temperature change). In *forced convection*, the mass transfer is forced by an actuator such as a fan or a pump.

2.5.2.2.1 Biot Number

Biot number (Bi) is a non-dimensional parameter given by the ratio:

$$\frac{[\text{Conductive resistance}]}{[\text{Convective resistance}]}$$

Now from equations (2.43) and (2.46), we get

$$\text{Bi} = \frac{R_k}{R_c} = \frac{\Delta x h_c A_c}{k A_k} \tag{2.47a}$$

Here we have allowed for the general case where the area of cross-section for the conduction (A_k) may be different from the area of cross-section for the convection (A_c). If the areas are equal, we have

$$\text{Bi} = \frac{h_c \Delta x}{k} \tag{2.47b}$$

The Biot (pronounce "bio") number may be used as the basis for approximating a distributed-parameter model for heat transfer (e.g., equation (2.44)) by an appropriate lumped-parameter model. Specifically, we divide the continuous conduction medium into slices of thickness Δx and use a lumped model for each slice. If for each slice, the Biot number ≤ 0.1 (as a rule of thumb), the used lumped-parameter model is adequate for that slice.

2.5.2.2.2 Rationale for Using Bi to Guide Lumped-Parameter Modeling:

The Biot number is an index of how the temperature varies within a body (through conduction) as heat is supplied into or taken out of the body through its outer surface (through convection). Specifically,

Smaller Bi → smaller temperature variation → better lumped-model approximation

This may be justified as follows: For a given convective resistance R_c and temperature gradient of convection (i.e., for a given heat transfer rate into or out of the body through convection), smaller Bi means smaller R_k. This in turn means smaller temperature gradient (i.e., smaller temperature variation) within the body, due to the heat transfer in conduction.

The rationale for using the expression $\frac{h_c \Delta x}{k}$ (equation (2.47b)) for Bi, which assumes the same area for both conductive and convective heat transfer, is as follows:

1. Smaller Δx → smaller length available for the variation in T (while the other parameters are unchanged) → better accuracy
2. Smaller h_c → smaller heat transfer into the body (while the other parameters are unchanged) → smaller variation in T → better accuracy
3. Larger k → smaller resistance for conduction heat transfer in the body (while the other parameters are unchanged) → smaller variation in T → better accuracy.

Example 2.2

A heat sink is a heat exchanger that is typically used for cooling electronic devices. Generally, it consists of a heat transfer base on which a bank of pins is mounted (see Figure 2.10a). The base of the heat sink is firmly attached to the object whose thermal energy is to be removed in order to

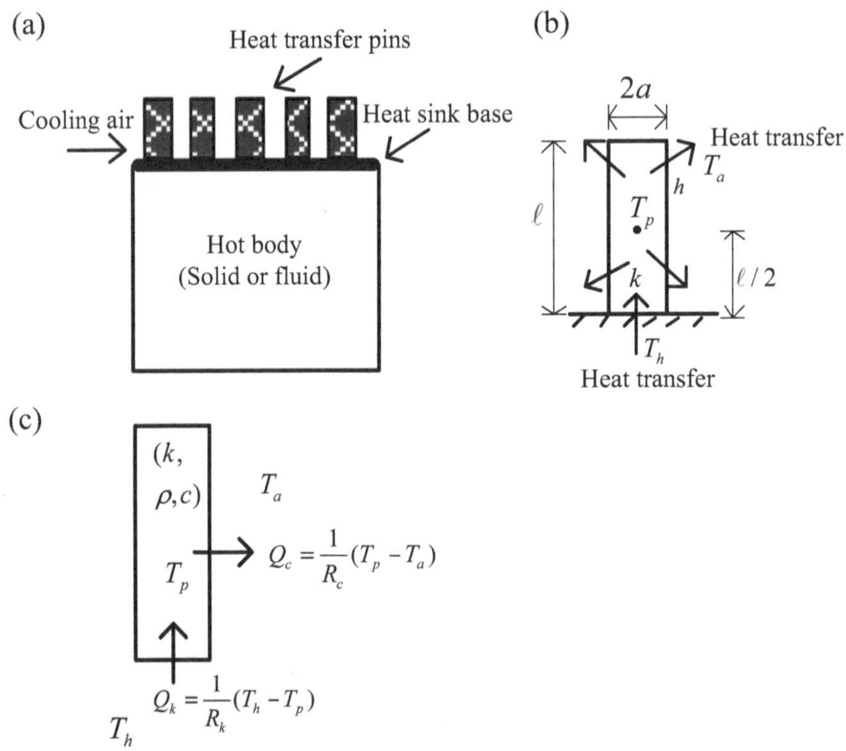

FIGURE 2.10 (a) A heat sink for cooling an object; (b) details of a pin of the heat sink; (c) heat transfer diagram for a pin.

avoid harmful consequences of high temperature in the object. Air (or some other fluid) is forced across the bank of pins to facilitate the heat transfer.

Assume that each pin is a uniform solid cylinder of length ℓ and radius of cross-section a (See Figure 2.10b). The following variables and parameters are defined:

T_h = temperature of the object to be cooled
T_p = average temperature inside a pin
T_a = ambient temperature (of the cross-flow air)
h = convective heat transfer coefficient of the pin surface exposed to air
k = thermal conductivity of a pin
ρ = mass density of a pin
c = specific heat of a pin

a. Obtain a lumped-parameter dynamic model for a single-pin system. What are the system inputs?
b. If ℓ=0.02 m, a=0.001 m, k=150 W/m/°C, and h=30 W/m²/°C, compute the appropriate Biot number for the single-pin system. On that basis, is the lumped model in Part (a) acceptable?

Solution

(a)

See Figure 2.10c.

Represent the temperature of the pin as the average value T_p at the centroid of the pin (*Note*: This is a common assumption in the modeling of thermal systems).

Conductive heat transfer into a pin (from the object):

$$Q_k = \frac{1}{R_k}(T_h - T_p) \tag{i}$$

where

$$R_k = \text{conductive thermal resistance} = \frac{\ell/2}{kA_k} \tag{ii}$$

$$A_k = \text{sectional area of conductive heat transfer} = \pi a^2 \tag{iii}$$

Convective heat transfer from the pin to the ambient air:

$$Q_c = \frac{1}{R_c}(T_p - T_a) \tag{iv}$$

where

$$R_c = \text{convective thermal resistance} = \frac{1}{hA_c} \tag{v}$$

$$A_c = \text{surface area of conductive heat transfer} = 2\pi a\ell + \pi a^2 \tag{vi}$$

$$\text{Mass of the pin } m_p = \pi a^2 \ell \rho \tag{vii}$$

$$\text{Thermal capacitance of the pin } C_t = m_p c \tag{viii}$$

Constitutive equation for the pin:

$$C_t \frac{dT_p}{dt} = Q_k - Q_c \tag{ix}$$

Substitute (i) and (iv) into (ix).

System equation:

$$C_t\frac{dT_p}{dt}=\frac{1}{R_k}\left(T_h-T_p\right)-\frac{1}{R_c}\left(T_p-T_a\right)$$

$$\rightarrow C_t\frac{dT_p}{dt}=-\left(\frac{1}{R_k}+\frac{1}{R_c}\right)T_p+\frac{1}{R_k}T_h+\frac{1}{R_c}T_a$$

T_p = state variable
T_h and T_a are the inputs.
C_t is given by (viii) and (vii).
R_k is given by (ii) and (iii).
R_c is given by (v) and (iv).

(b)

Biot number $\text{Bi}=\dfrac{R_k}{R_c}=\dfrac{\ell/2hA_c}{kA_k}\rightarrow \text{Bi}=\dfrac{lh\left(2\pi al+\pi a^2\right)}{2k\pi a^2}=\dfrac{lh(2l+a)}{2ka}$

Substitute numerical values.

$$\text{Bi}=\frac{0.02\times 30\times(2\times 0.02+0.001)}{2\times 150\times 0.001}=\frac{0.01}{5}\times(2\times 20+1)=0.082<0.1$$

⇒ The lumped model is acceptable.

2.5.2.3 Radiation

Besides conduction and convection, the heat transfer can take place from a higher temperature object (source) to a lower temperature object (receiver) through energy radiation, without needing a physical medium between the two objects (unlike in conduction and convection). This situation is shown in Figure 2.8c. The associated constitutive equation is the *Stefan–Boltzmann law:*

$$Q=\sigma c_e c_r A\left(T_s^4-T_r^4\right) \tag{2.48a}$$

Here,

T_s = temperature of the radiation source
T_r = temperature of the receiver
A = effective (normal) area of the receiver
c_e = effective *emissivity* of the source
c_r = shape factor of the receiver
σ = Stefan–Boltzmann constant (= 5.7×10^{-8} W/m²/°K⁴).

The relationship (2.48a) corresponds to a nonlinear thermal resistor. Heat transfer rate is measured in watts (W), the area in square meters (m²), and the temperature in degrees Kelvin (°K).

2.5.2.3.1 Linearized Radiation Resistor

The nonlinear relation (2.48a) may be linearized in several ways, as discussed in Chapter 4, resulting in the linear relationship

$$\left(T_s-T_r\right)=R_r Q \tag{2.48b}$$

where R_r=radiation thermal resistance.

As one approach, in equation (2.48a), write $T_s^4 - T_r^4 = \left(T_s^2 + T_r^2\right)(T_s + T_r)(T_s - T_r)$. Then, for small temperature changes about some operating condition $(\bar{T}_r, \bar{T}_s)$, we may use the following approximate expression for thermal resistance in equation (2.48b):

$$R_r = \frac{1}{\sigma c_e c_r A\left(\bar{T}_s^2 + \bar{T}_r^2\right)\left(\bar{T}_s + \bar{T}_r\right)} \tag{2.49}$$

Note: The over-bar denotes a representative operating condition.

TABLE 2.3
Some Linear Constitutive Relations

System	Constitutive Relations for		
	Energy Storage Elements		Energy Dissipating Element
Type	*A*-Type (Across) Element	*T*-Type (Through) Element	*D*-Type (Dissipative) Element
Mechanical (translatory): v=velocity f=force	Mass $m\frac{dv}{dt} = f$ (Newton's second law) m=mass	Spring $\frac{1}{k}\frac{df}{dt} = v$ (Hooke's law) k=stiffness	Viscous damper $v = \frac{1}{b}f$ b=damping constant
Electrical: v=voltage i=current	Capacitor $C\frac{dv}{dt} = i$ C=capacitance	Inductor $L\frac{di}{dt} = v$ L=inductance	Resistor $v = Ri$ R=resistance
Thermal: T=temperature difference Q=heat transfer rate	Thermal capacitor $C_t\frac{dT}{dt} = Q$ C_t=thermal capacitance	None	Thermal resistor $T = R_tQ$ R_t=thermal resistance
Fluid: P=pressure difference Q=volume flow rate	Fluid capacitor $C_f\frac{dP}{dt} = Q$ C_f=fluid capacitance	Fluid inertor (Inductor) $I_f\frac{dQ}{dt} = P$ I_f =inertance	Fluid resistor $P = R_fQ$ R_f=fluid resistance

TABLE 2.4
Force-Current Analogy

System Type	Mechanical	Electrical
System-response variables		
Through-variable	Force f	Current i
Across-variable	Velocity v	Voltage v
System parameters	m	C
	$1/k$	L
	$1/b$	R

2.6 DOMAIN ANALOGIES

Table 2.3 summarizes the linear constitutive relationships that describe the behavior of the basic translatory-mechanical, electrical, thermal, and fluid elements. In particular, the analogy used in Table 2.3 between mechanical and electrical elements is known as the *force-current analogy* and is given in Table 2.4. This follows from the fact that both force and current are through-variables, which do not change through an element. They are analogous as well to the fluid flow rate through a pipe and the heat transfer rate through a thermal element. Furthermore, both velocity and voltage are across-variables, which are measured across an element, with respect to one end (reference end), as in the case of fluid pressure along a pipe or temperature across a thermal element. This analogy is more logical than a force-voltage analogy because the structure of component interconnection is preserved across domains, through this analogy (i.e., a parallel connection between elements remains a parallel connection across domains, and a series connection between elements remains a series connection across domains). The correspondence between the parameter pairs given in Table 2.4 follows from the constitutive relations given in Table 2.3. A rotational (rotatory) mechanical element possesses constitutive relations between torque and angular velocity, which can be treated as a generalized force and a generalized velocity, respectively. Hence, a separate entry is not needed in Table 2.3 for rotatory mechanical elements.

2.6.1 Natural Oscillations

Mechanical systems can produce natural (free) oscillatory responses (or, free vibrations) because they can possess two different types of energy (kinetic and potential energies). When one type of stored energy is converted into the other type repeatedly and naturally, back and forth, the resulting response is a free oscillation. Of course, in practice, some of the energy will dissipate (through the dissipative mechanism of a D-type element or damper) and the free natural oscillations will decay as a result. Similarly, electrical circuits and fluid systems can exhibit free, natural oscillatory responses due to the presence of two distinct types of energy storage mechanism, where energy can "flow" back and forth repeatedly between the two types of elements. But, thermal systems have only one type of energy storage element (A-type) with only one type of energy (thermal energy). Hence, purely thermal systems cannot produce natural oscillations. Oscillations are possible, however, when forced by external means or integrated with other types of systems that can produce natural oscillations (e.g., integrating with a fluid system resulting in a thermo-fluid system).

2.7 SUMMARY SHEET

- **Lumped Elements**: Energy-storage elements, energy-dissipation elements, input elements (sources)
- **Across-Variable**: Measured across an element (e.g., velocity, voltage, pressure, temperature)
- **Through-Variable**: Passes through an element, unaltered (e.g., force, current, fluid flow rate, heat transfer rate)
- ***A*-type Element**: An element whose state variable is an across-variable (e.g., mass, electric capacitor, fluid capacitor, thermal capacitor)
- ***T*-type Element**: An element whose state variable is a through-variable (e.g., spring, inductor, inertor, or fluid inductor). There is no thermal T-type element
- ***D*-type Element**: An energy-dissipating element. It does not introduce a new state variable. State variables are introduced by the independent A-type and T-type elements connected to it
- ***A*-type Source**: An input element whose independent variable is an across-variable (e.g., velocity source, voltage source, pressure source, temperature source). *Note*: The independent variable of a source is not to be affected by the system dynamics
- ***T*-type Source**: An input element whose independent variable is a through-variable (e.g., force source, current source, flow source, heat transfer source)

- **Mass (Inertia Element)**: State is velocity v; Constitutive equation $m\frac{dv}{dt} = f \rightarrow v(t) = v(0^-) + \frac{1}{m}\int_{0^-}^{t} f\,dt$; Kinetic energy $E = \frac{1}{2}mv^2$; v cannot change instantaneously
- **Spring (Stiffness Element)**: State is force f; Constitutive equation $\frac{df}{dt} = kv \rightarrow f(t) = f(0^-) + k\int_{0^-}^{t} v\,dt$; potential (elastic) energy $E = \frac{1}{2}\frac{f^2}{k}$; f cannot change instantaneously
- **Damping (Dissipation Element)**: $f = bv$; no new state variables are introduced by it
- **Electric Capacitor**: State is voltage v; constitutive equation $C\frac{dv}{dt} = i \rightarrow v(t) = v(0^-) + \frac{1}{C}\int_{0^-}^{t} i\,dt$; electrostatic energy $E = \frac{1}{2}Cv^2$; v cannot change instantaneously
- **Electric Inductor**: State is current i; constitutive equation $L\frac{di}{dt} = v \rightarrow i(t) = i(0^-) + \frac{1}{L}\int_{0^-}^{t} v\,dt$; electromagnetic energy $E = \frac{1}{2}Li^2$; i cannot change instantaneously
- **Electric Resistor**: $Q = C_f\frac{dP}{dt}$; no new state variables are introduced
- **Fluid Capacitor or Accumulator (*A*-Type Element)**: $C_f\frac{dP}{dt} = Q$; capacitance due to: (a) liquid bulk modulus (β): $C_{\text{bulk}} = \frac{V}{\beta}$; (b) gas compressibility effect: isothermal $C_{\text{comp}} = \frac{V}{P} = \frac{mR_{\text{sp}}T}{P^2}$; adiabatic $C_{\text{comp}} = \frac{V}{kP} = \frac{mR_{\text{sp}}T}{kP^2}$, where ratio of specific heats at constant pressure and constant volume $k = c_P/c_V$; (c) container flexibility (stiffness k, area A): $C_{\text{flex}} = \frac{A^2}{k}$; (d) liquid gravity head (liquid column area A, mass density ρ): $C_{\text{grav}} = \frac{A}{\rho g}$
- **Inertor or Fluid Inductor**: $I_f\frac{dQ}{dt} = P$, inertance: $I_f = \int_0^L \frac{\rho}{A(x)}dx$, $I_f = \alpha\frac{\rho\Delta x}{A}$, α=2 for laminar flow with parabolic velocity profile; Reynolds number $\text{Re} = \frac{vL}{\upsilon} = \frac{\rho vL}{\mu}$, μ=absolute viscosity of fluid (dynamic viscosity), $\upsilon = \frac{\mu}{\rho}$=kinematic viscosity, v=fluid velocity along the pipe
- **Fluid Resistor**: $P = R_fQ$, for viscous, laminar flow: $R_f = 128\frac{\mu L}{\pi d^4}$ for a circular pipe of diameter d and length L; $R_f = 12\frac{\mu L}{wb^3}$, with $b \ll w$ for a pipe of length L and rectangular cross-section of width w and height b; for turbulent flow (i.e., high Re flow) $P = K_RQ^n$
- **Thermal Capacitor**: $C_h\frac{dT}{dt} = Q$, thermal capacitance $C_h = \rho Vc = mc$, c=specific heat
- **Thermal Resistor**: Conduction $Q = -kA\frac{\partial T}{\partial x}$, conductive resistance $R_k = \frac{\Delta x}{kA}$ with A=path area of X-section, k=thermal conductivity, Δx=conduction length; Laplace equation

$\frac{\partial^2 T}{\partial x^2}+\frac{\partial^2 T}{\partial y^2}+\frac{\partial^2 T}{\partial z^2}=\frac{1}{\alpha}\frac{\partial T}{\partial t}$, $\alpha=\frac{k}{\rho c}$=thermal diffusivity; convection $Q=h_cA(T_w-T_f)$ with h_c=convection heat transfer coefficient, convective resistance $R_c=\frac{1}{h_cA}$

- **Biot Number**: $\mathrm{Bi}=\frac{R_k}{R_c}=\frac{\Delta x h_c A_c}{kA_k}$; $\mathrm{Bi}=\frac{h_c\Delta x}{k}$ if $A_c=A_k$; For lumped-parameter approximation, we need $\mathrm{Bi}\le 0.1$ (a rule of thumb)
- **Radiation**: $Q=\sigma c_e c_r A(T_s^4-T_r^4)$, T_s=source temperature, T_r=receiver temperature, A= effective (normal) area of receiver, c_e=effective emissivity of source, c_r=shape factor of receiver, σ=Stefan–Boltzmann constant (=5.7×10^{-8} W/m²/°K⁴); Linearized radiation resistance $R_r=\frac{1}{\sigma c_e c_r A(\bar{T}_s^2+\bar{T}_r^2)(\bar{T}_s+\bar{T}_r)}$ or $R_r=\frac{1}{4KT_0^3}$ with $K=\sigma c_e c_r A$
- **Natural Oscillations**: Need two different types of energy storage elements (A-type and T-type. Thermal systems have only A-type energy storage elements.

PROBLEMS

2.1

a. Briefly explain and justify why: (i) voltage and not current is the natural state variable for an electrical capacitor and (ii) current and not voltage is the natural state variable for an electrical inductor.
b. List several advantages of using as state variables, the across-variables of independent A-type energy storage elements and through-variables of independent T-type energy storage elements, in the development of a state space model for an engineering system.
c. List several things to which the order of a dynamic system is equal.

2.2 What are the basic lumped elements of

i. A mechanical system
ii. An electrical system?

Indicate whether a distributed-parameter method is needed or a lumped-parameter model is adequate in the study of the following dynamic systems:

a. Vehicle suspension system (motion)
b. Elevated vehicle guideway (transverse motion)
c. Oscillator circuit (electrical signals)
d. Environmental (weather) system (temperature)
e. Aircraft (motion and stresses)
f. Large transmission cable (capacitance and inductance).

2.3

a. Why are analogies (i.e., analytical similarities among different physical domains) important in the modeling of dynamic systems?
b. In the force-current analogy, what mechanical element corresponds to an electrical capacitor?
c. Is the velocity-pressure analogy same as the force-current analogy? In the former analogy, is the fluid inertia element analogous to the mechanical inertia element?

2.4

a. What are through-variables in mechanical, electrical, fluid, and thermal system?
b. What are across-variables in mechanical, electrical, fluid, and thermal systems?
c. Can the velocity of a mass change instantaneously? Why?
d. Can the voltage across a capacitor change instantaneously? Why?
e. Can the force in a spring change instantaneously? Why?
f. Can the current in an inductor change instantaneously? Why?
g. Can purely thermal systems oscillate? Why?

2.5 A spherical thin shell is used as a liquid storage tank. The volume flow rate into the rank is Q and the internal pressure (with respect to the ambient condition—atmospheric pressure) is P (see Figure P2.5).

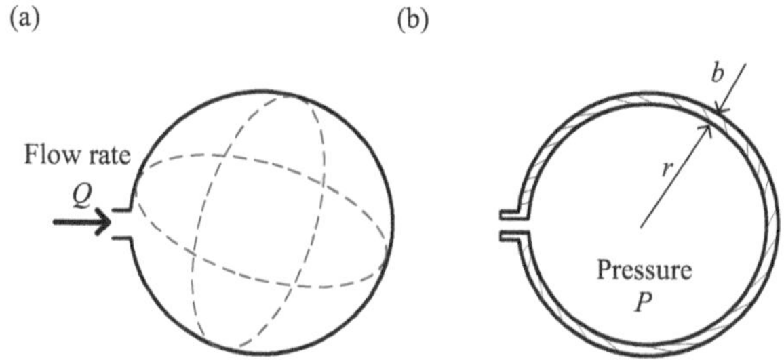

FIGURE P2.5 (a) Spherical shell for liquid storage; (b) cross-sectional details.

The following parameters are given:

r = internal radius of the tank
b = wall thickness of the tank
E = Young's modulus of the tank material
υ = Poisson's ratio of the tank material

Determine an expression for the fluid capacitance of the tank in terms of the given parameters.

Hints:

When the liquid pressure in the tank increases by ΔP, the internal radius of the rank increases by $\Delta r = \dfrac{r^2(1-\upsilon)}{2Eb}\Delta P$; the volume of a sphere of radius r is $V = \dfrac{4}{3}\pi r^3$.

2.6

a. Consider an overhead water storage tank of internal radius R at its base, as shown in Figure P2.6a. The inclination of the tank wall (wrt the base) is θ. The outlet valve is closed and the inlet valve provides an inflow at the rate Q (volume/s). At a given instant, the water level in the tank is h. Obtain an expression for the fluid capacitance C_f at that instant, in terms of h, R, θ, mass density ρ of water, and acceleration due to gravity (g).

Note: Here, C_f varies with h.

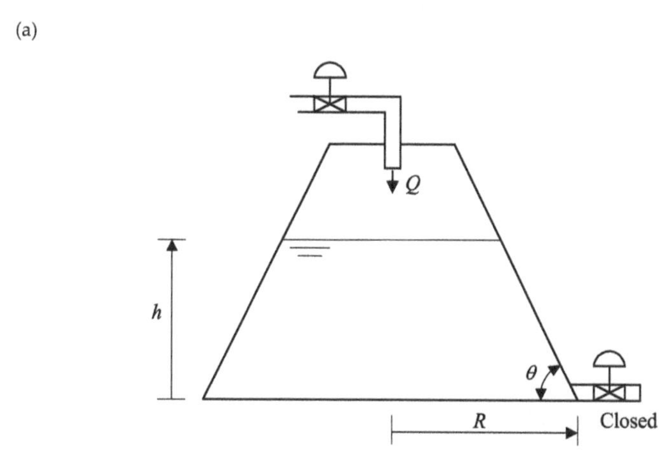

FIGURE P2.6 (a) Conical water tank with water level h; (b) cylindrical water tank with water level h.

(Continued)

(b)

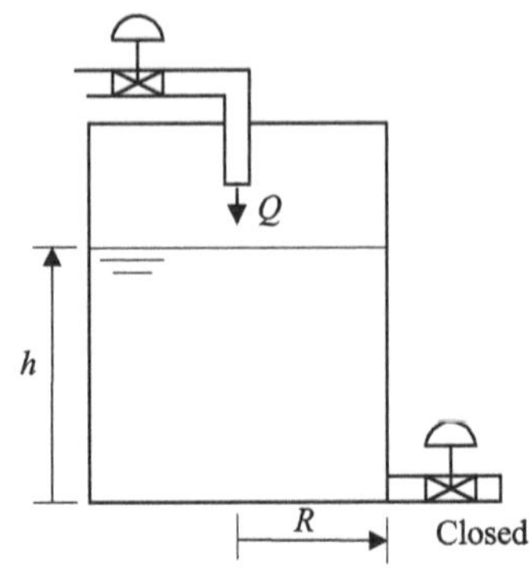

FIGURE P2.6 *(CONTINUED)* (a) Conical water tank with water level h; (b) cylindrical water tank with water level h.

Hint: Gage pressure (wrt the atmosphere pressure) of the water at the tank base, $P = \rho gh$
Let r = radius of the water surface (at height h)
If an incremental water volume δV enters the tank, the water level would rise by δh. Then, $\delta V = \pi r^2 \cdot \delta h$
Divide throughout by the incremental time $\delta t \Rightarrow \dfrac{\delta V}{\delta t} = \pi r^2 \dfrac{\delta h}{\delta t} \Rightarrow \dfrac{dV}{dt} = \pi r^2 \dfrac{dh}{dt}$

b. From your result in Part (a), show that for a cylindrical tank (i.e., $\theta = \dfrac{\pi}{2}$) of constant X-sectional radius R (see Figure P2.6b), we have the standard result $C_f = \dfrac{\pi R^2}{\rho g} = \dfrac{A}{\rho g} = \text{constant}$

Note: A =constant area of X-section of the tank

2.7

a. Suppose that the volume V of a fluid is a nonlinear function $V(P)$ of its pressure P. Show that its fluid capacitance C_f is given by $C_f = \dfrac{dV}{dP}$ which itself is a function of P
b. Using this, verify the result obtained in Problem 2.6, Part (a).

2.8 A cylindrical thin shell is used as a liquid storage tank. The volume flow rate into the tank is Q and the internal pressure of the cylinder (wrt the ambient condition—external pressure) is P. The tank has a uniform middle segment and two hemispherical ends. See Figure P2.8.

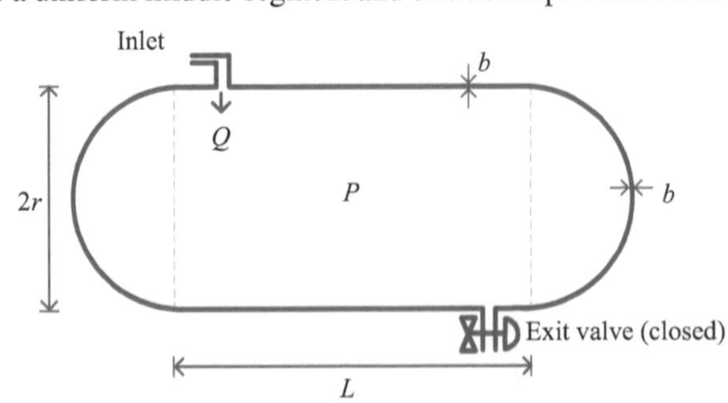

FIGURE P2.8 Sectional details of a thin-shell liquid storage tank.

The following parameters are given:

L = length of the uniform middle segment of the tank
r = internal radius of the end hemispheres (and the middle segment) of the tank
b = wall thickness of the tank
E = Young's modulus of the tank material
υ = Poisson's ratio of the tank material.

Determine an expression for the fluid capacitance C_f of the tank due to the flexibility of its walls, in terms of the given parameters. Ignore the compressibility of the liquid (the corresponding capacitance can be simply added—the two capacitors are in parallel).

Hint: when the pressure in the tank increases by ΔP, the internal radius of the tank increases by $\Delta r = \frac{r^2}{Eb}(1 - \upsilon / 2)\Delta P$ and the length of the uniform middle segment of the tank increases by $\Delta L = \frac{rL}{Eb}(\frac{1}{2} - \upsilon)\Delta P$.

2.9 In the electro-thermal analogy of thermal systems, where voltage is analogous to temperature and current is analogous to heat transfer rate, explain why there exists a thermal capacitor (*A*-type element) but not a thermal inductor (*T*-type element). What is a direct consequence of this fact, concerning the natural (free or unforced) response of a purely thermal system?

2.10 Consider a hollow cylinder of length l, inside diameter d_i, and the outside diameter d_o. If the conductivity of the cylinder material is k, what is the conductive thermal resistance of the cylinder in the radial direction? For the derivation, use the arrangement shown in Figure P2.10. It shows a cylinder whose ends are thermally insulated, so that the heat transfer is mainly in the radial direction. The inside (hollow core of the cylinder) temperature is assumed uniform at T_i, and the outside temperature is also assumed uniform at T_o.

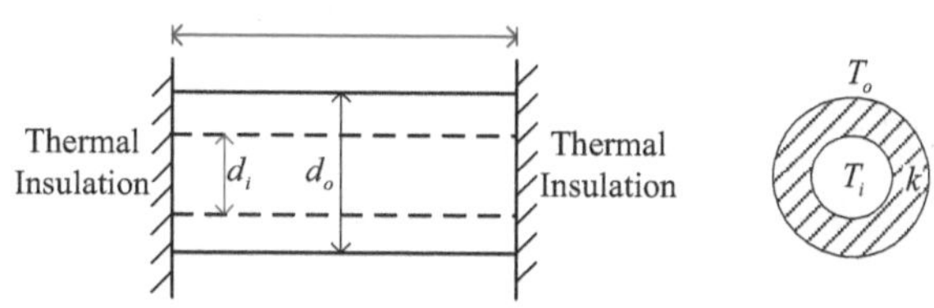

FIGURE P2.10 A cylinder conducting heat from the interior to the exterior.

Note: To facilitate your derivation, assume that $T_i > T_o$. But, the final result is the same regardless of this assumption.

2.11 A uniform metal bar length l and area of X-section A is heated to temperature T_b and placed in a thermally insulated compartment with one end exposed to the atmosphere of temperature $T_a < T_b$. This may be considered as a part of heat treatment process.

a. Obtain a differential equation (dynamic model) for the subsequent cooling process of the bar (see Figure P2.11).

Also given: ρ = mass density of the bar

c_p = specific heat (at constant pressure) of the bar
h_c = convective heat transfer coefficient at the end exposed to the atmosphere

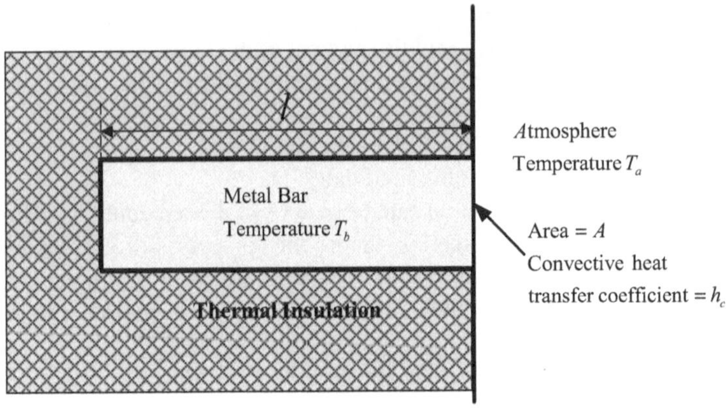

FIGURE P2.11 Heated metal bar in an insulated compartment.

b. Suppose that the following numerical values are known: $l = 0.2$ m, $A = 1.0$ cm^2, $h_c = 100.0$ W/m^2/°C, conductive heat transfer coefficient $k = 125.0$ W/m/°C.

Compute the Biot number Bi for the system.

From the result, determine whether the lumped-parameter model obtained in Part (a) is acceptable.

2.12

a. List four different types of fluid capacitance. Briefly indicate the physical process by which each of these four capacitances is generated.
b. Consider a water storage tank with gradually decreasing rectangular cross-section, as shown in Figure P2.12. Only one of the four sides of the tank wall is sloping (linearly) at an angle θ with the vertical. The other three sides are vertical (i.e., no slope). The internal dimensions of the base of the tank are $a \times b$, as shown. The outlet valve is kept closed and the inlet valve provides a water inflow at the rate Q (volume/s). At a given instant, the water level in the tank is h. Obtain an expression for the fluid capacitance C_f at that instant, due to the gravity head, in terms of a, b, θ, h, mass density ρ of water, and the acceleration due to gravity (g).

Note: C_f varies with h. Assume that the tank is rigid and the water is incompressible.

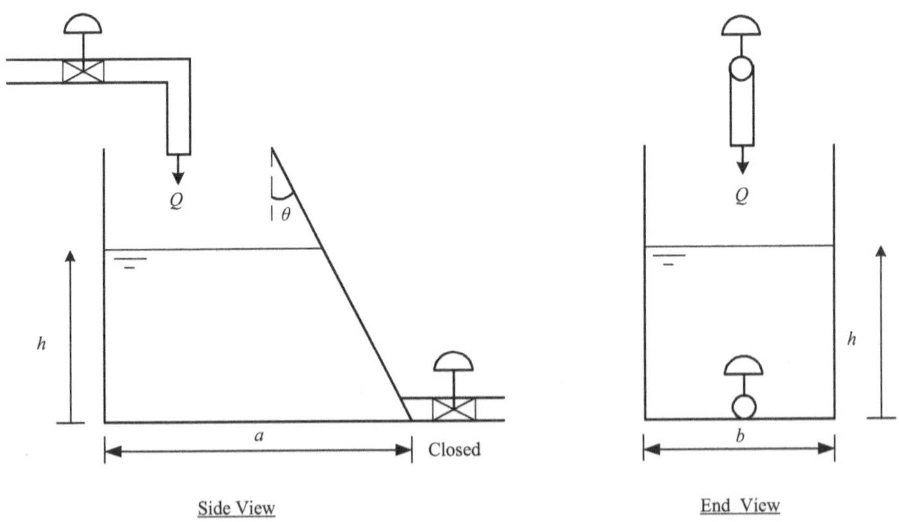

FIGURE P2.12 A sloping water tank.

3 Analytical Modeling

HIGHLIGHTS

- Types of Analytical Models
- Properties of Linear Systems (Principle of Superposition, etc.)
- Discrete-Time Models
- Lumped Model of a Distributed System (Heavy Spring)
- Analytical Model Development (Unified, Integrated, Systematic, Unique)
- State-Space Models and Input–Output Models
- Properties of State-Space Models
- Steps for State Model Development
- Mechanical, Electrical, Fluid, and Thermal Examples of State-Space Models

3.1 INTRODUCTION

The main focus of the present chapter is the unified, integrated, and systematic development of analytical models, leading to unique result. In Chapter 1, we identified several types of models, giving particular attention to practical applications of models. In Chapter 2, we presented the basic elements (components) in the mechanical, electrical, fluid, and thermal domains and their properties. In particular, we studied properties of these elements and established proper "state variables" for the (energy storage) elements in various physical domains. The properties of a dynamic model depend on the constituent basic components and the "structure" in which these components are interconnected. It is pertinent to note that the use of the independent across-variables and through-variables as the state variables to represent the behavior of the basic elements (or components) of a system (see Chapter 2) is quite advantageous in this regard. This is because these variables naturally represent the responses (outputs or states) of these components and also show how the components are interconnected (whether connected in parallel or in series). Specifically,

1. Components connected in parallel have a common across-variable
2. Components connected in series have a common through-variable.

Another advantage that stems from the use of these variables is the retention of the structural analogy across physical domains. For example, a parallel connection of mechanical elements will be analogous to a parallel connection (not a series connection) of the corresponding electrical elements.

In the present chapter, we concentrate on the integration (interconnection) of elements (components) of various physical domains to form analytical models. In particular, state-space models and input–output models are studied with examples in various physical domains. Prior to that, various types of analytical models are identified in a comparative manner. While comparing nonlinear models with their, approximated, linear models (more details are found in Chapter 4), some salient properties of linear models are discussed. Also, some important forms of discrete-time models, which are approximated (and computer-targeted) versions of continuous-time (or analog) models, are studied. In the comparison of distributed-parameter models with their approximated lumped-parameter models, some methods of approximating a distributed-parameter model by lumped-parameter models are presented, specifically through an illustrative example of a heavy spring. Since the main focus of the chapter is state-space models, the analytical formulation of a general state-space model and some definitions and key properties of them are discussed. The steps of converting a state-space model into an input–output models are presented, and illustrative examples are given.

DOI: 10.1201/9781003124474-3

The chapter concludes by presenting the systematic development of linear lumped-parameter state-space models in the four physical domains: mechanical, electrical, fluid, and thermal, using illustrative examples.

3.2 TYPES OF ANALYTICAL MODELS

In an analytical model, the response (output) to an applied excitation (input) may be expressed in:

1. The *time domain*, where the response value is expressed as a function of time (i.e., the independent variable is time t)
2. The *frequency domain*, where the amplitude and the phase angle of the response are expressed as functions of frequency (i.e., the independent variable is frequency ω).

Determination of the time-domain response generally involves the solution of a set of differential equations (e.g., state equations). The frequency-domain analysis is a special case of the Laplace transform analysis (in the Laplace domain) where the independent variable is the Laplace variable s. The corresponding analytical model is a set of transfer functions. A transfer function is the ratio of the Laplace transform of the output variable divided by the Laplace transform of the input variable. In the special case of the frequency domain, we have $s=j\omega$. The subject of frequency-domain models is studied in Chapters 6 and 7.

In summary, the main types of analytical models considered in this book are the following:

1. **Time-Domain Model**: It consists of differential equations whose independent variable is time t. Examples are state-space model (a set of first-order differential equations in time, called state equations) and input–output model (input–output differential equations in time)
2. **Transfer Function Model**: A transfer function is given by the Laplace transform of the pertinent output variable divided by the Laplace transform of the pertinent input variable. The corresponding model is a set of such transfer functions (algebraic equations with the Laplace variable s as the independent variable) for the pertinent input–output pairs. This is another type of input–output model
3. **Frequency-Domain Model**: Frequency transfer function (or *frequency response function*) is a special case of Laplace transfer function, where $s=j\omega$. This results from the fact that the Fourier transform is a special case of Laplace transform, where the independent variable is frequency ω rather than the Laplace variable s. The corresponding model is a set of frequency transfer functions. The model conversion between Laplace transfer function and frequency transfer function is straightforward, simply through the use of $s=j\omega$. In fact the conversion of them to and from the input–output differential equation is also very simple (see Chapter 6).

In considering various types of analytical models, it is particularly useful to consider them comparatively, in pairs, one type being a special (or approximate) version of the other. Typically, the approximate version is easier to analyze and implement. Such pairs of common model types are listed below.

1. Nonlinear models and linear models
2. Continuous-time models and discrete-time models
3. Distributed (continuous)-parameter models and lumped-parameter models
4. Time-varying (or non-stationary or non-autonomous) models and time-invariant (or stationary or autonomous) models
5. Random (or stochastic or probabilistic) models and deterministic models.

Some salient features of these classes of models are indicated in Table 3.1.

Next we consider several key types of systems (models) and explore their characteristics.

TABLE 3.1
Comparison of Different Types of Analytical Dynamic Models

Model Type	Key Properties
Nonlinear	Nonlinear differential equations in time; principle of superposition does not hold
Linear	Linear differential equations in time; principle of superposition holds
Continuous time	Differential equations in time; time variable is continuously defined; Laplace transform produces a transfer function model; Fourier transform produces a frequency transfer function model
Discrete time	Difference equations in time; time variable is defined by discrete values as a sequence of time points; z-transform produces a discrete transfer function (z-transform transfer function) which is equal to the z-transform of the discrete-time output divided by the z-transform of the discrete-time input
Distributed	Partial differential equations (PDEs) in time and space (dependent variables are functions of both time and space)
Lumped	Ordinary differential equations or ODEs (dependent variables are functions of time only, not space)
Time-varying	Differential equations with time-varying coefficients (model parameters vary with time)
Time-invariant	Differential equations with constant coefficients (model parameters are constants)
Stochastic	Stochastic differential equations in time; variables and/or parameters have some randomness and are governed by probability distributions; experiments repeated under same conditions do not produce identical results in general
Deterministic	Non-stochastic differential equations in time; there is no randomness in the system parameters or variables; experiments repeated under identical conditions produce identical results

3.2.1 Properties of Linear Systems

At the outset, it is important to emphasize that by "linear system" we mean "linear model" (i.e., a system that is represented by a linear model). All practical systems are nonlinear to some degree. If the nonlinearity is negligible, for the situation that is being considered (modeled), the system may be represented by a linear model. Since linear systems/models are far easier to handle (analyze, simulate, design, control, etc.) than nonlinear systems/models, a linear model of the system may be developed (or an available nonlinear model of the system may be linearized) for the purpose at hand. A linear model may be valid only for a limited range or set of conditions of operation. Linearization of nonlinear systems/models is treated in Chapter 4. Two key properties that are responsible for the analytical convenience of linear systems are outlined next.

- **Principle of Superposition**: All linear systems (linear models) satisfy the principle of superposition. Specifically, a system is linear if and only if the principle of superposition is satisfied. For a multi-input-multi-output (MIMO) system (or, *multi-variable system*) this principle states that, if $\mathbf{y}_1$ is the output vector (with many components) of the system when the input (vector) to the system is $\mathbf{u}_1$, and if $\mathbf{y}_2$ is the output vector of the system when the input vector to the system is $\mathbf{u}_2$, then $\alpha_1\mathbf{y}_1 + \alpha_2\mathbf{y}_2$ is the output vector when the input vector is $\alpha_1\mathbf{u}_1 + \alpha_2\mathbf{u}_2$, where α_1 and α_2 are any real constants. This property is graphically represented in Figure 3.1a. What is shown is the MIMO case where the inputs and outputs both are vectors (having multiple components) and hence represented by thick lines.

Example 3.1

An illustrative example of the principle of superposition in an MIMO system is shown in Figure 3.1b. Here the system has two (scalar) inputs (u_1 and u_2) and three (scalar) outputs y_1, y_2, and y_3. First suppose that only the input u_1 is applied. The corresponding three outputs are shown in the figure. Next suppose that only the input u_2 is applied. The corresponding three outputs are shown as well. Finally, if the linear combination $\alpha_1 u_1$ and $\alpha_2 u_2$ of the inputs is applied (where α_1 and α_2 are arbitrary real constants), the corresponding three outputs will have the same linear combination, as shown in Figure 3.1b.

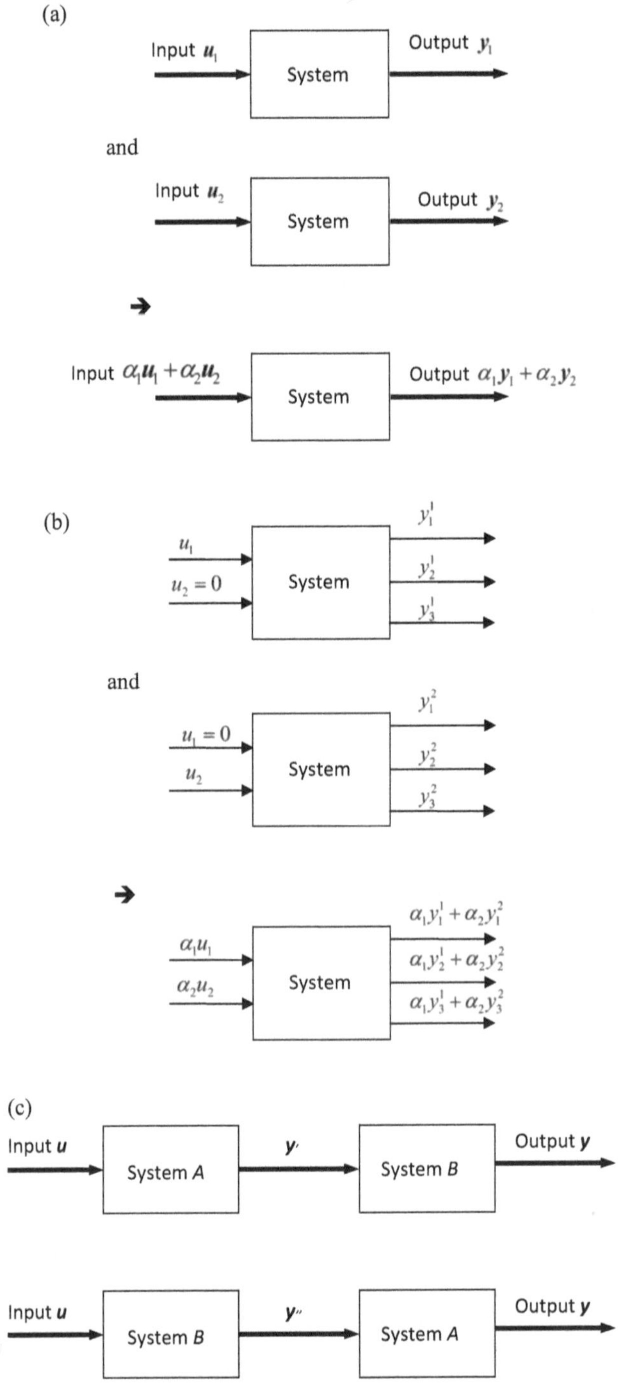

FIGURE 3.1 Properties of a linear system: (a) Principle of superposition (vector inputs); (b) a two-input example. (c) Interchangeability in series connection, for commutative systems.

- **Interchangeability in Series Connection**: The interchangeability (commutativity) in series connection is satisfied not always satisfied by general linear systems. Figure 3.1c shows the case of vector inputs and vector outputs (represented by thick lines and bold characteristics).

The two systems may be interchanged without affecting the output of the overall system for a given input if and only if the systems themselves are commutative (i.e., $\boldsymbol{AB} = \boldsymbol{BA}$). However, if the systems (and the input and the output) are scalar, the interchangeability is always valid.

Note: The interchangeability is not a property of just linear systems and is valid only for "commutative" systems. Interchangeability in parallel connection is a trivial fact, which is satisfied by both linear and nonlinear systems.

3.2.2 Discrete-Time Systems

A discrete-time signal is defined at a sequence of discrete time points of a continuous-time (analog) signal. Typically, it is obtained by discretizing (sampling) an analog signal, which is defined continuously in time (see Figure 3.2). Reading (or sampling) an analog signal at discrete time points is not the only way to generate a discrete-time signal. For example, a physical switching device may be activated in a discrete time sequence to generate discrete-time data. The time interval between two successive data points is the *sampling period* ΔT. In typical practical situations such as digital hardware (or digital computer), the sampling period is a constant. The *sampling rate* f_s is the inverse of the sampling period:

$$f_s = \frac{1}{\Delta T} \tag{3.1}$$

Note: An error is introduced when an analog signal is sampled (discretized). This is known as *aliasing* (i.e., sampling error). A discrete value is not necessarily a digital value. When a discrete value is represented in a digital form, a further error is introduced (i.e., *quantization* error).

- **Z-transform**: Consider an analog signal $x(t)$, which is sampled at constant the sampling period ΔT. We get the data sequence:

$$\{x_k\} = \{x_0, x_1, \ldots, x_k, x_{k+1}, \ldots\} \tag{3.2}$$

where $x_k = x(k \cdot \Delta T)$. This sequence can be represented by a polynomial function of the complex variable z as:

$$X(z) = \sum_{k=0}^{\infty} x_k z^{-k} \tag{3.3}$$

Then, $X(z)$ is termed the z-transform of the sequence $\{x_k\}$.

- **Difference Equations**: In a discrete-time system, the input signals and the output signals both are discrete-time sequences. The system equation (i.e., discrete-time model) can be represented

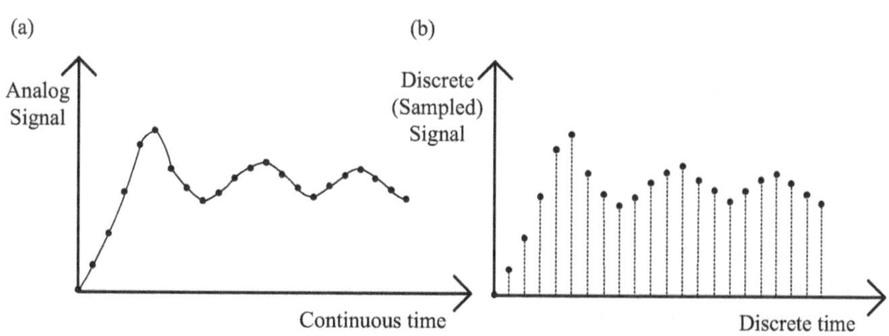

FIGURE 3.2 (a) An analog signal; (b) the corresponding discrete-time signal (sampled data).

by a set of *difference equations*. These difference equations relate the discrete output signals to the discrete input signals of the system. Consider, in particular, a single-input single-output (SISO) system. The input is the sequence $\{u_k\}$ and the output is the sequence $\{y_k\}$. The corresponding discrete-time model may be expressed by the *n*th order linear difference equation:

$$a_0 y_k + a_1 y_{k-1} + \ldots + a_n y_{k-n} = b_0 u_k + b_1 u_{k-1} + \ldots + b_m u_{k-m} \tag{3.4}$$

It is clear from equation (3.4) that, if the input sequence $\{u_k\}$ is known, the output sequence $\{y_k\}$ can be generated/computed starting with the first n values of the output sequence, which should be known. These initial n values are the *initial conditions*, which are required to determine the complete solution of a difference equation. In general, the model parameters a_i and b_i in equation (3.4) also depend on the sampling period ΔT.

- **Discrete Transfer Function**: The discrete-time model (3.4) may also be represented by the *discrete transfer function*:

$$G(z) = \frac{Y(z)}{U(z)} = \frac{b_0 + b_1 z^{-1} + \cdots + b_m z^{-m}}{a_0 + a_1 z^{-1} + \cdots + a_n z^{-n}} \tag{3.5}$$

This is analogous to representing a linear, input–output differential equation with respect to time (i.e., an analog system or *input–output model* in the time domain) by a transfer function (in the Laplace domain or frequency domain).
Note: By examining (3.4) and (3.5), it should be clear that z^{-1} corresponds to the "time delay" operation through ΔT.

- **Aliasing Distortion due to Signal Sampling**: Aliasing distortion (or, sampling error) is an important consideration when dealing with discrete-time data that are sampled from a continuous (analog) signal. Shannon's sampling theorem is relevant here, which is presented now.
- **Sampling Theorem**: If a time signal $x(t)$ is sampled at equal time periods of ΔT, no information regarding its frequency spectrum $X(f)$ can be obtained for frequencies higher than

$$f_c = \frac{1}{2\Delta T} = \frac{1}{2} f_s \tag{3.6}$$

This limiting (cutoff) frequency in the spectrum (of the sampled data) is called the *Nyquist frequency*.
It can be shown that signal sampling causes folding of the high-frequency segment of the signal frequency spectrum beyond the Nyquist frequency onto the low-frequency segment of the spectrum. The *aliasing error* is caused by this spectral folding, as illustrated in Figure 3.3. The aliasing error becomes more and more prominent for frequencies of the spectrum closer to the Nyquist frequency. In signal analysis, a sufficiently small sample step ΔT should be chosen in order to reduce the aliasing distortion in the frequency domain, depending on the highest frequency of interest in the analyzed signal. This, however, increases the amount of sampled data, which increases the signal processing time (computational load) and the computer storage requirements (and also the signal communication time, in a networked system), which are undesirable particularly in real-time analysis. It can also result in stability problems in numerical computations. The Nyquist sampling criterion requires that the sampling rate ($1/\Delta T$) for a signal should be at least twice the highest frequency of interest. Instead of making the sampling rate very high, a moderate value that satisfies the Nyquist sampling criterion is used in practice, together with an *antialiasing filter* to remove the frequency components in the original signal that would distort, through folding, the spectrum of the computed signal.

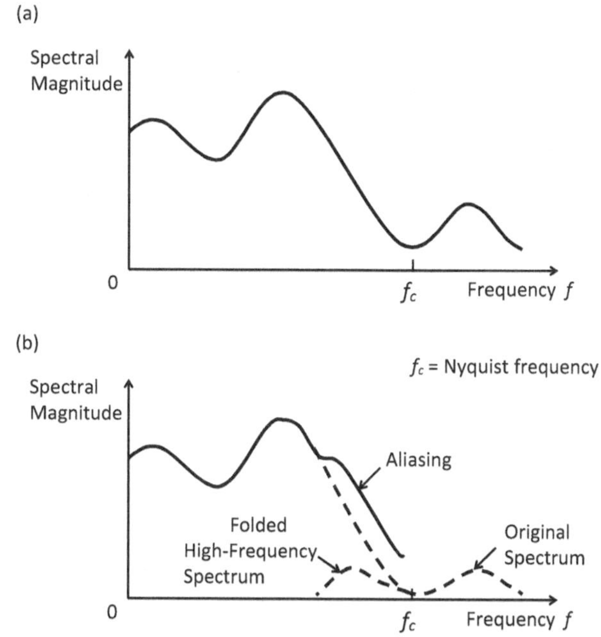

FIGURE 3.3 Aliasing distortion of a frequency spectrum: (a) Original spectrum, (b) distorted spectrum due to aliasing.

- **Antialiasing Filter**: It should be clear from Figure 3.3 that, if the original signal was low-pass filtered at a cutoff frequency equal to the Nyquist frequency, then the aliasing distortion caused by sampling would not occur. A filter of this type is called an *antialiasing filter*. Analog hardware filters may be used for this purpose. In practice, it is not possible to achieve perfect filtering. Hence, some aliasing could remain even after using an anti-aliasing filter, thereby further reducing the valid frequency range of the computed signal. Typically, the useful frequency limit of a sampled data set is $f_c/1.28$, and the last 20% of the spectral points near the Nyquist frequency should be neglected. Hence, typically, the filter cutoff frequency is chosen to be somewhat lower than the Nyquist frequency, for example, $f_c/1.28\ (\cong 0.8f_c)$. Then, the computed spectrum is accurate up to the filter cutoff frequency $0.8f_c$ and not up to the Nyquist frequency f_c.

3.2.3 Lumped Model of a Distributed System

As noted before, *lumped-parameter* models and *distributed-parameter* models are two broad categories of models for dynamic systems. In a lumped-parameter model, various characteristics of the system are limped into representative elements located at a discrete set of points in a geometric space. The corresponding analytical models are ordinary differential equations. Lumped-parameter model is the main focus of this book. In most physical systems, however, the properties (physical parameters such as mass, stiffness, inductance, and resistance) are continuously distributed in various components or segments; they are distributed-parameter (or *continuous*) components. To represent system parameters that are continuously distributed in space, we need spatial coordinates.

Since the distributed-parameter dynamic systems have time (t) and space coordinates (e.g., x, y, z) as the independent variables, the corresponding analytical models are partial differential equations with respect to time and space. For analytical convenience, we may attempt to approximate such distributed-parameter models into lumped-parameter ones. The accuracy of the model can be improved by increasing the number of discrete elements in such a model; for example, by using finite element techniques. In view of their convenience, lumped-parameter models are more commonly employed than continuous-parameter models. Furthermore, continuous-parameter components (e.g., beams, plates) may be included into otherwise lumped-parameter models in order to improve the model accuracy. Next we will address some pertinent issues by considering the example of a heavy spring.

3.2.3.1 Heavy Spring

A coil spring has a mass, an elastic (spring, flexibility) effect, and an energy-dissipation characteristic, each of which is distributed over the entire coil. The distributed mass of the spring has the capacity to store *kinetic energy* by acquiring velocity. Stored kinetic energy can be recovered as work done through a process of deceleration. Furthermore, in view of the distributed flexibility of the coil, each small element in the coil has the capacity to store *elastic potential energy* through reversible (elastic) deflection. If the coil was moving in the vertical direction, there would be changes in *gravitational potential energy,* but we can disregard this in dynamic-response studies if the deflections are measured from the static equilibrium position of the system (because in the static equilibrium configuration, the gravitational forces are balanced by the elastic forces). The coil will undoubtedly get warmer, make creaking noises, and over time will wear out at the joints, clear evidence of its capacity to dissipate energy. A further indication of damping is provided by the fact that when the coil is pressed and released, it will eventually come to rest: the work done by pressing the coil is completely dissipated. Even though these effects are distributed in the actual system, a discrete or lumped-parameter model is usually sufficient to predict the system response to a forcing function. Further approximations are possible under certain circumstances. For instance, if the maximum kinetic energy is small in comparison with the maximum elastic potential energy in general (particularly true for light stiff coils, and at low frequencies of oscillation), and if in addition the rate of energy dissipation is relatively small (determined with respect to the time span of interest), the coil can be modeled by just a discrete (lumped) stiffness (spring) element that is light (massless). These are modeling decisions.

In an analytical model, the individual distributed characteristics of inertia, flexibility, and dissipation of a heavy spring can be approximated (lumped) by a separate mass element, a spring element, and a damper element, which are interconnected in some parallel-series configuration (structure), thereby producing a lumped-parameter model. Since a heavy spring has its mass continuously distributed throughout its body, it has an infinite number of degrees of freedom. A single coordinate cannot represent its motion. But, for many practical purposes, a lumped-parameter approximation with just one or two lumped masses to represent the inertial characteristics of the spring would be sufficient. Such an approximation may be obtained by using one of several approaches. One is the energy approach. Another approach is equivalence of natural frequency. We will consider the energy approach next. Here we represent a distributed-parameter spring by a lumped-parameter "model" such that the original spring (distributed) and the model (lumped) have the same net kinetic energy and the same potential energy. This energy equivalence is used in deriving the lumped parameters in the model. In particular, kinetic energy equivalence is used to determine the mass parameters for the model. Even though damping (energy dissipation) is neglected in the present analysis, it is not difficult to incorporate that as well in the model (through, for example, energy-dissipation equivalence).

3.2.3.1.1 Potential Energy Equivalence

Consider a linear coil spring AB, as shown in Figure 3.4a. If an axial tensile force F is applied at end B while end A is fixed, the end B will stretch axially through distance δ. During that extension, the axial spring force will vary from 0 (in the beginning) to F (at the end of the extension). From considerations of equilibrium (i.e., force balance), the axial force in the spring will be the same (F) anywhere along it (*Note*: F is a through-variable). The work done by F during the extension of the spring is $\frac{1}{2}F\delta$ and this is indeed the *elastic potential energy* (PE) stored in the spring due to the work done. Also, the final spring force is $F = k\delta$ where k is the *stiffness* of the spring. Substituting, we have

$$\text{Elastic PE} = \frac{1}{2}k\delta^2 \tag{3.7}$$

Note: Stiffness k=force that is needed to push or pull the spring through a unit distance. It is clear that, even though the stiffness (or flexibility) is distributed along AB, the elastic PE of the spring is represented accurately by a single (lumped) stiffness parameter k. Hence, from the flexibility point of view, a distributed spring of overall stiffness k can be precisely represented by a lumped spring of the same stiffness k. Regardless of the velocity (dynamics, kinetic energy) of the spring during its deformation, the PE expression remains the same as given before. In other words, stiffness is a "static" property. Hence, for "mechanical" energy equivalence, only the kinetic energy equivalence needs to be considered.

3.2.3.2 Kinetic Energy Equivalence

Consider again the uniform, heavy spring shown in Figure 3.4a. One end (A) of the spring is fixed and the other end moving at velocity v.

Note: m_s=mass of spring; k=stiffness of spring; and l=length of spring.

Assume a linear distribution of the speed along the spring, with zero speed at the fixed end and the maximum speed v at the free end (Figure 3.4b). The local speed of an infinitesimal element δx of the spring is $\frac{x}{l}v$. Element mass$=\frac{m_s}{l}\delta x$. Hence, the element kinetic energy $KE=\frac{1}{2}\frac{m_s}{l}\delta x\left(\frac{x}{l}v\right)^2$

In the limit, we have $\delta x \to dx$. Accordingly, by performing the necessary integration, we get

$$\text{Total KE} = \int_0^l \frac{1}{2}\frac{m_s}{l}dx\left(\frac{x}{l}v\right)^2 = \frac{1}{2}\frac{m_s v^2}{l^3}\int_0^l x^2\,dx = \frac{1}{2}\frac{m_s v^2}{3}$$

Hence, for the energy (kinetic energy) equivalence,

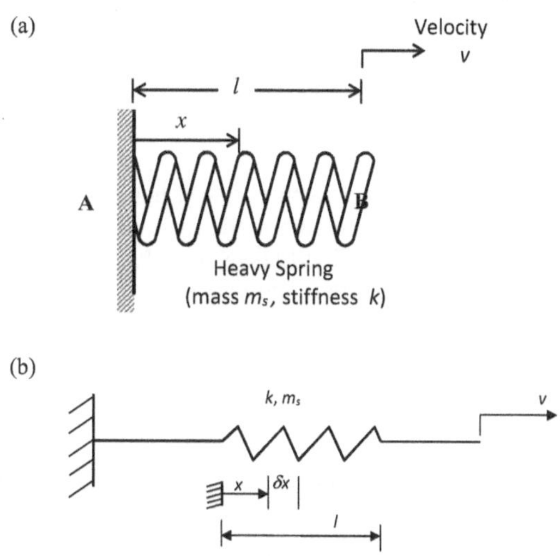

FIGURE 3.4 (a) A uniform heavy spring, (b) analytical representation.

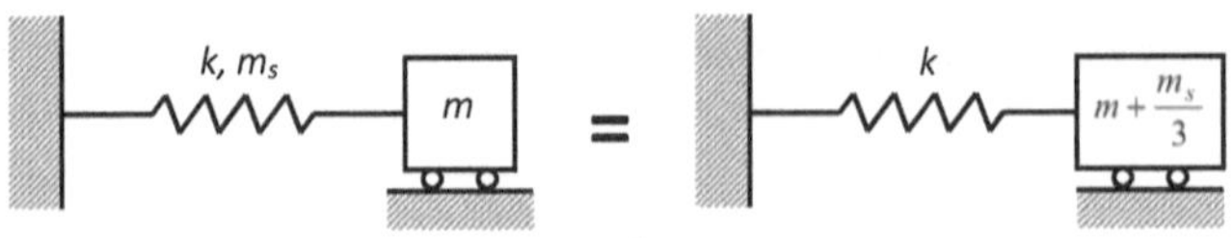

FIGURE 3.5 Lumped-parameter approximation for an oscillator with heavy spring.

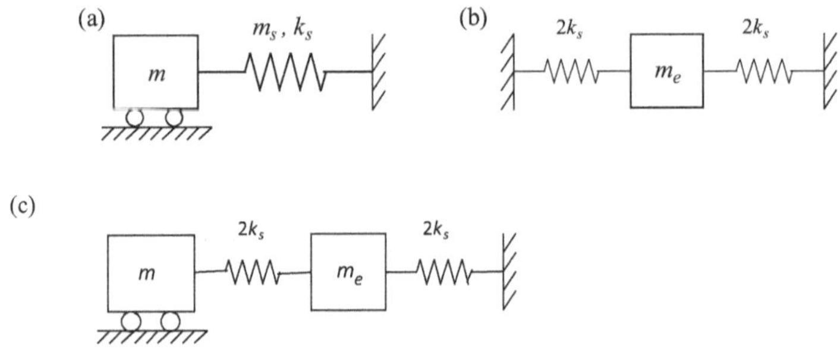

FIGURE 3.6 (a) A lumped mass with a distributed-parameter spring; (b) a lumped-parameter model of the system for large end mass; (c) overall lumped-parameter model.

$$\text{Equivalent lumped mass concentrated at the free end} = \frac{1}{3} \times \text{spring mass} \tag{3.8}$$

Note: This derivation assumes that one end of the spring is fixed, the conditions are uniform along the spring, and furthermore the spring velocity is distributed uniformly along the spring.

An example of utilizing this result is shown in Figure 3.5. Here a system with a heavy spring and a lumped mass is approximated by a light spring (having the same stiffness) and a lumped mass.

3.2.3.3 Natural Frequency Equivalence

Now consider the approach of natural frequency equivalence. Here we derive an equivalent lumped-parameter model by equating the fundamental (lowest) natural frequency of the distributed-parameter system to the lowest natural frequency of the lumped-parameter model. Next we illustrate this approach by using an example for the one-degree-of-freedom case. The method can be easily extended to multi-degree-of-freedom lumped parameter models as well (see the related problem at the end of the chapter).

Example 3.2

A heavy spring of mass m_s and stiffness k_s with one end fixed and the other end attached to a smoothly sliding mass m is shown in Figure 3.6a. If the mass m is sufficiently larger than m_s, then at relatively high frequencies, the mass will virtually stand still. Under these conditions, we have the configuration shown in Figure 3.6b, where the two ends of the spring are fixed. Also, we can approximate the distributed mass by an equivalent mass m_e at the midpoint of the spring. Then, each spring segment has double the stiffness of the original spring (why?). Hence the overall stiffness is $4k_s$. From the well-known fact, the natural frequency of the lumped model is

$$\omega_e = \sqrt{\frac{4k_s}{m_e}} \tag{3.9}$$

It is known from a complete analysis of a heavy (distributed mass) spring (which is beyond the present scope) that the exact natural frequency for the fixed-fixed configuration is

$$\omega_s = \pi n \sqrt{\frac{k_s}{m_s}} \tag{3.10}$$

where n is the mode number. Then, for the fundamental (first) mode (i.e., $n=1$), the natural frequency equivalence gives $\sqrt{\frac{4k_s}{m_e}} = \pi\sqrt{\frac{k_s}{m_s}}$

or,

$$m_e = \frac{4}{\pi^2} m_s \approx 0.4 m_s \tag{3.11}$$

The overall model, with the end mass included, is shown in Figure 3.6c. This result is valid when $m \gg m_s$.

Note: Since the effect of inertia decreases with increasing frequency, this model is valid for high frequencies as well, and it is not necessary to separately consider the case of high frequencies. However, the first mode approximation of the spring is valid only up to the first natural frequency of the first mode (i.e., equation (3.10) with $n=1$). Hence, the derived lumped approximation should not be used for arbitrarily high frequencies.

LEARNING OBJECTIVES

1. An alternative approach (natural frequency equivalence) for approximating a distributed-parameter model by a lumped-parameter one
2. Illustration that the result is not identical (but close) to what is obtained through energy equivalence
3. The significance of natural frequency of a system and its connection to excitation frequency
4. Lumped-parameter approximation of a heavy spring.

The natural frequency equivalence may be generalized as an *eigenvalue equivalence* (or *pole equivalence*) for any linear dynamic system in any physical domain (including electrical systems). In this general approach, the eigenvalues of the lumped parameter model are equated to the "corresponding eigenvalues" of the distributed-parameter system (which has an infinite number of eigenvalues), and the model parameters are determined accordingly.

3.3 ANALYTICAL MODEL DEVELOPMENT

We have made the following observations concerning analytical dynamic models:

- A dynamic model is a representation of a dynamic system
- It is useful in analysis, simulation, design, modification, instrumentation, monitoring (fault detection, diagnosis, prediction, etc.), and control of a system
- Analogies exist among mechanical, electrical, fluid, and thermal domains, in dynamic systems
- In view of the analogous multi-physics (multi-domain or mixed) nature of practical engineering systems, a *unified* development of models is possible and desirable. Then, suitable physical variables (specifically the across-variables of the independent A-type energy storage elements and the through-variables of the independent T-type energy storage elements) can be chosen as the state variables of the model, which choice leads to a *unique* (i.e., just one) model
- Also, it is desirable and possible to model all physical domains together (or concurrently). This is the basis of the *integrated* development of models

- The use of the *unified*, *integrated*, and *systematic* (with clear modeling steps) modeling process, leading to a unique model, is the essence of *mechatronic* modeling
- Incorporation of multi-functional devices (e.g., piezoelectric elements which work as both sensors and actuators) into the modeling framework is also facilitated by an integrated and unified approach to modeling.

Our systematic procedure for the development of a lumped-parameter analytical model of a dynamic system primarily involves the formulation of three types of equations:

1. Constitutive equations (physical laws) for the lumped elements
2. Continuity equations (or node equations or equilibrium equations) for the through-variables of the system
3. Compatibility equations (or loop equations or path equations) for the across-variables of the system.

Among these, the constitutive equations have been studied in Chapter 2.

A *continuity equation* is the equation written for the through-variables at a junction (i.e., a *node*) that connects several lumped elements in the system. It dictates the fact that there cannot be any accumulation (storage) or disappearance (dissipation) or generation (source) of the through-variables at a junction (i.e., what comes in must go out) because a node is not an element but a junction that connects elements. The summation of the forces (force balance or equilibrium) at a mechanical junction, the currents (Kirchhoff's current law) at a circuit node, the fluid flow rates (flow continuity equation) at a fluid junction, or the heat transfer rates at a thermal junction, and equating it to zero, provides a continuity equation. Clearly, all through-variables at a junction, including those of the source elements (input elements), should be included in writing a node equation.

A *compatibility equation* is the equation written for the across-variables around a closed path (i.e., a *loop*) connecting several lumped elements in the system. It dictates the fact that at a given instant, the value of the across-variable at a point in the system should be unique (i.e., it cannot have two or more different values). This guarantees the requirement that a closed path is indeed a closed path; there is no breakage of the loop (i.e., it is compatible). The summation of the velocities, or the voltages (Kirchhoff's voltage law), or the pressures, or the temperatures around a loop of elements in the system, and equating it to zero, provides a compatibility equation. Again, the across-variables of all the source elements in the loop should be included as well in writing a loop equation.

3.3.1 Steps of Model Development

The development of a suitable analytical model for a large and complex system particularly requires a systematic approach. Tools are available to aid this process. The process of modeling can be made simple by following a systematic sequence of steps. The desirable main steps of modeling are listed below:

1. Identify the system of interest by defining its *purpose*, *operation*, and the system *boundary* (physical or virtual)
2. Identify/determine the *variables* of interest. These include inputs (forcing functions or excitations), outputs (responses), and state variables (when possible)
3. Approximate (or model) various components (or processes or phenomena) in the system by *ideal elements*, which are suitably interconnected (which determines the model "structure")
4. Draw a structural diagram (e.g., electrical circuit, mechanical circuit, free-body diagram (FBD), linear graph, bond graph) showing the individual *elements* of the system and their interconnection (*structure*)
5. (a) Write the *constitutive equations* (physical laws) for the elements (not for the sources—input components)

(b) Write the *continuity* (or conservation) equations for the through-variables (equilibrium of forces or torques at joints; current balance at circuit nodes, fluid flow balance, etc.) at the nodes (junctions) of the system
(c) Write the *compatibility* equations for across-(potential or path) variables around the closed paths formed by the system elements. These are loop equations for velocities (geometric connectivity), voltage (potential balance), pressure drop, etc.
(d) Eliminate the *auxiliary* variables, which are redundant and not needed in the model.
6. Express the system *boundary conditions* (only for distributed-parameter models) and response to *initial conditions* using system variables.

These steps should be self-explanatory and should be integral with the particular modeling technique that is used. The associated procedures will be elaborated in the subsequent sections and chapters where many illustrative examples are provided as well.

3.4 STATE MODELS AND INPUT–OUTPUT MODELS

In this section, we study some key properties and characteristics of state-space models and input–output models, in the time domain. Also, we observe the relationship between these two types of models and learn how to convert one type of model into the other. Examples of analytical model development in various physical domains are presented in the final section of the chapter.

- **Input–Output Models**: More than one variable may be needed to represent the response (output) of a dynamic system. Furthermore, there may be more than one input variable in a system. Then we have a *multi-variable* system or an MIMO system. A time-domain analytical model of the system may be developed as a set of differential equations relating the response variables to the input variables. Specifically, this is a multi-variable *input–output model.* Generally, this set of system equations is coupled, so that more than one response variable appears in each differential equation, and each equation cannot be analyzed, solved, or computer-simulated separately/independently.
- **State-Space Models**: A particularly useful time-domain representation for a dynamic system is a state-space model. The state variables, as introduced in Chapter 2, are a minimal set of variables that can define the dynamic state of a system. In the state-space representation, the dynamics of an nth-order system is represented by n first-order differential equations, which generally are coupled. This is called a *state-space model* or simply a *state model.* An entire set of state equations can be written as a single vector-matrix state equation.

3.4.1 Properties of State-space Models

A state-space model (or simply a state model) is formulated using state variables. The choice of state variables is not unique: many choices are possible for a given system. Proper selection of state variables is crucial in developing an analytical model (state model) for a dynamic system. As presented in Chapter 2, in a unified, integrated, and systematic approach, leading to a single (unique) model, a proper choice is to use across-variables of the independent A-type (or, across-type) energy storage elements and the through-variables of the independent T-type (or, through-type) energy storage element as the state variables. Note that if any two elements are not independent (e.g., if two spring elements are directly connected in series or parallel), then only a single-state variable should be used to represent both elements. New state variables are not needed to represent D-type (dissipative) elements because their response can be represented in terms of the state variables of the energy storage elements (A-type and T-type) that are connected to it. In other words, D-type elements do not introduce new state variables. State-space models and their characteristics are discussed in a general sense now.

3.4.1.1 State Space

The concept of state was introduced in Chapter 2. The word "state" refers to the dynamic status or dynamic condition of a system. A complete description of the dynamic state of a system will require all the variables that are associated with the time evolution of the system response (i.e., both "magnitude" and "direction" of the response trajectory with respect to time). The state is a *vector.* The vector space in which all possible values of the state vector can be defined is the *state space.* As a dynamic system responds (to an input or an initial condition), its state vector traces out a trajectory in the state space. For example, a second-order system requires a two-dimensional (2-D or planar) state space, a third-order system requires a three-dimensional (3-D) state space, and so on.

3.4.1.2 Properties of State Models

A state vector $\boldsymbol{x}$ is a column vector, which contains a minimum set of state variables (x_1, x_2, ..., x_n), which completely determine the "dynamic state" of a dynamic system. The required minimum number of states variables (n), which is the order of the state vector, is also the *order* of the system.

3.4.1.2.1 Property 1

The state vector $\boldsymbol{x}(t_0)$ at time t_0 and the input (excitation) $\boldsymbol{u}[t_0, t_1]$ over the time interval $[t_0, t_1]$ will uniquely determine the state vector $\boldsymbol{x}(t_1)$ at any future time t_1. In other words, a transformation $\boldsymbol{g}$ can be defined such that

$$\boldsymbol{x}(t_1) = \boldsymbol{g}\left(t_0, t_1, \boldsymbol{x}(t_0), \boldsymbol{u}[t_0, t_1]\right) \tag{3.12}$$

By the *causality* property of a dynamic system, future states can be determined if all the inputs from the initial time up to that future time are known. This means, the transformation $\boldsymbol{g}$ is *nonanticipative* (i.e., inputs beyond t_1 are not needed to determine $\boldsymbol{x}(t_1)$). Each forcing function $\boldsymbol{u}[t_0, t_1]$ determines the corresponding "trajectory" of the state vector—the *state trajectory.* The n-dimensional vector space formed by all possible state trajectories is the *state space.*

3.4.1.2.2 Property 2

The state $\boldsymbol{x}(t_1)$ and the input $\boldsymbol{u}(t_1)$ at any time t_1 will uniquely determine the system output (or response) vector $\boldsymbol{y}(t_1)$ at that time. This can be expressed as

$$\boldsymbol{y}(t_1) = \boldsymbol{h}\left(t_1, \boldsymbol{x}(t_1), \boldsymbol{u}(t_1)\right) \tag{3.13}$$

This states that the system response (output) at time t_1 completely depends on the time, the input, and the state vector at that particular time.

The transformation $\boldsymbol{h}$ has no *memory*—the response at a previous time cannot be determined through the knowledge of the present state and the present input.

Note: In general, the system outputs ($\boldsymbol{y}$) are not identical to the system states ($\boldsymbol{x}$) even though the former can be uniquely determined by the latter.

A state model (in the time-domain) consists of a set of n first-order state equations. They are ordinary differential equations, which are coupled (i.e., inter-related or interacting). Also, a state model has m output equations. In the vector form, a complete state model is expressed as

$$\dot{\boldsymbol{x}} = \boldsymbol{f}(\boldsymbol{x}, \boldsymbol{u}, t) \tag{3.14}$$

$$\boldsymbol{y} = \boldsymbol{h}(\boldsymbol{x}, \boldsymbol{u}, t) \tag{3.15}$$

Equation (3.14) represents the n *state equations* (first-order ordinary differential equations) and equation (3.15) represents the m *algebraic output equations.* If $\boldsymbol{f}$ is a nonlinear vector function, then the state model is nonlinear, which is the general case.

Summarizing:

- A state model is a set of n first-order differential equations (which are coupled) using n state variables (which is an nth-order system)
- State equations define the dynamic state of a system at any time
- Required minimum set of state variables $x_1, x_2, \ldots x_n$ forms the *state vector* $\boldsymbol{x}$
- Many choices are possible for state variables (i.e., the choice is not unique). This book provides a justifiable unique choice of state variables
- The state vector traces out a *trajectory* in the *state space*
- To complete the state model, it must include the *output equations* (a set of m algebraic equations relating the output vector to the state vector). Sometimes, these output equations contain input variables as well, indicating the presence of a "feedforward nature" in the system.

3.4.2 Linear State Equations

Nonlinear state models are difficult to analyze and simulate. Often linearization is necessary, through various forms of approximations and assumptions. Linearization is studied in Chapter 4. An nth-order linear, state model is given by the state equations (differential):

$$\begin{aligned}
\dot{x}_1 &= a_{11}x_1 + a_{12}x_2 + a_{13}x_3 + \cdots + a_{1n}x_n + b_{11}u_1 + b_{12}u_2 + \cdots + b_{1r}u_r \\
\dot{x}_2 &= a_{21}x_1 + a_{22}x_2 + a_{23}x_3 + \cdots + a_{2n}x_n + b_{21}u_1 + b_{22}u_2 + \cdots + b_{2r}u_r \\
&\vdots \\
&\vdots \\
\dot{x}_n &= a_{n1}x_1 + a_{n2}x_2 + a_{n3}x_3 + \cdots + a_{nn}x_n + b_{n1}u_1 + b_{n2}u_2 + \cdots + b_{nr}u_r
\end{aligned} \tag{3.16a}$$

and the output equations (algebraic):

$$\begin{aligned}
y_1 &= c_{11}x_1 + c_{12}x_2 + c_{13}x_3 + \cdots + c_{1n}x_n + d_{11}u_1 + d_{12}u_2 + \cdots + d_{1r}u_r \\
y_2 &= c_{21}x_1 + c_{22}x_2 + c_{23}x_3 + \cdots + c_{2n}x_n + d_{21}u_1 + d_{22}u_2 + \cdots + d_{2r}u_r \\
&\vdots \\
y_m &= c_{m1}x_1 + c_{m2}x_2 + c_{m3}x_3 + \cdots + c_{mn}x_n + d_{m1}u_1 + d_{m2}u_2 + \cdots + d_{mr}u_r
\end{aligned} \tag{3.17a}$$

Here, $\dot{x}_i = dx_i/dt$; $x_1, x_2, \ldots, x_n$ are the n *state variables*; $u_1, u_2, \ldots, u_r$ are the r *input variables*; and $y_1, y_2, \ldots, y_m$ are the m *input variables*. Equation (3.16) simply states that a change in any of the n state variables and the r inputs of the system will affect the rate of change of any given state variable. Both the state equations given by (3.16a) and the output equations given by (3.17a) are needed for a complete representation of the state model. The output equations in general contain the input variables, as indicated in equation (3.17a), which happen when the system has "feedforward" characteristics (i.e., one or more inputs are directly exerted on the output). More often, however, the input variables are not present in this set of output equations (i.e., the coefficients d_{ij} are all zero in equation (3.17a)).

This state model may be rewritten in the compact vector-matrix form as

$$\dot{\boldsymbol{x}} = \boldsymbol{A}\boldsymbol{x} + \boldsymbol{B}\boldsymbol{u} \tag{3.16b}$$

$$\boldsymbol{y} = \boldsymbol{C}\boldsymbol{x} + \boldsymbol{D}\boldsymbol{u} \tag{3.17b}$$

Conventionally, a boldface upper-case letter represents a *matrix*, and a boldface lower-case letter represents a *vector*, typically a column vector. Specifically, we have

$$\boldsymbol{x} = \begin{bmatrix} x_1 \\ x_2 \\ \vdots \\ x_n \end{bmatrix}; \dot{\boldsymbol{x}} = \begin{bmatrix} \dot{x}_1 \\ \dot{x}_2 \\ \vdots \\ \dot{x}_n \end{bmatrix}; \boldsymbol{A} = \begin{bmatrix} a_{11} & a_{12} & \cdots & a_{1n} \\ a_{21} & a_{22} & \cdots & a_{2n} \\ \vdots & \vdots & & \vdots \\ a_{n1} & a_{n2} & \cdots & a_{nn} \end{bmatrix}; \boldsymbol{B} = \begin{bmatrix} b_{11} & b_{12} & \cdots & b_{1r} \\ b_{21} & b_{22} & \cdots & b_{2r} \\ \vdots & \vdots & & \vdots \\ b_{n1} & b_{n2} & \cdots & b_{nr} \end{bmatrix};$$

$$\boldsymbol{C} = \begin{bmatrix} c_{11} & c_{12} & \cdots & c_{1n} \\ c_{21} & c_{22} & \cdots & c_{2n} \\ \vdots & \vdots & & \vdots \\ c_{m1} & c_{m2} & \cdots & c_{mn} \end{bmatrix}; \boldsymbol{D} = \begin{bmatrix} d_{11} & d_{12} & \cdots & d_{1r} \\ d_{21} & d_{22} & \cdots & d_{2r} \\ \vdots & \vdots & & \vdots \\ d_{m1} & d_{m2} & \cdots & d_{mr} \end{bmatrix}$$

Also,

$$\boldsymbol{x} = \begin{bmatrix} x_1 & x_2 & \cdots & x_n \end{bmatrix}^T = \text{state vector } (n\text{th order})$$

$$\boldsymbol{u} = \begin{bmatrix} u_1 & u_2 & \cdots & u_r \end{bmatrix}^T = \text{input vector } (r\text{th order})$$

$$\boldsymbol{y} = \begin{bmatrix} y_1 & y_2 & \cdots & y_m \end{bmatrix}^T = \text{output vector } (m\text{th order})$$

$\boldsymbol{A}$ = system matrix ($n \times n$)
$\boldsymbol{B}$ = input distribution matrix ($n \times r$)
$\boldsymbol{C}$ = output (or measurement) gain matrix ($m \times n$)
$\boldsymbol{D}$ = feedforward input gain matrix ($m \times r$)

Note: $[\ ]^T$ denotes the transpose of a matrix or vector. The system matrix $\boldsymbol{A}$ tells us how the system responds naturally (freely) in the absence of any external input, but it is needed as well to determine the system response in the presence of an external input. The input distribution matrix $\boldsymbol{B}$ tells us how the input $\boldsymbol{u}$ affects (i.e., how it is amplified and distributed when reaching) the system.

3.4.2.1 Time-Invariant Systems

The functions $\boldsymbol{f}$ and $\boldsymbol{h}$ in equations (3.14) and (3.15) show their explicit dependence on time t. Then, the corresponding dynamic system (state model) is said to be *time-variant* (or *nonstationary* or *non-autonomous*). That means, the system parameters (such as mass, stiffness, capacitance, inductance) vary with time. In the time-variant case, the system behavior depends on the time origin, for a given initial state and input function. If the system parameters do not change with time, the functions $\boldsymbol{f}$ and $\boldsymbol{h}$ will not have an explicit t in its representation. Then, the corresponding system (state model) is said to be *time-invariant*, or *stationary*, or *autonomous*. A linear system is time-invariant if (and only if) the matrices $\boldsymbol{A}$, $\boldsymbol{B}$, $\boldsymbol{C}$, and $\boldsymbol{D}$ (in equations (3.16) and (3.17)) are constants.

3.4.3 Derivation of Input–Output Models from State-Space Models

Once a state-space model is obtained, it is easy to convert it into a time-domain input–output model. In the general, nonlinear case, suppose the equations (3.15) and their derivatives are substituted into (3.14) to eliminate $\boldsymbol{x}$ and $\dot{\boldsymbol{x}}$, and get a set of differential equations for $\boldsymbol{y}$ (with $\boldsymbol{u}$ and its derivatives present). The result is the input–output model, in the time domain. If these input–output differential equations are nonlinear, then the system (or strictly, the input–output model) is nonlinear.

The analytical procedure of achieving this for a linear, time-invariant state model is as follows:

1. Successively differentiate the equations (3.17) (e.g., $\dot{\boldsymbol{y}} = \boldsymbol{C}\dot{\boldsymbol{x}} + \boldsymbol{D}\dot{\boldsymbol{u}}$)
2. Eliminate $\dot{\boldsymbol{x}}$ by substituting equations (3.16) (e.g., $\dot{\boldsymbol{y}} = \boldsymbol{CAx} + \boldsymbol{CBu} + \boldsymbol{D}\dot{\boldsymbol{u}}$)
3. Use the resulting equations to eliminate the remaining state variables.

A matrix equation cannot be given for the last step. The final I-O equations have to be determined by using the actual resulting equations (see Example 3.4).

Example 3.3

The concepts of state, output, and order of a system, and the importance of the system's initial state can be shown using a simple example. Consider the rectilinear (translatory) motion of a particle of mass m subject to an input force $u(t)$ (Figure 3.7).

By Newton's second law, its velocity v can be expressed as the first-order differential equation:

$$m\frac{dv}{dt} = u(t) \text{ or } \dot{v} = \frac{1}{m}u \tag{i}$$

This is a first-order model, which is completely represented by a single *state equation* with the state variable v. This equation is also the *input–output model* of this system. Furthermore, it satisfies the formal definition of a state (state vector in general, but a scalar in the present example) as stated before. Specifically, by integrating (i), we get

$$v = v_0 + \frac{1}{m}\int_0^t u \cdot dt \tag{ii}$$

where v_0=velocity at t=0.

Equation (ii) is consistent with the formal definition of state because it says that we can determine v at time t by knowing the v at time 0 and the input u from time 0 to t.

Next we can show that the position x of the mass does not constitute a complete state of it. To show that, we integrate (ii), again from t=0 to t. We get

$$x = x_0 + v_0 t + \frac{1}{m}\int_0^t \int_0^\tau u \cdot d\tau \cdot dt \tag{iii}$$

Note: We have introduced the dummy variable τ for time, for the first integration, which is independent of the second integration.

Equation (ii) says that to determine x at time t, we need to know both x and v at time 0 (not just x at time 0) and the input u from time 0 to t. This is not consistent with the formal definition of state, and hence, x alone cannot completely represent the state of a point mass. If we need

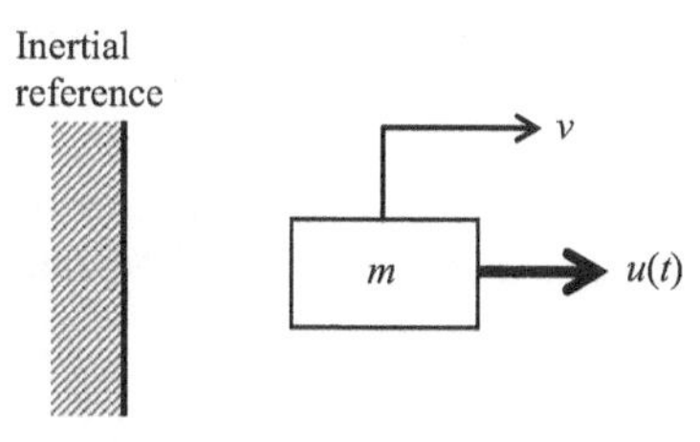

FIGURE 3.7 Point mass in rectilinear motion under a force.

to know x as an output, we have to use both x and v as states. Then we have a second-order state vector:

$$\mathbf{x} = \begin{bmatrix} x & v \end{bmatrix}^T$$

The corresponding two state equations are:

$$\dot{x} = v$$

$$\dot{v} = \frac{1}{m} u(t)$$

The associated model matrices are:

$$\mathbf{A} = \begin{bmatrix} 0 & 1 \\ 0 & 0 \end{bmatrix}; \quad \mathbf{B} = \begin{bmatrix} 0 \\ 1/m \end{bmatrix}; \quad \mathbf{C} = \begin{bmatrix} 1 & 0 \end{bmatrix}$$

Note: The use of x as well is not a natural choice of states for this system (a mass particle in translatory motion) since we know that, naturally, the system is first order, not second order. Because we seek an "unnatural" output (position x) for the mass, we get the unnatural result of a second-order model for a first-order system (which has only a single energy storage element). This further supports our previous assertion (see Chapter 2) that the natural state variable for a mass (an A-type element) is its across-variable, which is v.

LEARNING OBJECTIVES

1. Simple investigation into the concepts of state and order of a system/model
2. The natural choice of state variable may have to be modified depending on the required output
3. The system order may have to be modified depending on the required output.

3.4.3.1 System Order

The "order" of a dynamic system (model) indicates the dynamic "size" of it. The order may be defined/indicated in many ways, but all of them indicate the same thing. The following are some common indicators of the model order:

1. Number of independent energy storage elements
2. Number of state variables (i.e., order of the state vector)
3. Order of the system matrix $\mathbf{A}$ (in the linear case)
4. Number of initial conditions required to determine the system response
5. Order of the input–output differential equation (time-domain I-O model)
6. Order of the denominator polynomial of the system transfer function (i.e., the frequency-domain I-O model—see Chapter 6)
7. Number of eigenvalues of $\mathbf{A}$ (or poles—roots of the denominator equation of the transfer function—see Chapter 6).

Some of these indicators are general (e.g., 1, 2, 4, 5) while some others assume a linear or linear and time-invariant system (e.g., 3 assumes a linear system while 6 and 7 assume a linear time-invariant system).

Example 3.4

Convert the following five state-space models into input–output models:

$$\text{(a) } \begin{aligned}\dot{x}_1 &= 2x_1 + 3x_2 + u_1\\ \dot{x}_2 &= x_1 - x_2 - u_2\\ y &= x_1 + 2x_2\end{aligned} \;;\quad \text{(b) } \begin{aligned}\dot{x}_1 &= 2x_1 + 3x_2 + u\\ \dot{x}_2 &= x_1 - x_2\\ y &= x_1 + 2x_2\end{aligned} \;;\quad \text{(c) } \begin{aligned}\dot{x}_1 &= 2x_1 + 3x_2\\ \dot{x}_2 &= x_1 - x_2\\ y &= x_1 + 2x_2\end{aligned} \;;$$

$$\text{(d) } \begin{aligned}\dot{x}_1 &= 2x_1 + u\\ \dot{x}_2 &= x_1 - x_2\\ y &= x_1\end{aligned} ;\; \text{(e) } \begin{aligned}\dot{x}_1 &= 2x_1 + 3x_2 + u_1\\ \dot{x}_2 &= x_1 - x_2 - u_2\\ y &= x_1 + 2x_2 + 3u_1\end{aligned}$$

Solution

(a)

$$\begin{aligned}\dot{x}_1 &= 2x_1 + 3x_2 + u_1\\ \dot{x}_2 &= x_1 - x_2 - u_2\\ y &= x_1 + 2x_2\end{aligned} \qquad \text{(i)}$$

Differentiate (i): $\dot{y} = \dot{x}_1 + 2\dot{x}_2$

Substitute state equations:

$$\begin{aligned}\dot{y} &= 2x_1 + 3x_2 + u_1 + 2(x_1 - x_2 - u_2)\\ &= 4x_1 + x_2 + u_1 - 2u_2\end{aligned} \qquad \text{(ii)}$$

Differentiate (ii):

$$\ddot{y} = 4\dot{x}_1 + \dot{x}_2 + \dot{u}_1 - 2\dot{u}_2$$

Substitute state equations:

$$\begin{aligned}\ddot{y} &= 4(2x_1 + 3x_2 + u_1) + x_1 - x_2 - u_2 + \dot{u}_1 - 2\dot{u}_2\\ &= 9x_1 + 11x_2 + 4u_1 - u_2 + \dot{u}_1 - 2\dot{u}_2\end{aligned} \qquad \text{(iii)}$$

Substitute (i) to eliminate x_1:

$$\text{(ii)}: \quad \dot{y} = 4(y - 2x_2) + x_2 + u_1 - 2u_2 = 4y - 7x_2 + u_1 - 2u_2$$

$$\begin{aligned}\text{(iii)}: \quad \ddot{y} &= 9(y - 2x_2) + 11x_2 + 4u_1 - u_2 + \dot{u}_1 - 2\dot{u}_2\\ &= 9y - 7x_2 + 4u_1 - u_2 + \dot{u}_1 - 2\dot{u}_2\end{aligned}$$

Subtract the first result from the second to eliminate x_2:

$$\begin{aligned}\ddot{y} - \dot{y} &= 9y + 4u_1 - u_2 + \dot{u}_1 - 2\dot{u}_2 - 4y - u_1 + 2u_2\\ &= 5y + 3u_1 + u_2 + \dot{u}_1 - 2\dot{u}_2\\ &\Rightarrow \ddot{y} - \dot{y} - 5y = 3u_1 + u_2 + \dot{u}_1 - 2\dot{u}_2\end{aligned}$$

Note: This is a "two-input single-output" system. We can consider one input at a time (u_1 or u_2), which will result in two SISO systems. Then, in view of the system linearity, we can apply the "principle of superposition" with the two SISO results, to determine the output (y) when both inputs are present.

(b)

$$\begin{aligned} \dot{x}_1 &= 2x_1 + 3x_2 + u \\ \dot{x}_2 &= x_1 - x_2 \\ y &= x_1 + 2x_2 \end{aligned} \tag{i}$$

Differentiate (i): $\dot{y} = \dot{x}_1 + 2\dot{x}_2$

Substitute state equations:

$$\begin{aligned} \dot{y} &= 2x_1 + 3x_2 + u + 2(x_1 - x_2) \\ &= 4x_1 + x_2 + u \end{aligned} \tag{ii}$$

Differentiate (ii):

$$\begin{aligned} \ddot{y} &= 4\dot{x}_1 + \dot{x}_2 + \dot{u} = 4(2x_1 + 3x_2 + u) + (x_1 - x_2) + \dot{u} \\ &= 9x_1 + 11x_2 + 4u + \dot{u} \end{aligned} \tag{iii}$$

Substitute (i) to eliminate x_1:

$$\begin{aligned} \text{(ii)}: \quad \dot{y} &= 4(y - 2x_2) + x_2 + u = 4y - 7x_2 + u \\ \text{(iii)}: \quad \ddot{y} &= 9(y - 2x_2) + 11x_2 + 4u + \dot{u} \\ &= 9y - 7x_2 + 4u + \dot{u} \end{aligned}$$

Eliminate x_2 (Subtract the first result from the second result):

$$\begin{aligned} \ddot{y} - \dot{y} &= 9y + 4u + \dot{u} - 4y - u \\ &= 5y + 3u + \dot{u} \\ &\Rightarrow \ddot{y} - \dot{y} - 5y = 3u + \dot{u} \end{aligned}$$

This is a SISO system.

Note: This result can be obtained from the result of Part (a) by simply setting $u_1 = u$ and $u_2 = 0$.

(c)

$$\begin{aligned} \dot{x}_1 &= 2x_1 + 3x_2 \\ \dot{x}_2 &= x_1 - x_2 \\ y &= x_1 + 2x_2 \end{aligned} \tag{i}$$

Differentiate (i):

$$\begin{aligned} \dot{y} &= \dot{x}_1 + 2\dot{x}_2 = 2x_1 + 3x_2 + 2(x_1 - x_2) \\ &= 4x_1 + x_2 \end{aligned} \tag{ii}$$

Differentiate (ii):

$$\begin{aligned} \ddot{y} &= 4\dot{x}_1 + \dot{x}_2 = 4(2x_1 + 3x_2) + (x_1 - x_2) \\ &= 9x_1 + 11x_2 \end{aligned} \tag{iii}$$

Substitute (i) to eliminate x_1:

$$\text{(ii)}: \quad \dot{y} = 4[y - 2x_2] + x_2 = 4y - 7x_2$$

$$\text{(iii)}: \quad \ddot{y} = 9(y - 2x_2) + 11x_2 = 9y - 7x_2$$

Subtract the first result from the second to eliminate x_2 :

$$\ddot{y} - \dot{y} = 9y - 4y = 5y$$
$$\Rightarrow \ddot{y} - \dot{y} - 5y = 0$$

This represents the free (natural) response of the system (i.e., no input).

Note: This result can be obtained from Part (a) or Part (b) by simply setting the inputs to zero.

(d)

$$\begin{aligned} \dot{x}_1 &= 2x_1 + u \\ \dot{x}_2 &= x_1 - x_2 \\ y &= x_1 \end{aligned} \tag{i}$$

$$\text{Differentiate (i)}: \quad \dot{y} = \dot{x}_1 = 2x_1 + u \tag{ii}$$

Substitute (i):

$$\dot{y} = 2y + u$$
$$\Rightarrow \dot{y} - 2y = u$$

This is a "first-order" system. It is not coupled with the subsystem denoted by $\dot{x}_2 = x_1 - x_2$ and is a completely independent first-order system. However, the subsystem denoted by $\dot{x}_2 = x_1 - x_2$ is influenced by the first (uncoupled) subsystem.

Note: If the output depends on both x_1 and x_2, then, the second subsystem has to be included in the model, and the overall system becomes "second order."

$$\text{E.g. 1}: \text{ Let } y = x_1 + 2x_2 \tag{i}$$

Differentiate and substitute state equations

$$\begin{aligned} \dot{y} &= \dot{x}_1 + 2\dot{x}_2 = 2x_1 + u + 2(x_1 - x_2) \\ &= 4x_1 - 2x_2 + u \end{aligned} \tag{ii}$$

$$\begin{aligned} \ddot{y} &= 4\dot{x}_1 - 2\dot{x}_2 + \dot{u} = 4(2x_1 + u) - 2(x_1 - x_2) + \dot{u} \\ &= 6x_1 + 2x_2 + 4u + \dot{u} \end{aligned} \tag{iii}$$

Substitute (i) to eliminate x_1:

$$\text{(ii)}: \quad \dot{y} = 4(y - 2x_2) - 2x_2 + u = 4y - 10x_2 + u$$

$$\text{(iii)}: \quad \ddot{y} = 6(y - 2x_2) + 2x_2 + 4u + \dot{u} = 6y - 10x_2 + 4u + \dot{u}$$

Subtract first from the second (to eliminate x_2):

$$\ddot{y}-\dot{y}=6y+4u+\dot{u}-4y-u=2y+3u+\dot{u}$$
$$\Rightarrow \ddot{y}-\dot{y}-2y=3u+\dot{u}$$

Note: Clearly, this is a second-order system.

$$\text{E.g. 2: Let } y=x_2 \tag{i}$$

Differentiate and substitute state equations

$$\dot{y}=\dot{x}_2=x_1-x_2\text{(ii)}$$

$$\begin{aligned}\ddot{y}&=\dot{x}_1-\dot{x}_2=2x_1+u-(x_1-x_2)\\&=x_1+x_2+u\end{aligned} \tag{iii}$$

Substitute (i) to eliminate x_2:

$$\text{(ii)}:\quad \dot{y}=x_1-y$$
$$\text{(iii)}:\quad \ddot{y}=x_1+y+u$$

Subtract first from the second (to eliminate x_1):

$$\ddot{y}-\dot{y}=y+u+y$$
$$\Rightarrow \ddot{y}-\dot{y}-2y=u$$

Note: This is also a second-order system (the LHS is identical to before, representing common "free" dynamics).

(e)

$$\begin{aligned}\dot{x}_1&=2x_1+3x_2+u_1\\\dot{x}_2&=x_1-x_2-u_2\\y&=x_1+2x_2+3u_1\end{aligned} \tag{i}$$

Differentiate (i):

$$\begin{aligned}&\dot{y}=\dot{x}_1+2\dot{x}_2+3\dot{u}_1=2x_1+3x_2+u_1+2(x_1-x_2-u_2)+3\dot{u}_1\\&\Rightarrow \dot{y}=4x_1+x_2+u_1-2u_2+3\dot{u}_1\end{aligned} \tag{ii}$$

Differentiate (ii):

$$\begin{aligned}\ddot{y}&=4\dot{x}_1+\dot{x}_2+\dot{u}_1-2\dot{u}_2+3\ddot{u}_1\\&=4(2x_1+3x_2+u_1)+x_1-x_2-u_2+\dot{u}_1-2\dot{u}_2+3\ddot{u}_1\\\Rightarrow \ddot{y}&=9x_1+11x_2+4u_1-u_2+\dot{u}_1-2\dot{u}_2+3\ddot{u}_1\end{aligned} \tag{iii}$$

Substitute (i) in (ii) and (iii) to eliminate x_1:

$$\text{(ii)}: \quad \dot{y} = 4(y - 2x_2 - 3u_1) + x_2 + u_1 - 2u_2 + 3\dot{u}_1$$
$$= 4y - 7x_2 - 11u_1 - 2u_2 + 3\dot{u}_1$$
$$\text{(iii)}: \quad \ddot{y} = 9(y - 2x_2 - 3u_1) + 11x_2 + 4u_1 - u_2 + \dot{u}_1 - 2\dot{u}_2 + 3\ddot{u}_1$$
$$= 9y - 7x_2 - 23u_1 - u_2 + \dot{u}_1 - 2\dot{u}_2 + 3\ddot{u}_1$$

Subtract to eliminate x_2:

$$\ddot{y} - \dot{y} = 9y - 23u_1 - u_2 + \dot{u}_1 - 2\dot{u}_2 + 3\ddot{u}_1 - 4y + 11u_1 + 2u_2 - 3\dot{u}_1$$
$$= 5y - 12u_1 + u_2 - 2\dot{u}_1 - 2\dot{u}_2 + 3\ddot{u}_1$$
$$\Rightarrow \ddot{y} - \dot{y} - 5y = -12u_1 + u_2 - 2\dot{u}_1 - 2\dot{u}_2 + 3\ddot{u}_1$$

In this model, the highest order of the input side (i.e., $\ddot{u}_1$ term) and of the output side (i.e., $\ddot{y}$ term) is the same. This is a property of a system with the "feedforward" characteristic.

Note: The above result can be obtained from the result of Part (a) as follows:

Define a new output variable as:

$$y' = y - 3u_1 \quad \text{(iv)}$$

Then the output equation becomes: $y' = x_1 + 2x_2$

This transformed problem is identical to that of Part (a) albeit with the new output variable y'. Hence, the input–output model of the present problem can be stated as (from the result of Part (a)):

$$\ddot{y}' - \dot{y}' - 5y' = 3u_1 + u_2 + \dot{u}_1 - 2\dot{u}_2$$

Transform back the original problem using (iv):

$$(\ddot{y} - 3\ddot{u}_1) - (\dot{y} - 3\dot{u}_1) - 5(y - 3u_1) = 3u_1 + u_2 + \dot{u}_1 - 2\dot{u}_2$$

On simplifying we get: $\ddot{y} - \dot{y} - 5y = -12u_1 + u_2 - 2\dot{u}_1 - 2\dot{u}_2 + 3\ddot{u}_1$

This result is identical to what we obtained before.

3.5 MODELING EXAMPLES

From the foregoing discussions and illustrative examples, the following statements can be made:

- State vector is a least (minimal) set of variables that completely determine the dynamic state of system. Hence, a state variable cannot be expressed as a linear combination of the remaining state variables
- State vector is not unique. Many choices are possible for a given system
- Output (response) variables can be completely determined from the state variables
- State variables may or may not have a physical interpretation.

3.5.1 Systematic Development of a State Model

The steps for developing analytical models have been identified. At this stage, let us focus on state-space models. The key steps in our systematic approach for formulating a state-space model are given now.

Note: We assume that the inputs (***u***) and the outputs (***y***) of the system are specified or identified.

Step 1: Identify the physical elements in the system (e.g., energy storage elements, energy dissipative elements, input/source elements) and draw a structural diagram (e.g., circuit diagram, linear graph—Chapter 5, bond graph, etc.) to show how they are interconnected

Step 2: Identify the state variables (vector $\boldsymbol{x}$). These are the across-variables of the independent A-type energy storage elements and the through-variables of the independent T-type energy storage elements. Two energy storage elements (e.g., springs) are dependent if they can be combined into (represented by) a single energy storage element of the same type

Step 3: Write the constitutive equations for all the elements in the system (including the energy storage elements, energy dissipative elements, and two-port elements—see Chapter 5), while excluding the source (input) elements (a source has an independent variable, which is not affected by the system dynamics)

Step 4: Write the continuity equations for the through-variables for all the nodes, less 1 (node is a point that connects two or more system components. If there are n nodes in the system, only $n-1$ node equations are independent. The other node equation can be formed by combining the first $n-1$ node equations)

Step 5: Write the compatibility equations for the across-variables around all the independent loops (a loop is a closed path that is formed by two or more system components. It is independent if it cannot be formed by combining the remaining loops)

Step 6: Form the state-space shell. It is formed by the constitutive equations for the independent energy storage elements. Eliminate the auxiliary variables in the state-space shell (by using the remaining equations from steps 3, 4, and 5)

Note: In the state equations, we keep only the state variables and the input variables. Any other variable is an auxiliary (unnecessary/redundant) variable in the shell equations

Step 7: Express the outputs in terms of the state variables.

Note: Sometimes (i.e., in the presence of the feedforward character), an output equation may have to include one or more input variables as well.

Now we present several examples in the mechanical, electrical, fluid, and thermal domains for the systematic development of a state-space model. In some examples, the conversion from a state-space model to an input–output model is illustrated as well.

3.5.2 Modeling in the Mechanical Domain

In the mechanical domain, the modeling approach is facilitated by the use of a structural diagram similar to the circuit diagram that is used in the electrical domain. Furthermore, sometimes, the loop equations may be automatically satisfied in a mechanical system (if the across-variables are not introduced separately, and if some of them are defined in terms of other existing absolute variables). Also, (1) for an inertia element, the inertial reference (with zero velocity) is always one end of the element; and (2) for a mechanical source, one end is always fixed (with zero velocity). Hence, their across-variables are "absolute" variables.

Example 3.5

A rigid mechanical object of mass m is cushioned by a flexible pad represented by two springs of stiffness k_1 and k_2 and a damper of damping constant b. It is excited by a source with force $f(t)$. A sketch of the device is shown in Figure 3.8. The velocity of the mass is v. The across-variables of (change in velocity across) the other elements are also shown in the figure. The output of the system is the velocity v of the object. The three independent loops of the system are shown as well, to assist the solution of the problem.

a. Sketch a circuit diagram showing the nodes of the system, and mark suitable through-variables (forces) at the nodes.

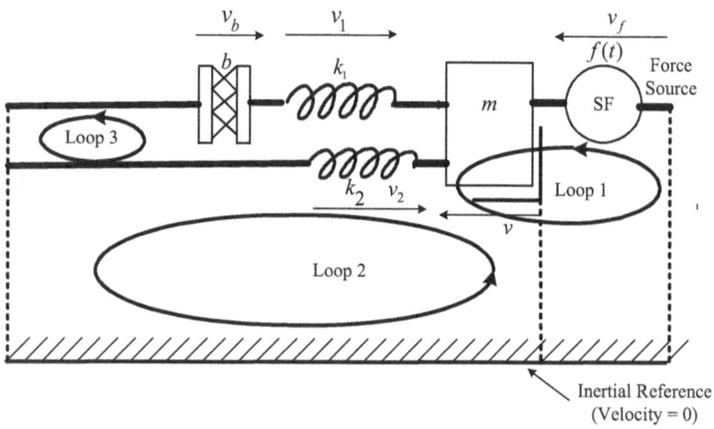

FIGURE 3.8 A mechanical device with a force source.

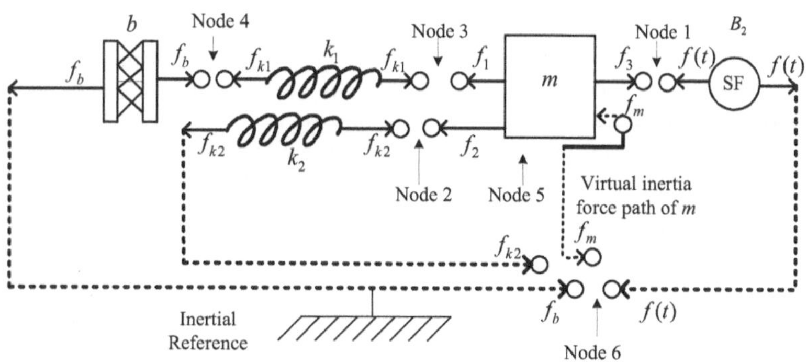

FIGURE 3.9 Mechanical circuit with the node details.

b. Using the detailed systematic approach, develop the state-space model for this system. Why are the loop equations not automatically satisfied?
c. Express the model, including the output equation, in the vector-matrix form.

Solution

(a)

See Figure 3.9.

(b)

Constitutive Equations

State-Space Shell

To formulate the state-space shell, we write the constitutive equations of the independent energy storage elements. In the present system, there is a mass element and two spring elements, all of which are independent. The associated three equations are written now.

$$\text{For } k_1\text{:} \quad \frac{df_{k1}}{dt} = k_1 v_1$$

$$\text{For } k_2\text{:} \quad \frac{df_{k2}}{dt} = k_2 v_2$$

$$\text{For } m\text{:} \quad m\frac{dv}{dt} = f_m$$

Remaining Constitutive Equation

$$\text{For } b\text{:} \quad f_b = bv_b$$

Loop Equations

To write the loop equations, we use Figure 3.8. It shows the three *primary loops*. We use the sign convention where "velocity increase," shown by an arrow, is taken positive.

$$\text{Loop 1: } v_f - v = 0$$

$$\text{Loop 2: } v - v_2 = 0$$

$$\text{Loop 3: } -v_1 - v_b + v_2 = 0$$

It is seen that the loop equations are not automatically satisfied here because we had defined separate across-variables (velocity changes) for all the elements in the system, not the absolute velocities at the joints.

Node Equations

To write the node equations, we use the circuit diagram shown in Figure 3.9, which separates the elements of the system, showing its *nodes*. There are six nodes in this system. Only five node equations should be written, however, because the equation of the remaining node is determined from the equations of the first five nodes. Using the nomenclature for the forces, as indicated in Figure 3.9, and taking the forces acting to the right as positive, we write the following node equations.

$$\text{Node 1: } f_3 - f(t) = 0$$

$$\text{Node 2: } f_{k2} - f_2 = 0$$

$$\text{Node 3: } f_{k1} - f_1 = 0$$

$$\text{Node 4: } f_b - f_{k1} = 0$$

$$\text{Node 5: } f_3 - f_1 - f_2 - f_m = 0 \quad (\textit{Note}\text{: This is the node representing all the forces applied to the mass } m.\ f_m \text{ is the inertia force})$$

As a check, we write the equation for node 6: $f_{k2} + f_b + f_m - f(t) = 0$.

This equation is obtained simply by combining the previous five node equations, and hence, it is not independent.

Auxiliary Variables

We eliminate the three auxiliary variables in the state-space shell, by using:

$$v_1 = v_2 - v_b = -\frac{T_{d_1}}{B_1} = v - \frac{f_b}{b} = v - \frac{1}{b} f_{k1}$$

(from loop 2 and 3 equations+damper constitutive equation+node 4 equation)

$$v_2 = v$$

(from loop 2 equation)

$$f_m = f_3 - f_1 - f_2 = T - T_{d2} = f(t) - f_{k1} - f_{k2}$$

(from nodes 1, 2, and 3 equations)

State Equations

On substituting for the auxiliary variables in the state-space shell, we get the final state-space equations

$$\frac{df_{k1}}{dt} = k_1 v - \frac{k_1}{b} f_{k1}$$

$$\frac{df_{k2}}{dt} = k_2 v$$

$$\frac{dv}{dt} = \frac{1}{m}\left[f(t) - f_{k1} - f_{k2}\right]$$

Output equation: $y = v$

(c)

The vector-matrix form of the state equations and the algebraic output equation can be easily obtained from the obtained model.

State vector $\boldsymbol{x} = \left[f_{k1} \;\; f_{k2} \;\; v\right]^T$; input $\boldsymbol{u} = [f(t)]$; output $\boldsymbol{y} = [v]$

The corresponding matrices are

$$\boldsymbol{A} = \begin{bmatrix} k_1 & -\dfrac{k_1}{b} & 0 \\ k_2 & 0 & 0 \\ 0 & -\dfrac{1}{m} & -\dfrac{1}{m} \end{bmatrix}; \quad \boldsymbol{B} = \begin{bmatrix} 0 \\ 1/m \end{bmatrix}; \quad \boldsymbol{C} = [0 \;\; 0 \;\; 1]$$

LEARNING OBJECTIVES

1. Proper choice of state variables
2. Systematic development of a state-space model
3. Modeling of mechanical systems
4. Implications of the across-variable choice on the loop equations.

Important Comments

1. Some of the continuity equations (node equations) and compatibility equations (loop equations) may be automatically satisfied by the particular choice of variables. Then, the corresponding equations are redundant (not in the above example)
2. Some of the node equations and/or loop equations may not be needed for the elimination of the auxiliary variable in the state-space shell
3. The reference for the inertia elements and the input elements is the ground (inertial reference, with zero velocity).

3.5.3 Modeling in the Electrical Domain

In the electrical domain, the modeling approach is facilitated by the use of a circuit diagram. Then, the usual modeling steps are followed. The element equations introduced in Chapter 2 are used in writing the constitutive equations. The node equations and loop equations are easy to write, through the use of the circuit diagram. These facts are illustrated in the following example.

Example 3.6

A bridge network is shown in Figure 3.10. v_a and v_b are the voltage inputs to the circuit. Using the voltage v_C across the capacitor (with capacitance C) and the current i_L through the inductor (with

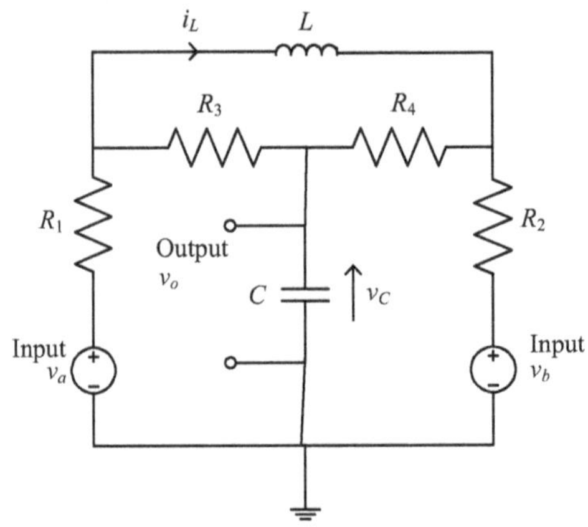

FIGURE 3.10 A bridge network.

inductance L) as the state variables (which corresponds to the "unique" choice in the learned modeling approach), obtain a complete state-space model. The circuit output is the voltage across the capacitor.

How would you determine the corresponding input–output model (differential equation)?

Note: R_1, R_2, R_3, R_4 are the resistances of the four arms of the bridge; L: "extra" bringing element; C: load (output). If $\frac{R_1}{R_2} = \frac{R_3}{R_4}$, we have a balanced bridge.

Solution

In Figure 3.11, we have indicated the important variables and other useful information of the circuit. It identifies some auxiliary variables, three "independent" nodes, and three "independent" loops.

Constitutive Equations

$$\left.\begin{aligned} C: C\frac{dv_C}{dt} &= i_C \quad \text{(i)} \\ L: L\frac{di_L}{dt} &= v_L \quad \text{(ii)} \end{aligned}\right\} \text{state-space shell}$$

$$R_1 : v_1 = R_1 i_1 \quad \text{(iii)}$$

$$R_2 : v_2 = R_2 i_2 \quad \text{(iv)}$$

$$R_3 : v_3 = R_3 i_3 \quad \text{(v)}$$

$$R_4 : v_4 = R_4 i_4 \quad \text{(vi)}$$

Node Equations

$$i_1 - i_3 - i_L = 0 \quad \text{(vii)}$$

$$i_3 - i_C + i_4 = 0 \quad \text{(viii)}$$

$$i_L - i_4 + i_2 = 0 \quad \text{(ix)}$$

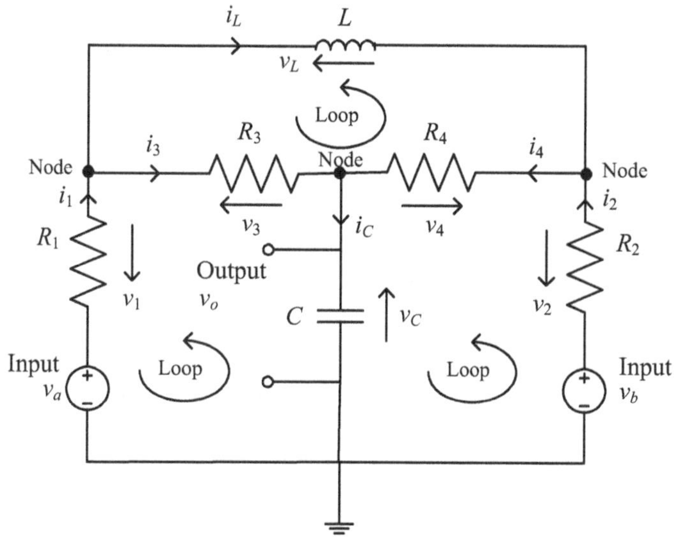

FIGURE 3.11 Circuit variables

Loop Equations

$$-v_C - v_3 - v_1 + v_a = 0 \tag{x}$$

$$-v_b + v_2 + v_4 + v_C = 0 \tag{xi}$$

$$v_3 - v_4 - v_L = 0 \tag{xii}$$

In the node equations and the loop equations, we have used the standard sign convention.

Eliminate the auxiliary variables in the state-space shell.

$$\text{(viii)}: i_C = i_3 + i_4 = \frac{v_3}{R_3} + \frac{v_4}{R_4} \tag{xiii}$$

$$\text{(xii)}: v_L = v_3 - v_4 \tag{xiv}$$

Determine v_3 and v_4 in terms of the state and input variables (i.e., variables that need to be kept) and substitute in (xiii) and (xiv):

$$\text{(vii)} \frac{v_1}{R_1} - \frac{v_3}{R_3} - i_L = 0 \tag{xv}$$

Substitute (x) in (xv):

$$\frac{1}{R_1}(-v_C - v_3 + v_a) - \frac{v_3}{R_3} - i_L = 0$$

$$\Rightarrow \left(\frac{1}{R_1} + \frac{1}{R_3}\right) v_3 = -\frac{v_C}{R_1} + \frac{v_a}{R_1} - i_L \tag{xvi}$$

$$\Rightarrow \frac{v_3}{R_3} = \frac{1}{(R_1 + R_3)}(-v_C + v_a) - \frac{R_1}{(R_1 + R_3)} i_L$$

$$\text{(ix)}: i_L - \frac{v_4}{R_4} + \frac{v_2}{R_2} = 0$$

Substitute (xi) in this result: $i_L - \frac{v_4}{R_4} + \frac{1}{R_2}(v_b - v_4 - v_c) = 0 \Rightarrow \left(\frac{1}{R_2} + \frac{1}{R_4}\right)v_4 = -\frac{v_C}{R_2} + \frac{v_b}{R_2} + i_L$

$$\Rightarrow \frac{v_4}{R_4} = \frac{1}{(R_2 + R_4)}(-v_C + v_b) + \frac{R_2}{(R_2 + R_4)} i_L \qquad \text{(xvii)}$$

Substitute (xiii) and (xiv) with (xvi) and (xvii) into (i) and (ii):

$$C\frac{dv_C}{dt} = \frac{1}{(R_1 + R_3)}(-v_C + v_a) - \frac{R_1}{(R_1 + R_3)} i_L + \frac{1}{(R_2 + R_4)}(-v_C + v_b) + \frac{R_2}{(R_2 + R_4)} i_L$$

$$L\frac{di_L}{dt} = \frac{R_3}{(R_1 + R_3)}(-v_C + v_a) - \frac{R_1 R_3}{(R_1 + R_3)} i_L - \frac{R_4}{(R_2 + R_4)}(-v_C + v_b) - \frac{R_2 R_4}{(R_2 + R_4)} i_L$$

On grouping the like terms, we get the state equations:

$$C\frac{dv_C}{dt} = -\left[\frac{1}{(R_1 + R_3)} + \frac{1}{(R_2 + R_4)}\right]v_C - \left[\frac{R_1}{(R_1 + R_3)} - \frac{R_2}{(R_2 + R_4)}\right]i_L + \frac{1}{(R_1 + R_3)}v_a + \frac{1}{(R_2 + R_4)}v_b$$

$$L\frac{di_L}{dt} = -\left[\frac{R_3}{(R_1 + R_3)} - \frac{R_4}{(R_2 + R_4)}\right]v_C - \left[\frac{R_1 R_3}{(R_1 + R_3)} + \frac{R_2 R_4}{(R_2 + R_4)}\right]i_L + \frac{R_3}{(R_1 + R_3)}v_a - \frac{R_4}{(R_2 + R_4)}v_b$$

Output Equation: $v_o = v_C$

State vector=$\begin{bmatrix} v_C & i_L \end{bmatrix}^T$

Input vector=$\begin{bmatrix} v_a & v_b \end{bmatrix}^T$

Output vector=$\begin{bmatrix} v_o \end{bmatrix}$

Model Matrices

$$\mathbf{A} = \begin{bmatrix} -\left[\frac{1}{(R_1 + R_3)} + \frac{1}{(R_2 + R_4)}\right] & -\left[\frac{R_1}{(R_1 + R_3)} - \frac{R_2}{(R_2 + R_4)}\right] \\ -\left[\frac{R_3}{(R_1 + R_3)} - \frac{R_4}{(R_2 + R_4)}\right] & -\left[\frac{R_1 R_3}{(R_1 + R_3)} + \frac{R_2 R_4}{(R_2 + R_4)}\right] \end{bmatrix};$$

$$\mathbf{B} = \begin{bmatrix} \frac{1}{(R_1 + R_3)} & \frac{1}{(R_2 + R_4)} \\ \frac{R_3}{(R_1 + R_3)} & -\frac{R_4}{(R_2 + R_4)} \end{bmatrix}; \quad \mathbf{C} = \begin{bmatrix} 1 & 0 \end{bmatrix}; \quad \mathbf{D} = \begin{bmatrix} 0 & 0 \end{bmatrix}$$

To determine the I-O model, eliminate i_L from the two state equations.

Step 1: Substitute the first state equation and its derivative into second state equation.
Step 2: Replace v_C by v_o.

3.5.4 Modeling in the Fluid Domain

First we present some important concepts in the development of lumped-parameter fluid models. Next we indicate several special considerations that can be made when modeling pneumatic systems, which have elements that are more compressible than hydraulic elements. We end the treatment with an illustrative example.

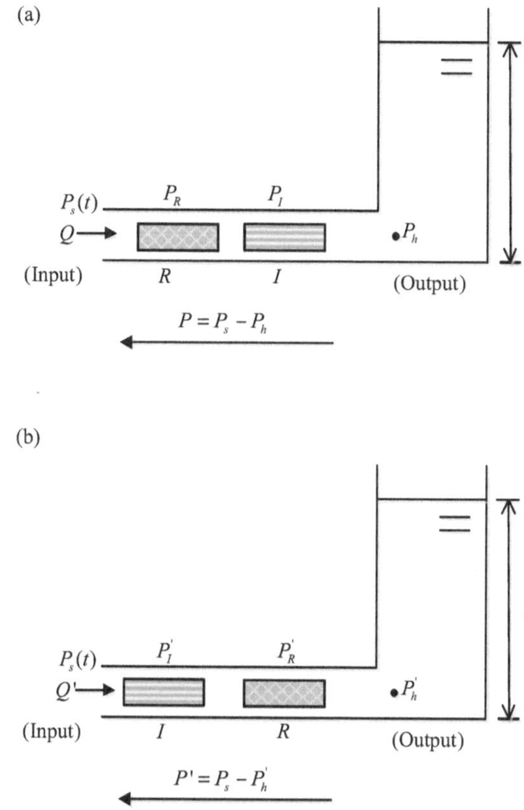

FIGURE 3.12 Lumped model of fluid flow through a pipe into a tank. (a) Case 1; (b) case 2 (with elements interchanged).

3.5.4.1 Commutativity of Series Resistor and Inertor Elements

Consider the case of a pipe flow represented by the two lumped elements: Fluid resistor (R) and fluid inertor (I), driven by the pressure source $P_s(t)$, feeding into a uniform storage tank, as shown in Figure 3.12a. This is called Case 1. When the two lumped elements are interchanged, we get the model in Figure 3.12b. This is called Case 2, where the conditions change to those marked with (').

Note: The variable P represents the pressure drop (an across-variable), not the absolute pressure.

Element (Constitutive) Equations for Case 1:

$$RQ = P_R$$

$$I\frac{dQ}{dt} = P_I$$

Add: $RQ + I\dfrac{dQ}{dt} = P_R + P_I = P = P_s - P_h$

$$\Rightarrow \quad RQ + I\frac{dQ}{dt} = P_s - P_h \tag{3.18}$$

$$\text{with } C_{\text{grv}}\frac{dP_h}{dt} = Q \tag{3.19}$$

By substituting (3.19) into (3.18), we get the input–output system equation (a second-order differential equation in P_h):

$$IC_{\text{grv}}\frac{d^2P_h}{dt^2}+RC_{\text{grv}}\frac{dP_h}{dt}+P_h=P_s(t) \tag{3.20}$$

Element (Constitutive) Equations for Case 2:

$$I\frac{dQ'}{dt}=P_I'$$

$$RQ'=P_R'$$

$$\text{Add:}\quad RQ'+I\frac{dQ'}{dt}=P_I'+P_R'=P_s-P_h'=P'$$

$$\Rightarrow\quad RQ'+I\frac{dQ'}{dt}=P_s-P_h' \tag{3.18'}$$

$$\text{with}\quad C_{\text{grv}}\frac{dP_h'}{dt}=Q' \tag{3.19'}$$

By substituting (3.19') into (3.18'), we get the input–output system equation (a second-order differential equation in P_h'):

$$IC_{\text{grv}}\frac{d^2P_h'}{dt^2}+RC_{\text{grv}}\frac{dP_h'}{dt}+P_h'=P_s(t) \tag{3.20'}$$

From (3.20) and (3.20'), it is seen that we get the same overall system equation (input–output model) in the two cases.

The inertor and resistor elements are commutative (interchangeable), and the two systems shown in Case 1 and Case 2 are equivalent.

In particular, for the same input $P_s(t)$, we will get the same output $\Rightarrow P_h'=P_h$ and $Q'=Q$

Furthermore, since the element parameters (R or I) and the fluid volume flow rate (Q) are the same in the two cases, we must have the pressure drops across the corresponding elements in the two cases: $P_R'=P_R$ and $P_I'=P_I$

Note: By induction, we can prove that this property of commutativity is satisfied by any number of R and I elements that are in series (i.e., having a common through-variable).

3.5.4.2 Extension of Commutativity

The R and I elements in the system of Figure 3.13a may be regrouped as in Figure 3.13b. This is accomplished simply by interchanging the elements I_v and R_h, which are commutative according to the result obtained before. Then, the like elements may be combined to give the equivalent parameters:

$$R_v+R_h=R$$

$$I_v+I_h=I$$

General steps for combining series R and I elements

1. Group together the elements of the same type, by considering two elements at a time
2. Add the resistances to give the equivalent resistance R_{eq}
3. Add the inertances to give the equivalent intertance I_{eq}

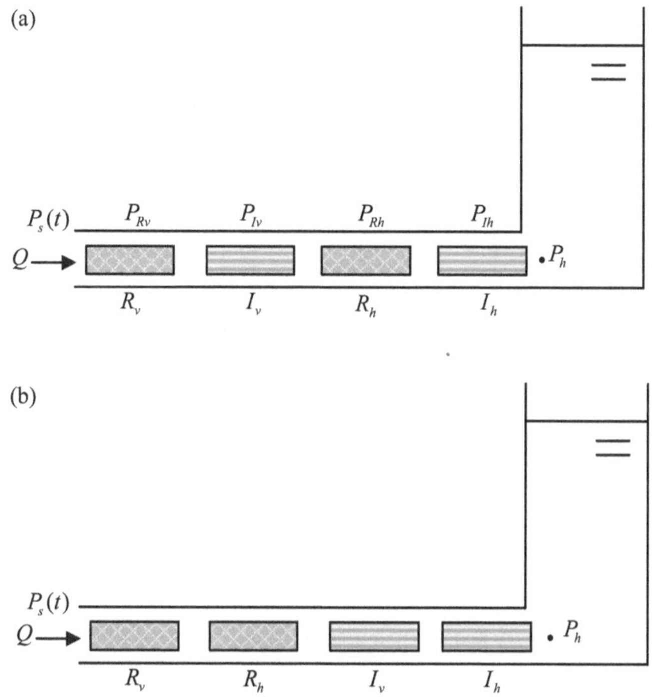

FIGURE 3.13 An example of commutativity of series elements. (a) Original model; (b) equivalent model.

3.5.4.3 Generalization

The generalization of the commutativity to any number of cascaded pairs of series R and I elements can be proved "by induction." This is done now.

We have proved commutativity for the case of one series pair of R and I elements (i.e., for $n=1$).

To prove the general result (i.e., for any n), first assume: Commutativity is true for n. Then we will show that it is true for $n+1$. Hence, by induction, it is true for any n, since it was shown to be true for $n=1$.

Step 1: Since commutativity is assumed for n pairs, group all n number of R-elements together (hence, the n number of I-elements are also grouped together).

Step 2: Add the n + first series I and R elements to the grouped n pair.

Step 3: Using the result for $n=1$ (already proved), interchange the added two elements, if necessary → Grouped $n+1$ number of R-elements and grouped $n+1$ number of I-elements → Commutativity holds for $n+1$ pairs → Proof is complete.

3.5.4.4 Pressure Head in a Vertical Pipe

Consider the vertical pipe segment as shown in Figure 3.14. For the present discussion, neglect inertance and resistance.

If we ignore the capacitance due to gravity head of the vertical pipe segment (because the area of cross-section of the pipe can be considerably small compared to that of the tank; *Note*: Gravity head capacitance is $\frac{A}{\rho g}$ —see Chapter 2), then the flow rate $Q=0$ (from the equation of a fluid capacitor). Consequently, the pressure drop in the vertical pipe segment: $P_s - P_h$ = constant.

Then, we can include this constant pressure drop (gravity head) as a constant pressure "input" (source) rather than a fluid capacitance.

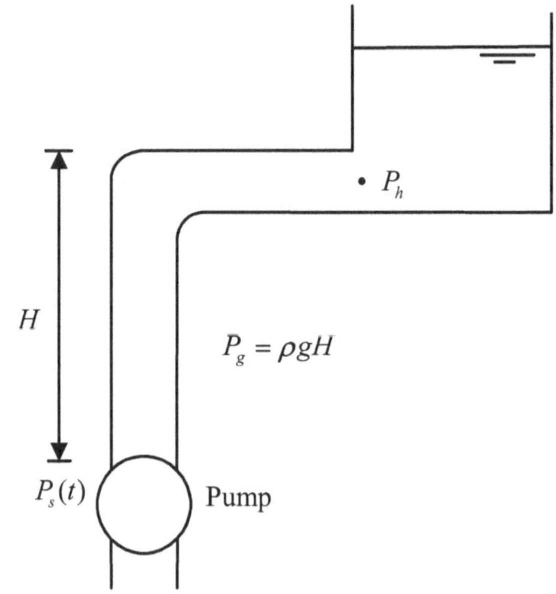

FIGURE 3.14 Effect of the gravity head from a vertical pipe.

Specifically, with $P_g = \rho gH$, we modify (reduce) the pressure source as $P_s - P_g$, where H=height of the pipe segment.

Note: Nevertheless, any pressure drops that are due to fluid resistance or inertance have to be included separately, if they are present in the actual pipe segment.

3.5.4.5 Parallel Connection of Fluid Capacitors

Consider the situation of compressible fluid flow into a tank of flexible (deformable) wall, as shown in Figure 3.15.

Here, the overall fluid flow (volume) rate Q into the tank is the result of three separate flow rates, due to:

1. Compression of the fluid bulk
2. Expansion of the tank
3. Gravity head of the fluid column.

However, we cannot use $C_{\text{bulk}}\dfrac{dP}{dt} = Q_1$ for effect 1, neither can we use $C_{\text{elastic}}\dfrac{dP}{dt} = Q_2$ for effect 2 because the pressure is not uniform within the fluid body in the tank, unless h is relatively very small.

Nevertheless, we can use $C_{\text{grav}}\dfrac{dP}{dt} = Q_3$ for effect 3 because here, P represents the pressure at the bottom of the tank (wrt the atmospheric pressure).

3.5.4.6 Thin Fluid Layer Approximation

Suppose that h is very small. Then, the pressure difference $P = P_1 - P_a$ represents the pressure in the entire fluid (wrt the ambience), since this fluid quantity is small, and P is common to all three flows listed above, where

P_1=pressure at the bottom of the tank
P_a=ambient pressure

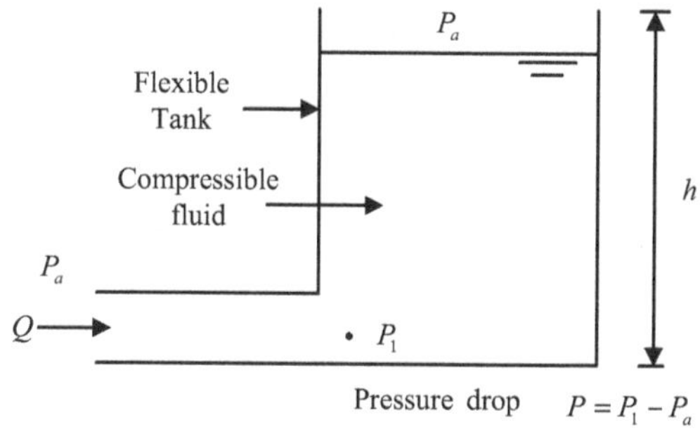

FIGURE 3.15 Flow of compressible liquid into a flexible tank.

Adding the individual flow components, we have $C_{\text{bulk}}\dfrac{dP}{dt} + C_{\text{elastic}}\dfrac{dP}{dt} + C_{\text{grav}}\dfrac{dP}{dt} = Q_1 + Q_2 + Q_3 = Q$.

This is a case of parallel connection (pressure is common and flow rate is additive).

$$\text{Hence} \quad C_{\text{eq}}\frac{dP}{dt}$$

$$\text{with} \quad C_{\text{eq}} = C_{\text{bulk}} + C_{\text{elastic}} + C_{\text{grav}}$$

Note: If h is not small, we have to partition the fluid in the tank into many layers and write equations for each thin layer separately, in this manner (or, integrate the effects of infinitesimal layers). This case is considered under linear graph modeling, in Chapter 5. The situation is further complicated by the fact that due to the tank flexibility, the area of cross-section of the tank changes with the fluid pressure.

Note: An example of two fluid capacitors in series is the case of an incompressible liquid column (C_{grav}) in a rigid tank with a flexible (spring-loaded) cap or accumulator (C_{elastic}). Then the equivalent capacitance (C_{eq}) is given by $\dfrac{1}{C_{\text{eq}}} = \dfrac{1}{C_{\text{elastic}}} + \dfrac{1}{C_{\text{grav}}}$ (see Problems 3.13 and 3.14 at the end of the chapter).

3.5.4.7 Pneumatic Systems

For pneumatic (gas) systems, pressure is the across-variable (as for hydraulic systems). However, in view of the relatively high compressibility of the flowing fluid (gas), it is desirable to use the mass flow rate Q_m as the through-variable (unlike in hydraulic systems where the volume flow rate Q is the through-variable). This modified representation of the through-variable, in pneumatic systems, is considered now.

***A*-type Pneumatic Element (Fluid Capacitor):**

$$\text{For a hydraulic system}: C_f\frac{dP}{dt} = Q$$

$$\Rightarrow \quad \rho C_f\frac{dP}{dt} = \rho Q = Q_m = \text{mass flow rate}$$

$$\text{For a pneumatic system we use: } C_{\text{mf}}\frac{dP}{dt} = Q_m \tag{3.21}$$

$$\text{Hence, pneumatic capacitance } C_{\text{mf}} = \rho C_f \tag{3.22}$$

T-type Pneumatic Element (Fluid Inertor)

$$\text{For a hydraulic system}: I_f \frac{dQ}{dt} = P \Rightarrow \frac{1}{\rho} I_f \frac{dQ_m}{dt} = P$$

$$\text{For a pneumatic system we use}: I_{\text{mf}} \frac{dQ_m}{dt} = P \tag{3.23}$$

$$\text{Hence, pneumatic inertance}: I_{\text{mf}} = \frac{1}{\rho} I_f \tag{3.24}$$

Note: $\frac{dQ_m}{dt} = \frac{d(\rho Q)}{dt} = \rho \frac{dQ}{dt} + Q \frac{d\rho}{dt}$. Hence, the equation $\frac{dQ}{dt} = \frac{1}{\rho}\frac{dQ_m}{dt}$ is strictly valid only when the mass density ρ is a constant. However, if we assume that the rate of change of ρ is much smaller than the rate of change Q, this equation still holds.

Another way to approach this issue is by treat equation (3.23) as the constitutive equation for a pneumatic inertor (linear) to begin with and not using the constitutive equation of a hydraulic inertor. Then, however, the relation $I_{\text{mf}} = \frac{1}{\rho} I_f$ does not generally hold.

D-type Pneumatic Element (Fluid Resistor):

$$\text{For a hydraulic system}: P = R_f Q = \frac{R_f}{\rho} Q_m$$

$$\text{For a pneumatic system we use}: P = R_{\text{mf}} Q_m \tag{3.25}$$

$$\text{Hence, pneumatic resistance}: R_{\text{mf}} = \frac{1}{\rho} R_f \tag{3.26}$$

Example 3.7

(a)

Consider two water tanks joined by a horizontal pipe with an on-off valve. With the valve closed, the water levels in the two tanks were initially maintained unequal. When the valve was suddenly opened, some oscillations were observed in the water levels of the tanks. Suppose that the system is modeled as two gravity-type capacitors linked by a fluid resistor. Would this model exhibit oscillations in the water levels when subjected to an initial-condition excitation? Clearly explain your answer.

A centrifugal pump is used to pump water from a well into an overhead tank. This fluid system is schematically shown in Figure 3.16a. The pump is considered as a pressure source $P_s(t)$ and the water level h in the overhead tank is the system output. The ambient pressure is denoted by P_a. The following system parameters are given:

L_v, d_v = length and internal diameter of the vertical segment of the pipe
L_h, d_h = length and internal diameter of the horizontal segment of the pipe
A_t = area of cross-section of the overhead tank (uniform)
ρ = mass density of water
μ = dynamic viscosity of water
g = acceleration due to gravity.

Suppose that this fluid system is approximated by the lumped-parameter model shown in Figure 3.15b. Rationalize this approximation.

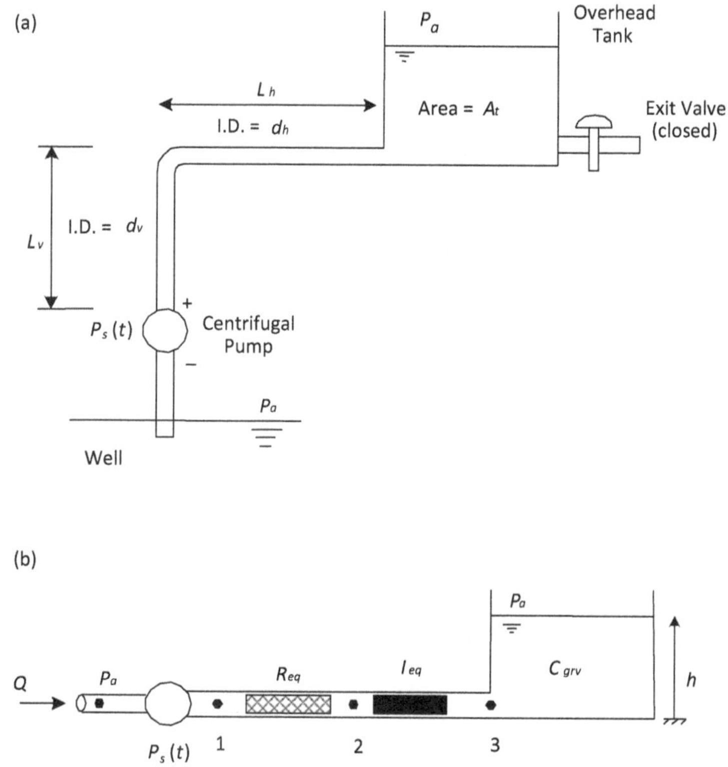

FIGURE 3.16 (a) A system for pumping water from a well into an overhead tank; (b) a lumped parameter model of the fluid system.

(b) Give expressions for the equivalent linear fluid resistance of the overall pipe (i.e., combined vertical and horizontal segments) R_{eq}, the equivalent fluid inertance within the overall pipe I_{eq}, and the gravitational fluid capacitance of the overhead tank C_{grv}, in terms of the given system parameters.

(c) Treating $\mathbf{x} = \begin{bmatrix} P_{3a} & Q \end{bmatrix}^T$ as the state vector (which is our preferred choice), where P_{3a}=pressure head of the overhead tank and Q=volume flow rate through the pipe, develop a complete state-space model for the system. Specifically, obtain the matrices ***A***, ***B***, ***C***, and ***D***.

(d) Obtain the input–output differential equation of the system.

Solution

(a)

If the inertia effects in the system are neglected, and only fluid capacitors are used as the energy storage elements, there exists only one type of energy in this system. Then, the model cannot provide an oscillatory response to an initial condition excitation (i.e., natural oscillations are not possible). But, the actual physical system has fluid inertia, and hence, the system can exhibit a free (natural) oscillatory response, to an initial condition (unequal liquid levels).

(b)

The pressure of the vertical pipe segment in Figure 3.16a may be included in the pressure source. Then, after replacing $P_s(t)$ by $P_s(t) - \rho g l_v$, we can use the model shown in Figure 3.16b. In the following analysis, we assume that $P_s(t)$ has this modified form.

Assuming a parabolic velocity profile (viscous, laminar flow), the fluid inertance in a pipe of uniform cross-section A and length L is given by $I = \dfrac{2\rho L}{A}$, with the usual notation (see Chapter 2).

Since the same volume flow rate Q is common to both segments of piping (i.e., continuity), we have, for series connection, $I_{eq} = \frac{2\rho L_v}{\frac{\pi}{4} d_v^2} + \frac{2\rho L_h}{\frac{\pi}{4} d_h^2} = \frac{8\rho}{\pi}\left[\frac{L_v}{d_v^2} + \frac{L_h}{d_h^2}\right]$.

The linear fluid resistance (viscous, laminar flow) in a circular pipe is $R = \frac{128\mu L}{\pi d^4}$ (see Chapter 2) where d=internal diameter. Again, since the same Q exists in both segments of the series-connected pipe, $R_{eq} = \frac{128\mu}{\pi}\left[\frac{L_v}{d_v^4} + \frac{L_h}{d_h^4}\right]$

Also, $C_{grv} = \frac{A_t}{\rho g}$

(c)

State-Space Shell:

$$C_{grv}\frac{dP_{3a}}{dt} = Q$$

$$I_{eq}\frac{dQ}{dt} = P_{23}$$

Remaining constitutive equation: $P_{12} = R_{eq}Q$

Note: Continuity (node) equations are automatically satisfied.

Compatibility (Loop) Equations: $P_{1a} = P_{12} + P_{23} + P_{3a}$ with $P_{1a} = P_s(t)$ and $P_{3a} = \rho g h$

Now eliminate the auxiliary variable P_{23} in the state-space shell, using the remaining equations. We get $P_{23} = P_{1a} - P_{12} - P_{3a} = P_s(t) - R_{eq}Q - P_{3a}$

Hence, the state-space model is given as follows.

State Equations:

$$\frac{dP_{3a}}{dt} = \frac{1}{C_{grv}}Q \quad \text{(i)}$$

$$\frac{dQ}{dt} = \frac{1}{I_{eq}}\left[P_s(t) - P_{3a} - R_{eq}Q\right] \quad \text{(ii)}$$

$$\text{Output Equation}: h = \frac{1}{\rho g}P_{3a} \quad \text{(iii)}$$

Corresponding matrices are:

$$\mathbf{A} = \begin{bmatrix} 0 & 1/C_{grv} \\ -1/I_{eq} & -R_{eq}/I_{eq} \end{bmatrix}; \quad \mathbf{B} = \begin{bmatrix} 0 \\ 1/I_{eq} \end{bmatrix}; \quad \mathbf{C} = \begin{bmatrix} \frac{1}{\rho g} & 0 \end{bmatrix}; \quad \mathbf{D} = 0$$

(d)

Substitute equation (i) into (ii):

$$I_{eq}C_{grv}\frac{d^2P_{3a}}{dt^2} = P_s(t) - P_{3a} - R_{eq}C_{grv}\frac{dP_{3a}}{dt}$$

Now substitute equation (iii) for P_{3a}:

$$I_{eq}C_{grv}\frac{d^2h}{dt^2} + R_{eq}C_{grv}\frac{dh}{dt} + h = \frac{1}{\rho g}P_s(t)$$

This is the input–output model.

LEARNING OBJECTIVES

1. Modeling of fluid systems
2. Compensation for gravity head in a vertical pipe
3. Proper choice of state variables
4. Development of a state-space model
5. Development of an input–output model

3.5.5 Modeling in the Thermal Domain

In the thermal domain, modeling is somewhat simplified by the fact that there is no thermal inductor (there exist only thermal capacitor and thermal resistor). Now we present an illustrative example for modeling in the thermal domain.

Example 3.8

A simplified model of an active chilled beam air-conditioning system is sketched in Figure 3.17.

The heat exchanger is represented by a copper slab of thermal capacitance C_h. Its average temperature is T_h. The air that is pumped into the system (possibly recirculated from the cooled space) enters the heat exchanger at temperature T_{ai} and leaves it (into the cooled space) at temperature T_{ao}.

The water that is pumped into the system from the chiller enters the heat exchanger at temperature T_{wi} and leaves it at temperature T_{wo}. The following parameters are given:

$\dot{m}_a$ = mass flow rate of air
$\dot{m}_w$ = mass flow rate of water
c_a = specific heat of air (at constant pressure)
c_w = specific heat of water

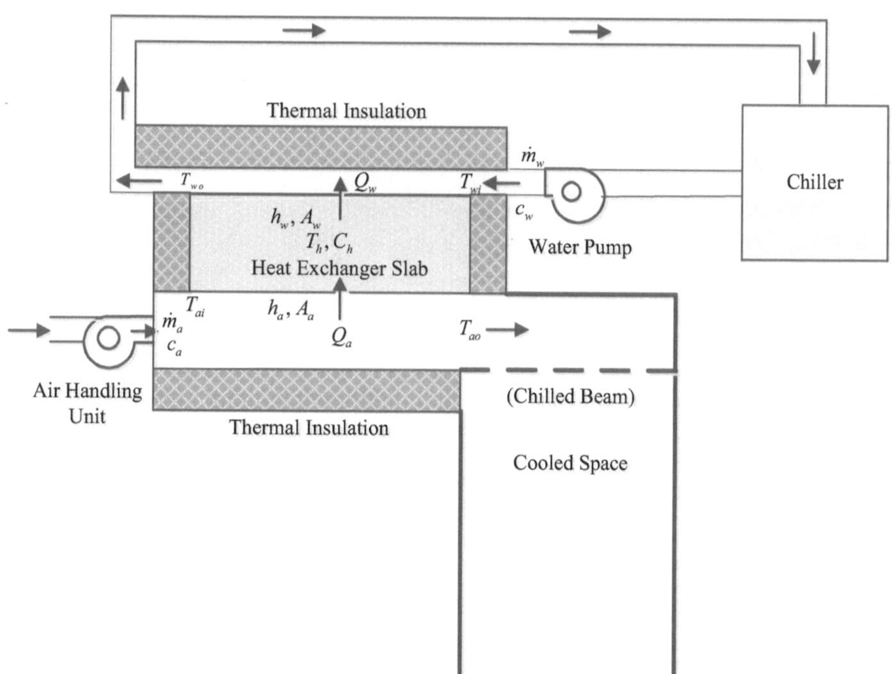

FIGURE 3.17 An active chilled beam air-conditioning system.

C_h = thermal capacitance of the heat exchanger
h_a = convective heat transfer coefficient at the air-side of the heat exchanger, with effective area A_a
h_w = convective heat transfer coefficient at the water-side of the heat exchanger, with effective area A_w

a. What are the inputs and what are the outputs of the system?
b. Develop a complete state-space model for the system. Indicate some characteristics of this model.

Solution

a. The inputs of the system are T_{wi} (a deliberate input) and T_{ai} (an unintentional or disturbance input).

   The outputs of the system are T_{ao} (the desired output), T_{wo}, and T_h. Of course, it is up to the engineer (model developer) to decide which outputs among these should be included in the model.
b. The constitutive equations are given below.

   Heat exchanger slab (thermal capacitor):

$$C_h \frac{dT_h}{dt} = Q_a - Q_w \quad \text{(i)}$$

where
Q_a = heat transfer rate from the air into the heat exchanger
Q_w = heat transfer rate from the heat exchanger into the water

Air-side wall (thermal resistor):

$$Q_a = h_a A_a (T_a - T_h) \quad \text{(ii)}$$

with $T_a = \frac{T_{ai} + T_{ao}}{2}$ (This is a common approximation in modeling HVAC—heating ventilation and air-conditioning system equations)

Thermal resistance $R_a = \frac{1}{h_a A_a}$

Water-side wall (thermal resistor):

$$Q_w = h_w A_w (T_h - T_w) \quad \text{(iii)}$$

with $T_w = \frac{T_{wi} + T_{wo}}{2}$ (as commonly approximated in this type of models).

Thermal resistance $R_w = \frac{1}{h_w A_w}$

Note: In general heat exchangers, $A_a \neq A_w$

Node equation (thermal energy conservation) for air flow:

$$\dot{m}_a c_a T_{ai} - \dot{m}_a c_a T_{ao} = Q_a \quad \text{(iv)}$$

Node equation (thermal energy conservation) for water flow:

$$\dot{m}_w c_w T_{wi} + Q_w = \dot{m}_w c_w T_{wo} \quad \text{(v)}$$

Now, we eliminate the auxiliary variables to obtain the state equation (first-order system) from the state-space shell equation (i).

Substitute (ii) and (iii) into (i):

$$C_h \frac{dT_h}{dt} = \frac{1}{R_a}(T_a - T_h) - \frac{1}{R_w}(T_h - T_w) \;\Rightarrow\; C_h \frac{dT_h}{dt} + \left(\frac{1}{R_a} + \frac{1}{R_w}\right)T_h = \frac{1}{R_a}T_a + \frac{1}{R_w}T_w$$

$$= \frac{1}{R_a} \times \frac{T_{ai} + T_{ao}}{2} + \frac{1}{R_w} \times \frac{T_{wi} + T_{wo}}{2} \qquad \text{(i)}^*$$

Substitute (ii) into (iv):

$$T_{ao} = T_{ai} - \frac{Q_a}{\dot{m}_a c_a} = T_{ai} - \frac{h_a A_a}{\dot{m}_a c_a}(T_a - T_h) = T_{ai} - p_a\left[\frac{T_{ai} + T_{ao}}{2} - T_h\right]$$

where $p_a = \dfrac{h_a A_a}{\dot{m}_a c_a}$

$$\Rightarrow \quad (1 + p_a/2)T_{ao} = (1 - p_a/2)T_{ai} + p_a T_h$$

$$\rightarrow \quad T_{ao} = \frac{(1 - p_a/2)}{(1 + p_a/2)}T_{ai} + \frac{p_a}{(1 + p_a/2)}T_h \qquad \text{(vi)}$$

Note: This is indeed an output equation.

Substitute (iii) into (v):

$$T_{wo} = T_{wi} + \frac{Q_w}{\dot{m}_w c_w} = T_{wi} + \frac{h_w A_w (T_h - T_w)}{\dot{m}_w c_w} = T_{wi} + p_w\left[T_h - \frac{T_{wi} + T_{wo}}{2}\right]$$

where $p_w = \dfrac{h_w A_w}{\dot{m}_w c_w}$

$$\Rightarrow \quad (1 + p_w/2)T_{wo} = (1 - p_w/2)T_{wi} + p_w T_h$$

$$\rightarrow \quad T_{wo} = \frac{(1 - p_w/2)}{(1 + p_w/2)}T_{wi} + \frac{p_w}{(1 + p_w/2)}T_h \qquad \text{(vii)}$$

Note: This is another output equation.

Substitute (vi) and (vii) into (i)* to eliminate the auxiliary variables T_{ao} and T_{wo}.

$$C_h \frac{dT_h}{dt} + \left(\frac{1}{R_a} + \frac{1}{R_w}\right)T_h = \frac{1}{2R_a}\left[\frac{(1 - p_a/2)}{(1 + p_a/2)}T_{ai} + \frac{p_a}{(1 + p_a/2)}T_h\right] + \frac{1}{2R_w}\left[\frac{(1 - p_w/2)}{(1 + p_w/2)}T_{wi} + \frac{p_w}{(1 + p_w/2)}T_h\right] + \frac{T_{ai}}{2R_a} + \frac{T_{wi}}{2R_w}$$

$$\Rightarrow C_h \frac{dT_h}{dt} + \left[\frac{1}{R_a} + \frac{1}{R_w} - \frac{p_a}{2R_a(1 + p_a/2)} - \frac{p_w}{2R_w(1 + p_w/2)}\right]T_h = \left[\frac{(1 - p_a/2)}{(1 + p_a/2)} + 1\right]\frac{1}{2R_a}T_{ai} + \left[\frac{(1 - p_w/2)}{(1 + p_w/2)} + 1\right]\frac{1}{2R_w}T_{wi}$$

We now have the state-space equation:

$$C_h \frac{dT_h}{dt} + 2\left[\frac{1}{R_a(2 + p_a)} + \frac{1}{R_w(2 + p_w)}\right]T_h = \frac{2}{R_a(2 + p_a)}T_{ai} + \frac{2}{R_w(2 + p_w)}T_{wi}$$

Note the following:

1. There are three outputs and the corresponding three algebraic output equations
2. Two of the output equations are (vi) and (vii). These equations have the input variables in them. Hence the model has the "feedforward" characteristic
3. The state variable T_h is also an output. The corresponding output equation is trivial.

LEARNING OBJECTIVES

1. Modeling of thermal systems
2. Proper formulation of constitutive equations and continuity equations
3. The number of outputs can be greater than the number of states
4. The presence of "feedforward" terms in the output equations.

3.6 SUMMARY SHEET

- **Types of Analytical Models**: Time-domain model (differential equations with time t as the independent variable, e.g., state state-space model—a set of first-order differential equations in time, and input–output model—input–output differential equations in time); transfer function model (this is an input–output model. Set of transfer functions: [Laplace transform of output]/[Laplace transform of input], independent variable is Laplace variable s); frequency-domain model (frequency transfer function or frequency response function. A special case of Laplace transfer function, with $s=j\omega$. Independent variable is frequency ω). Discrete-time forms of these models are used in digital computer implementation and simulation. Notable: Difference equations (discretized differential equations) and corresponding z-transform transfer functions
- **Analytical Model Simplification**: 1. Nonlinear → linear (principle of superposition holds; and interchangeability in series connection does not hold in general, unless the system matrices are commutative or the system is SISO; 2. continuous time → discrete time; 3. distributed (continuous) parameter → lumped parameter; 4. time-varying (or non-stationary or non-autonomous) → time-invariant (or stationary or autonomous); 5. random (or stochastic or probabilistic) → deterministic models
- **Discrete-Time Systems**: Sampling rate $f_s = \dfrac{1}{\Delta T}$, sampling period $=\Delta T$; sampled data at constant $\Delta T : \{x_k\} = \{x_0, x_1, \ldots, x_k, x_{k+1}, \ldots\}$, Its z-transform $X(z) = \sum_{k=0}^{\infty} x_k z^{-k}$; nth-order *linear difference equation*: $a_0 y_k + a_1 y_{k-1} + \cdots + a_n y_{k-n} = b_0 u_k + b_1 u_{k-1} + \cdots + b_m u_{k-m}$ → *discrete transfer function* (z-transform transfer function): $G(z) = \dfrac{Y(z)}{U(z)} = \dfrac{b_0 + b_1 z^{-1} + \cdots + b_m z^{-m}}{a_0 + a_1 z^{-1} + \cdots + a_n z^{-n}}$, $u=$input, $y=$output.
- **Sampling Theorem**: In sampled data, no information regarding frequency spectrum $X(f)$ can be found for frequencies > *Nyquist* (cutoff) *frequency* $f_c = \dfrac{1}{2\Delta T} = \dfrac{1}{2} f_s$
- **Aliasing Error**: Caused by sampling → folding of the high-frequency segment of the frequency spectrum beyond the Nyquist frequency onto the low-frequency segment. Method of reducing it: 1. Increase the sampling frequency (increased data), 2. use an antialiasing filter of cutoff at $< \dfrac{1}{2} f_s$

- **Lumped Model of Heavy Spring**: One end fixed and the equivalent lumped mass placed at the other end. 1. Energy (kinetic) equivalence $\rightarrow$ lumped mass$=\frac{1}{3}\times$ spring mass; 2. natural frequency equivalence $\rightarrow 0.4\times$ spring mass
- **Properties of State Models**: 1. State $\boldsymbol{x}(t_0)$ at time t_0 and input $\boldsymbol{u}[t_0,t_1]$ over time interval $[t_0,t_1]$ will uniquely determine state $\boldsymbol{x}(t_1)$ at time $t_1 \Rightarrow$ **Nonanticipative** $\boldsymbol{g}$: $x(t_1)=g\big(t_0,t_1,x(t_0),u[t_0,t_1]\big)$; that is, inputs beyond t_1 are not needed; 2. $\boldsymbol{x}(t_1)$ and $\boldsymbol{u}(t_1)$ at t_1 will uniquely determine output $\boldsymbol{y}(t_1) \Rightarrow$ **Memory-less** $\boldsymbol{h}$: $y(t_1)=h\big(t_1,x(t_1),u(t_1)\big)$; that is, past responses cannot be determined from the present state and the present input
- **Nonlinear State Model**: $\dot{\boldsymbol{x}}=\boldsymbol{f}(\boldsymbol{x},\boldsymbol{u},t)$, $\boldsymbol{y}=\boldsymbol{h}(\boldsymbol{x},\boldsymbol{u},t)$
- **Linear State Model**: $\dot{\boldsymbol{x}}=\boldsymbol{A}\boldsymbol{x}+\boldsymbol{B}\boldsymbol{u}$, $\boldsymbol{y}=\boldsymbol{C}\boldsymbol{x}+\boldsymbol{D}\boldsymbol{u}$
- **Properties**: 1. State model: n first-order differential equations (coupled) using n state variables (nth-order); 2. determines dynamic state $\boldsymbol{x}$ of the system; 3. n state variables $x_1,x_2,\ldots,x_n$ are the required minimum and maximum number (minimal set) $\Rightarrow$ a state variable cannot be expressed as a linear combination of the remaining state variables; 4. $\boldsymbol{x}$ traces a *trajectory* in the *state space* (vector space of x); 5. $\boldsymbol{x}$ is not unique. Many choices are possible; 6. $\boldsymbol{x}$ may or may not have a physical interpretation; 7. output (response) variables can be completely determined from any choice of state variables. To complete the state model, include the algebraic *output equations* (m algebraic equations relating the outputs to the states; sometimes they may contain inputs as well—feedforward case)
- **Steps of Developing of a State Model**: Assume that inputs ($\boldsymbol{u}$) and outputs ($\boldsymbol{y}$) are specified/identified. 1. Identify the physical elements of the system (e.g., energy storage elements, energy dissipative elements, and input/source elements). Draw a *structural diagram* (circuit diagram, linear graph, bond graph, etc.) to show their interconnection; 2. state variable ($\boldsymbol{x}$) selection: Across-variables of independent A-type energy storage elements and through-variables of independent T-type energy storage elements. This choice will result in a realistic and unique state-space model; 3. write constitutive (physical) equations for all elements in the system (energy storage elements, energy dissipative elements, two-port elements), except the source elements; 4. write continuity equations for through-variables at each node (points connecting two or more components), less one; 5. write compatibility equations for across-variables around each primary (independent) loop (closed path that is formed by two or more system components); 6. state-space shell: Constitutive equations for the independent energy storage elements. Eliminate the auxiliary (redundant/unwanted) variables in the state-space shell (by using the remaining equations); 7. express the outputs in terms of state variables (if the system has the feedforward character, the input variables have to be included in output equations)
- **Comments**: 1. Some continuity (node) equations and compatibility (loop) equations may be automatically satisfied by the choice of the corresponding variables; 2. all remaining equations may not be needed for the elimination of the auxiliary variables in the state-space shell; 3. equations having the dependent variables of the sources are not useful; 4. compatibility equations may be automatically satisfied depending on the use of the across-variables for elements (e.g., the use of separate variables of different elements or the use of absolute across-variables at the joints); 5. the inertial reference (ground, with zero velocity) is always at one end (reference end) of an inertia element or an input element
- **Input–Output Model from State Model**: 1. Successively differentiate output equations, e.g., $\dot{\boldsymbol{y}}=\boldsymbol{C}\dot{\boldsymbol{x}}+\boldsymbol{D}\dot{\boldsymbol{u}} \rightarrow \dot{\boldsymbol{y}}=\boldsymbol{C}\boldsymbol{A}\boldsymbol{x}+\boldsymbol{C}\boldsymbol{B}\boldsymbol{u}+\boldsymbol{D}\dot{\boldsymbol{u}}$; 2. eliminate the derivatives of the states (e.g., $\dot{\boldsymbol{x}}$) by substituting the state equations; 3. use the resulting equations to eliminate $\boldsymbol{x}$.

PROBLEMS

3.1 Indicate whether a distributed-parameter model is needed or a lumped-parameter model is adequate in the study of the following dynamic systems:

a. Vehicle suspension system (motion)
b. Elevated vehicle guideway (transverse motion)
c. Oscillator circuit (electrical signals)
d. Environment (weather) system (temperature)
e. Aircraft (motion and structural stresses)
f. Large transmission cable (electrical capacitance and inductance).

If a distributed model is desirable, indicate what part of the system needs a distributed model.
Note: The quantities of primary interest are indicated in parentheses.

3.2 What are some of the limitations in using the equivalent lumped-mass model, through energy equivalence (specifically, kinetic energy equivalence, because potential energy equivalence is automatically satisfied through the stiffness parameter, which is the same for the heavy spring and its lumped model), for a heavy spring (with distributed mass)?

3.3 One end of a heavy, uniform, helical spring of mass m_s and stiffness k is fixed and the other end is attached to an object of mass M, which is free to move on rollers (see Figure P3.3a). Using the method of natural frequency equivalence determine equivalent lumped masses, one to be located at the midpoint of the spring and the other at the free end (see Figure P3.3b) in an equivalent lumped-parameter model.
Hint: The natural frequencies of a heavy helical spring with one end fixed and the other end free are given by $\omega_n = \frac{\pi}{2}(2n-1)\sqrt{\frac{k}{m_s}}$, where n is the mode number. Use only the first two modes (i.e., $n=1$ and 2) in this example.

3.4 Answer the following questions true or false:

a. A state-space model is uniqueh.
b. The number of state variables in a state vector is equal to the order of the systemi.
c. The outputs of a system are always identical to the state variables
d. Outputs can be expressed in terms of state variablesk.
e. State model is a time-domain model.

3.5 List three things to which the order of an electromechanical dynamic system is equal. Write down the order of the systems shown in Figure P3.5.

3.6 Real systems are nonlinear. Under what conditions a linear model is sufficient in studying a real system?
Consider the following system equations:

a. $\ddot{y} + (2\sin\omega t + 3)\dot{y} + 5y = u(t)$
b. $3\ddot{y} - 2y = u(t)$

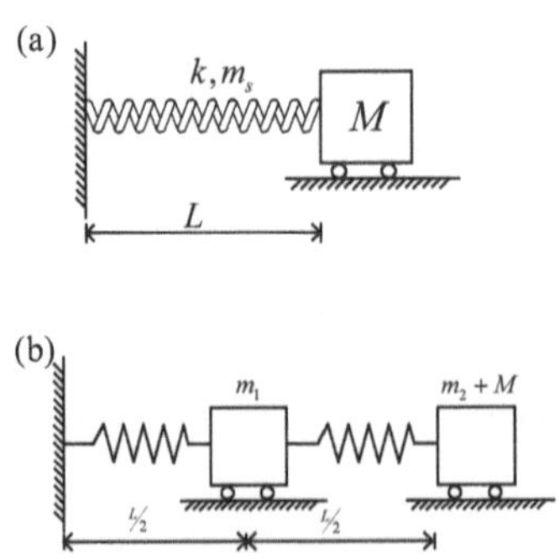

FIGURE P3.3 (a) A system with a uniform heavy spring; (b) an approximate lumped-parameter model.

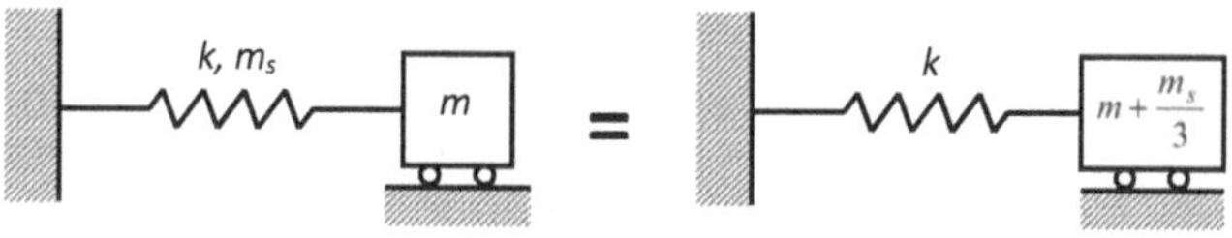

FIGURE P3.5 (a–d) Models of four mechanical systems.

c. $3\ddot{y}+2\dot{y}^3+y=u(t)$
d. $5\ddot{y}+2\dot{y}+3y=5u(t)$
   i. Which ones of these are linear?
   ii. Which ones are nonlinear?
   iii. Which ones are time-variant?

3.7 Identify some dynamic characteristics in the flowing five state-space models. Also tabulate the matrices ***A***, ***B***, ***C***, and ***D*** for the five models.

a. $\dot{x}_1=2x_1+3x_2+u_1$

$\dot{x}_2=x_1-x_2-u_2$

$y=x_1+2x_2$

b. $\dot{x}_1=2x_1+3x_2+u$

$\dot{x}_2=x_1-x_2$

$y=x_1+2x_2$

c. $\dot{x}_1=2x_1+3x_2$

$\dot{x}_2=x_1-x_2$

$y=x_1+2x_2$

d. $\dot{x}_1=2x_1+u$

$\dot{x}_2=x_1-x_2$

$y=x_1$

e. $\dot{x}_1=2x_1+3x_2+u_1$

$\dot{x}_2=x_1-x_2-u_2$

$y=x_1+2x_2+3u_1$

3.8 Consider a system represented by the state equations:

$$\dot{x}_1=x_1+2x_2$$

$$\dot{x}_2=-x_1+2u$$

in which x_1 and x_2 are the state variables and u is the input variable. Suppose that the output y is given by: $y=2x_1-x_2$.

a. Write this state-space model in the vector-matrix form:
   $\dot{\boldsymbol{x}}=\boldsymbol{A}\boldsymbol{x}+\boldsymbol{B}u$
   $y=\boldsymbol{C}\boldsymbol{x}$
   Specifically, identify the elements of the matrices ***A***, ***B***, and ***C***
b. What is the order of the system?

3.9 Consider the mass-spring system shown in Figure P3.9.

The mass m is supported by a spring of stiffness k and is excited by the dynamic force $f(t)$.

a. Taking $f(t)$ as the input, and the position and the speed of the mass as the two outputs, obtain a state-space model for the system.
b. What is the order of the system?
c. Repeat the problem, this time taking the compression force in the spring as the only output.
d. How many initial conditions are needed to determine the complete response of the system?

3.10 A torsional dynamic model of a pipeline segment is shown in Figure P3.10a. The FBD in Figure P3.10b shows the internal torques acting at sectioned inertia junctions, for free motion.

Develop a state model for this system by using the generalized velocities (angular velocities Ω_i) of the independent inertia elements and the generalized forces (torques T_i) of the independent elastic (torsional spring) elements as the state variables. As it is known, a minimum set of states that is required for a complete representation of the system dynamics determines the order of the system/model.

3.11 A one-quarter model of a vehicle is shown in Figure P3.11a. Since the gravitational forces (of the masses m_1 and m_2) are balanced by the static deflections of the springs, they can be ignored in the model, assuming that the deflections are measured with respect to the static positions of the springs. The corresponding horizontal model is shown in Figure P3.11b.

a. For a systematic and unified (domain-independent) procedure in the development of a state-space model of the system, identify the nodes and the loops of the model and indicate the relevant through-variables and across-variables
b. Write: the loop equations (and show that they are automatically satisfied); the node equations; and the constitutive equations
c. By eliminating the auxiliary variables in the state-space shell, determine a state-space model for the system.

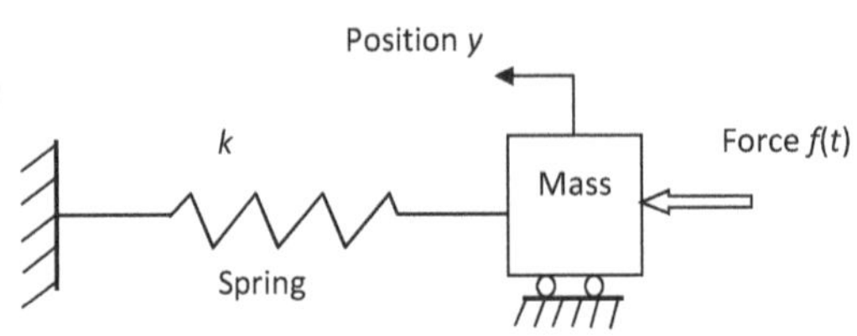

FIGURE P3.9 A mechanical oscillator.

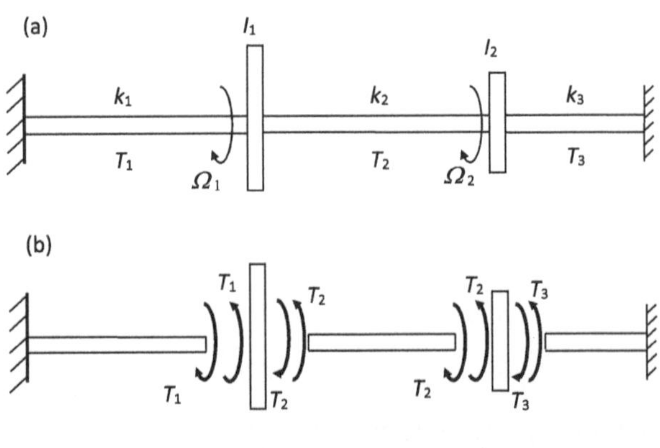

FIGURE P3.10 (a) Dynamic model of a pipeline segment; (b) free-body diagram.

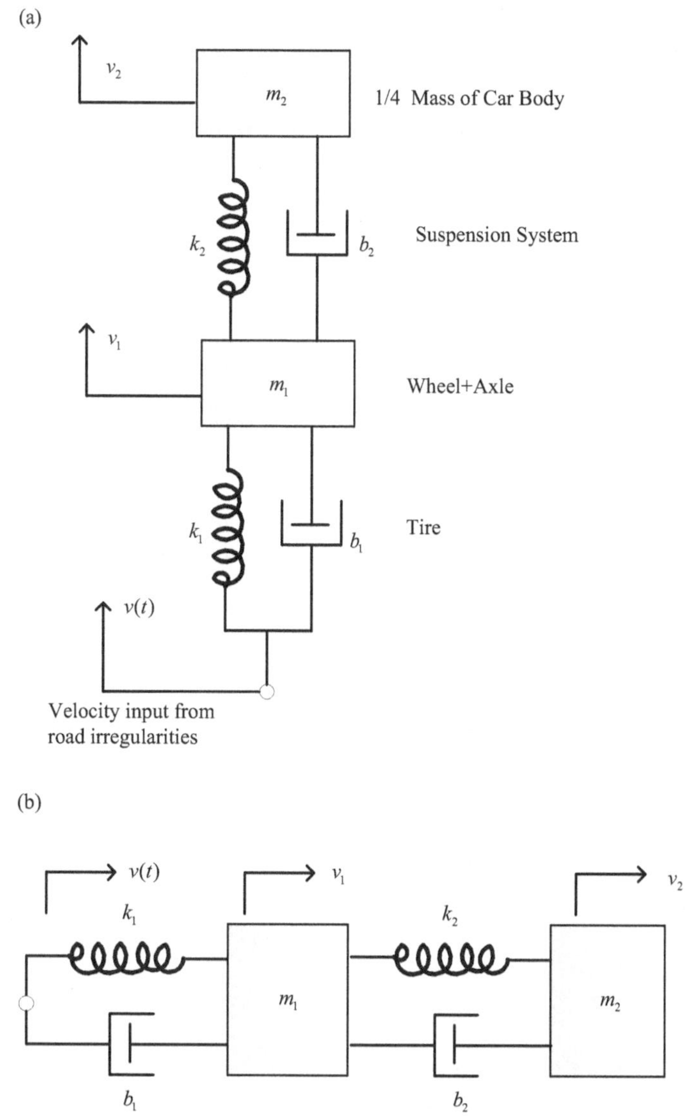

FIGURE P3.11 (a) Quarter model of a vehicle; (b) horizontal model (gravity balanced by static deflection of springs).

Note: Take the velocities of the independent masses and the compressive forces in the independent springs as the state variables (this is the preferred choice). The input to the vehicle model is the velocity $v(t)$ that is imparted on the tires in the vertical direction, due to the irregularities of the road surface, as the vehicle moves at some speed.

d. Complete the state-space model by developing the output equations for the three cases:
   1. The outputs (two) are the heave velocity of the passenger compartment and the suspension force (spring and damper combined)
   2. The output is the vertical (heave) displacement of the car body with respect to the wheel axle
   3. The output is the absolute vertical displacement of the car body.

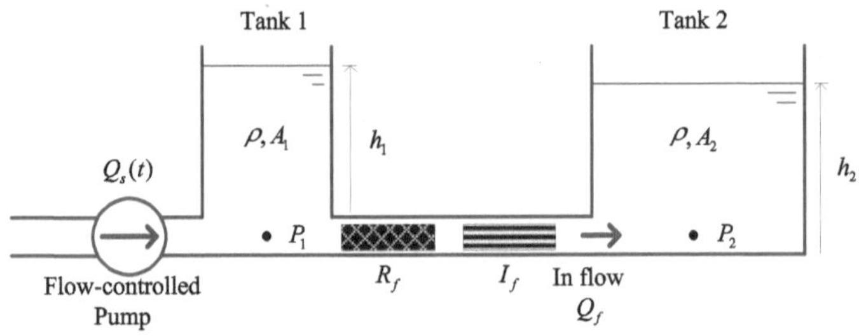

FIGURE P3.12 A pump-controlled two-tank system.

3.12

a. It is desirable to have a modeling approach that is integrated, unified, systematic, and unique. Briefly explain the meanings of these four terms in the present context of our approach to modeling.

b. A flow-controlled pump supplies a liquid to the bottom of a uniform tank. The bottom of this tank is connected to the bottom of a second uniform tank through a long uniform pipe. See Figure P3.12.

Given:

Mass density of the liquid = ρ

Volume flow rate of the supply pump (system input) = $Q_s(t)$
Area of cross-section of the first tank = A_1
Area of cross-section of the second tank = A_2
Gauge pressure at the bottom of the first tank = P_1
Gauge pressure at the bottom of the second tank = P_2
Volume flow rate of the liquid in the connecting pipe = Q_f
Fluid resistance in the pipe = R_f
Fluid inertance in the pipe = I_f

Use

$\boldsymbol{u} = [Q_s(t)]$ as the input

$\boldsymbol{x} = \begin{bmatrix} P_1 & Q_f & P_2 \end{bmatrix}^T$ as the state vector

$\boldsymbol{y} = \begin{bmatrix} h_1 & h_2 \end{bmatrix}^T$ as the output vector

Determine a complete state-space model for the system. In particular, determine the matrices $\boldsymbol{A}$, $\boldsymbol{B}$, $\boldsymbol{C}$, and $\boldsymbol{D}$.

Acceleration due to gravity = g

Note 1: Express your results only in terms of the given system parameters A_1, A_2, ρ, I_f, R_f.

Note 2: Gauge pressure is the pressure that is measured with respect to the ambient pressure.

3.13 Figure P3.13 shows a spring-loaded accumulator for liquid. The following parameters are given:

A = area of cross-section (uniform) of the accumulator cylinder
k = spring stiffness of the accumulator piston
ρ = mass density of the liquid.

Assume that the liquid is incompressible. The following variables are important:

$P_{21} = P_2 - P_1$ = pressure at the inlet of the accumulator with respect to that at the top (flexible) wall (P_1)

$P_{1r} = P_1 - P_r$ = pressure at the top of the accumulator (has a flexible wall) with respect to the ambient reference (P_r)

Q = volume flow rate of liquid into the accumulator

h = height of the liquid column in the accumulator.

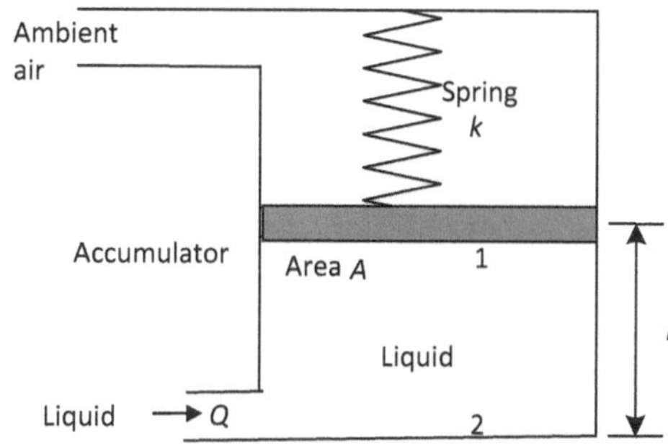

FIGURE P3.13 Model of a liquid accumulator.

Note: The piston (wall) of the accumulator can move against the spring, thereby varying h. Considering the effects of the movement of the spring-loaded accumulator wall and also the gravity head of the liquid in the accumulator, obtain an expression for the equivalent fluid capacitance C_{eq} of the accumulator in terms of k, A, ρ, and g. Are the two capacitors that contribute to C_{eq} (i.e., due to stretching of the accumulator wall and the gravity head of the liquid column) connected in parallel or in series?

Note: Neglect the effect of the bulk modulus (compressibility) of the liquid.

3.14

a. Briefly explain why a purely thermal system typically cannot have a natural (free) oscillatory response whereas a fluid system can.

b. Figure P3.14 shows a pressure-regulated system that can provide a high-speed jet of liquid. Its applications may include spray painting, material coating, and cleaning. The system consists of a pump, a spring-loaded accumulator, and a fairly long section of piping, which ends with a nozzle. The pump is considered as a flow source of value Q_s. The following parameters are important:

   A = area of cross-section (uniform) of the accumulator cylinder
   k = spring stiffness of the accumulator piston
   L = length of the section of piping from the accumulator to the nozzle
   A_p = area of cross-section (uniform, circular) of the piping
   A_o = discharge area of the nozzle
   C_d = discharge coefficient of the nozzle
   ρ = mass density of the liquid.

   Assume that the liquid is incompressible. The following variables are important:

   $P_{1r} = P_1 - P_r$ = pressure at the inlet of the accumulator wrt the ambient (reference) pressure P_r
   Q = volume flow rate through the nozzle
   h = height of the liquid column in the accumulator.

Note: The piston (wall) of the accumulator can move against the spring, thereby varying h.

   i. Considering the effects of the movement of the spring-loaded accumulator wall and also the gravity head of the liquid, obtain an expression for the equivalent fluid capacitance C_a of the accumulator in terms of k, A, ρ, and g.

Note: Neglect the effect of the bulk modulus (compressibility) of the liquid.

   ii. Considering the capacitance C_a, the inertance I of the fluid in the piping (length L and cross-section area A_p), and the resistance of the nozzle only, develop a nonlinear state-space model for the system. The state vector $\boldsymbol{x} = \begin{bmatrix} P_{1r} & Q \end{bmatrix}^T$ and the input $\boldsymbol{u} = [Q_s]$.

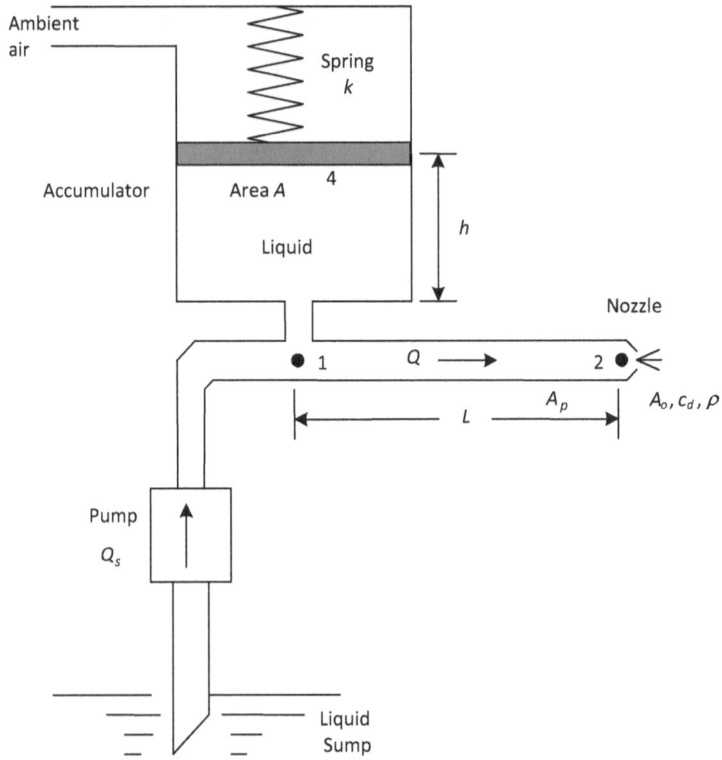

FIGURE P3.14 Pressure-regulated liquid jet system.

Assume that the liquid flow in the (circular) pipe is viscous laminar, with a parabolic velocity profile, and the corresponding inertance $I = \dfrac{2\rho L}{A_p}$. The volume rate of fluid discharge through the nozzle is $Q = A_o c_d \sqrt{\dfrac{2P_{2r}}{\rho}}$, in which

P_{2r}=pressure inside the nozzle with respect to the outside reference (P_r)
c_d=discharge coefficient.

3.15 A model for the automatic gage control (AGC) system of a steel rolling mill is shown in Figure P3.15. The rollers are pressed using a single-acting hydraulic actuator, which is controlled by the valve displacement u. As a result, the rollers are displaced through y, thereby pressing the steel that is being rolled. For a given y, the rolling force F is completely known from the steel parameters. Identify the inputs and the controlled variable in this control system.

i. In terms of the variables and the system parameters indicated in Figure P3.15, write dynamic equations for the system, while including valve nonlinearities.
ii. What is the order of the system? Identify the response variables.
iii. What variables would you measure (and feedback through a suitable controller) in order to improve the performance of the control system?

3.16 An integrated-circuit (IC) package consists primarily of a wafer of crystalline silicon substrate on which a film of minute amounts of silicon dioxide, etc., is deposited. It is heat treated at high temperature as an intermediate step in the production of IC chips. An approximate model of the heating process is shown in Figure P3.16.

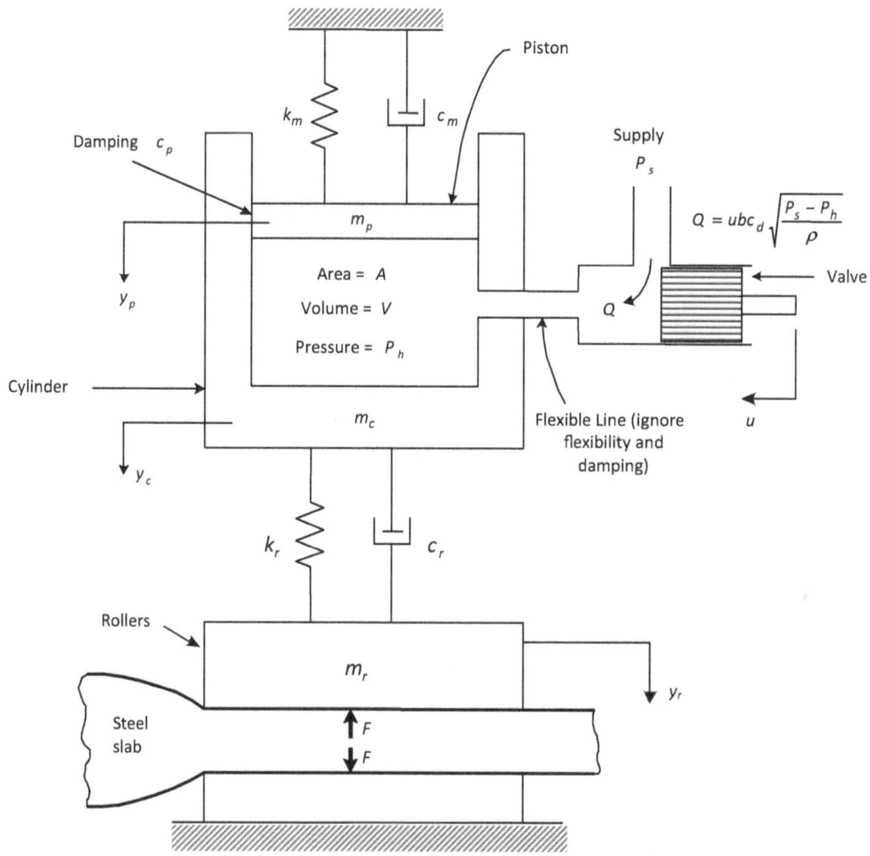

FIGURE P3.15 Automatic gage control (AGC) system of a steel rolling mill.

The package is placed inside a heating chamber whose walls are uniformly heated by a distributed heating element. The associated heat transfer rate into the wall is Q_i. The interior of the chamber contains a gas of mass m_c and specific heat c_c and is maintained at uniform temperature T_c. The temperature of silicon chip is T_s and that of the chamber wall is T_w. The outside environment is maintained at temperature T_o. The specific heats of the silicon package and the wall are denoted by c_s and c_w, respectively, and the corresponding masses are denoted by m_s and m_w, as shown. The convective heat transfer coefficient at the interface of silicon and gas inside the chamber is h_s, and the effective surface area is A_s. Similarly, h_i and h_o denote the convective heat transfer coefficients at the inside and the outside surfaces of the chamber wall, and the corresponding surface areas are A_i and A_o, respectively.

a. Using T_s, T_c, and T_w as the state variables, write state equations for the process.
b. Express these equations in terms of the parameters $C_{hs} = m_s c_s$, $C_{hc} = m_c c_c$, $C_{hw} = m_w c_w$, $R_s = \frac{1}{h_s A_s}$, $R_i = \frac{1}{h_i A_i}$, and $R_o = \frac{1}{h_o A_o}$. Explain the electrical analogy and the physical significance of these parameters.
c. What are the inputs to the process? If T_s is the output of importance, obtain the matrices $\boldsymbol{A}$, $\boldsymbol{B}$, $\boldsymbol{C}$, and $\boldsymbol{D}$ of the state-space model.
d. Comment on the accuracy of the model in the context of the actual physical process of producing IC chips.

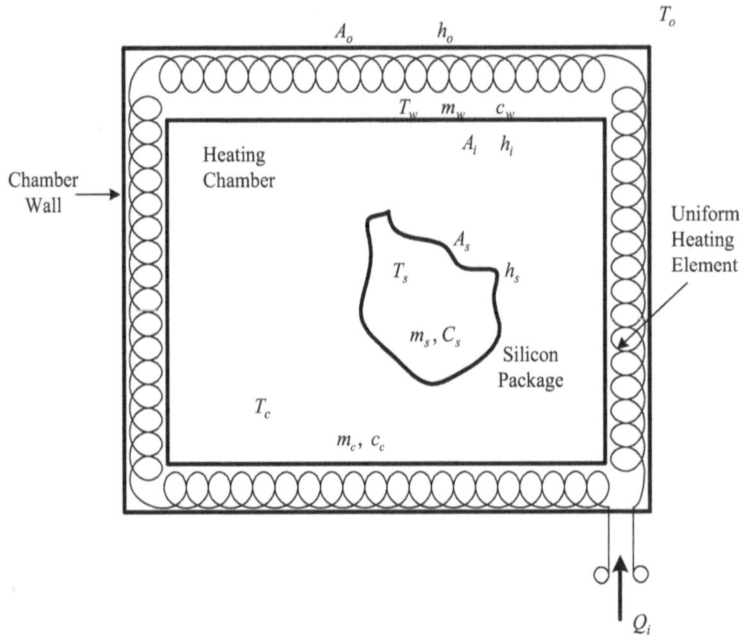

FIGURE P3.16 A model for the heat treatment of an IC package of silicon.

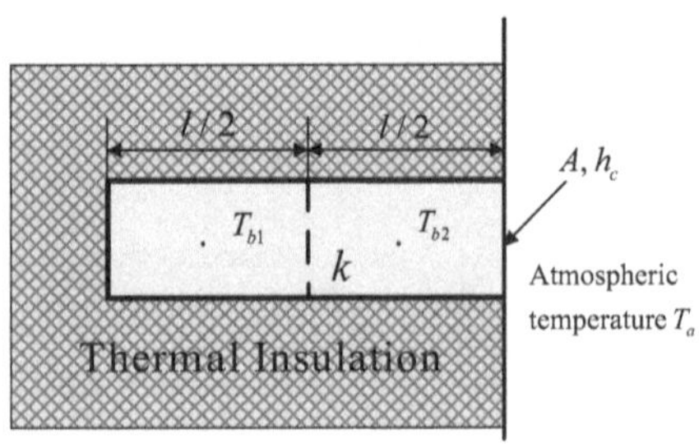

FIGURE P3.17 Improved model for the cooling bar.

3.17 A uniform metal bar of length l and area of cross-section A is heated to temperature T_b and placed in a thermally insulated compartment with one end exposed to the atmosphere of temperature $T_a < T_b$. Consider this as a part of a heat treatment process.
Given:
ρ=mass density of the bar
c_p=specific heat (as constant pressure) of the bar
h_c=convective heat transfer coefficient at the end exposed to the atmosphere.
Divide the metal bar into two equal segments of length $l/2$. Taking the temperatures in these two segments to be uniform at T_{b1} and T_{b2} (see Figure P3.17) obtain a state-space model for the cooling process of the bar. Is this lumped-parameter model sufficiently accurate? Confirm using the Biot number.
The following numerical values are known:
$l = 0.2$ m, $A = 1.0$ cm^2, $h_c = 100.0$ W/m^2/°C, conductive heat transfer coefficient k = 125.0 W/m/°C

3.18 A simplified version of a counter-flow heat-exchanger tube is shown in Figure P3.18. Its outside is thermally insulated. The tubular core duct of radius r_a and length L carries air, to

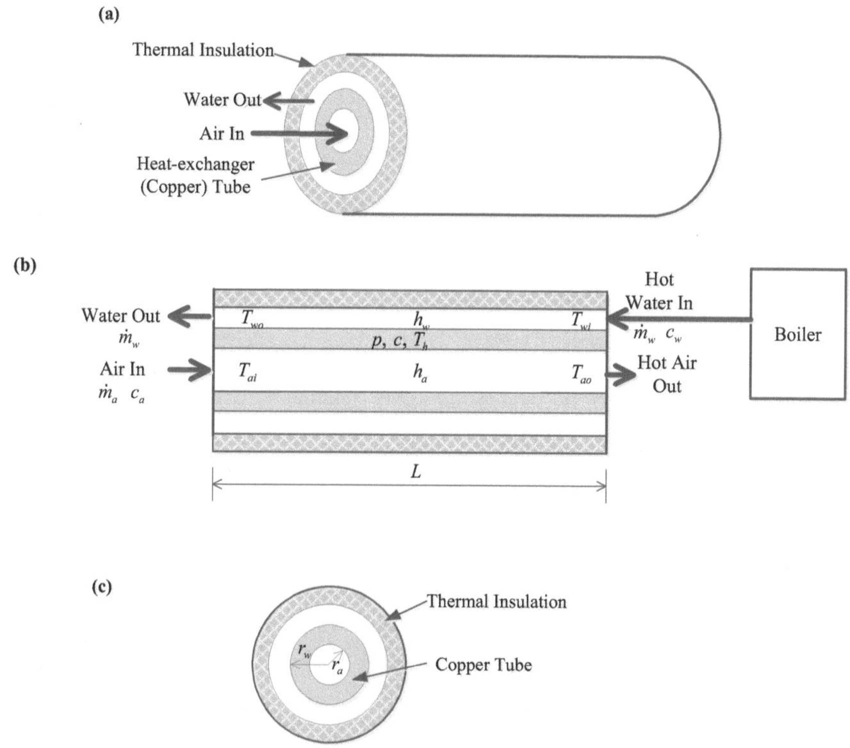

FIGURE P3.18 A counter-flow heat-exchanger unit. (a) Perspective view; (b) sectional side view; (c) sectional end view.

be heated, in one direction, at the steady mass flow rate $\dot{m}_a$. Thick copper tubing of exterior radius r_w (interior radius r_a) and length L separates the air flow (core) from the counter-flow of water at the steady mass flow rate $\dot{m}_w$. A thermal insulation tube, which is placed outside the copper tube, forms an annular path for the water flow.

Hot water from the boiler enters the heat exchanger tube at temperature T_{wi} and leaves it at temperature T_{wo}. The air flow enters the heat exchanger tube at temperature T_{ai} and leaves it at temperature T_{ao}.

In addition, the following parameters are given (known):

ρ = mass density of the heat exchanger (copper) tube

c = specific heat of the copper tube

c_w = specific heat of water

c_a = specific heat of air (at constant pressure)

h_w = convective heat transfer coefficient at the water-side wall of the heat-exchanger tube

h_a = convective heat transfer coefficient at the air-side wall of the heat-exchanger tube.

Also, let T_h be the average temperature of the heat exchanger tube (copper).

a. In terms of the given parameters, obtain expressions for:
   i. Thermal capacitance C_h of the heat exchanger (copper) tube
   ii. Convective thermal resistance R_w at the water-side wall of the heat-exchanger tube
   iii. Convective thermal resistance R_a at the air-side wall of the heat-exchange tube.
b. What are the inputs and possible outputs of this system?
c. Taking T_h as the state variable, develop a complete state-space model (including the output equations) for this system.

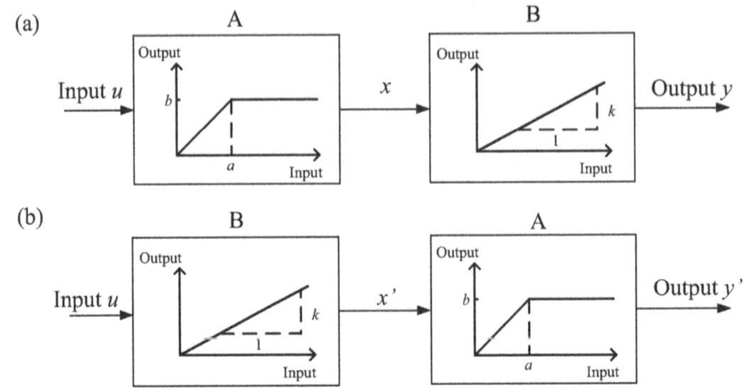

FIGURE P3.20 Possible violation of interchangeability (commutativity) of SISO systems in series connection.

3.19 Identify some dynamic characteristics in the flowing five state-space models. Also, tabulate their matrices ***A***, ***B***, ***C***, and ***D***.

a. $\dot{x}_1 = 2x_1 + 3x_2 + u_1$
$\dot{x}_2 = x_1 - x_2 - u_2$
$y = x_1 + 2x_2$

b. $\dot{x}_1 = 2x_1 + 3x_2 + u$
$\dot{x}_2 = x_1 - x_2$
$y = x_1 + 2x_2$

c. $\dot{x}_1 = 2x_1 + 3x_2$
$\dot{x}_2 = x_1 - x_2$
$y = x_1 + 2x_2$

d. $\dot{x}_1 = 2x_1 + u$
$\dot{x}_2 = x_1 - x_2$
$y = x_1$

e. $\dot{x}_1 = 2x_1 + 3x_2 + u_1$
$\dot{x}_2 = x_1 - x_2 - u_2$
$y = x_1 + 2x_2 + 3u_1$

3.20 Consider the system in Figure P3.20.

i Assume $u > a$ and $k > 1$. Check whether the SISO systems **A** and **B** are interchangeable (commutative).

ii Assume $u < a$ and $k < 1$. Check whether the SISO systems **A** and **B** are interchangeable (commutative).

What conclusions can you arrive at, based on these results?

3.21 Consider a particle of mass m subjected to an input force $u(t)$. By Newtonå's second law, its position x can be expressed by the second-order differential equation:

$$m\frac{d^2x}{dt^2} = u(t) \text{ or } m\ddot{x} - u = 0$$

Develop state models and input–output (I-O) models for the following three cases:
Case 1: Position x is the output
Case 2: Velocity $\dot{x} = v$ is the output
Case 3: Both position x and velocity $\dot{x} = v$ are outputs.

3.22 The rigid output shaft of a diesel engine prime mover is running at known angular velocity $\Omega(t)$. It is connected through a frictional clutch to a flexible shaft, which in turn drives

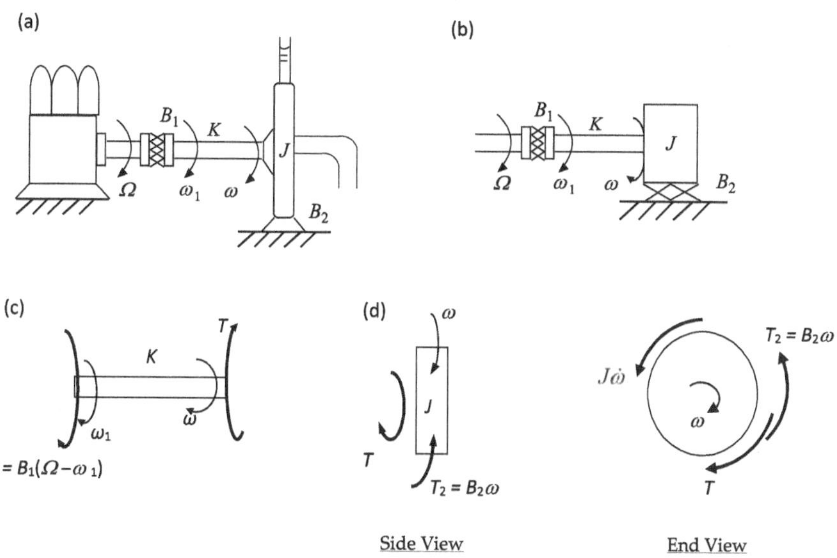

FIGURE P3.22 (a) Diesel engine; (b) linear model; (c) free-body diagram of the shaft; (d) free-body diagram of the wheel.

a hydraulic pump (see Figure P3.22a). A linear model for this system is schematically shown in Figure P3.22b. The clutch is represented by a viscous rotatory damper of damping constant B_1 (units: torque/angular velocity). The stiffness of the flexible shaft is K (units: torque/angle of twist). The pump is represented by a wheel of moment of inertia J (units: torque/angular acceleration) and viscous damping constant B_2.

a. Write the two state equations, where the state variables are: T=torque in flexible shaft and ω=pump speed, and input=Ω, through a simple and intuitive approach by using the: 1. FBD for the shaft, shown in Figure P3.22c, where ω_1 is the angular speed at the left end of the shaft. To get one state equation, write the "torque balance" and the "constitutive" relations for the shaft, and eliminate ω_1; and the 2. FBD for the wheel J shown in Figure P3.21d. To get the other state equation, use D'Alembert's principle.
b. Express the state equations in the vector-matrix form.
c. To complete the state-space model, determine the output equation for: (i) output=ω; (ii) output=T; (iii) output=ω_1
d. Which one of the translatory systems in Figure P3.22e is analogous to the system in Figure P3.22b?
e. Comment on why the compatibility equations and continuity equations are not explicitly used in this development of the state equations.
f. Present a more detailed and systematic approach, as presented in this book, for the development of the state-space model in this example.

3.23

a. Figure P3.23a shows an electrical circuit having a resistor of resistance R and a capacitor of capacitance C.
v_i=input voltage
v_o=output voltage
Using the capacitor voltage v_C as the state variable, obtain a complete state model. From it obtain the input–output model.
b. Figure P3.23b shows the circuit formed by cascading two circuit modules of the type in Figure 3.23a. Again,
v_i=input voltage
v_o=output voltage

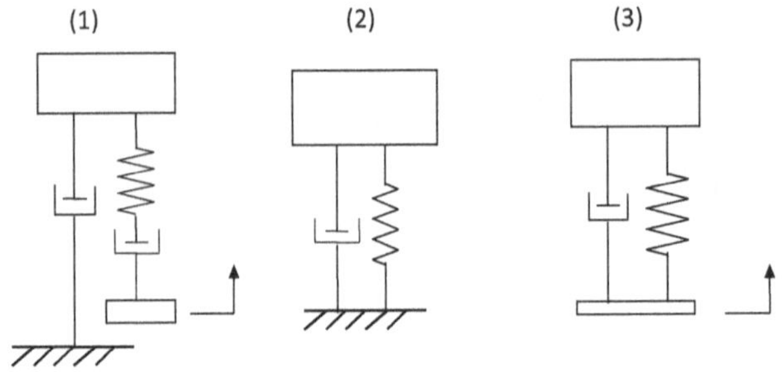

FIGURE P3.22E Three translatory mechanical systems.

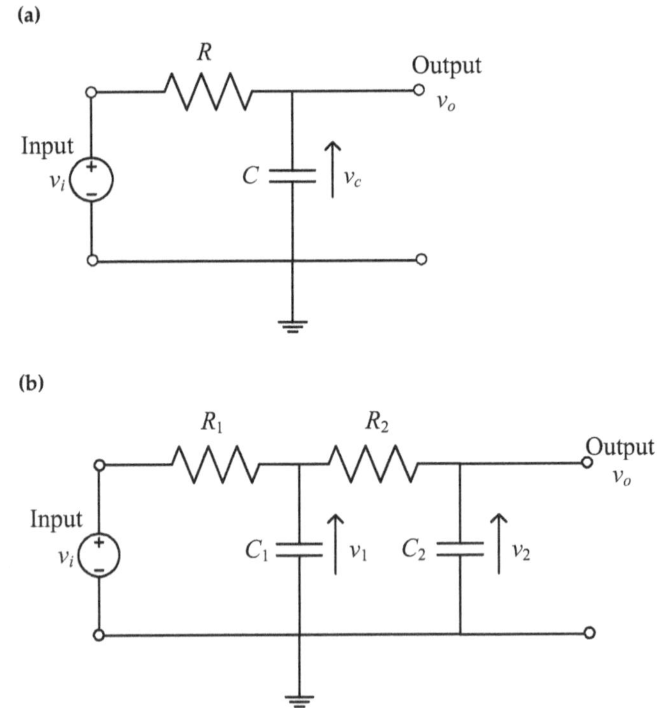

FIGURE P3.23 (a) A circuit module; (b) cascade circuit.

Using the voltages v_1 and v_2 across the capacitors C_1 and C_2 as the state variables, determine a complete state model. From it obtain the I-O model.

Show that this I-O model cannot be obtained by simply cascading two I-O models of Part (a). Specifically, the second circuit model "loads" the first circuit module (i.e., changes the conditions of the first circuit module).

3.24 A bridged-T *RC* network is shown in Figure P3.24. The T-circuit is formed by the two capacitors C_1 and C_2 and the resistor R. The resistor R_b directly connects the input and the output and forms the "bridge." Here,

v_i = input voltage

v_o = output voltage

This circuit has many practical applications such as forming an integral part of a resistance-tuned oscillator (in which case $C_1 = C_2$).

Determine a complete state model for the system using the capacitor voltages v_1 and v_2 as the state variables. Comment on the nature of the system.

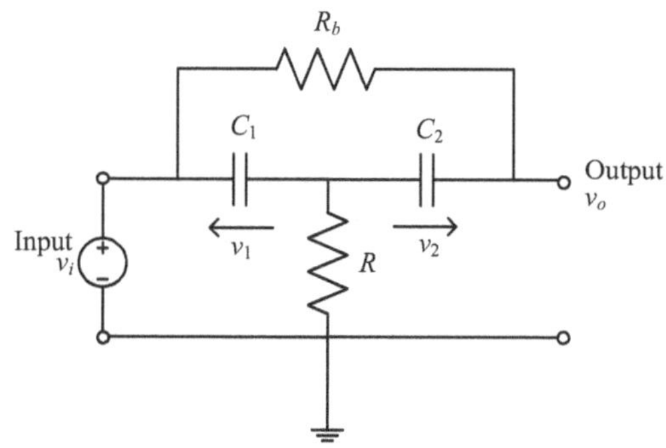

FIGURE P3.24 A bridged-T *RC* network.

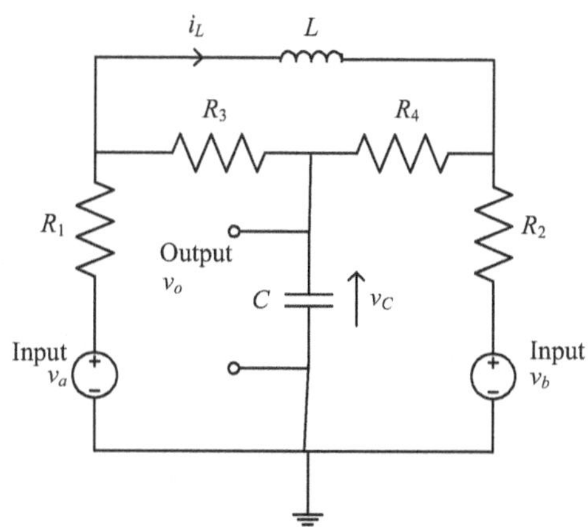

FIGURE P3.25 A bridged network.

Note: Another form of this network is the bridged-T *RL* network where inductors are used in place of capacitors.

3.25 A bridged network is shown in Figure P3.25. Here, v_a and v_b are the voltage inputs to the network. Using the voltage v_C across the capacitor C and the current i_L through the inductor as the state variables, obtain a complete state model for the system. Output is the voltage across the capacitor.
Note: If $R_1 = R_2$ and $R_3 = R_4$, we have a balanced bridge.

3.26 Consider the pneumatic system shown in Figure P3.26a. The pressure source (pressure-controlled pump) of gauge pressure P_s generates a gauge pressure P_l at a distant location, which drives a load. Suppose that the load is not defined yet. There is a bellows unit of gauge pressure P_b, which acts as a pneumatic capacitor (due to the compressibility of the gas and the flexibility of the bellows). The pipeline from the pump to the bellows has pneumatic inertance I_{mf} and pneumatic resistance R_{mf}. The branch from the bellows to the load has pneumatic resistance R_{ml}. The absolute pressures at various locations of the pneumatic system are indicated in Figure P3.26a. Also, Q_m=mass flow rate of gas from the pump; P_a=ambient pressure.

a. Taking Q_m and P_b as the state variables, obtain a state-space model for the pneumatic system. What are the system inputs?
b. Suppose that the load is a pneumatic cylinder of area A, which drives a mechanical system represented by mass m, spring of stiffness k, and viscous damper of damping constant b, as shown in Figure P3.26(b). Enhance the state-space model to incorporate

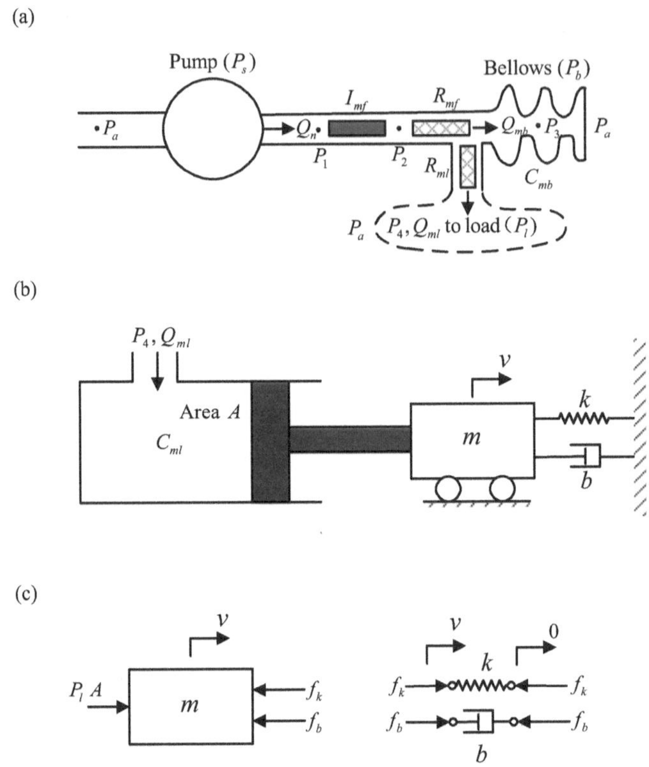

FIGURE P3.26 (a) Pneumatic system; (b) details of the load; (c) free-body diagrams of the load elements.

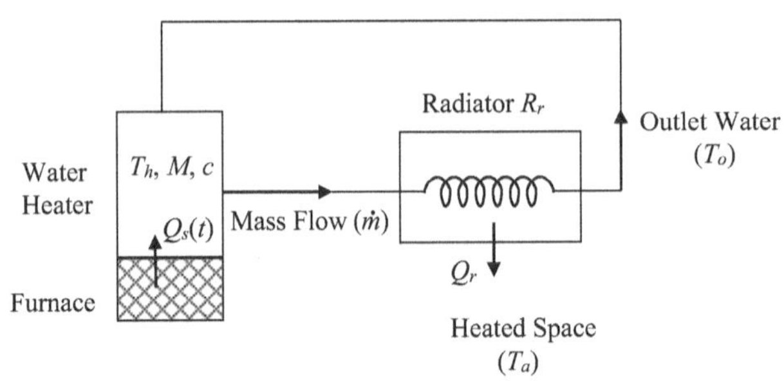

FIGURE P3.27 A household heating system.

this specific load, by including the additional state variables: P_l, v = velocity of the piston, and f_k = stiffness of the spring.

3.27 A simplified model of a hot water heating system is shown in Figure P3.27.
Pertinent nomenclature is as follows:
Q_s = rate of heat supplied by the furnace to the water heater (=1000 kW)
T_a = ambient temperature (°C)
T_h = temperature of the water in the water heater—assumed uniform (°C)
T_o = temperature of the water leaving the radiator (°C)
Q_r = rate of heat transfer from the radiator to the ambience (kW)
M = mass of water in the water heater (=500 kg)
$\dot{m}$ = mass flow rate of water through the radiator (=25 kg/min)
c = specific heat of water (=4200 J/kg/°C).

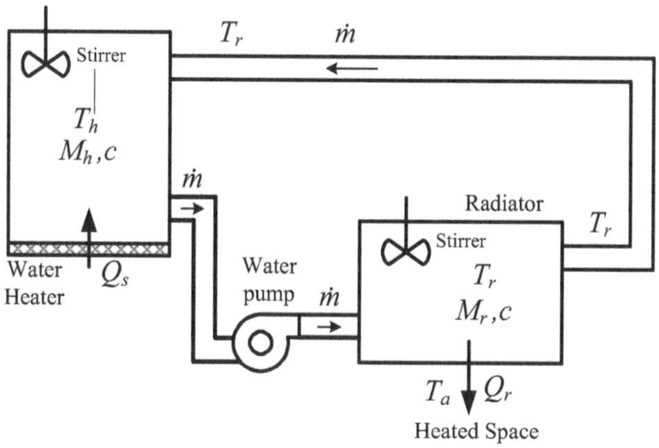

FIGURE P3.28 Hot water-based space heating system.

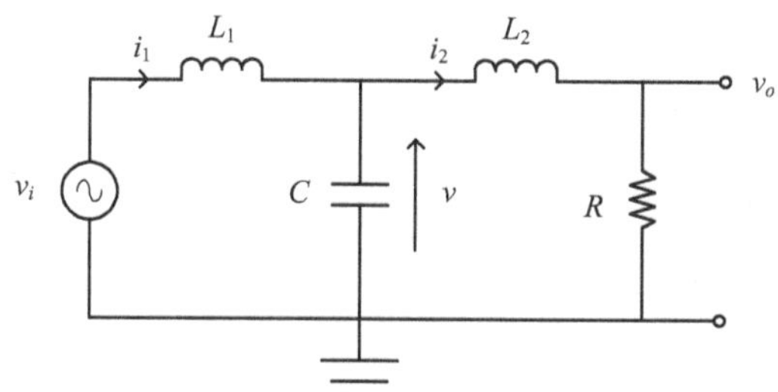

FIGURE P3.29 A passive three-pole Butterworth filter circuit.

The radiator satisfies the equation:

$$T_h - T_a = R_r Q_r$$

where R_r=thermal resistance of the radiator (2×10^{-3} °C/kW)

a. What are the inputs to the system?
b. Using T_h as a state variable, develop a state-space model for the system
c. Give the output equations (algebraic) for Q_r and T_o. Discuss their characteristics.

3.28 Hot water-based space heating system is shown in Figure P3.28. Water in the hot water heater receives heat (thermal energy) at the rate Q_s (**input**) from a natural gas burner. Water in the heater tank is maintained at uniform temperature T_h using a stirrer. The pump circulates the hot water from the tank at the mass flow rate $\dot{m}$ through the radiator back into the water heater. In the radiator, the water temperature is maintained uniform at T_r (using another stirrer). The radiator transfers heat from the water into the heated space (**output**). Given: M_h=mass of water in water heater; M_r=mass of water in radiator; c=specific heat of water; R=thermal resistance between radiator and ambience (heated space). Also,
T_a=temperature of the heated space (**another input, why?**)
Q_r=heat transfer rate from the radiator into the heated space (**output**)
Develop a state-space model for the system. **Assume:** steady conditions for water flow.

3.29 Consider the passive filter circuit shown in Figure P3.29. It is a three-pole, low-pass, Butterworth filter. As shown, it consists of two inductors (L_1 and L_2), one capacitor (C), and one resistor (R). The circuit input is the voltage $v_i(t)$ and the circuit output is the voltage v_o (across the resistor R).

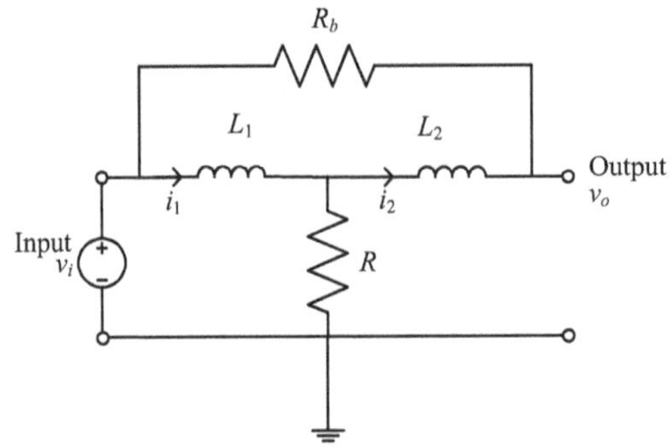

FIGURE P3.30 A bridged-T *RL* network.

a. On the same circuit diagram, indicate the voltages across and the currents through all four circuit branches, and the useful (primary) nodes of the circuit. What is the order of the system?
b. Systematically derive a complete state-space model for the circuit. Determine the model matrices, $\boldsymbol{A}$, $\boldsymbol{B}$, $\boldsymbol{C}$, and $\boldsymbol{D}$
c. From the state-space model, determine the input–output differential equation of the system.

3.30 A bridge-T *RL* network is shown in Figure P3.30. The circuit has two inductors with inductance L_1 and L_2 and two resistors with resistance R and R_b. In particular, R_b is the bridging resistance. The voltage input to the circuit is v_i, and v_o is the voltage output of the circuit.

a. Systematically (clearly indicating all the key steps) determine a complete state-space model for this circuit.

   Hint: Use the currents i_1 and i_2 through the inductors L_1 and L_2, respectively, as the state variables.
b. Indicate an important characteristic of this model.

4 Model Linearization

HIGHLIGHTS

- Common Nonlinearities and Properties
- Nonlinear Electrical Elements
- Analytical Linearization Using Local Slope
- Operating Condition and Equilibrium State
- Linearization of State-Space Model
- Linearization of Input–Output Model
- Reduction of System Nonlinearities
- Linearization Using Experimental Data
- Experimental Model for Actuator Control
- Calibration Curve Method
- Equivalent Model Approach
- Describing Function Method
- Linearization Using Analog Hardware
- Feedback Linearization

4.1 INTRODUCTION

Real systems are nonlinear and may be represented by nonlinear analytical models. Linear systems (models) are in fact idealized representations, and they are represented by linear differential equations in the time domain or by analytical transfer functions (ratio of polynomials) in the frequency domain. Clearly, it is far more convenient to analyze, simulate, design, and use/implement linear models. Also, in some systems, the degree of nonlinearity may not be significant, particularly depending on the objectives of modeling. For such reasons, nonlinear systems are often "approximated" by linear models.

It is not possible to represent a highly nonlinear system by a single linear model in its entire range of operation. For small "changes" in response about some operating condition, a linear model may be used, which is valid in the neighborhood of the operating condition. Commonly, linearization about an operating condition is done based on the "local slope" of the nonlinearity at the operating condition. Such linearization is not always feasible or satisfactory depending on the nature of the nonlinearity, for reasons which we will highlight later. Then, problem-specific and ad hoc approaches may have to be used to deal with system nonlinearities. In this chapter, we study linearization of a nonlinear system (i.e., of a nonlinear model). The studied topics are the following:

1. Slope-based analytical linearization over a limited range of operations about an operating condition (for both state-space models and input–output models)
2. Slope-based linearization using experimental input–output data (leading to experimental models)
3. Static linearization through recalibration or rescaling
4. Linearization based on an equivalent model (using some criterion of equivalence such as energy)
5. Describing function method of linearization (this method also uses a criterion of equivalence—fundamental frequency component of output)
6. Linearization using analog hardware
7. Feedback linearization.

Examples are given to illustrate these methods of linearization.

DOI: 10.1201/9781003124474-4

4.2 PROPERTIES OF NONLINEAR SYSTEMS

In analytical and dynamic sense, a device is considered linear if it can be modeled by a set of linear differential equations with time t as the independent variable or by a set of analytical transfer functions (ratios of polynomials) with frequency ω (or the Laplace variable s) as the independent variable. The use of an analytical transfer function to represent a device implicitly assumes the device to be linear and constant parameter (i.e., time-invariant). A useful property of a linear system is the satisfaction of the *principle of superposition*, as discussed in Chapter 3. Conversely, a nonlinear system does not obey the principle of superposition. In this section, we discuss some characteristic properties of nonlinear systems/devices/models, which distinguish them from linear systems/devices/models.

4.2.1 Static Nonlinearity

If the input–output relation of a device is a nonlinear "algebraic" equation, it represents a *static nonlinearity*. Such nonlinearity can be handled (or linearized) simply by using a proper calibration or rescaling curve. If, on the other hand, the input–output relation of a device is a nonlinear differential equation of time, it represents a *dynamic nonlinearity*. Then the analysis and linearization usually become more complex.

According to industrial and commercial terminology, a "linear" device (e.g., a measuring instrument) has its output proportional to the input (e.g., a *measured value* varies linearly with the value of the *measurand*—the variable that is measured). This is consistent with the definition of static linearity and is appropriate because for those commercial devices it is typically required that the operating range be limited to the "static region" where the dynamics do not appreciably affect the device output.

4.2.2 Nonlinear Characteristics of Practical Devices

All physical devices are nonlinear to some degree. Broadly speaking, the nonlinear behavior in devices may be classified into two types:

1. Geometric (including kinematic) nonlinearity, specifically in a "mechanical" device
2. Physical (including kinetic) nonlinearity.

Geometric nonlinearity of a mechanical device stems primarily from large deflections or large motions of the device resulting in the introduction of nonlinear terms (e.g., trigonometric terms such as sine, cosine, and tan) in the representation of its input–output behavior. Kinematic relations of a robotic arm are an example of geometric nonlinearity. Physical nonlinearity of a device results from the deviation from the linear (ideal) behavior of its physical relations due to such causes as electrical and magnetic saturation, deviation from Hooke's law in elastic elements, plasticity, Coulomb and Stribeck friction, creep at joints, aerodynamic damping, backlash in gears and other loose components, and component wear out. Nonlinear Newton's second law equation (i.e., kinetic nonlinearity) in a mechanical system also falls into the category of physical nonlinearity.

Nonlinearities (particularly, physical nonlinearities) in devices are often manifested as some peculiar characteristics. Some examples are given now.

- **Saturation**: Nonlinear devices may exhibit saturation (see Figure 4.1a). When saturated, the output of the device remains unchanged even when the input changes. This may be the result of causes such as magnetic saturation, which is common in magnetic-induction devices, transformer-like devices and other electromagnetic devices with a ferromagnetic core, differential transformers (in the current-magnetic field curve, the magnetic field strength saturates); electronic saturation, as in amplifiers (in the input–output behavior, output saturates); elastoplastic behavior in material (in the strain–stress curve, stress

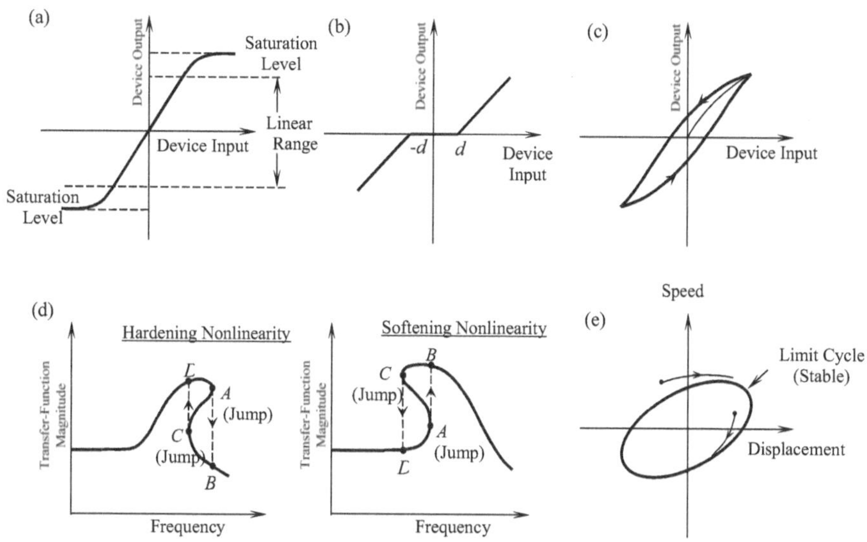

FIGURE 4.1 Common manifestations of nonlinearity in devices: (a) Saturation, (b) dead zone, (c) hysteresis, (d) jump phenomenon, (e) limit cycle response.

saturates as the strain increases); and in mechanical components like nonlinear springs (in the displacement-force curve, spring force saturates).

- **Ideal Relay**: A special, ideal case of saturation is the "two-state *switching function*" or an *ideal relay*. In this case, the device saturates at two different (usually opposite) states and does not have a linear (or variable) region in between. Hence, the device can only switch between these two states (a relay).
- **Dead Zone**: A dead zone is a region in which a device would not respond to an excitation (input). Stiction in mechanical devices with Coulomb friction is a good example (where x-axis is frictional force, y-axis is displacement or speed). Because of stiction, a component will not move until the applied force (input) reaches some minimum value. Once the motion is initiated, subsequent behavior can be either linear or nonlinear. Another example is the backlash in loose components such as gear wheels that do not mesh perfectly (where the rotation is the input and the transmitted torque is the output). Bias signal in electronic devices is a third example. In them, until the bias signal reaches a specific level, the circuit action will not take place, as in reverse bias of a diode until breakdown (here, input is bias voltage, output is transmitted current). A dead zone with subsequent linear behavior is shown in Figure 4.1b.
- *Note*: In the case of stiction or Coulomb friction, if the input of the device is a motion (displacement or velocity) and the output is the corresponding force in the device, then the behavior corresponds to an ideal relay (see under saturation) rather than a dead zone.
- **Hysteresis**: Nonlinear devices may produce hysteresis. In hysteresis, the value of the input–output curve at a particular point is different depending on the direction of the input (see Figure 4.1c), resulting in a hysteresis loop. This behavior is common in loose components such as gears, which have backlash; in components with nonlinear damping, such as Coulomb friction; and in magnetic devices or magnetic circuits (here, input is magnetizing current, output is the magnetic field strength) with ferromagnetic media and various dissipative mechanisms (e.g., eddy current dissipation). For example, consider a coil wrapped around a ferromagnetic core. If a dc current is passed through the coil, a magnetic field is generated. As the current is increased from zero, the field strength will also increase. Now, if the current is decreased back to zero, the field strength will not return to zero because of the residual magnetism in the ferromagnetic core. A negative current has to be applied to demagnetize the core. It follows that the field strength vs. current curve looks somewhat like Figure 4.1c. This is magnetic hysteresis.

- The presence of a hysteresis loop alone does not imply that the device is nonlinear. For example, linear viscous damping also exhibits a hysteresis loop in its force–displacement curve. This is a property of any mechanical component that dissipates energy (area within the hysteresis loop=energy dissipated in one cycle of motion). This topic is further explored in some examples and problems, later in the chapter. In general, if force in a device depends on the displacement (as in the case of a spring) and the velocity (as in the case of a damping element), the value of force at a given value of displacement will change depending on the direction of the velocity. In particular, the force when the component is moving in one direction (say positive velocity) will be different from the force at the same location when the component is moving in the opposite direction (negative velocity), thereby producing a hysteresis loop in the force–displacement plane. If the relationship of the displacement and velocity to the force is linear (as in viscous damping), the hysteresis effect is linear. If on the other hand, the relationship is nonlinear (as in Coulomb damping and aerodynamic damping), the resulting hysteresis is nonlinear.
- **Hysteresis Loop and Energy Dissipation**: When the two axes represent "force" and "displacement," it is known that the area of the hysteresis loop gives the net work done (or dissipated energy) in one cycle of movement. In this case, the input is the force and the output is the displacement. Note the loop arrows in Figure 4.1c. Since "work done" is given by the integral of "force"×"incremental displacement," the area projected onto the y-axis (i.e., output axis or displacement axis) gives the work done. It is clear from Figure 4.1c that this area is greater in the forward direction of the loop (forward arrow) than in the backward movement. Hence, the net area is positive and is equal to the area of the hysteresis loop, indicating an overall network done (or energy dissipation).
- **Jump Phenomenon**: Some nonlinear devices exhibit an instability known as the jump phenomenon (or *fold catastrophe*) in the frequency response (transfer) function curve that is determined experimentally. This is shown in Figure 4.1d. As the frequency increases, the jump occurs from A to B; and as the frequency decreases, it occurs from C to D. In particular, note the bending of the resonant peak, corresponding to either a *hardening* device (resonant frequency increases from the linear value; the peak bends forward) or a *softening* device (resonant frequency decreases from the linear value; the peak bends backward). Furthermore, the experimentally determined transfer function of a nonlinear device may depend on the magnitude of the input excitation. (The experimental transfer function of a linear device does not depend on the magnitude of the input.) Also, see under "describing function method" later in this chapter.
- *Note*: It should be clear that the segment AC in the nonlinear frequency response curve cannot be determined experimentally (it has to be determined either analytically or by interpolation of the experimental curve).
- **Limit Cycles**: A notable property of a nonlinear system is that its stability may depend on the system inputs and/or initial conditions. In particular, nonlinear devices may produce limit cycles. An example is given in Figure 4.1e on the phase plane (2-D) of velocity versus displacement. A limit cycle is a closed trajectory in the state space that corresponds to sustained oscillations at a specific frequency and amplitude, without decay or growth. The amplitude of these oscillations does not depend on the initial location from which the response started (unlike in a linear device). In addition, an external input is not needed to sustain a limit-cycle oscillation. In the case of a *stable limit cycle*, the response will move onto the limit cycle irrespective of the location in the neighborhood of the limit cycle from which the response was initiated (see Figure 4.1e). In the case of an *unstable limit cycle*, the response will move away from it with the slightest disturbance, and an unstable limit cycle cannot be determined experimentally.
- **Frequency Creation**: A linear device, when excited by a sinusoidal signal, will generate a response at the same frequency as the excitation, at steady state. On the other hand, at steady state, a nonlinear device may create frequencies that are not present in the excitation signals. These created frequencies might be *harmonics* (integer multiples of the excitation frequency), *subharmonics* (integer fractions of the excitation frequency), or *nonharmonics* (usually rational fractions of the excitation frequency). An example for this behavior is given next.

Example 4.1

Consider a nonlinear device that is modeled by the differential equation $\left\{\frac{dy}{dt}\right\}^{1/2} = u(t)$, where $u(t)$ is the input and y is the output. Show that this device creates frequency components that are different from the excitation frequencies.

Solution

First, we express the given system equation as $y = \int_0^t u^2(t)\,dt + y(0)$.

Now, for an input given by $u(t) = a_1 \sin\omega_1 t + a_2 \sin\omega_2 t$, straightforward integration using properties of trigonometric functions gives the following response:

$$y = \left(a_1^2 + a_2^2\right)\frac{t}{2} - \frac{a_1^2}{4\omega_1}\sin 2\omega_1 t - \frac{a_2^2}{4\omega_2}\sin 2\omega_2 t$$
$$+ \frac{a_1 a_2}{2(\omega_1 - \omega_2)}\sin(\omega_1 - \omega_2)t - \frac{a_1 a_2}{2(\omega_1 + \omega_2)}\sin(\omega_1 + \omega_2)t - y(0)$$

Note: $\sin^2\theta = (1 - \cos 2\theta)/2$; $2\sin\theta_1 \sin\theta_2 = \cos(\theta_1 - \theta_2) - \cos(\theta_1 + \theta_2)$

It is seen that the discrete frequency components $2\omega_1$, $2\omega_2$, $(\omega_1 - \omega_2)$, and $(\omega_1 + \omega_2)$ are created by the nonlinear device. Additionally, there is a continuous spectrum contributed by the linear function of t that is present in the response (but not in the input).

4.2.3 Nonlinear Electrical Elements

The three lumped-parameter passive elements in an electrical system are capacitor (an A-type element with the *across-variable* voltage as the state variable); inductor (a T-type element with the *through-variable* current as the state variable); and resistor (a D-type element representing energy dissipation; a new state variable is not introduced by it). The linear versions of these elements are discussed in Chapter 2. Now let us briefly look into the general, nonlinear versions of these elements.

4.2.3.1 Capacitor

Electrical charge (q) is a function of the voltage (v) across a capacitor, as given by the nonlinear constitutive equation:

$$q = q(v) \tag{4.1a}$$

For the linear case we have,

$$q = Cv \tag{4.1b}$$

where C is the *capacitance*. Then the current i, as given by $\frac{dq}{dt}$, is obtained by differentiating equation (4.1a) as,

$$i = \frac{\partial q}{\partial v}\frac{dv}{dt} \tag{4.2a}$$

which gives a nonlinear capacitance, $C(v) = \frac{\partial q}{\partial v}$

or by differentiating equation (4.1b) as

$$i = C\frac{dv}{dt} + v\frac{dC}{dt} \tag{4.2b}$$

Here, we have allowed for a time-varying (but still linear) capacitance. If C is assumed constant, we have the familiar linear constitutive equation:

$$i = C\frac{dv}{dt} \tag{4.2c}$$

4.2.3.2 Inductor

Magnetic flux linkage (λ) of an inductor is a function of the current (i) passing through the inductor, as given by the nonlinear constitutive equation:

$$\lambda = \lambda(i) \tag{4.3a}$$

For the linear case, we have

$$\lambda = Li \tag{4.3b}$$

where L is the *inductance*.

The voltage v induced in an inductor is equal to the rate of change of the flux linkage $\left(v = \frac{d\lambda}{dt}\right)$. Hence, by differentiating equation (4.3a), we get

$$v = \frac{\partial\lambda}{\partial i}\frac{di}{dt} \tag{4.4a}$$

which gives a nonlinear inductance, $L(i) = \frac{\partial\lambda}{\partial i}$

or, by differentiating equation (4.3b), we get

$$v = L\frac{di}{dt} + i\frac{dL}{dt} \tag{4.4b}$$

Assuming that the inductance is constant, we have the familiar linear constitutive equation:

$$v = L\frac{di}{dt} \tag{4.4c}$$

4.2.3.3 Resistor

In general, the voltage across a (nonlinear) resistor is a function of the current through the resistor, as given by,

$$v = v(i) \tag{4.5a}$$

This gives nonlinear resistance $R(i) = \frac{\partial v}{\partial i}$.

In the linear case, we have the familiar Ohm's law:

$$v = Ri \tag{4.5b}$$

where R is the *resistance*, which can be time-varying in general. In most cases, however, we assume R to be a constant.

4.3 ANALYTICAL LINEARIZATION USING LOCAL SLOPES

Real systems are nonlinear and may be more accurately represented by nonlinear analytical models. Analytical techniques (e.g., response analysis, frequency-domain analysis, eigenvalue problem analysis, simulation, control) commonly employ linear models, which are far more convenient to use. Linear models may be adequate for a problem at hand, for several reasons: (a) The degree of nonlinearity in the actual system is not considerable; (b) the physical system operates, in practice, in a steady state with only minor deviations from that state; (c) several linear models (or a single linear model with variable parameters), in different operating conditions, may be adequate to represent the actual nonlinear system, in the considered range of operation; (d) the required accuracy of the analysis is not high (i.e., only an approximate, preliminary analysis is adequate); and (e) only some aspects of the actual system are of interest, and they are by and large linear.

Popularly, nonlinear devices are linearized by considering small excursions about an operating point. In other words, a "linear local model" is used. If a single such model has sufficient accuracy over the entire operating range of the device, it is an indication that the device is linear (over the entire operating range). If not, a series of local linear models may have to be used over the operating range.

A nonlinear analytical model may contain one or more nonlinear terms. The approach taken in the local slope method is to linearize each nonlinear term by using the first-order Taylor series approximation, which involves only the first derivative of the nonlinear term (or, the "slope" in its graphical representation). A nonlinear term may be a function of more than one independent variable. Then, the first "partial derivatives" with respect to all its independent variables (i.e., the slopes along all orthogonal directions of the coordinate axes, which represent the independent variables) are needed in the linearization process. In this section, we study "analytical' linearization of a system using local derivatives (slopes) of the nonlinear terms in the system. In a subsequent section, we consider local linearization through experimental data (by determining the local slopes of experimental data that correspond to the nonlinear terms).

4.3.1 Analytical Linearization about an Operating Point

In the approach described now, linearization is carried out by determining the partial derivatives of the nonlinear terms, with respect to the independent variables, at the considered operating point. Typically, this linearizing point is the normal operating condition of the system. Of necessity, the normal operating condition has to be in the steady state or the equilibrium state.

4.3.1.1 Equilibrium State

In a steady state, by definition, the rates of changes of the system variables are zero. Hence, the steady state (equilibrium state) is determined by setting the time-derivative terms in the system equations to zero and then solving the resulting algebraic equations. This may lead to more than one solution, since the steady-state (algebraic) equations themselves are nonlinear. The real (i.e., non-complex) steady-state (equilibrium) solutions will correspond to one of the following three types:

1. Stable (here, given a slight shift, the system response eventually returns to the original steady state)
2. Unstable (here, given a slight shift, the system response continues to move in the direction of the shift, away from the original steady state)
3. Neutral (here, given a slight shift, the system response will remain in the shifted condition).

These three types of equilibrium are schematically shown in Figure 4.2.

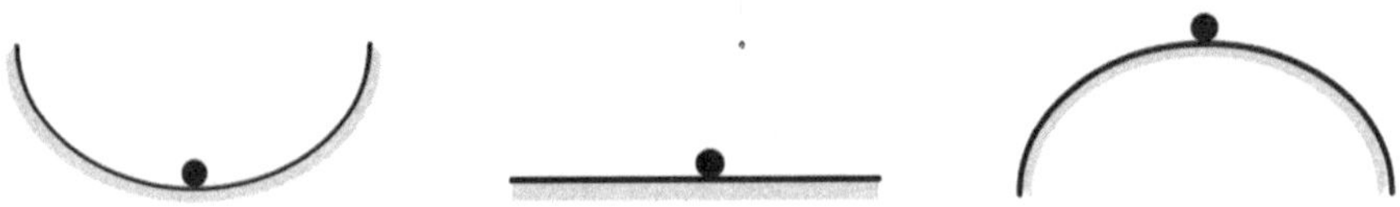

FIGURE 4.2 Equilibrium types. (a) Stable equilibrium; (b) neutral equilibrium; (c) unstable equilibrium.

4.3.2 Nonlinear Function of One Variable

Consider a nonlinear function $f(x)$ of the independent variables x. Its Taylor series approximation about an operating point $(\cdot)_o$, up to the first derivative, is given by

$$f(x) \approx f(x_o) + \frac{df(x_o)}{dx}\delta x \quad \text{with} \quad x = x_o + \delta x \tag{4.6a}$$

Here, δx represents a small change from the operating point.

Alternatively, denote the operating condition by $(\bar{\ })$ and the small increment about that condition by $(\hat{\ })$. We have,

$$f(\bar{x}+\hat{x}) \approx f(\bar{x}) + \frac{df(\bar{x})}{dx}\hat{x} \tag{4.6b}$$

From equation (4.6), it is seen that the increment of the function, due to the increment in its independent variable, is given by,

$$\delta f = f(x) - f(x_o) \approx \frac{df(x_o)}{dx}\delta x \tag{4.7a}$$

or

$$\hat{f} = f(\bar{x}+\hat{x}) - f(\bar{x}) \approx \frac{df(\bar{x})}{dx}\hat{x} \tag{4.7b}$$

Equation (4.7) is a linear relationship between the increment of the independent variable ($\hat{x}$) and the increment of the function ($\hat{f}$). In other words, we have linearized the nonlinear function, about an operating point (denoted by $f(x_o)$ or $f(\bar{x})$. A graphical illustration of this linearization approach is shown in Figure 4.3.

The error resulting from this linear approximation is

$$\text{Error } e = f(\bar{x}+\hat{x}) - [f(\bar{x}) + \frac{df(\bar{x})}{dx}\hat{x}] \tag{4.8}$$

This error decreases:

1. If the nonlinear function is more linear
2. By making the increment $\hat{x}$ from the operating point smaller

Note: If the function is already linear, we have: $f = ax$

where a is a constant coefficient. The corresponding incremental relation is

$$\delta f = a\delta x \tag{4.9a}$$

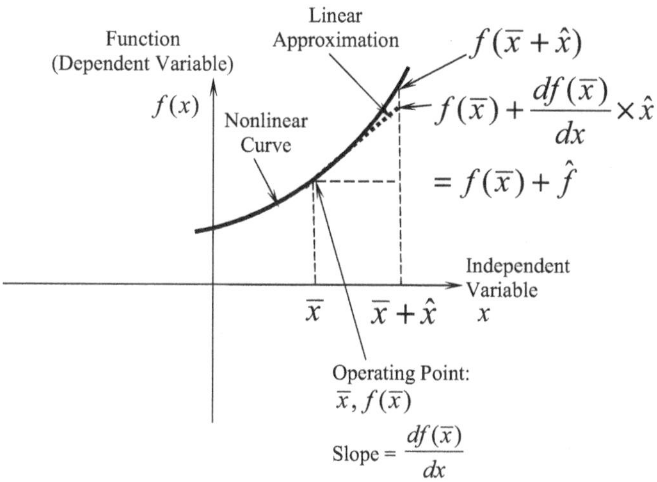

FIGURE 4.3 Linearization about an operating point.

or

$$\hat{f} = a\hat{x} \tag{4.9b}$$

As expected, the incremental relation is also linear because the original relation is linear. In this case, however, no error is introduced through the process of linearization.

4.3.2.1 Commutativity of Increment and Derivative

The increment of a time derivative is equal to the time derivative of the increment. Hence,

$$\delta\dot{x} = \frac{d\hat{x}}{dt} = \dot{\hat{x}} \tag{4.10}$$

Now, if we set $x = \dot{z}$ in (4.10), we get $\delta(\ddot{z}) = \frac{d\hat{\dot{z}}}{dt} = \frac{d\dot{\hat{z}}}{dt} = \frac{d^2\hat{z}}{dt^2}$. The second equality is obtained in view of (4.10). Then, by replacing z by x in the final result, we have

$$\delta\ddot{x} = \frac{d^2\hat{x}}{dt^2} = \ddot{\hat{x}} \tag{4.11}$$

4.3.2.2 Further Useful Information: Derivatives of Nonlinear Functions

Signum (sign) function:

$$\operatorname{sgn}\omega_c = 1 \quad \text{when } \omega_c > 0$$
$$= -1 \quad \text{when } \omega_c < 0$$

Hence, it is a constant in each of the two ranges of $\omega_c \rightarrow$ Its time derivative is zero (except at $\omega_c = 0$, which is not important here as it corresponds to the static condition when the function value is zero). Also, since it is a constant, it can be taken out when differentiating a product of it with another quantity.

Useful Derivatives:

$$|\omega_c| = \omega_c \operatorname{sgn}\omega_c$$

$$\frac{d}{d\omega_c}(|\omega_c|) = \frac{d}{d\omega_c}(\text{sgn}\,\omega_c \times \omega_c) = \text{sgn}\,\omega_c \times 1 = \text{sgn}\,\omega_c$$

$$\frac{d}{d\omega_c}(|\omega_c|\omega_c) = \frac{d}{d\omega_c}(\text{sgn}\,\omega_c \times \omega_c^{\,2}) = \text{sgn}\,\omega_c \times 2\omega_c = 2|\omega_c|$$

$$\frac{d}{dt}(|\omega_c|\omega_c) = \frac{d}{dt}(\text{sgn}\,\omega_c \times \omega_c^2) = \text{sgn}\,\omega_c \times 2\omega_c\dot{\omega}_c = 2|\omega_c|\dot{\omega}_c$$

$$\frac{d^2}{dt^2}(|\omega_c|\omega_c) = \frac{d}{dt}(2|\omega_c|\dot{\omega}_c) = 2\,\text{sgn}\,\omega_c\frac{d}{dt}(\omega_c\dot{\omega}_c) = 2\,\text{sgn}\,\omega_c \times (\dot{\omega}_c^2 + \omega_c\ddot{\omega}_c) = 2\,\text{sgn}(\omega_c)\dot{\omega}_c^2 + 2|\omega_c|\ddot{\omega}_c$$

4.3.3 Nonlinear Function of Two Variables

The process of linearization of a function, as presented for one independent variable, can be easily extended to functions of more than one independent variable. For illustration, consider a nonlinear function $f(x, y)$ of two independent variables x and y. Its first-order Taylor series approximation is

$$f(x,y) \approx f(x_o, y_o) + \frac{\partial f(x_o, y_o)}{\partial x}\delta x + \frac{\partial f(x_o, y_o)}{\partial y}\delta y \quad \text{with} \quad x = x_o + \delta x, \quad y = y_o + \delta y \tag{4.12a}$$

or

$$f(\bar{x}+\hat{x}, \bar{y}+\hat{y}) \approx f(\bar{x},\bar{y}) + \frac{\partial f(\bar{x},\bar{y})}{\partial x}\hat{x} + \frac{\partial f(\bar{x},\bar{y})}{\partial y}\hat{y} \tag{4.12b}$$

where ($\bar{}$) denotes the operating condition and ($\hat{}$) denotes a small increment about that condition, as indicated before.

From equation (4.12), it is seen that the increment of the nonlinear function, due to the increments of its independent variables, is given by,

$$\delta f = f(x,y) - f(x_o, y_o) \approx \frac{\partial f(x_o, y_o)}{\partial x}\delta x + \frac{\partial f(x_o, y_o)}{\partial y}\delta y \tag{4.13a}$$

or

$$\hat{f} = f(\bar{x}+\hat{x}, \bar{y}+\hat{y}) - f(\bar{x},\bar{y}) \approx \frac{\partial f(\bar{x},\bar{y})}{\partial x}\hat{x} + \frac{\partial f(\bar{x},\bar{y})}{\partial y}\hat{y} \tag{4.13b}$$

Equation (4.13) is a linear relationship between the increments of the independent variables ($\hat{x}$ and $\hat{y}$) and the increment of the function ($\hat{f}$). In other words, we have linearized the nonlinear function, about an operating point (denoted by $f(x_o, y_o)$ or $f(\bar{x},\bar{y})$). In the present case, for the process of linearization, we need to use two local slopes (partial derivatives) $\frac{\partial f(\bar{x},\bar{y})}{\partial x}$ and $\frac{\partial f(\bar{x},\bar{y})}{\partial y}$ along the two orthogonal directions of the independent variables x and y.

From these illustrations, it should be clear that linearization of a nonlinear system (model) is carried out by replacing each term in the system equation by its increment, about an operating point. We summarize below the steps of local slope-based analytical linearization of a system (model) about an operating point:

1. Select the operating point (or, reference condition). This is typically a steady state, which can be determined by setting the time-derivative terms in the system equations to zero and solving the resulting nonlinear algebraic equations

2. Determine the slopes (first-order derivatives) of each nonlinear term (function) in the systems equation at the operating point, with respect to (i.e., along) each independent variable
3. Consider each term in the system equation. If a term is nonlinear, replace it by its slope (at the operating point) times the corresponding incremental variable. If a term is linear, replace it by its original coefficient (which is indeed the constant slope of the linear term) times the corresponding incremental variable

4.4 NONLINEAR STATE-SPACE MODELS

Consider a general nonlinear, time-variant, nth-order system represented by n first-order nonlinear differential equations, which generally are coupled, as given by,

$$\begin{aligned} \frac{dq_1}{dt} &= f_1(q_1, q_2, \ldots, q_n, r_1, r_2, \ldots, r_m, t) \\ \frac{dq_2}{dt} &= f_2(q_1, q_2, \ldots, q_n, r_1, r_2, \ldots, r_m, t) \\ &\vdots \\ \frac{dq_n}{dt} &= f_n(q_1, q_2, \ldots, q_n, r_1, r_2, \ldots, r_m, t) \end{aligned} \tag{4.14a}$$

The state vector is

$$\boldsymbol{q} = [q_1, q_2, \ldots, q_n]^T \tag{4.15}$$

and the input vector is

$$\boldsymbol{r} = [r_1, r_2, \ldots, r_m]^T \tag{4.16}$$

Equation (4.14a) may be written in the vector notation:

$$\dot{\boldsymbol{q}} = \boldsymbol{f}(\boldsymbol{q}, \boldsymbol{r}, t) \tag{4.14b}$$

Note: The explicit presence of the time variable "t" as an argument in the function f indicates that the parameters of the function vary with time (i.e., it is a *time-variant* model).

4.4.1 Linearization of State Model

Now we linearize the nonlinear state-space model given by equation (4.14) about its equilibrium state (an operating point). An equilibrium state of the state-space model (4.14) corresponds to the condition when the rates of changes of the state variables are all zero. Specifically, we set

$$\dot{\boldsymbol{q}} = \boldsymbol{0} \tag{4.17}$$

This is true because in an equilibrium state the system response remains steady and hence its rate of change is zero. Consequently, the equilibrium states $\bar{\boldsymbol{q}}$ are obtained by solving the set of n nonlinear algebraic equations:

$$\boldsymbol{f}(\boldsymbol{q}, \boldsymbol{r}, t) = \boldsymbol{0} \tag{4.18}$$

for a particular steady input $\bar{\boldsymbol{r}}$.

Note: Since the algebraic equations (4.18) are nonlinear, they can have more than one "real" (i.e., non-complex) roots.

Usually, under steady conditions, a system operates at one of its "stable" equilibrium states. To study the stability of various equilibrium states of a nonlinear dynamic system, it is first necessary to linearize the system model about a general equilibrium state given by a solution of equation (4.18).

As noted before, equation (4.14) can be linearized for small variations $\delta\boldsymbol{q}$ and $\delta\boldsymbol{r}$ of the states and the inputs, about an equilibrium point $(\bar{\boldsymbol{q}},\bar{\boldsymbol{r}})$, by employing up to only the first derivative term (i.e., O(1) term) in the Taylor series expansion of the nonlinear function $\boldsymbol{f}$. The higher-order terms are negligible for small $\delta\boldsymbol{q}$ and $\delta\boldsymbol{r}$. As explained before, this method yields the linear model:

$$\delta\dot{\boldsymbol{q}} = \frac{\partial \boldsymbol{f}}{\partial \boldsymbol{q}}(\bar{\boldsymbol{q}},\bar{\boldsymbol{r}},t)\delta\boldsymbol{q} + \frac{\partial \boldsymbol{f}}{\partial \boldsymbol{r}}(\bar{\boldsymbol{q}},\bar{\boldsymbol{r}},t)\delta\boldsymbol{r} \tag{4.19}$$

Denote the state vector and the input vector of the linearized model by the "incremental" vectors,

$$\delta\boldsymbol{q} = \boldsymbol{x} = \left[x_1, x_2, \ldots, x_n\right]^T \tag{4.20}$$

$$\delta\boldsymbol{r} = \boldsymbol{u} = \left[u_1, u_2, \ldots, u_m\right]^T \tag{4.21}$$

This results in the linear model,

$$\dot{\boldsymbol{x}} = \boldsymbol{A}\boldsymbol{x} + \boldsymbol{B}\boldsymbol{u} \tag{4.22}$$

The linear system matrix $\boldsymbol{A}(t)$ and the input distribution (gain) matrix $\boldsymbol{B}(t)$ are given by

$$\boldsymbol{A}(t) = \frac{\partial \boldsymbol{f}}{\partial \boldsymbol{q}}(\bar{\boldsymbol{q}},\bar{\boldsymbol{r}},t) \tag{4.23a}$$

$$\boldsymbol{B}(t) = \frac{\partial \boldsymbol{f}}{\partial \boldsymbol{r}}(\bar{\boldsymbol{q}},\bar{\boldsymbol{r}},t) \tag{4.24a}$$

The elements of these matrices are explicitly given by,

$$\boldsymbol{A} = \begin{bmatrix} \frac{\partial f_1}{\partial q_1} & \frac{\partial f_1}{\partial q_2} & \cdots & \cdots & \frac{\partial f_1}{\partial q_n} \\ \frac{\partial f_2}{\partial q_1} & \frac{\partial f_2}{\partial q_2} & \cdots & \cdots & \frac{\partial f_2}{\partial q_n} \\ \cdots & \cdots & \cdots & \cdots & \cdots \\ \cdots & \cdots & \cdots & \cdots & \cdots \\ \frac{\partial f_n}{\partial q_1} & \frac{\partial f_n}{\partial q_2} & \cdots & \cdots & \frac{\partial f_n}{\partial q_n} \end{bmatrix} \tag{4.23b}$$

$$\boldsymbol{B} = \begin{bmatrix} \frac{\partial f_1}{\partial r_1} & \cdots & \cdots & \frac{\partial f_1}{\partial r_m} \\ \frac{\partial f_2}{\partial r_1} & \cdots & \cdots & \frac{\partial f_2}{\partial r_m} \\ \cdots & \cdots & \cdots & \cdots \\ \cdots & \cdots & \cdots & \cdots \\ \frac{\partial f_n}{\partial r_1} & \cdots & \cdots & \frac{\partial f_n}{\partial r_m} \end{bmatrix} \tag{4.24b}$$

If it is a constant-parameter (i.e., *stationary* or *time-invariant*) dynamic system, or if it can be assumed as such for the time period of interest, then $\boldsymbol{A}$ and $\boldsymbol{B}$ are constant matrices.

4.4.2 Mitigation of System Nonlinearities

Under steady conditions (i.e., in the static case), system nonlinearities can be "precisely" removed through recalibration. Under dynamic conditions, however, the task becomes far more difficult. In general, the following approaches may be used to remove nonlinearities:

1. Recalibration and rescaling (e.g., log) of the output (in the static case)
2. Use of linearizing elements (e.g., resistors, amplifiers in bridge circuits) to neutralize the nonlinear effects
3. Use of nonlinear feedback (feedback linearization).

These approaches are illustrated in examples of this chapter.

The following precautions can be taken to reduce nonlinear behavior in dynamic systems:

1. Avoid operating the device over a wide range of inputs or signal levels
2. Avoid operation over a wide frequency band
3. Use devices that do not generate large deformations (e.g., deviation from Hooke's law—a physical nonlinearity) or large mechanical motions (e.g., introduction of trigonometric terms—a geometric or kinematic nonlinearity)
4. Minimize nonlinear processes of the system (e.g., Coulomb and Stribeck friction, stiction, wear and tear) through practical means (e.g., proper lubrication)
5. Avoid loose joints, gear coupling, etc., which can cause backlash (e.g., use direct-drive mechanisms, harmonic drives, etc.)
6. Minimize sensitivity to undesirable influences (e.g., environmental influences such as temperature).

Next we illustrate the analytical linearization and operating condition analysis using two practical examples. The first example involves a state-space model and an input–output model in two physical domains (mechanical and electrical).

Example 4.2

The robotic spray-painting system of an automobile assembly plant employs an induction motor and pump combination to supply paint at an overall peak rate of 15 gal/min to a cluster of spray-painting heads in several painting booths. The painting booths are an integral part of the production line in the plant. The pumping and filtering stations are in the ground level of the building and the painting booths are in an upper level. Not all booths or painting heads operate at a given time. The pressure in the paint supply line is maintained at a desired level (approximately 275 psi or 1.8 MPa) by controlling the pump speed, which is achieved through a combination of voltage control and frequency control of the induction motor. An approximate model for the paint pumping system is shown in Figure 4.4a.

The induction motor is linked to the pump through a gear transmission of efficiency η and speed ratio 1:r (typically, a speed reducer is used, with $r<1$) and a flexible shaft of torsional stiffness k_p. The moments of inertia of the motor rotor and the pump impeller are denoted by J_m and J_p, respectively. *Note*: The pump inertia J_p may include the "added-mass effect" of the paint. The gear inertia is neglected (or lumped with J_m). The mechanical dissipation in the motor and its bearings is modeled as a linear viscous damper of damping constant b_m. The load on the pump (i.e., much of the paint load plus any mechanical dissipation) is also modeled by a viscous damper, and

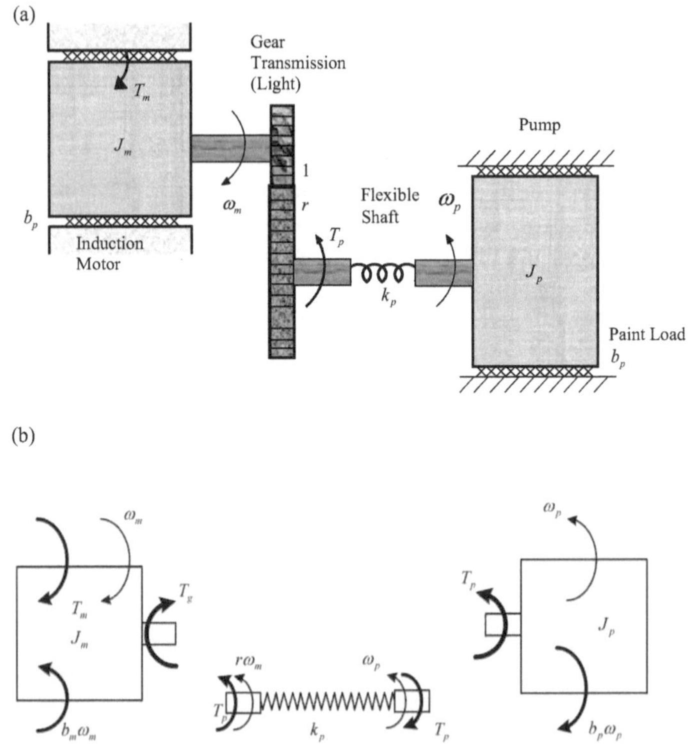

FIGURE 4.4 (a) A model of a paint pumping system in an automobile assembly plant; (b) free-body diagrams of the energy storage elements.

its equivalent damping constant is b_p. The magnetic torque T_m generated by the induction motor is given by,

$$T_m = \frac{T_0 q \omega_0 (\omega_0 - \omega_m)}{\left(q\omega_0^2 - \omega_m^2\right)} \tag{4.25}$$

Here, ω_m is the motor speed; the parameter T_0 depends directly (quadratically) on the phase (ac) voltage supplied to the motor; the second parameter ω_0 is the speed of the rotating magnetic field of the motor and is directly proportional to the line frequency of the ac supply; the third parameter q is positive and greater than unity, and this parameter is assumed constant in the control system.

a. Comment on the accuracy of the model shown in Figure 4.4a.
b. Taking the motor speed ω_m, the pump-shaft torque T_p, and the pump speed ω_p as the state variables, systematically derive the three state equations for this (nonlinear) model. Clearly explain all steps involved in the derivation. What are the inputs to the system?
c. What do the motor parameters ω_0 and T_0 represent, with regard to the motor behavior? Obtain the partial derivatives $\frac{\partial T_m}{\partial \omega_m}$, $\frac{\partial T_m}{\partial T_0}$, and $\frac{\partial T_m}{\partial \omega_0}$. Verify that the first of these three expressions is negative and the other two are positive. *Note*: Under normal operating conditions, for an induction motor (i.e., with "slip" in the motor rotation with respect to the rotating magnetic field), $0 < \omega_m < \omega_0$.
d. Consider the steady-state operating condition of the system, where the motor speed is steady at $\bar{\omega}_m$. Obtain expressions for $\bar{\omega}_p$, $\bar{T}_p$, and $\bar{T}_0$ (at this operating point), in terms of $\bar{\omega}_m$ and $\bar{\omega}_0$.

e. Suppose that $\frac{\partial T_m}{\partial \omega_m} = -b$, $\frac{\partial T_m}{\partial T_0} = \beta_1$, and $\frac{\partial T_m}{\partial \omega_0} = \beta_2$ in the operating condition given in Part (d). *Note*: Voltage control is achieved by varying T_0 and frequency control by varying ω_0. Linearize the state model obtained in Part (b) about the operating condition and express it in terms of the incremental variables $\hat{\omega}_m$, $\hat{T}_p$, $\hat{\omega}_p$, $\hat{T}_0$, and $\hat{\omega}_0$. Suppose that the (incremental) output variables are the incremental pump speed $\hat{\omega}_p$ and the incremental angle of twist of the pump shaft. Express the linear state-space model in the usual form and obtain the associated matrices ***A***, ***B***, ***C***, and ***D***.

f. For the case of frequency control alone (i.e., $\hat{T}_0=0$) obtain the input–output model (differential equation) relating the incremental output $\hat{\omega}_p$ to the incremental input $\hat{\omega}_0$. Using this equation, show that if $\hat{\omega}_0$ is suddenly changed by a step of $\Delta\hat{\omega}_0$, then $\frac{d^3\hat{\omega}_p}{dt^3}$ will instantaneously change by a step of $\frac{\beta_2 r k_p}{J_m J_p}\Delta\hat{\omega}_0$, but the lower derivatives of $\hat{\omega}_p$ will not change instantaneously.

Solution

(a)

- Backlash and inertia of the gear transmission have been neglected in the model. This assumption is not quite valid in general. Also, the gear efficiency η, which is assumed constant here, usually varies with the gear speed
- Usually there is some flexibility in the shaft (coupling) that connects the gear to the drive motor
- Energy dissipation (in the pump load and in various bearings) has been lumped into a single linear viscous-damping element. In practice, this energy dissipation is nonlinear and distributed
- At least part of the pump load may be included as an "added inertia" to the pump rotor (usually this is not a constant). Alternatively, the pump load may be more accurately represented by a torque versus speed curve.

(b)

Let T_g=reaction torque on the motor from the gear. Output speed of the gear transmission is $r\omega_m$. Also, power=torque×speed. Hence, by definition, the gear efficiency is given by

$$\eta = \frac{\text{Output Power}}{\text{Input Power}} = \frac{T_p r\,\omega_m}{T_g\,\omega_m}$$

This gives

$$T_g = \frac{r}{\eta}T_p \tag{i}$$

The free-body diagrams for the energy storage elements are shown in Figure 4.4b. There are two independent A-type elements and a T-type element. For them, the following three constitutive equations are written:

Newton's second law (torque=inertia×angular acceleration) for the motor:

$$J_m\dot{\omega}_m = -b_m\omega_m + T_m - T_g = -b_m\omega_m + T_m - \frac{r}{\eta}T_p \tag{ii}$$

Note: We have substituted (i).

Hooke's law (torque rate=torsional stiffness×twisting speed) for the flexible shaft:

$$\dot{T}_p = k_p\left(r\omega_m - \omega_p\right) \tag{iii}$$

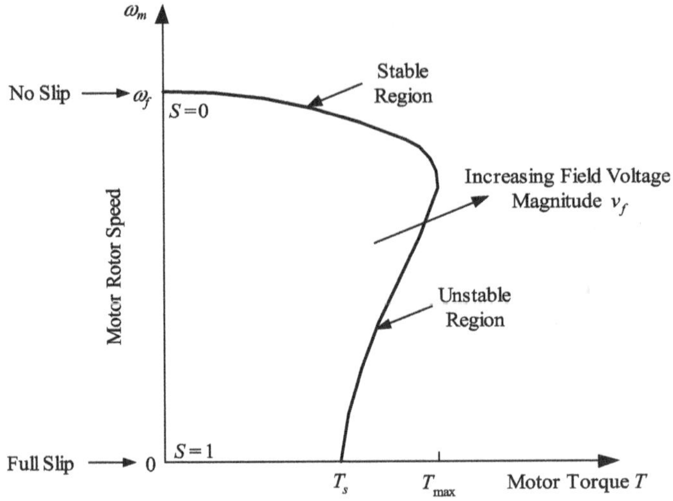

FIGURE 4.5 Torque–speed characteristic curve of an induction motor.

Newton's second law for the pump:

$$J_p \dot{\omega}_p = T_p - b_p \omega_p \tag{iv}$$

Equations (ii)–(iv) are the three state equations, with the state vector $\begin{bmatrix} \omega_m & T_p & \omega_p \end{bmatrix}^T$.

These equations provide a "linear" state-space model whose input is T_m. Strictly, the system is nonlinear with two inputs, in view of the torque–speed characteristic curve of the motor as given by equation (4.25) and sketched in Figure 4.5. It also indicates the fractional slip S of the induction motor, as given by,

$$S = \frac{(\omega_0 - \omega_m)}{\omega_0} \tag{4.26}$$

here, ω_0=speed of the rotating magnetic field of the motor (proportional to the line frequency).

Actually there are two inputs into the system: ω_0 and T_0. The latter depends quadratically on the phase voltage.

Note: Varying ω_0 corresponds to *frequency control* and varying T_0 corresponds to *voltage control*.

(c)

From (4.25), we note the following:

When $\omega_m = 0$, we have $T_m = T_0$. Hence, T_0=starting torque of the motor.

When $T_m = 0$, we have $\omega_m = \omega_0$. Hence, ω_0=no-load speed. This is the synchronous speed—Under no-load conditions, there is no slip in the induction motor (i.e., actual speed of the motor is equal to the speed ω_0 of the rotating magnetic field).

Differentiate (4.25) separately with respect to T_0, ω_0, and ω_m. We get

$$\frac{\partial T_m}{\partial T_0} = \frac{q\omega_0(\omega_0 - \omega_m)}{(q\omega_0^2 - \omega_m^2)} = \beta_1 \text{ (say)} \tag{4.27}$$

$$\frac{\partial T_m}{\partial \omega_0} = \frac{T_0 q\left[(q\omega_0^2 - \omega_m^2)(2\omega_0 - \omega_m) - \omega_0(\omega_0 - \omega_m)2q\omega_0\right]}{(q\omega_0^2 - \omega_m^2)^2}$$

$$= \frac{T_0 q\omega_m\left[(\omega_0 - \omega_m)^2 + (q-1)\omega_0^2\right]}{(q\omega_0^2 - \omega_m^2)^2} = \beta_2 \text{ (say)} \tag{4.28}$$

$$\frac{\partial T_m}{\partial \omega_m} = \frac{T_0 q\omega_0\left[\left(q\omega_0^2 - \omega_m^2\right)(-1) - (\omega_0 - \omega_m)(-2\omega_m)\right]}{\left(q\omega_0^2 - \omega_m^2\right)^2}$$

$$= -\frac{T_0 q\omega_0\left[(q-1)\omega_0^2 + (\omega_0 - \omega_m)^2\right]}{\left(q\omega_0^2 - \omega_m^2\right)^2} = -b_e \text{ (say)} \tag{4.29}$$

Since $q > 1$, we notice that $\beta_1 > 0$, $\beta_2 > 0$, and $b_e > 0$.

Note: b_e = electrical damping constant of the motor.

(d)

In a steady-state operating condition, the rates of changes of the state variables are zero. Hence, set $\dot{\omega}_m = 0 = \dot{T}_p = \dot{\omega}_p$ in equations (ii)–(iv). We get

$$0 = \bar{T}_m - b_m\bar{\omega}_m - \frac{r}{\eta}\bar{T}_p$$

$$0 = k_p\left(r\bar{\omega}_m - \bar{\omega}_p\right)$$

$$0 = \bar{T}_p - b_p\bar{\omega}_p$$

Hence,

$$\bar{\omega}_p = r\bar{\omega}_m \tag{v}$$

$$\bar{T}_p = b_p r\bar{\omega}_m \tag{vi}$$

$$\bar{T}_m = b_m\bar{\omega}_m + r^2 b_p\bar{\omega}_m/\eta = \frac{\bar{T}_0 q\bar{\omega}_0\left(\bar{\omega}_0 - \bar{\omega}_m\right)}{\left(q\bar{\omega}_0^2 - \bar{\omega}_m^2\right)} \quad \text{(from (4.25))}$$

or

$$\bar{T}_0 = \frac{\bar{\omega}_m\left(b_m + r^2 b_p/\eta\right)\left(q\bar{\omega}_0^2 - \bar{\omega}_m^2\right)}{q\bar{\omega}_0\left(\bar{\omega}_0 - \bar{\omega}_m\right)} \tag{vii}$$

Strictly, we need to express the operating values of the state variables: $[\bar{\omega}_m \quad \bar{T}_p \quad \bar{\omega}_p]^T$ in terms of the operating values (known) of the two inputs: $\left[\bar{T}_0 \quad \bar{\omega}_0\right]^T$. In theory, this can be done because we have three equations (v)–(vii) in the three unknowns. This is not straightforward, however, in view of the complex nature of the expression (vii). The specific procedure to accomplish objective is as follows:

1. Solve the "cubic equation" (vii) to express $\bar{\omega}_m$ in terms of the known quantities (steady-state inputs) $\bar{T}_0$ and $\bar{\omega}_0$. We should get at least one stable solution—see Figure 4.5. Pick the stable solution
2. Once we know $\bar{\omega}_m$, we can express $\bar{\omega}_p$ and $\bar{T}_p$ in terms of the known input quantities $\bar{T}_0$ and $\bar{\omega}_0$ (using (v) and (vi)).

Hint: The cubic equation in this case is of the form $ax^3 - bx + c = 0$ with $x = \bar{\omega}_m$. Its three roots are:

$$x_1 = \left[\left(\frac{c^2}{4a^2} - \frac{b^3}{27a^3}\right)^{\frac{1}{2}} - \frac{c}{2a}\right]^{\frac{1}{3}} + \frac{b}{3a\left[\left(\frac{c^2}{4a^2} - \frac{b^3}{27a^3}\right)^{\frac{1}{2}} - \frac{c}{2a}\right]^{\frac{1}{3}}}$$

$$x_2 = -\frac{1}{2}\left[\left(\frac{c^2}{4a^2} - \frac{b^3}{27a^3}\right)^{\frac{1}{2}} - \frac{c}{2a}\right]^{\frac{1}{3}} - \frac{b}{6a\left[\left(\frac{c^2}{4a^2} - \frac{b^3}{27a^3}\right)^{\frac{1}{2}} - \frac{c}{2a}\right]^{\frac{1}{3}}}$$

$$+\frac{j\sqrt{3}}{2}\left\{\left[\left(\frac{c^2}{4a^2} - \frac{b^3}{27a^3}\right)^{\frac{1}{2}} - \frac{c}{2a}\right]^{\frac{1}{3}} - \frac{b}{3a\left[\left(\frac{c^2}{4a^2} - \frac{b^3}{27a^3}\right)^{\frac{1}{2}} - \frac{c}{2a}\right]^{\frac{1}{3}}}\right\}$$

$$x_3 = -\frac{1}{2}\left[\left(\frac{c^2}{4a^2} - \frac{b^3}{27a^3}\right)^{\frac{1}{2}} - \frac{c}{2a}\right]^{\frac{1}{3}} - \frac{b}{6a\left[\left(\frac{c^2}{4a^2} - \frac{b^3}{27a^3}\right)^{\frac{1}{2}} - \frac{c}{2a}\right]^{\frac{1}{3}}}$$

$$-\frac{j\sqrt{3}}{2}\left\{\left[\left(\frac{c^2}{4a^2} - \frac{b^3}{27a^3}\right)^{\frac{1}{2}} - \frac{c}{2a}\right]^{\frac{1}{3}} - \frac{b}{3a\left[\left(\frac{c^2}{4a^2} - \frac{b^3}{27a^3}\right)^{\frac{1}{2}} - \frac{c}{2a}\right]^{\frac{1}{3}}}\right\}$$

It is seen that the first root is real and positive, and it is the only valid root for $\bar{\omega}_m$. The other two roots are complex conjugates and are not valid.

MATLAB® Code:

```
syms a b c x
eqn = a*x^3-b*x+c==0
solve(eqn, x)
```

(e)

The linearized nonlinear expression (the derivatives) for the motor torque has been determined before. Specifically, the increment of the motor torque from the operating point is:

$$\hat{T}_m = \frac{\partial T_m}{\partial \omega_m}\hat{\omega}_m + \left[\frac{\partial T_m}{\partial T_0}\right]\hat{T}_0 + \left[\frac{\partial T_m}{\partial \omega_0}\right]\hat{\omega}_0 = -b_e\hat{\omega}_m + \beta_1\hat{T}_0 + \beta_2\hat{\omega}_0 \tag{viii}$$

Take the increments of the state equations (ii)–(iv) and substitute (viii). We get the following linear state-space model:

$$J_m\dot{\hat{\omega}}_m = -(b_m + b_e)\hat{\omega}_m - \frac{r}{\eta}\hat{T}_p + \beta_1\hat{T}_0 + \beta_2\hat{\omega}_0 \tag{ix}$$

$$\dot{\hat{T}}_p = k_p\left(r\hat{\omega}_m - \hat{\omega}_p\right) \tag{x}$$

$$J_p\dot{\hat{\omega}}_p = \hat{T}_p - b_p\hat{\omega}_p \tag{xi}$$

State vector $\boldsymbol{x} = \begin{bmatrix} \hat{\omega}_m & \hat{T}_p & \hat{\omega}_p \end{bmatrix}^T$

Input vector $\boldsymbol{u} = \begin{bmatrix} \hat{T}_0 & \hat{\omega}_0 \end{bmatrix}^T$

Output vector $\boldsymbol{y} = \begin{bmatrix} \hat{\omega}_p & \hat{T}_p/k_p \end{bmatrix}^T$

Note: The second output is the angle of twist of the pump shaft. Its incremental value is $\hat{T}_p/k_p$.
We have the corresponding model matrices

$$\boldsymbol{A} = \begin{bmatrix} -(b_e + b_m)/J_m & -r/(\eta J_m) & 0 \\ k_p r & 0 & -k_p \\ 0 & 1/J_p & -b_p/J_p \end{bmatrix}; \quad \boldsymbol{B} = \begin{bmatrix} \beta_1/J_m & \beta_2/J_m \\ 0 & 0 \\ 0 & 0 \end{bmatrix};$$

$$\boldsymbol{C} = \begin{bmatrix} 0 & 0 & 1 \\ 0 & 1/k_p & 0 \end{bmatrix}; \quad \boldsymbol{D} = 0$$

Note:

b_e = electrical damping constant of the motor
b_m = mechanical damping constant of the motor

(f)

In frequency control, we have $\hat{T}_0 = 0$.

To get the linear I-O differential equation, we should eliminate the state variables $\hat{\omega}_m$ and $\hat{T}_p$ from the state equations (ix)–(xi). The general procedure for this is as follows:

1. Pick an equation where a variable to be eliminated occurs by itself, only in a single term
2. Substitute that equation into the remaining equations to eliminate the variable
3. Repeat the above two steps for the remaining variables that need to be eliminated

In the present example, we use this approach as follows:

1. Substitute (x) into (ix) in order to eliminate $\hat{\omega}_m$
2. Substitute (xi) into the above result in order to eliminate $\hat{T}_p$.

On simplification, we get the following input–output model (differential equation):

$$J_m J_p \frac{d^3\hat{\omega}_p}{dt^3} + \left[J_m b_p + J_p(b_m + b_e)\right]\frac{d^2\hat{\omega}_p}{dt^2} + \left[k_p\left(J_m + \frac{r^2 J_p}{\eta}\right) + b_p(b_m + b_e)\right]\frac{d\hat{\omega}_p}{dt} + k_p\left(\frac{r^2 b_p}{\eta} + b_m + b_e\right)\hat{\omega}_p$$
$$= \beta_2 r k_p \hat{\omega}_0 \qquad \text{(xii)}$$

This is a third-order differential equation, as expected, since the system is third order. Also, as we have seen, the state-space model is also third order.

Observation From (xii)

When $\hat{\omega}_0$ is changed by the "finite" step $\Delta\hat{\omega}_0$, the RHS of (xii) will change by a finite amount. Hence, the LHS also must change by a finite amount.

In this process, suppose that the lowest order term $\hat{\omega}_p$ instantaneously changes by a finite amount. That means, the higher-order terms (higher derivatives) $\frac{d\hat{\omega}_p}{dt}$ and $\frac{d^2\hat{\omega}_p}{dt^2}$ have to change by "infinite" amounts, instantaneously (*Note*: The derivative of a step is an impulse—infinite). Then the LHS will change by an "infinite" amount, which violates the equation.

Hence, only the highest derivative ($\frac{d^3\hat{\omega}_p}{dt^3}$) will change instantaneously. The lower derivatives will not change instantaneously.

Further Verification: In fact $\hat{\omega}_p$, $\frac{d\hat{\omega}_p}{dt}$, and $\frac{d^2\hat{\omega}_p}{dt^2}$ constitute another choice of state variables, for this system (not our preferred choice, however). We know that state variables cannot change instantaneously (see Chapter 2) as that will violate causality.

For the instantaneous change, set all the lower order terms on the LHS of (xii) to zero. Then, we can determine the change of the highest order term $\frac{d^3\hat{\omega}_p}{dt^3}$ as: $\frac{\beta_2 r k_p}{J_m J_p}\Delta\hat{\omega}_0$

The following somewhat general observations can be made from this example:

1. Mechanical damping constant b_m comes from bearing friction and other mechanical sources
2. Electrical damping constant b_e comes from the electromagnetic interactions in the motor
3. The two damping parameters occur together (and should be treated together, in analysis, simulation, design, control, etc.). For example, whether the response is underdamped or overdamped depends on the sum and not the individual damping parameters. This is a consequence of electromechanical coupling.

Note: If the characteristic curve corresponding to (4.25) is experimentally determined, the damping parameter "*b*" that is determined from the curve will contain mechanical damping (e.g., in bearings) as well since the torque is measured outside the bearings.

LEARNING OBJECTIVES

1. The effect of electromechanical coupling on damping
2. Mechanical damping and electrical damping should be treated together. System behavior depends on the combined effect. This is a case for integrated analysis, simulation, design, control, etc., as suggested before.
3. System nonlinearity can come from nonlinear coupling of inputs and state variables
4. A state variable cannot change instantaneously
5. When an input changes instantaneously, the highest derivative of the output in the I-O differential equation changes with it (instantaneously); and the lower derivatives of the output do not change.

Note: These lower derivatives may be taken as state variables of the system, in an alternative choice.

Example 4.3

Radiation heat transfer may be expressed by the nonlinear relationship $Q = K\left(T_1^4 - T_2^4\right)$

Here,

Q = heat transfer rate (watts)
T_1 = heat source temperature (°K)
T_2 = heat receiver temperature (°K)
K = a system constant

Consider the reference condition Q_0, T_{10}, and T_{20}, which satisfies $Q_0 = K\left(T_{10}^4 - T_{20}^4\right)$
Change in heat transfer rate from the reference value is given by

$$\delta Q = Q - Q_0 = K\left(T_1^4 - T_2^4\right) - K\left(T_{10}^4 - T_{20}^4\right)$$

Non-dimensional "change" in the heat transfer rate:

$$q = \frac{\delta Q}{Q_0} = \frac{K\left(T_1^4 - T_2^4\right) - K\left(T_{10}^4 - T_{20}^4\right)}{K\left(T_{10}^4 - T_{20}^4\right)} = \frac{T_1^4 - T_2^4}{T_{10}^4 - T_{20}^4} - 1 \tag{4.30}$$

Several approaches may be used to express the change δQ of the heat transfer rate from the reference value (Q_0) as the temperature changes by δT from the reference condition. Specifically consider the following *four* approaches:

(a)

i. (**Approach 1, Source** temperature changes): Let $\delta T = T_1 - T_{10};\ T_2 = T_{20};$ define the non-dimensional temperature change (of source), $r = \dfrac{\delta T}{T_{10} - T_{20}}$.
   Problem: Study the variation of q wrt r near T_{10} and T_{20}.

ii. (**Approach 2, Receiver** temperature changes): Let $\delta T = T_{20} - T_2;\ T_1 = T_{10};$ define the non-dimensional temperature change (of receiver), $r = \dfrac{\delta T}{T_{10} - T_{20}}$.
   Problem: Study the variation of q wrt r near T_{10} and T_{20}

(b)

i. (**Approach 3, Source** temperature changes): Local slope $\dfrac{\partial Q}{\partial T_1} = 4KT_1^3$ for source temperature variation $\delta T = T_1 - T_{10};\ \delta Q = \dfrac{\partial Q}{\partial T}\delta T \rightarrow$

$$\frac{\delta Q}{Q_0} \cong \frac{4KT_{10}^3}{K(T_{10}^4 - T_{20}^4)} \cdot \delta T = \frac{4T_{10}^3(T_{10} - T_{20})}{(T_{10}^4 - T_{20}^4)} \cdot \frac{\delta T}{T_{10} - T_{20}} = \frac{4T_{10}^3}{(T_{10} + T_{20})(T_{10}^2 + T_{20}^2)} \cdot \frac{\delta T}{(T_{10} - T_{20})} \rightarrow$$

$$q \cong \frac{4T_{10}^3}{(T_{10} + T_{20})(T_{10}^2 + T_{20}^2)} r$$

Problem: Study the variation of q wrt r near T_{10} and T_{20}

ii. **(Approach 4, Receiver** temperature changes**):** Local slope $\dfrac{\partial Q}{\partial T_2} = -4KT_2^3$ for receiver temperature variation $\delta T = T_{20} - T_2;$ near reference: $\dfrac{\delta Q}{Q_0} \cong \dfrac{4K \cdot T_{20}^3}{K\left(T_{10}^4 - T_{20}^4\right)} \cdot \delta T \rightarrow$

$$q \cong \frac{4T_{20}^3}{(T_{10} + T_{20})(T_{10}^2 + T_{20}^2)} r$$

Problem: Study the variation of q wrt r near T_{10} and T_{20}

(c)
Compare these four approaches for thermal resistance (near the reference condition).

Solution

(a)

$$q = \frac{\delta Q}{Q_0} = \frac{T_1^4 - T_2^4}{T_{10}^4 - T_{20}^4} - 1, \quad r = \frac{\delta T}{T_{10} - T_{20}}$$

Case (i):

$$\delta T = T_1 - T_{10}; \quad T_2 = T_{20}$$

$$\Rightarrow q = \frac{\left(T_{10} + \delta T\right)^4 - T_{20}^4}{\left(T_{10}^4 - T_{20}^4\right)} - 1 = \frac{\left[T_{10} + \left(T_{10} - T_{20}\right)r\right]^4 - T_{20}^4}{\left(T_{10}^4 - T_{20}^4\right)} - 1$$

$$= \frac{\left[T_{10}{}^4 + 4T_{10}^3\left(T_{10} - T_{20}\right)r + 6T_{10}^2\left(T_{10} - T_{20}\right)^2 r^2 + 4T_{10}\left(T_{10} - T_{20}\right)^3 r^3 + \left(T_{10} - T_{20}\right)^4 r^4 - T_{20}^4\right]}{\left(T_{10}^4 - T_{20}^4\right)}$$

$$\Rightarrow q = Ar + Br^2 + Cr^3 + Dr^4 \tag{4.31}$$

with $A=\frac{4T_{10}^3(T_{10}-T_{20})}{(T_{10}^4-T_{20}^4)}; B=\frac{6T_{10}^2(T_{10}-T_{20})^2}{(T_{10}^4-T_{20}^4)}; C=\frac{4T_{10}(T_{10}-T_{20})^3}{(T_{10}^4-T_{20}^4)}; D=\frac{(T_{10}-T_{20})^4}{(T_{10}^4-T_{20}^4)}$

Note: $A>0, B>0, C>0, D>0$

Reference point: $q=0, r=0$

Non-dimensional thermal resistance $=\frac{1}{dq/dr}=\frac{1}{A}$ at the reference point.

Thermal resistance $R_T=\frac{1}{dQ/dT}$

$$=\frac{(T_{10}-T_{20})}{Q_0}\frac{1}{dq/dr}=\frac{(T_{10}-T_{20})}{Q_0 A}=\frac{(T_{10}-T_{20})\times(T_{10}^4-T_{20}^4)}{K(T_{10}^4-T_{20}^4)4T_{10}^3(T_{10}-T_{20})}=\frac{1}{4KT_{10}^3}$$

at the reference point.

Case (ii):

$\delta T=T_{21}-T_2;\ T_1=T_{10}$

$$\Rightarrow q=\frac{T_{10}^4-(T_{20}-\delta T)^4}{(T_{10}^4-T_{20}^4)}-1=\frac{T_{10}^4-[T_{20}-(T_{10}-T_{20})r]^4}{(T_{10}^4-T_{20}^4)}-1$$

$$=\frac{\{T_{20}^4-[T_{10}{}^4-4T_{20}^3(T_{10}-T_{20})r+6T_{20}^2(T_{10}-T_{20})^2r^2-4T_{20}(T_{10}-T_{20})^3r^3+(T_{10}-T_{20})^4r^4]\}}{(T_{10}^4-T_{20}^4)}-1$$

$$\Rightarrow q=A'r-B'r^2+C'r^3-D'r^4 \tag{4.32}$$

with

$$A'=\frac{4T_{20}^3(T_{10}-T_{20})}{(T_{10}^4-T_{20}^4)}=\left(\frac{T_{20}}{T_{10}}\right)^3A;\quad B'=\frac{6T_{20}^2(T_{10}-T_{20})^2}{(T_{10}^4-T_{20}^4)}=\left(\frac{T_{20}}{T_{10}}\right)^2B;$$

$$C'=\frac{4T_{20}(T_{10}-T_{20})^3}{(T_{10}^4-T_{20}^4)}=\left(\frac{T_{20}}{T_{10}}\right)C;\quad D'=\frac{(T_{10}-T_{20})^4}{(T_{10}^4-T_{20}^4)}=D$$

Note: $A'>0, B'>0, C'>0, D'>0$

As in Case (i),

Non-dimensional thermal resistance $=\left.\frac{1}{dq/dr}\right|_{ref}=\frac{1}{A'}$

Thermal resistance $R_T=\left.\frac{1}{dQ/dT}\right|_{ref}$

$$=\frac{(T_{10}-T_{20})}{Q_0}\left.\frac{1}{dq/dr}\right|_{ref}=\frac{(T_{10}-T_{20})}{Q_0 A'}=\frac{(T_{10}-T_{20})\times(T_{10}^4-T_{20}^4)}{K(T_{10}^4-T_{20}^4)4T_{20}^3(T_{10}-T_{20})}=\frac{1}{4KT_{20}^3}$$

(b)

Case (i):

From the result in (a)(i), the linearized relation is

$$q=Ar \tag{4.33}$$

TABLE 4.1
Expressions for Thermal Resistance

Case	Thermal Resistance
(a)(i)	$\frac{1}{4KT_{10}^3}$
(a)(ii)	$\frac{1}{4KT_{20}^3}$
(b)(i)	$\frac{1}{4KT_{10}^3}$
(b)(ii)	$\frac{1}{4KT_{20}^3}$

with $A = \frac{4T_{10}^3(T_{10} - T_{20})}{(T_{10}^4 - T_{20}^4)} = \frac{4T_{10}^3}{(T_{10} + T_{20})(T_{10}^2 + T_{20}^2)}$

Non-dimensional thermal resistance $= \frac{1}{A}$

Thermal resistance $R_T = \frac{1}{4KT_{10}^3}$

Case (ii):
From the result in (a)(ii), the linearized relation is

$$q = A'r \tag{4.34}$$

with $A' = \frac{4T_{20}^3(T_{10} - T_{20})}{(T_{10}^4 - T_{20}^4)} = \frac{4T_{20}^3}{(T_{10} + T_{20})(T_{10}^2 + T_{20}^2)}$

Non-dimensional thermal resistance $= \frac{1}{A'}$

Thermal resistance $R_T = \frac{1}{4KT_{20}^3}$

(c)
For the four approaches, the thermal resistances at the reference point (operating point) are given in Table 4.1.

The curves of q versus r (non-dimensional) are plotted in Figure 4.6 for the operating condition $T_{10} = 310\,°\text{K}, T_{20} = 290\,°\text{K}$.

Comparison:

1. Cases (a)(i) and (ii) are general and are not linearized relations. The corresponding thermal resistances are determined by the "inverse" of the slope at the reference point $q = 0, r = 0$
2. Cases (b)(i) and (ii) are the linearized versions of (a)(i) and (ii), respectively. Hence, the corresponding thermal resistances are the same as those for Part (a), at the reference point. *Note*: Thermal resistance is defined locally (linear)
3. The larger the T_{10}, the smaller the thermal resistance, because for a given T_{20}, a larger T_{10} produces a greater heat transfer rate

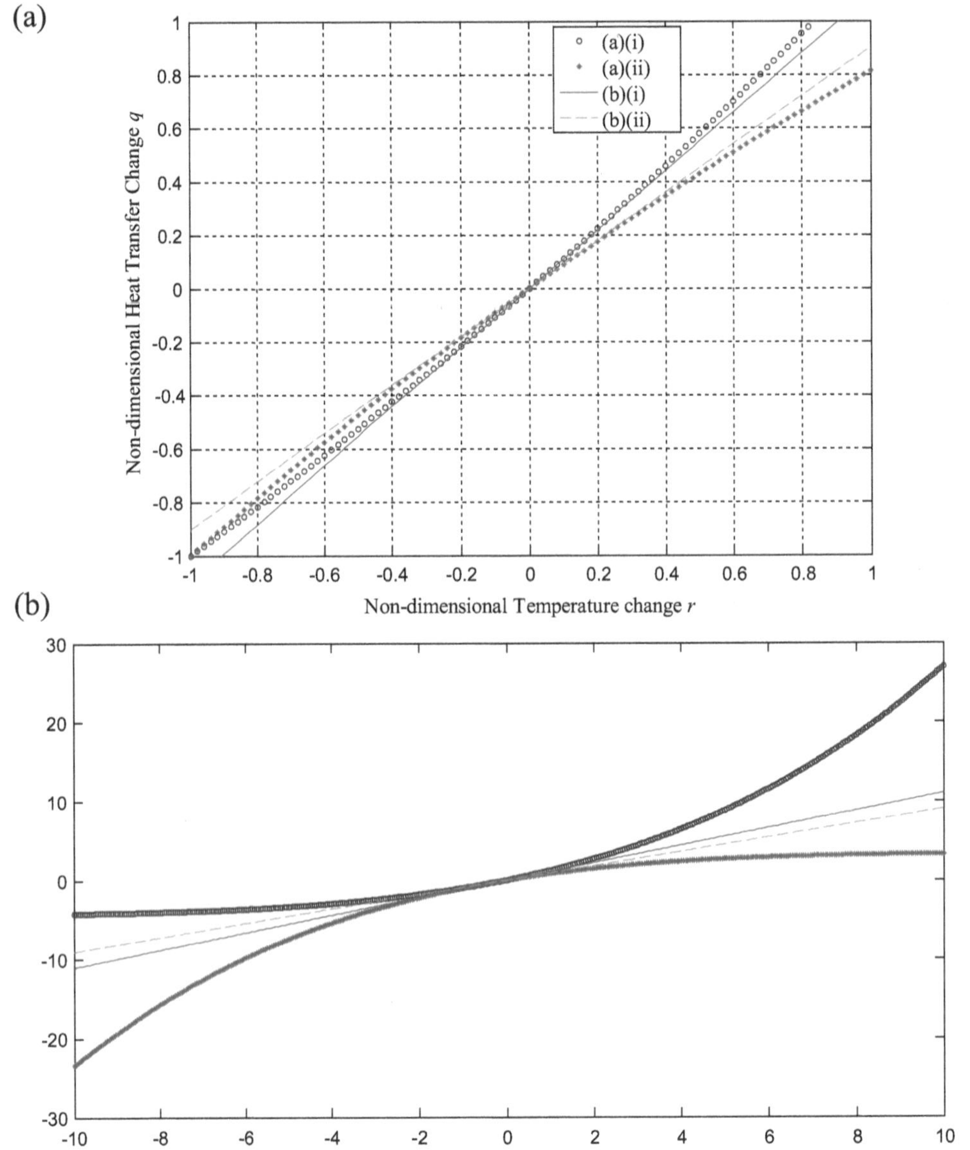

FIGURE 4.6 Non-dimensional curves of q versus r for the four cases: (a) Local details; (b) global view.

4. The larger the T_{20}, the larger the thermal resistance, because for a given T_{10}, a larger T_{20} produces a smaller heat transfer rate.

LEARNING OBJECTIVES

1. Modeling of a nonlinear thermal system
2. Characteristics of radiation heat transfer
3. Linearization of radiation heat transfer
4. The local concept of thermal resistance.

4.5 LINEARIZATION USING EXPERIMENTAL DATA

Linearization of nonlinear analytical models was studied in the previous sections. In some situations, an accurate analytical model may not be readily available for an existing physical system. This could be due to reasons such as: The physics of some processes of the system is not clearly known; the system is highly complex for an analytical formulation; and the limited skillset of the project personnel. Furthermore, as noted before, slope-based local linearization may not be possible with some analytical models (e.g., the local slope may be zero, infinite, or indeterminate). If a prototype of the system is available, another option for developing a model is through the use of experimental data. This is called *experimental modeling* (or in the language of control engineers, *model identification* or *system identification*) and may be achieved in several ways including:

1. Model identification of the entire dynamic system using input–output test data. This is what is commonly known as system identification. It can be a somewhat complicated process, particularly for complex and high order systems, because the data must be fitted to all important dynamic characteristics of the system
2. Testing only a key part or component of the system, which may be nonlinear, under *steady-state* conditions. Since steady-state test data are used, even if the entire system is tested, some characteristics of the dynamics (e.g., inertia forces/torques) will not be present in the data. Hence, some dynamics (e.g., inertia) of the system have to be integrated into the identified model separately (and analytically).

It is the second approach (which is a special case of the first approach) that we particularly focus on in the present section. In this approach, experiments are conducted on the specific part of the system at steady state, to determine the corresponding *operating curves*. These operating curves, which typically are nonlinear, are used in deriving a model. This approach is discussed now, to derive a "linear" model first taking an electric motor as the example system and then extending it to an example in the thermo-fluid domain. The approach is not limited to a particular physical domain, however. The approach is as follows:

1. Linearization is done by determining the local slopes (derivatives with respect to the independent variables) of the experimental characteristic curves
2. For reasons of experimental feasibility, the curves are determined by varying one independent variable at a time (keeping the other independent variables constant at some operating condition)
3. On a two-axis coordinate frame, only two such independent variables can be represented
4. Typically, the experiments are conducted at steady state. Dynamics (particularly, inertial dynamics) is incorporated separately into the model, using the corresponding analytical terms.

4.5.1 Torque–Speed Curves of Motors

The speed versus torque curves of motors under steady conditions (i.e., steady-state operating curves) are available from the motor manufacturers (data sheets). These curves have the characteristic shape that they decrease slowly up to a point and then drop rapidly to zero. One simple reason for this characteristic is the power supply constraint, which requires a drop in speed in order to achieve a greater torque. We have already discussed an example that uses the nonlinear characteristic curve of an ac induction motor (see Figure 4.5). The operating curves of dc motors take a similar, but not identical, characteristic form. Figure 4.7 presents three characteristic curves of a dc motor that has both stator windings (field windings) and rotor windings (armature windings). In order to be powered by a single input (dc voltage), the two sets of windings have to be connected together. Commonly, they are connected

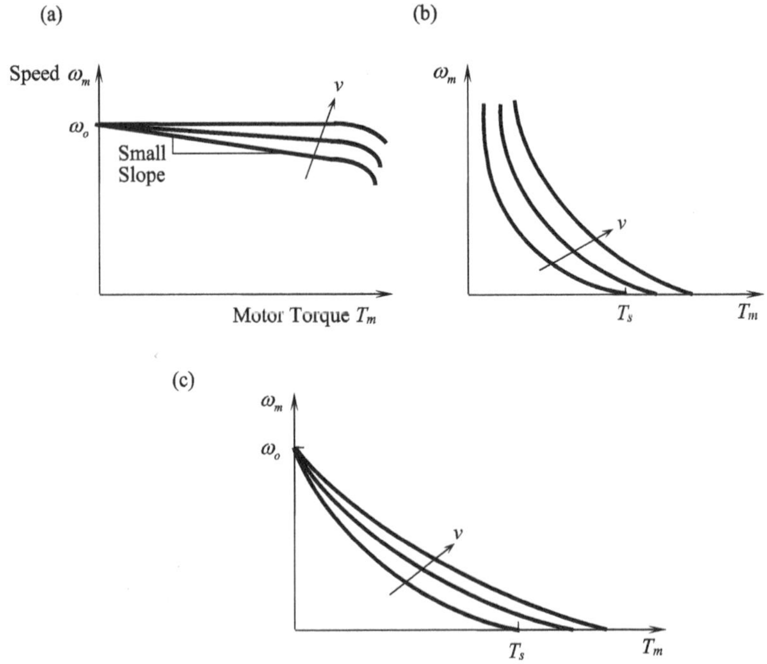

FIGURE 4.7 Torque–speed operating curves of dc motors. (a) Shunt-wound; (b) series-wound; (c) compound-wound.

1. In parallel (i.e., shunt-wound)
2. In series (i.e., series-wound)
3. With part of the stator windings connected in series and the remaining part connected in parallel with the rotor windings (i.e., compound-wound).

It is seen from Figure 4.7 that the interconnection arrangement between the stator windings and the rotor windings determines the shape of the characteristic curves.

The torque at zero speed is the *braking torque* or *starting torque* or *stalling torque*. The speed at zero torque is the *no-load speed*, which, for an ac induction motor, is also the *synchronous speed* (when the motor speed equals the speed of the rotating magnetic field, giving the condition of *no slip*).

Typically, the measurement of the characteristic curves of a dc motor with a common input voltage for exciting both stator and rotor is done as follows:

1. Keep the supply voltage to the motor windings at a known constant value
2. Apply a constant known load (torque) to the motor shaft
3. Once the conditions of the motor become steady (i.e., reaches a constant speed) measure the motor speed
4. Repeat steps 2 and 3 for increments of torques within an appropriate practical range. This gives one operating curve, for a specified supply voltage
5. Go to Step 1, change the supply voltage by a suitable increment, to another constant value and repeat steps 2 through 4, to get another operating curve
6. Repeat steps 1 through 5 to get the required number of operating curves.

It should be noted that the motor speed is maintained steady in these experiments as they are carried out in "steady" operating conditions. That means the motor inertia (inertial torque) is not present

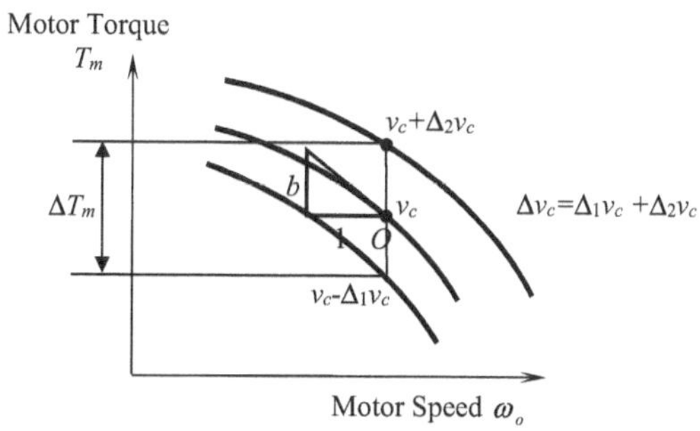

FIGURE 4.8 Two steady-state operating curves of a motor at constant input voltage.

when obtaining these curves, while the mechanical damping is. Hence, motor inertia has to be introduced separately when using these curves to determine a "dynamic" model for a motor. Since mechanical damping is included in the measurements, because the torque is measured in a shaft segment outside the motor bearings, it should not be added again in the model again. Also, electrical damping is inherent in the operating curves. Of course, if the motor is connected to an external load, the damping, inertia, and flexibility of the load all have to be included separately, by a suitable analytical model, when using experimental operating curves of motors in developing models for motor-integrated dynamic systems.

Note: Some characteristics of the load also may be experimentally determined as its operating curves and then converted into an analytical model (as for the load).

4.5.2 Experimental Linear Model of a Motor

Consider an experimental set of steady-state operating curves for a motor, each obtained at a constant supply/control voltage. In particular, one curve (primary curve), which passes through the normal operating point O, is measured at voltage v_c. Two other curves adjacent to (enclosing) this curve are measured at voltages $v_c - \Delta_1 v_c$ and $v_c + \Delta_2 v_c$, where $\Delta v_c = \Delta_1 v_c + \Delta_2 v_c$ is the voltage increment between these adjacent curves at O and ΔT_m is the corresponding torque increment, as shown in Figure 4.8.

Draw a tangent to the primary curve at the normal operating point O and determine its slope. This slope is negative, as shown. Its magnitude b is given by

$$\text{Damping constant } b = -\left.\frac{\partial T_m}{\partial \omega_m}\right|_{v_c=\text{constant}} = \text{slope magnitude at } O \tag{4.35}$$

The parameter b represents the equivalent rotary damping constant (torque/angular speed) in the operating condition of the motor and includes both electromagnetic and mechanical damping effects in the motor since typically the motor torque is measured from a motor shaft segment outside the motor bearings. The included mechanical damping comes primarily from the friction of the motor bearings and aerodynamic effects. Since a specific load is not considered in the operating curve, the load damping is not included in b.

Draw a vertical line through the operating point O to intersect the two adjacent operating curves whose voltage difference is Δv_c. From that, we get:

$$\Delta T_m = \text{torque intercept between the two operating curves,}$$

$$\text{Voltage gain } k_v = \left.\frac{\partial T_m}{\partial v_c}\right|_{\omega_m=\text{ constant}} \simeq \frac{\Delta T_m}{\Delta v_c} \tag{4.36}$$

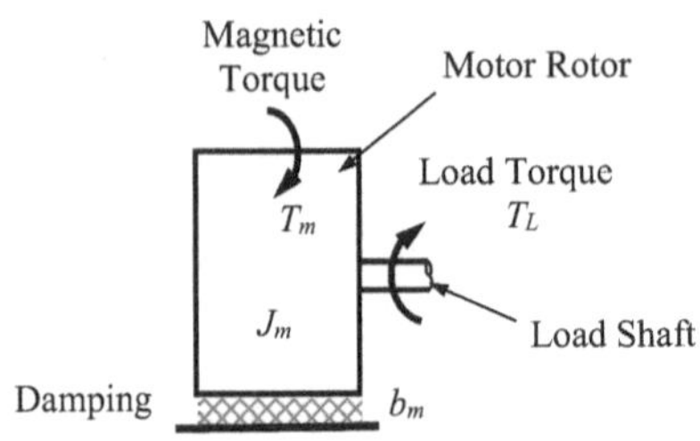

FIGURE 4.9 Mechanical system of the motor.

Note: A vertical line is a constant-speed line.

Since the motor torque T_m is a function of both motor speed ω_m and the input voltage v_c (i.e., $T_m = T_m(\omega_m, v_c)$), we have from basic calculus:

$$\delta T_m = \left.\frac{\partial T_m}{\partial \omega_m}\right|_{v_c} \delta\omega_m + \left.\frac{\partial T_m}{\partial v_c}\right|_{\omega_m} \delta v_c \tag{4.37a}$$

or

$$\delta T_m = -b\delta\omega_m + k_v\delta v_c \tag{4.37b}$$

where the motor damping constant b and the voltage gain k_v are given by equations (4.35) and (4.36), respectively.

Equation (4.37) represents a linearized model of the motor. The torque that is needed to accelerate/decelerate the motor rotor inertia is not included in this equation (because steady-state curves are used in determining the model parameters). The motor inertia term, as given by the Newton's second law, should be explicitly included in the linearized (incremental) model of the motor rotor (see Figure 4.9):

$$J_m \frac{d\delta\omega_m}{dt} = \delta T_m - \delta T_L \tag{4.38}$$

where J_m = moment of inertia of the motor rotor and T_L = load torque (equivalent torque applied on the motor by the load that is driven by the motor).

Note: Mechanical damping (with damping constant b_m) of the motor and its bearings, as shown in Figure 4.9, is not explicitly included in equation (4.38) because it (along with electromagnetic damping of the motor) is already included in equations (4.35) and (4.37), as indicated before.

4.5.3 Experimental Linear Model of a Nonlinear System

The method of obtaining the experimental model of a motor and its linearization are presented in the previous section. This approach can be extended for any nonlinear dynamic system. Experimentally obtaining the steady operating curves of the system and determining a linear model of the system from them are outlined now. Assuming that only two independent variables are present in the characteristic function, the relevant steps are as follows:

1. Keep one independent variable constant at some operating condition, vary the other independent variable in steps, and successively measure the system response
2. Plot the steady-state characteristic curve of the system using the measured data
3. Repeat steps 1 and 2 by changing the operating condition of the first independent variable by a small increment. Obtain a series of characteristic curves in this manner

4. Linearization: For a specific operating condition, determine the slope of the curve that passes through it (in which the first variable is kept constant); also find the vertical increment of two adjacent characteristic curves that enclose the operating point (the second variable is constant along a vertical line)
5. Slopes with respect to the two independent variables are obtained from the two values determined in step 4.

Once the local slopes of the nonlinear characteristic function with respect to the two independent variables are determined at the operating condition in this manner, they can be easily incorporated into a linear model (as done in the motor example). An example is given now.

Example 4.4

A gas turbine drives a centrifugal pump through a long shaft, as shown in Figure 4.10a. The variables and parameters of the system are identified in the figure.

For Gas Turbine:
Fuel input rate=Q (gal/s)
Moment of inertia of the rotor=$J_t = 0.1\ \text{kg}\cdot\text{m}^2$
Mechanical damping constant (at bearings, etc.) = $b_t = 0.5\ \text{N}\cdot\text{m/rad/s}$

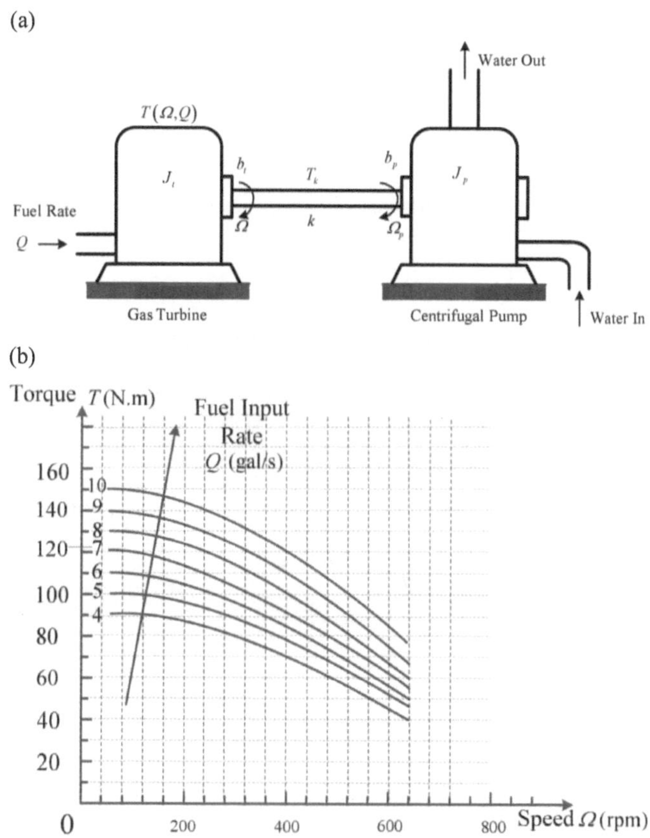

FIGURE 4.10 (a) Gas turbine operating a water pump; (b) steady-state operating curves of the gas turbine.

For Shaft:
Torsional stiffness$=k = 20.0\ \text{N}\cdot\text{m}/\text{rad}$
Torque in the shaft$=T_k$

For Centrifugal Pump:
Moment of inertia of the rotor (may include the "added mass" due to the fluid load)

$$= J_p = 0.05\ \text{kg}\cdot\text{m}^2$$

Mechanical damping constant (linear viscous) of pump (may partly include the fluid load)

$$= b_t = 3.0\ \text{N}\cdot\text{m/rad/s}$$

The torque versus speed curves of the turbine, at steady state, for different values of the fuel input rate are shown in Figure 4.10b.

Note: Torque of the turbine is measured on the turbine shaft prior to the bearings (unlike in the motor example). Speed Ω is given in rpm, not rad/s.

The system output is the pump speed.

a. Determine a complete state-space model for the system. $T(\Omega,Q)$ is the generated torque of the turbine; Ω, T_k and Ω_p are the state variables; Q is the input variable; and Ω_p is the output. *Note*: This model is nonlinear because of the term $T(\Omega,Q)$
b. Linearize the model about an operating point, for incremental variables of the states and the input, given by,

$$\delta\Omega = \omega,\ \delta T_k = \tau_k,\ \delta\Omega_p = \omega_p, \text{ and } \delta Q = q$$

Note: First formulate the linear model using given symbols for the model parameters. Next substitute the numerical values that are given and those extracted from the experimental curves. The operating point is given by $Q = 8$ gal/s and $\Omega = 400$.

Solution

(a)
Consider the free-body diagrams of the turbine rotor, shaft, and the pump rotor, as shown in Figure 4.11a.

Constitutive Equations:

$$\text{Turbine Rotor}: J_t\dot{\Omega} = T(\Omega,Q) - b_t\Omega - T_m \tag{i*}$$

$$\text{Shaft}: \dot{T}_k = k \tag{ii}$$

$$\text{Pump Rotor}: J_p\dot{\Omega}_p = T_p - b_p\Omega_p \tag{iii*}$$

Node equations at the motor–shaft interface and the shaft–pump interface:

$$T_m - T_k = 0 \tag{iv}$$

$$T_k - T_p = 0 \tag{v}$$

Substitute (iv) into (i*):

$$J_t\dot{\Omega} = T(\Omega,Q) - b_t\Omega - T_k \tag{i}$$

Substitute (v) into (iii*):

$$J_p\dot{\Omega}_p = T_k - b_p\Omega_p \tag{iii}$$

State Equations: (i) through (iii), with
State vector $\mathrm{X} = \left[\Omega, T_k, \Omega_p\right]^T$

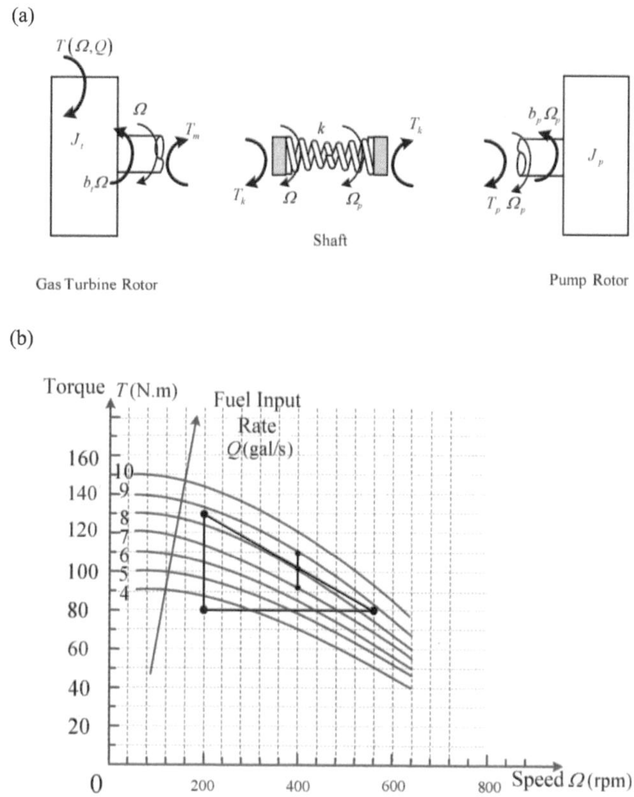

FIGURE 4.11 (a) Free-body diagrams; (b) linear parameter estimation using the experimental curves.

Input vector $\boldsymbol{U} = [Q]$
Output vector $Y = \left[\Omega_p\right]$

(b)

Linearize the nonlinear model (i)–(iii) by taking increments:

$$J_t\delta\dot{\Omega} = \delta T - b_t\delta\Omega - \delta T_k$$

$$\delta\dot{T}_k = k\left(\delta\Omega - \delta\Omega_p\right)$$

$$J_p\delta\dot{\Omega}_p = \delta T_k - b_p\delta\Omega_p$$

From calculus, $\delta T = \left.\frac{\partial T}{\partial \Omega}\right|_0 \delta\Omega + \left.\frac{\partial T}{\partial Q}\right|_0 \delta Q = -b\delta\Omega + k_q\delta Q == -b\omega + k_q q$

where

$$b = -\left.\frac{\partial T}{\partial \Omega}\right|_o \tag{vi}$$

and

$$k_q = \left.\frac{\partial T}{\partial Q}\right|_o \tag{vii}$$

Substitute the given notation. We get the linear state equations:

$$J_t\dot{\omega} = -b\omega + k_q q - b_t\omega - \tau_k$$
$$\dot{\tau}_k = k(\omega - \omega_p) \qquad \text{(viii)}$$
$$J_p\dot{\omega}_p = \tau_k - b_p\omega_p$$

where the incremental state vector and the incremental input vector are

$$\mathbf{x} = \begin{bmatrix} \omega & \tau_k & \omega_p \end{bmatrix}^T; \mathbf{u} = [q]$$

In the vector-matrix form, we have

$$\dot{\mathbf{x}} = \mathbf{A}\mathbf{x} + \mathbf{B}\mathbf{u}$$
$$\mathbf{y} = \mathbf{C}\mathbf{x} + \mathbf{D}\mathbf{u}$$

$$\text{with, } \mathbf{A} = \begin{bmatrix} -(b+b_t)/J_t & -1/J_t & 0 \\ k & 0 & -k \\ 0 & 1/J_p & -b_p/J_p \end{bmatrix}; \quad \mathbf{B} = \begin{bmatrix} k_q/J_t \\ 0 \\ 0 \end{bmatrix}; \quad \mathbf{C} = [\,0 \quad 0 \quad 1\,]; \quad \mathbf{D} = [0]$$

Now we estimate b and k_q (according to (vi) and (vii)) from the given operating curves, as shown in Figure 4.11b:

b = slope at the operating point on the $Q = 8$ gal/s curve

$$= \frac{(130-80)(\text{N}\cdot\text{m})}{(560-200)\times\frac{2\pi}{60}(\text{rad/s})} = \frac{50}{360\times\frac{2\pi}{60}} = 1.326\ \text{N}\cdot\text{m/rad/s}$$

k_q = increment of T over the change in Q from 7 to 9 gal/s with Ω kept constant at 400 rpm $= \dfrac{(110-92)(\text{N}\cdot\text{m})}{(9-7)(\text{gal/s})} = 9.0\ \text{N}\cdot\text{m/gal/s}$

Substitute the given parameter values:

$$\frac{b+b_t}{J_t} = \frac{1.326+0.5}{0.1} = 18.26\ \text{s}^{-1}; \quad \frac{1}{J_t} = 10.0\ (\text{kg}\cdot\text{m}^2)^{-1}; \quad \frac{1}{J_p} = \frac{1}{0.05} = 20.0\ (\text{kg}\cdot\text{m}^2)^{-1}$$

$$\frac{b_p}{J_p} = \frac{3.0}{0.05} = 60.0\ \text{s}^{-1}; \quad \frac{k_q}{J_t} = \frac{9.0\ (\text{N}\cdot\text{m/gal/s})}{0.1\ (\text{kg}\cdot\text{m}^2)} = 90.0\ \text{gal/s}^3$$

$$\text{We have } \mathbf{A} = \begin{bmatrix} -18.26 & -10.0 & 0 \\ 20.0 & 0 & -20.0 \\ 0 & 20.0 & -60.0 \end{bmatrix}; \quad \mathbf{B} = \begin{bmatrix} 90.0 \\ 0 \\ 0 \end{bmatrix}; \quad \mathbf{C} = [\,0 \quad 0 \quad 1\,]; \quad \mathbf{D} = [0]$$

LEARNING OBJECTIVES

1. Use of test data in model development of a dynamic system
2. Modeling of a multi-physics (thermo-fluid and mechanical) system
3. Model linearization using test data
4. Integration of analytical and experimental models
5. Incorporation of system dynamics using steady-state characteristics.

4.6 OTHER METHODS OF MODEL LINEARIZATION

As we have studied, popular methods of linearization of a nonlinear device employ the local behavior of the device over a small operating range. This local linearization is straightforward but is not generally applicable. In particular, local slope-based linearization (whether analytical or experimental) of a dynamic system may fail (or may not be feasible) due to such reasons as the following:

1. The operating conditions can change considerably and a single local slope may not be valid over the entire range
2. Multiple local slopes may exist at the same operating condition (e.g., switching from one slope value to another at the same value of the independent variable)
3. The local slope may not exist, may be infinite, or may be insignificant (practically, zero) compared to the O(2) terms of the Taylor series expansion (e.g., Coulomb friction)
4. In some nonlinear systems, the use of local slopes (e.g., negative damping in a control law) may lead to undesirable consequences (e.g., instability).

In such situations, other problem-specific or ad hoc approaches may have to be used. For example, several different linear models may be used over the operating range of the system.

On the other hand, methods are available to reduce or eliminate the nonlinear behavior in devices. They include the use of *calibration curves* (in the static case), *linearizing elements* (e.g., resistors and amplifiers in bridge circuits) to neutralize the nonlinear effects, and nonlinear feedback (*feedback linearization*) to eliminate the nonlinear characteristics in the system. Also, methods other than the local slope-based methods are available for developing linear models. They include the use of an equivalent linear model based on some criterion of equivalence such as energy per operating cycle and the describing function method (which extends the transfer function method to nonlinear systems and is based on the equivalence of the fundamental frequency response). Several of these methods are outlined now.

4.6.1 Calibration Curve Method

A static nonlinearity of a device corresponds to a nonlinear algebraic (rather than differential equation) relationship between the input and the output of the device. Typically, this "static nonlinearity" (rather than "dynamic nonlinearity") occurs under steady-state conditions. A significant consequence of a static nonlinearity is the algebraic distortion of the output. This can be removed (linearized) by *recalibration* or *rescaling*, without introducing any error, at least in principle. This approach is explained now using an example.

Example 4.5

Suppose that the steady-state input–output behavior of a device is given by $y = ke^{pu}$, where u is the input and y is the output. Consider a sinusoidal input $u = u_0 \sin \omega t$. The corresponding output is $y = ke^{pu_0 \sin \omega t}$, which is far from sinusoidal.

Note: In a linear system, the steady-state output also will be sinusoidal at the same frequency (ω).

Now let us "transform" the problem as $\log(y) = pu + \log(k)$. In this manner, the input–output relationship is "exactly" linearized by simply using a log scale for the output and also adding a constant offset of $-\log(k)$. In this recalibrated form, the output for a sinusoidal input will be purely sinusoidal at the same frequency (ω) with the amplitude magnified by p. The phase angle does not change. No error has been introduced by this approach.

To illustrate this behavior numerically, use the parameter values: k=2.0, p=1.5, u_0=2.0, and ω=1.0. We use the following MATLAB function to determine the input–output behavior of this numerical example (Figure 4.12a) and the corresponding two signals (Figure 4.12b):

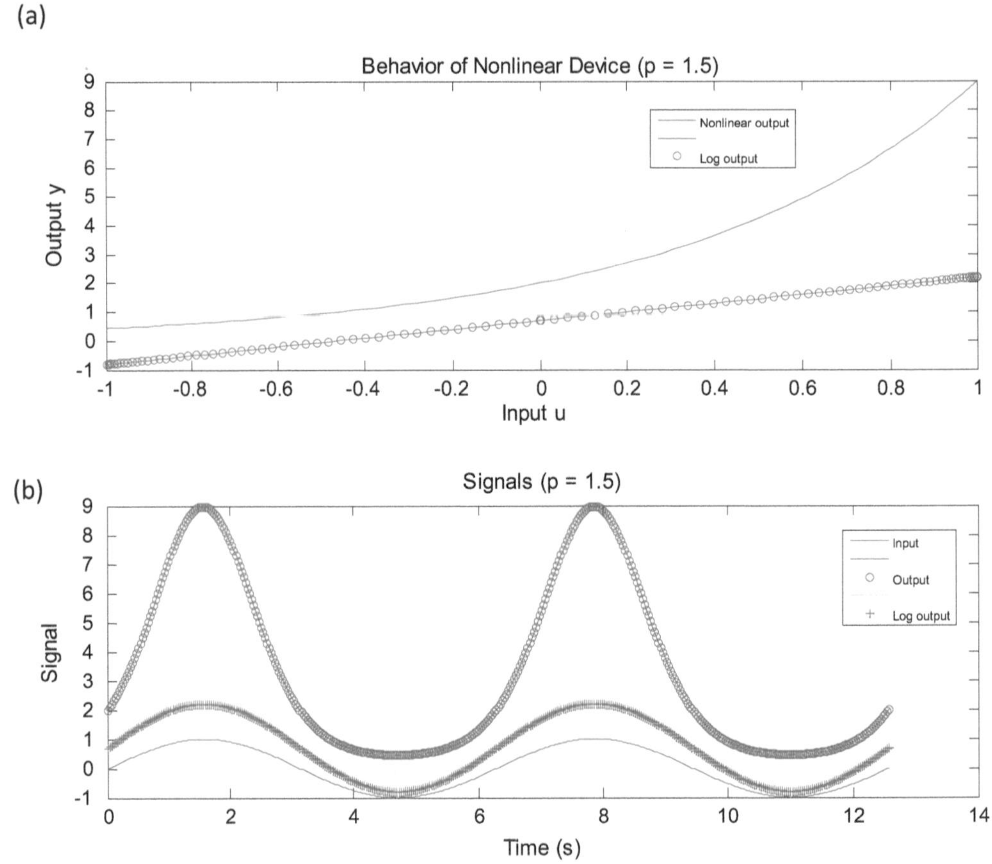

FIGURE 4.12 Static sine response of a nonlinear device. (a) Input–output behavior; (b) signals.

```
% Response of nonlinear device
u0=2.0;k=2.0;p=1.5;
t=0:0.01*pi:4*pi;
u=sin(t);
y=k*exp(p*u);
y2=log(y);
% plot the results
plot(u, y, '-', u, y2,'-', u, y2,'o')
plot(t, u, '-', t, y, '-', t, y, 'o!, t, y2,'-', t, y2,'+')
```

It is seen that the actual nonlinear function considerably distorts the input sine signal, and the output is not sinusoidal at all. The use of a log scale for the output, however, conveniently and accurately linearizes the behavior, giving a sinusoidal output. Furthermore, by using the log output shown in Figure 4.12a, we can extract the two parameters p and k from the slope and the y-intercept of the linearized (scaled) input–output curve. Specifically, $p=\text{slope}=3.0/2.0=1.5$; Log $k=0.7 \rightarrow k=2.0$.

LEARNING OBJECTIVES

1. Removal of static nonlinearity
2. Calibration (rescaling) method
3. Parameter extraction from static input–output data.

4.6.2 Equivalent Model Approach of Linearization

Another way to linearize a nonlinear model is through some criterion of equivalence. Specifically, a linear model that is equivalent to the nonlinear model is determined, based on some criterion. Approximating a distributed-parameter model by a lumped-parameter model by using energy equivalence is addressed in Chapter 3. Similarly, energy equivalence is a practical and convenient criterion of equivalence for linearizing a nonlinear model. Specifically, the energy absorbed (or energy dissipated, or work done) in an operating cycle is used as the criterion for equating the two models. This approach is illustrated now using an example.

Example 4.6

Consider the nonlinear damper (e.g., fluid damper) model shown in Figure 4.13a. The constitutive relation of the damper is

$$f = c\dot{x}\,|\,\dot{x}\,| \tag{4.39}$$

where

f = damping force
x = relative displacement of the damper
$\dot{x}$ = relative velocity of the damper
c = damping parameter.

Suppose that a harmonic (sinusoidal) force of frequency ω is applied to the damper. At steady state, the dominant sinusoidal component of the displacement x will be $x = x_0 \sin(\omega t + \phi)$
where

x_o = displacement amplitude

ϕ = phase angle of the displacement x with respect to the force f

a. Determine the energy dissipation of this nonlinear damper in one cycle of motion (i.e., in the period $T = 2\pi / \omega$). Express your result in terms of c, x_0 and ω.
b. Now consider the linear viscous damper model shown in Figure 4.13b. Here, the constitutive relation is

$$f = b\dot{x} \tag{4.40}$$

where b = viscous damping coefficient. Following the same procedure as for the nonlinear damper, determine the energy dissipation in one cycle of motion of the linear damper. Express your result in terms of b, x_0 and ω.

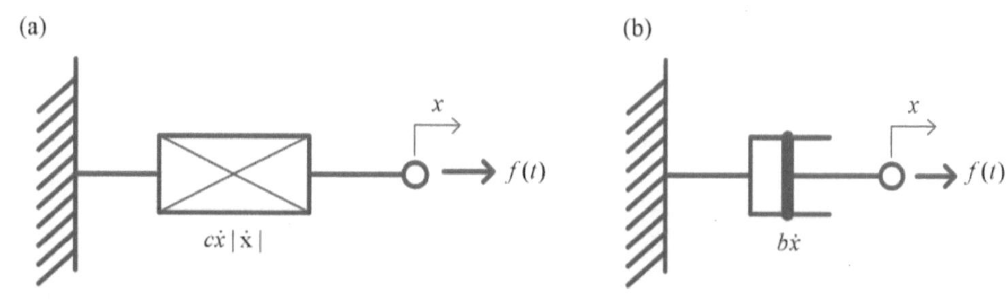

FIGURE 4.13 (a) A nonlinear fluid damper; (b) a linear viscous damper.

c. From the two results, give an equivalent linear viscous damping coefficient b_{eq} that results in the same energy dissipation per cycle as for the nonlinear damper. Express your result in terms of c, x_0 and ω.

Note: We assume that the equivalent linear damper (equivalent model) will operate at the same frequency (ω), which is the excitation frequency, as the nonlinear damper and has the same displacement amplitude (x_0) as the nonlinear damper. These two added criteria of equivalence (in addition to the energy dissipation per cycle) can be enforced on the equivalent linear system, because it is a representative "model" that is developed in any desirable way, and it is not a model of the actual system (the given nonlinear model).

d. Comment on the limitations in the use of this method of linearization.

Solution

(a)

$$f = c\dot{x}|\dot{x}| \tag{4.39}$$

$$x = x_0 \sin(\omega t + \phi) \tag{i}$$

$$\dot{x} = x_0\omega \cos(\omega t + \phi) \tag{ii}$$

Energy dissipation per cycle of excitation, $\Delta u = \int\limits_{cycle} f\, dx = \int\limits_{cycle} f \frac{dx}{dt} dt$

Substitute (4.39) and (ii): $\Delta u = c \int\limits_{cycle} \dot{x}^2 |\dot{x}|\; dt = cx_0^3\omega^3 \int\limits_{-\phi/\omega}^{(2\pi-\phi)\omega} \cos^2(\omega t + \phi)|\cos(\omega t+\phi)|\, dt$

Change variables: $\theta = \omega t + \phi$. Then,

$$\text{At}\quad t = -\frac{\phi}{\omega},\quad \theta = 0$$

$$\text{At}\quad t = \frac{(2\pi - \phi)}{\omega},\quad \theta = 2\pi$$

$$\text{Also,}\quad d\theta = \omega dt$$

Substitute

$$\Delta u = cx_0^3\omega^2 \int\limits_0^{2\pi} \cos^2\theta |\cos\theta|\, d\theta = 4cx_0^3\omega^2 \int\limits_0^{\pi/2} \cos^3\theta d\theta \qquad \left(\textit{Note}: \cos\theta \ge 0 \text{ when } \theta = 0 \text{ to } \frac{\pi}{2}\right)$$

$$= 4cx_0^3\omega^2 \int\limits_0^{\pi/2} \cos\theta\left(1 - \sin^2\theta\right)d\theta = 4cx_0^3\omega^2\left[\sin\theta - \frac{1}{3}\sin^3\theta\right]_0^{\frac{\pi}{2}} = 4cx_0^3\omega^2\left(1 - \frac{1}{3}\right)$$

$$\Delta u = \frac{8}{3}cx_0^3\omega^2 \tag{4.41}$$

(b)

$$f = b\dot{x} \tag{4.40}$$

As before,

$$\Delta u' = b\int_{\text{cycle}} \dot{x}^2\,dt = bx_0^2\omega^2 \int_{-\phi/\omega}^{(2\pi-\phi)\omega} \cos^2(\omega t+\phi)dt = bx_0^2\omega\int_0^{2\pi}\cos^2\theta d\theta = bx_0^2\omega\int_0^{2\pi}\frac{1}{2}[1+\cos 2\theta]d\theta$$

$$= \frac{1}{2}bx_0^2\omega\left[\theta+\frac{1}{2}\sin 2\theta\right]_0^{2\pi} = \frac{1}{2}bx_0^2\omega\times[2\pi+0]$$

$$\Delta u' = \pi b x_0^2\omega \tag{4.42}$$

Note: Alternatively, as in Part (a), we could have used

$$\Delta u' = 4\times bx_o{}^2\omega\int_0^{\pi/2}\frac{1}{2}[1+\cos 2\theta]d\theta$$

$$= \frac{4}{2}bx_o{}^2\omega\left[\theta+\frac{1}{2}\sin 2\theta\right]_0^{\pi/2} = 2bx_o{}^2\omega\left[\frac{\pi}{2}+0\right] = \pi bx_o{}^2\omega$$

We get the same answer.

(c)

For the equivalence of energy dissipation in one cycle, we must satisfy $\Delta u = \Delta u'$ with $b = b_{eq}$.

$$\pi b_{eq}x_0^2\omega = \frac{8}{3}cx_0^3\omega^2$$

$$b_{eq} = \frac{8}{3\pi}cx_0\omega \tag{4.43}$$

(d)

It is seen from result (4.43) that the equivalent damping constant depends on both the excitation frequency and the response amplitude (hence on the excitation amplitude). This is a common characteristic of a nonlinear system, as pointed out before, and also seen in the describing function method, which is considered next. Therefore, the equivalent damping constant has to be changed during the use of this linear model, depending on the excitation parameters, unless the excitation signal is sinusoidal.

Also we require that the equivalent linear model has the same displacement amplitude and frequency as the nonlinear damper, which is not a limitation, as pointed out earlier.

Note: At steady state (i.e., as $\omega \to 0$), $b_{eq} \to 0$, and the model will be able to provide very large speeds even with a very small excitation force. At very high frequencies (i.e., $\omega \to \infty$), $b_{eq} \to \infty$, and the model will not move, which is quite realistic.

The nature of the damping modules of Figure 4.13 is shown in Figure 4.14.

It is clear that the nonlinear damping (Figure 4.14a) has an infinite slope at zero force (and at zero speed) unlike the linear viscous damping (Figure 4.14b), which has a finite slope.

Note: Infinite slope means zero damping constant because slope in Figure 4.14 is $\frac{\partial \dot{x}}{\partial f}$, while the damping constant is $\frac{\partial f}{\partial \dot{x}}$.

Hence, local linearization (based on the local slope) is not practical in this case of nonlinear damping, particularly for low speeds (and low damping forces).

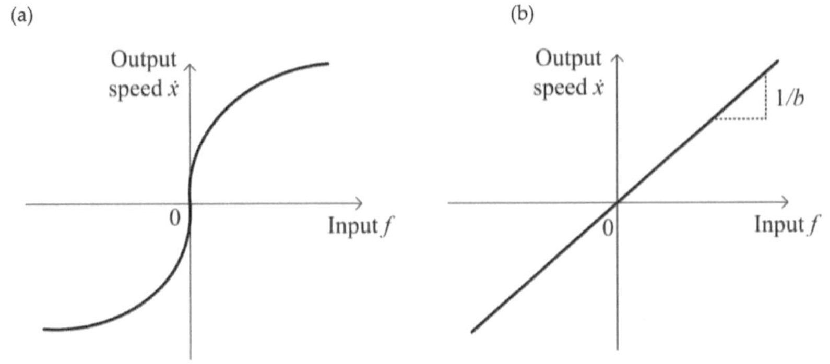

FIGURE 4.14 (a) Nonlinear damping; (b) linear viscous damping.

LEARNING OBJECTIVES

1. Why the local slope method of linearization is not feasible in some cases of nonlinearity
2. Energy-equivalence method of linearization
3. Dependence of the linearized model on the excitation amplitude
4. Limitations of energy-equivalence linearization
5. Change of variables to simplify analysis.

Depending on the situation, other criteria of equivalence may be used for determining the equivalent linear model. The describing function method of linearization, for example, uses the criterion of "fundamental frequency equivalence," which is presented next.

4.6.3 Describing Function Method

Nonlinear systems can be analyzed in the frequency domain by using the *describing function* approach. This is similar to the use of *transfer functions* (or *frequency transfer functions* or *frequency response functions*) for linear systems, but the usage is restrictive. As observed before, when a sinusoidal input (at a specific frequency) is applied to a nonlinear device, the resulting output at steady state will have a component at this fundamental frequency and also components, typically harmonics, at other frequencies (as a result of the property of "frequency creation" by the nonlinear device). The response may be represented by a *Fourier series*, which will have a signal component at this input frequency (fundamental frequency) and also at integer multiples of the input frequency. The describing function approach neglects all the higher frequency components (harmonics) in the output and retains only the fundamental component (i.e., the component at the input frequency). This output component, when divided by the input, produces the describing function of the device. In this manner, we obtain a "transfer function model" called the *describing function* for the nonlinear device. But, unlike for a linear device, the gain and the phase shift of a describing function will vary with the input amplitude, in general.

Clearly, the describing function method of linearization is a method of using an *equivalent model* (see previous section). Specifically, for the describing function method, the equivalence criterion is the fundamental frequency component of the output (response). The obtained equivalent model is a frequency-domain model (a transfer function—see Chapter 6), albeit with limitations, not a time-domain model. In particular, as in the case of energy equivalence, the equivalent model represented by a describing function will depend in general on the excitation amplitude and the excitation frequency. We now illustrate this method using an example.

Example 4.7

The ideal relay is a two-state switching function. Its input–output behavior is given by the analytical relationship,

$$y = y_0 \operatorname{sgn}(u) \tag{4.44}$$

where "sgn" denotes the *signum* (sign) function. Specifically, if the input takes a positive value, the output switches to y_0, and if the input takes a negative value, the output switches to $-y_0$. When $u=0$, the output value is not strictly defined, but we will take it to be 0 as well. Consider a sinusoidal input, given by $u = u_0 \sin \omega t$.

The output is a pulse with the same frequency (same period) as the input, as shown in Figure 4.15a.

Determine a describing function N for this ideal relay, which can be represented by the *block diagram* in Figure 4.15b.

Solution

First we need to determine the signal component of frequency ω in the output signal. This can be obtained by the *Taylor series expansion* of the output signal. We have

$$y = y_0 \operatorname{sgn}(\sin \omega t) = b_0 + \sum_{i=1}^{\infty} (a_i \sin i\omega t + b_i \cos i\omega t) \tag{i}$$

First, at $t=0$ in (i), we note that $b_0 = 0$. Then, we only need to determine a_1 and b_1, one of which can be made zero (without loss of generality) from the nature of the nonlinear function. To determine

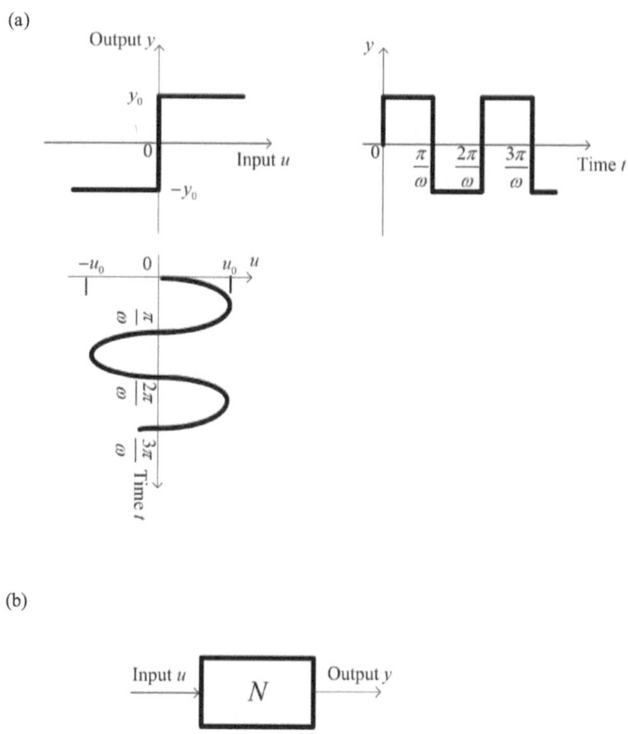

FIGURE 4.15 (a) The output of an ideal relay for a sine input; (b) the describing function representation of a nonlinear device.

a_1, multiply both sides of (i) by $\sin\omega t$ and integrate wrt time t, over a full cycle of the signal (i.e., from 0 to $2\pi/\omega$). It is easy to see that all the terms on the RHS of (i), except the term containing a_1, go to zero because they have identical positive and negative halves over the integration period. Then we have

$$\int_0^{2\pi/\omega} y_0 \sin\omega t \,\mathrm{sgn}(\sin\omega t)dt = \int_0^{2\pi/\omega} (\sin\omega t)(a_1 \sin\omega t)dt$$

$$2y_0 \int_0^{\pi/\omega} \sin\omega t dt = a_1 \int_0^{2\pi/\omega} \sin^2 \omega t dt = a_1 \int_0^{2\pi/\omega} \frac{1}{2}(1-\cos 2\omega t)dt$$

$$\frac{2y_0}{\omega}[-\cos\omega t]_0^{\pi/\omega} = a_1 \frac{1}{2}\left[t - \frac{1}{2\omega}\sin 2\omega t\right]_0^{2\pi/\omega}$$

$$\frac{4y_0}{\omega} = a_1 \frac{\pi}{\omega}$$

Hence, $a_1 = \dfrac{4y_0}{\pi}$.

In the same manner, we can show that $b_1 = 0$. This means the fundamental component of the output y is $a_1 \sin\omega t = \dfrac{4y_0}{\pi}\sin\omega t$.

We get the describing function by dividing this output component by the input signal $u = u_0 \sin\omega t$. Specifically,

$$N = \frac{4y_0}{\pi u_0} \tag{4.45}$$

Clearly, this function depends on the input amplitude u_0, which is a characteristic of a nonlinear system.

More details and examples of the describing function approach can be found in textbooks on nonlinear control systems.

4.6.4 Linearization Using Analog Hardware

Hardware, particularly those that employ operational amplifiers (op-amps), may be used to linearize a nonlinear device, often without sacrificing the device accuracy. We now illustrate this approach by using resistance bridge (Wheatstone bridge) as an example.

A bridge circuit is commonly used in applications of instrumentation, to make some form of measurement. Typical measurements include change in resistance, change in inductance, change in capacitance (or, generally, change in impedance), oscillating frequency, or some variable (stimulus) that causes these changes. There are many types of bridge circuits. Figure 4.16a shows a Wheatstone bridge. It is a resistance bridge with a constant dc voltage supply (i.e., it is a constant-voltage resistance bridge). A Wheatstone bridge is particularly useful in strain-gage measurements, and consequently in force, torque, and tactile sensors that employ strain-gage techniques. Since a Wheatstone bridge is used primarily in the measurement of small changes in resistance, it could be used in other types of sensing applications as well. For example, in resistance temperature detectors (RTDs), the change in resistance in a metallic (e.g., platinum) element, as caused by a change in temperature, is measured using a bridge circuit.

In Figure 4.16a, suppose that $R_1=R_2=R_3=R_4=R$. Then, the bridge is balanced, and the bridge output v_o will be zero. Now increase the resistance R_1 by δR. For example, R_1 may represent the only active strain gage, while the remaining three elements in the bridge are identical dummy elements. It can be shown that the change in the bridge output due to the change δR is given by

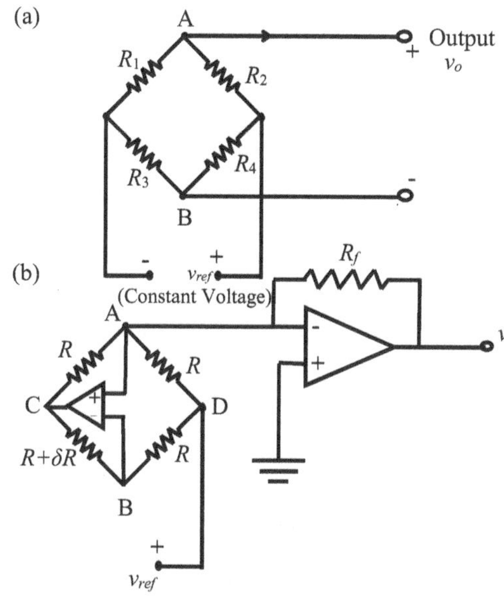

FIGURE 4.16 (a) Wheatstone bridge (constant-voltage resistance bridge); (b) a linearized bridge.

$$\frac{\delta v_o}{v_{\text{ref}}} = \frac{\delta R / R}{(4 + 2\delta R / R)} \tag{4.46}$$

It is clear that the output is nonlinear, with respect to the change in resistance of the active elements. Such a nonlinear bridge can be linearized using hardware; particularly op-amp elements. To illustrate this approach, modify the bridge circuit by connecting two op-amp elements, as shown in Figure 4.16b. The output amplifier has a feedback resistor R_f. The output equation for this circuit can be obtained by using the properties of an op-amp, in the usual manner. In particular, the potentials at the two input leads must be equal and the current through these leads must be zero. From these properties, the output of the hardware-modified bridge can be determined as

$$\frac{\delta v_o}{v_{\text{ref}}} = \frac{R_f}{R}\frac{\delta R}{R} \tag{4.47}$$

This relationship is linear.

4.6.5 Feedback Linearization

Feedback linearization is an "active linearization" technique. It is used in a control system to make a nonlinear plant (i.e., the *process* or the system that is being controlled) behave like a linear plant. This is done primarily to make a linear control method effective in a nonlinear system. However, in this method, the nonlinearities and dynamic coupling in the system must be compensated for faster than the control speed. In the *feedback linearization technique (FLT)*, this is accomplished by implementing a linearizing and decoupling controller "inside" the direct control loops. Feedback linearization of a nonlinear and coupled mechanical dynamic systems (e.g., robotic manipulator) is outlined now.

Consider a mechanical dynamic system (plant) given by

$$\boldsymbol{M}(\boldsymbol{q})\frac{d^2\boldsymbol{q}}{dt^2} = \boldsymbol{n}(\boldsymbol{q},\dot{\boldsymbol{q}}) + \boldsymbol{f}(t) \tag{4.48}$$

in which

$$\boldsymbol{f} = \begin{bmatrix} f_1 \\ f_2 \\ \vdots \\ f_r \end{bmatrix} = \text{vector of input forces at various locations of the system}$$

$$\boldsymbol{q} = \begin{bmatrix} q_1 \\ q_2 \\ \vdots \\ q_r \end{bmatrix} = \text{vector of response variables (e.g., positions) at the forcing locations of the system}$$

$\boldsymbol{M}(\boldsymbol{q})$ = inertia matrix (nonlinear)

$\boldsymbol{n}(\boldsymbol{q},\dot{\boldsymbol{q}})$ = a vector of nonlinear effects in the system (e.g., damping, backlash, Coriolis and centrifugal accelerations, gravitational effects)

Now suppose that we can model $\boldsymbol{M}$ by $\hat{\boldsymbol{M}}$ and $\boldsymbol{n}$ by $\hat{\boldsymbol{n}}$. Then, we can compute $\hat{\boldsymbol{M}}$ using the model $\hat{\boldsymbol{M}}$ and the online measurement $\boldsymbol{q}$. Similarly, we can compute $\hat{\boldsymbol{n}}$ using the model $\boldsymbol{n}$ and the online measurements $\boldsymbol{q}$ and $\dot{\boldsymbol{q}}$. With the use of these computed values of $\hat{\boldsymbol{M}}$ and $\hat{\boldsymbol{n}}$, the following linearizing feedback controller can be implemented:

$$\boldsymbol{f} = \hat{\boldsymbol{M}}\boldsymbol{K}\left[\boldsymbol{e} + \boldsymbol{T}_i^{-1}\int \boldsymbol{e}\,dt - \boldsymbol{T}_d\frac{d\boldsymbol{q}}{dt}\right] - \hat{\boldsymbol{n}} \tag{4.49}$$

in which,

$$\boldsymbol{e} = \boldsymbol{q}_d - \boldsymbol{q} = \text{error (correction) vector}$$

$$\boldsymbol{q}_d = \text{desired response}$$

and $\boldsymbol{K}$, $\boldsymbol{T}_i$, and $\mathbf{T}_d$ are constant control parameter matrices. This control scheme is sown in Figure 4.17.

Note: Since the computed values are used in the generation of the control action $\boldsymbol{f}$, this is an "active" scheme of linearization.

To show that this is indeed a linearizing controller, substitute the controller equation (4.49) into the plant equation (4.48). We get

$$\boldsymbol{M}\frac{d^2\boldsymbol{q}}{dt^2} = \boldsymbol{n} - \hat{\boldsymbol{n}} + \hat{\boldsymbol{M}}\boldsymbol{K}\left[\boldsymbol{e} + \boldsymbol{T}_i^{-1}\int \boldsymbol{e}\,dt - \boldsymbol{T}_d\frac{d\boldsymbol{q}}{dt}\right] \tag{4.50}$$

If our models are exact, we have $\boldsymbol{M} = \hat{\boldsymbol{M}}$ and $\boldsymbol{n} = \hat{\boldsymbol{n}}$. Then, because the inverse of matrix $\hat{\boldsymbol{M}}$ exists in general (because the inertia matrix *is positive definite*), we get

$$\frac{d^2\boldsymbol{q}}{dt^2} = \boldsymbol{K}\left[\boldsymbol{e} + \boldsymbol{T}_i^{-1}\int \boldsymbol{e}\,dt - \boldsymbol{T}_d\frac{d\boldsymbol{q}}{dt}\right] \tag{4.51}$$

Clearly, equation (4.51) represents a linear, constant-parameter system with proportional-integral-derivative (PID) control. In other words, the controller has linearized the system. The proportional control parameters are the elements of the gain matrix $\boldsymbol{K}$, the integral control parameters are the elements of the matrix $\boldsymbol{T}_i$, and the derivative control parameters are the elements of the matrix $\boldsymbol{T}_d$. We are free to select these parameters so as to achieve the desired response. In particular, if these three parameter matrices are chosen to be diagonal, then the control system, as given by equation

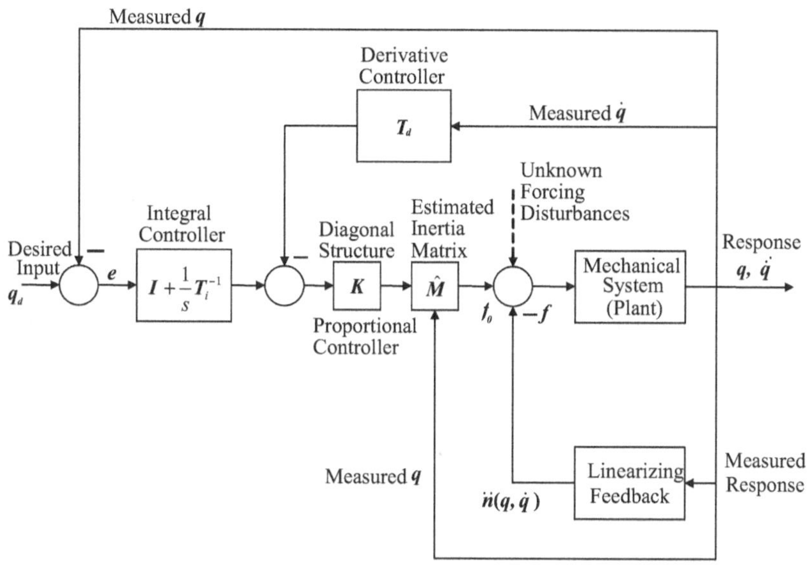

FIGURE 4.17 The structure of a linearizing feedback control system.

(4.51) and shown in Figure 4.17, is *uncoupled* (i.e., one input affects only one output) and will not have *dynamic interactions*. In summary, this controller both linearizes and decouples the system. Its main limitation is that accurate models are needed for the nonlinearities.

Instead of using analytical modeling, the parameters in $\hat{M}$ and $\hat{n}$ may be obtained through the measurement of various input–output pairs. This is the method of *model identification*, as discussed before. Strictly, nonlinear model identification is needed. It can cause further complications with regard to the instrumentation and data processing speed, particularly because the system is nonlinear and also some of the model parameters must be estimated in real time.

4.7 SUMMARY SHEET

- **Useful Methods of Linearization**: 1. Slope-based analytical, local linearization; 2. slope-based linearization using experimental models; 3. static linearization through recalibration or rescaling; 4. linearization based on an equivalent model; 5. describing function method; 6. linearization using analog hardware, and 7. feedback linearization
- **Static Nonlinearity**: Nonlinear, algebraic, steady-state, input–output relation
- **Geometric Nonlinearity**: Nonlinear trigonometric (kinematic) terms (cos, sine, tan, etc.) due to large deflections or large motions of a mechanical device
- **Physical Nonlinearity**: Nonlinear physical relations (e.g., nonlinear kinetic relation—Newton's second law with nonlinear accelerations; nonlinear stress–strain relation)
- **Manifestations of Physical Nonlinearity**: Saturation (output remains unchanged when the input changes); dead zone (region with zero output to an input)); hysteresis (loop in the input–output curve); jump phenomenon (frequency response function jumps from one value to another at a particular frequency); limit cycles (sustained oscillations at a specific frequency and amplitude, independent of the initial condition); frequency creation (at steady-state creates frequency components that are not present in the input)
- **Nonlinear Electrical Elements**: Capacitor: $q = q(v)$; inductor: $\lambda = \lambda(i)$; resistor: $v = v(i)$; q=charge, λ = flux linkage, v=voltage, i=current
- **Steady-State Operating Condition**: Set the rates of changes of system variables to zero. Solve the resulting algebraic equations; stable (given a slight shift, the system returns to

original state); unstable (given a slight shift, the system continues to move away); neutral (given a slight shift, the system remains in the shifted condition)

- **Slope-based Linearization**: For one independent variable, $\hat{f} = f(\bar{x}+\hat{x}) - f(\bar{x}) \approx \frac{df(\bar{x})}{dx}\hat{x}$, Error $e = f(\bar{x}+\hat{x}) - \left[f(\bar{x}) + \frac{df(\bar{x})}{dx}\hat{x}\right]$; for two independent variables, $\hat{f} = f(\bar{x}+\hat{x}, \bar{y}+\hat{y}) - f(\bar{x},\bar{y}) \approx \frac{\partial f(\bar{x},\bar{y})}{\partial x}\hat{x} + \frac{\partial f(\bar{x},\bar{y})}{\partial y}\hat{y}$; $\delta\dot{x} = \frac{d\hat{x}}{dt} = \dot{\hat{x}}$, $\delta\ddot{x} = \frac{d^2\hat{x}}{dt^2} = \ddot{\hat{x}}$
- **Slope-Based Linearization Procedure**: 1. Select operating condition (steady state: set time derivatives to zero and solve the nonlinear algebraic equations); 2. determine the first-order derivatives of each nonlinear term wrt each independent variable; 3. replace the nonlinear term by its slope×incremental variable, and linear term by its coefficient×incremental variable
- **Nonlinear State-Space Model**: $\frac{dq_i}{dt} = f_i(q_1,q_2,\ldots,q_n,r_1,r_2,\ldots,r_m,t)$, $i = 1,2,\ldots,n$ or $\dot{\boldsymbol{q}} = \boldsymbol{f}(\boldsymbol{q},\boldsymbol{r},t)$; state vector $\boldsymbol{q} = [q_1,q_2,\ldots,q_n]^T$, input vector $\boldsymbol{r} = [r_1,r_2,\ldots,r_m]^T$; "$t$" in argument → time-variant system
- **Linearization**: $\dot{\boldsymbol{q}} = \boldsymbol{0}$, solve $\boldsymbol{f}(\boldsymbol{q},\boldsymbol{r},t)=\boldsymbol{0}$ → Equilibrium states; incremental state equations: $\delta\dot{\boldsymbol{q}} = \frac{\partial \boldsymbol{f}}{\partial \boldsymbol{q}}(\bar{\boldsymbol{q}},\bar{\boldsymbol{r}},t)\delta\boldsymbol{q} + \frac{\partial \boldsymbol{f}}{\partial \boldsymbol{r}}(\bar{\boldsymbol{q}},\bar{\boldsymbol{r}},t)\delta\boldsymbol{r}$; $\delta\boldsymbol{q} = \boldsymbol{x} = [x_1,x_2,\ldots,x_n]^T$, $\delta\boldsymbol{r} = \boldsymbol{u} = [u_1,u_2,\ldots,u_m]^T$; or $\dot{\boldsymbol{x}} = \boldsymbol{A}\boldsymbol{x} + \boldsymbol{B}\boldsymbol{u}$ with $\boldsymbol{A}(t)=\frac{\partial \boldsymbol{f}}{\partial \boldsymbol{q}}(\bar{\boldsymbol{q}},\bar{\boldsymbol{r}},t)$ and $\boldsymbol{B}(t)=\frac{\partial \boldsymbol{f}}{\partial \boldsymbol{r}}(\bar{\boldsymbol{q}},\bar{\boldsymbol{r}},t)$;

$$\boldsymbol{A} = \begin{bmatrix} \frac{\partial f_1}{\partial q_1} & \frac{\partial f_1}{\partial q_2} & \cdots & \cdots & \frac{\partial f_1}{\partial q_n} \\ \frac{\partial f_2}{\partial q_1} & \frac{\partial f_2}{\partial q_2} & \cdots & \cdots & \frac{\partial f_2}{\partial q_n} \\ \cdots & \cdots & \cdots & \cdots & \cdots \\ \cdots & \cdots & \cdots & \cdots & \cdots \\ \frac{\partial f_n}{\partial q_1} & \frac{\partial f_n}{\partial q_2} & \cdots & \cdots & \frac{\partial f_n}{\partial q_n} \end{bmatrix};\ \boldsymbol{B} = \begin{bmatrix} \frac{\partial f_1}{\partial r_1} & \cdots & \cdots & \frac{\partial f_1}{\partial r_m} \\ \frac{\partial f_2}{\partial r_1} & \cdots & \cdots & \frac{\partial f_2}{\partial r_m} \\ \cdots & \cdots & \cdots & \cdots \\ \cdots & \cdots & \cdots & \cdots \\ \frac{\partial f_n}{\partial r_1} & \cdots & \cdots & \frac{\partial f_n}{\partial r_m} \end{bmatrix};$$

linear time-invariant system → $\boldsymbol{A}$ and $\boldsymbol{B}$ are constant (time-invariant) matrices

- **Mitigation of System Nonlinearities**: Calibrate/rescale (e.g., log) the output; use linearizing elements (e.g., resistors, amplifiers in bridge circuits); use feedback linearization; avoid operation over wide input ranges, signal levels, or frequency bands; avoid large deformations (e.g., deviation from Hooke's law—a physical nonlinearity) or large mechanical motions (e.g., trigonometric terms—a geometric or kinematic nonlinearity); minimize nonlinear friction (e.g., Coulomb, Stribeck), stiction, wear and tear (e.g., through proper lubrication); avoid loose joints, gear coupling, etc., which cause backlash (e.g., use direct-drive mechanisms, harmonic drives, etc.); minimize sensitivity to undesirable influences (e.g., environmental influences such as temperature)
- **Linearization Using Experimental Data**: 1. Determine steady-state characteristic curves by varying one independent variable at a time (keeping the other variables constant); 2. determine local slopes (with respect to independent variables) of the experimental characteristic curves at operating point; 3. dynamics (particularly, inertial dynamics) are incorporated separately into the model, using analytical terms

- **Experimental Linear Model for Motor Control**: 1. Determine a set of characteristic curves $T_m = T_m(\omega_m, v_c)$ for motor; 2. draw a tangent to primary curve at operating point O. Damping constant $b = -\left.\frac{\partial T_m}{\partial \omega_m}\right|_{v_c=\text{constant}}$ = slope magnitude at O (includes both electromagnetic damping and mechanical damping, if torque T is measured outside the motor bearings); 3. draw vertical line through O to intersect two other adjacent operating curves. ΔT_m = torque intercept between the two curves, Δv_c = voltage increment corresponding to the two adjacent curves. Voltage gain $k_v = \left.\frac{\partial T_m}{\partial v_c}\right|_{\omega_m=\text{constant}} = \frac{\Delta T_m}{\Delta v_c}$; 4. $\delta T_m = \left.\frac{\partial T_m}{\partial \omega_m}\right|_{v_c} \delta\omega_m + \left.\frac{\partial T_m}{\partial v_c}\right|_{\omega_m} \delta v_c \rightarrow$ $\delta T_m = -b\delta\omega_m + k_v\delta v_c$; 5. linear dynamic model $\delta T_m = -b\delta\omega_m + k_v\delta v_c$; motor torque = T_m, motor speed = ω_m, input voltage = v_c, moment of inertia of motor rotor = J_m, load torque = T_L
- **Linearization by Calibration**: Example: $y = ke^{pu}$ (nonlinear input u-output y relation); take log $\rightarrow \log(y) = pu + \log(k)$ (linear)
- **Linearization through Energy Equivalence**: Example: Nonlinear damper $f = c\dot{x}\,|\,\dot{x}\,| \rightarrow$ equivalent linear damper $f = b_{eq}\dot{x}$ where, $b_{eq} = \frac{8}{3\pi} c x_0 \omega$ (for same energy dissipation in a cycle; *Note*: Dependence on excitation amplitude; f = damping force, x = relative displacement of damper, $\dot{x}$ = relative velocity of damper, ω = frequency of oscillation, c = nonlinear damping parameter, b_{eq} = equivalent linear viscous damping constant
- **Describing Function**: A frequency-domain equivalent model of a nonlinear system. Uses the equivalence criterion: fundamental frequency component in the output. Example: Ideal relay $y = y_0 \,\text{sgn}(u)$. "sgn" is the signum (sign) function. Output switches between y_0 and $-y_0$ depending on the input sign. Input $u = u_0 \sin\omega t$ gives output y, a pulse of amplitude y_0 and same frequency. Describing function of relay: $N = \frac{4y_0}{\pi u_0}$ (depends on the input amplitude u_0)
- **Linearization Using Analog Hardware**: Connect special analog hardware into the circuit, to obtain precise linear behavior (e.g., use of op-amps in a bridge circuit)
- **Feedback Linearization**: This is a method of "active" linearization. Use feedback (active control signals) to remove the nonlinear effects in a dynamic system. After linearization, linear control techniques can be used without loss of accuracy. Example: Mechanical dynamic system $\boldsymbol{M}(\boldsymbol{q})\ddot{\boldsymbol{q}} = \boldsymbol{n}(\boldsymbol{q},\dot{\boldsymbol{q}}) + \boldsymbol{f}(t)$, $\boldsymbol{f}$ = vector of input forces, $\boldsymbol{q}$ = vector of movements at the forcing locations, $\boldsymbol{M}(\boldsymbol{q})$ = inertia matrix (nonlinear in $\boldsymbol{q}$), $\boldsymbol{n}(\boldsymbol{q},\dot{\boldsymbol{q}})$ = other nonlinear effects in the system. Linearizing feedback controller: $\boldsymbol{f} = \hat{\boldsymbol{M}}\boldsymbol{K}\left[\boldsymbol{e} + \boldsymbol{T}_i^{-1}\int \boldsymbol{e}\,dt - \boldsymbol{T}_d \frac{d\boldsymbol{q}}{dt}\right] - \hat{\boldsymbol{n}}$;
- $\boldsymbol{e} = \boldsymbol{q}_d - \boldsymbol{q}$ = error (correction) vector, $\boldsymbol{q}_d$ = desired response, $\hat{\boldsymbol{M}}$ = model of $\boldsymbol{M}$, $\hat{\boldsymbol{n}}$ = model of $\boldsymbol{n}$; $\boldsymbol{K}$, $\boldsymbol{T}_i$, and $\boldsymbol{T}_d$ are proportional, integral, derivative control parameter matrices.
- $\boldsymbol{M}\ddot{\boldsymbol{q}} = \boldsymbol{n} - \hat{\boldsymbol{n}} + \hat{\boldsymbol{M}}\boldsymbol{K}\left[\boldsymbol{e} + \boldsymbol{T}_i^{-1}\int \boldsymbol{e}\,dt - \boldsymbol{T}_d \frac{d\boldsymbol{q}}{dt}\right] \rightarrow \frac{d^2\mathbf{q}}{dt^2} = \boldsymbol{K}\left[\boldsymbol{e} + \boldsymbol{T}_i^{-1}\int \boldsymbol{e}\,dt - \boldsymbol{T}_d \frac{d\boldsymbol{q}}{dt}\right]$, if models are exact (i.e., $\boldsymbol{M} = \hat{\boldsymbol{M}}$ and $\boldsymbol{n} = \hat{\boldsymbol{n}}$)—a shortcoming of the method.

PROBLEMS

4.1 Read about the followings nonlinear phenomena:

i. Saturation
ii. Hysteresis
iii. Jump phenomena
iv. Frequency creation
v. Limit cycle
vi. Dead band.

Often, the local slopes (derivatives) of a nonlinear function with respect to its independent variables are used in linearizing the function. Indicate situations where this approach is not appropriate.

4.2 What precautions may be taken in developing and operating a mechanical system, in order to reduce nonlinearities in the system?

Two types of nonlinearities are shown in Figure P4.2.

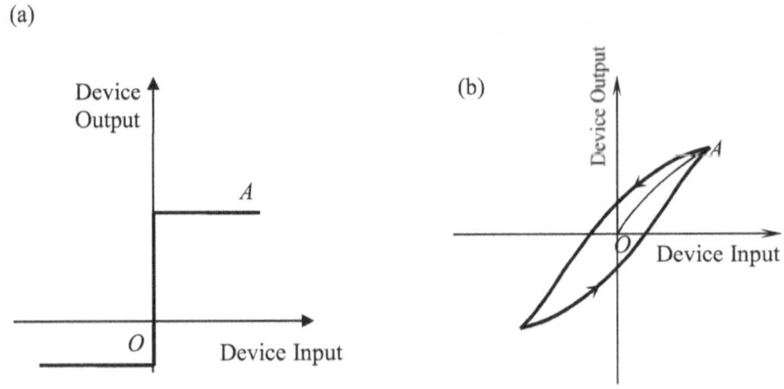

FIGURE P4.2 Two types of nonlinearities: (a) Ideal saturation; (b) hysteresis.

In each case indicate the difficulties of developing an analytical model for operation near:

i. Point O
ii. Point A.

4.3 An excitation was applied to a system and its response was observed. Next, the excitation was doubled. It was found that the response also doubled as a result. Is the system linear? What kind of useful conclusion may be reached from a single test of this nature?

4.4

a. Determine the derivative $\dfrac{d}{dx}x|x|$.
b. Determine an expression for the value of each of the following terms at a point that is away from the steady-state operating point $x = \bar{x} = 2$ by a small increment $\delta x = \hat{x}$:

(i) $3x^3$ (ii) $|x|$ (iii) $\dot{x}^2$

4.5 A nonlinear device obeys the relationship $y=y(u)$ and has an operating curve (characteristic curve) as shown in Figure P4.5.

i. Is this device a dynamic system?

A linear model of the form $y=ku$ is to be determined for operation of the device:

ii. In a small neighborhood of point B
iii. Over the entire range from A to B.

Suggest a suitable value for k in each case.

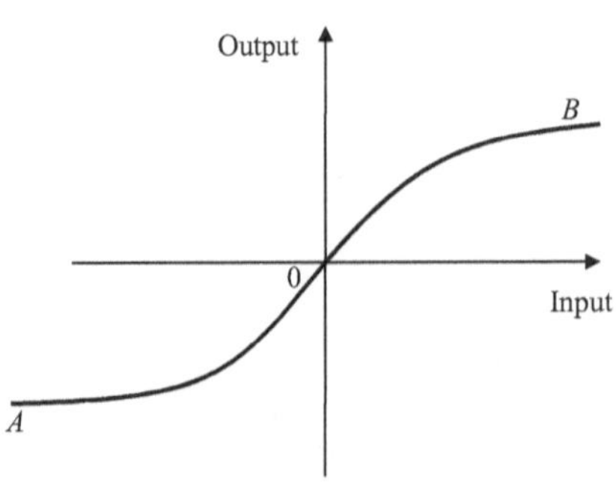

FIGURE P4.5 The characteristic curve of a nonlinear device.

4.6 A nonlinear damper is connected to a mechanical system as shown in Figure P4.6. The force f, which is exerted by the damper on the system, is $c(v_2 - v_1)^2$ where c is a constant parameter.

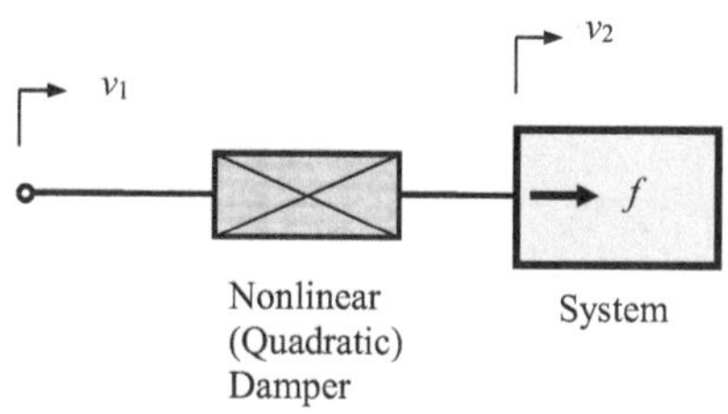

FIGURE P4.6 A nonlinear mechanical system.

xii. Give an analytical expression for f in terms of v_1, v_2, and c, which will be generally valid.
xiii. Give an appropriate linear model.
xiv. If the operating velocities v_1 and v_2 are equal, what will be the linear model about (in the neighborhood of) this operating point?

4.7 Suppose that a system is in equilibrium under the forces F_i and F_o as shown in Figure P4.7. If the point of application of F_i is given a small "virtual" displacement x in the same direction as the force, suppose that the location of F_o moves through $y = k\,x$ in the direction opposite to F_o.

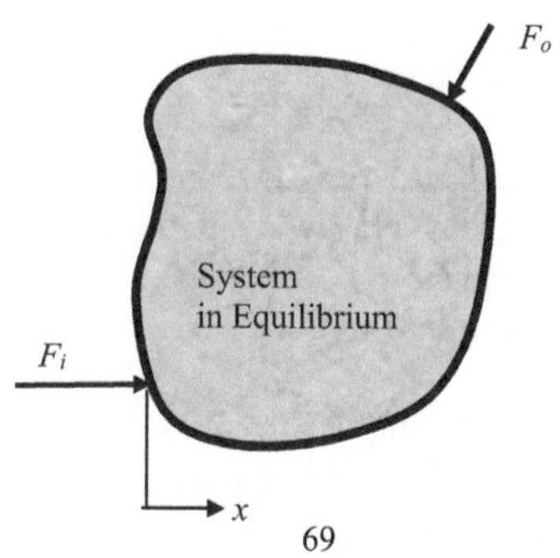

FIGURE P4.7 Virtual displacement of a system in equilibrium.

i. Determine F_o in terms of F_i (This method uses the "principle of virtual work").
ii. What is the relationship between small changes $\hat{F}_i$ and $\hat{F}_o$, about the operating conditions $\bar{F}_i$ and $\bar{F}_o$, assuming that the system is in equilibrium?

4.8 A simplified model of an elevator is shown in Figure P4.8.
The model parameters are:

J = moment of inertia of the cable pulley
r = radius of the pulley
k = stiffness of the cable
m = mass of the car and its occupants.

a. Which system parameters are variable? Explain.
b. Suppose that the damping torque $T_d(\omega)$ at the bearing of the pulley is a nonlinear function of the angular speed ω of the pulley. Let:

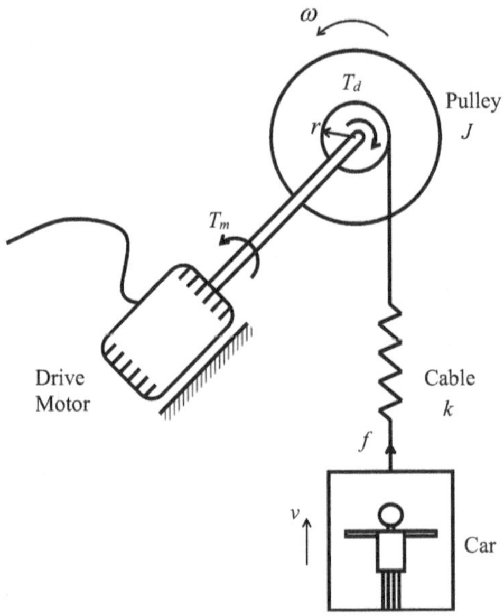

FIGURE P4.8 A simplified model of an elevator.

State vector $\boldsymbol{x} = [\ \omega \quad f \quad v\ ]^T$, with f = tension force in the cable; v = velocity of the car (taken positive upwards),

Input vector $\boldsymbol{u} = [T_m]$, with T_m = torque applied by the motor to the pulley (positive in the direction indicated in Figure P4.8)

Output vector $\boldsymbol{y} = [v]$

Obtain a complete, nonlinear, state-space model for the system.

c. With T_m as the input and v as the output, convert the state-space model into a nonlinear input–output differential equation model. What is the order of the system?
d. Give an equation whose solution provides the steady-state operating speed $\bar{v}$ of the elevator car.
e. Linearize the nonlinear input–output differential equation model obtained in Part (c), for small changes $\hat{T}_m$ of the input and $\hat{v}$ of the output, about an operating point.

   Note: $\bar{T}_m$ = steady-state operating-point torque of the motor (known).

   Hint: Denote $\dfrac{dT_d}{d\omega}$ by $b(\omega)$.
f. Linearize the state-space model obtained in Part (b) and give the model matrices $\boldsymbol{A}$, $\boldsymbol{B}$, $\boldsymbol{C}$, and $\boldsymbol{D}$ in the usual notation. Obtain the linear input–output differential equation from this state-space model and verify that it is identical to what was obtained in Part (e).

4.9 A rocket-propelled spacecraft of mass m is fired vertically up (in the Y-direction) from the earth's surface (see Figure P4.9). The vertical distance of the centroid of the spacecraft, measured from the earth's surface, is denoted by y. The upward thrust force of the rocket is $f(t)$. The gravitational pull on the spacecraft is given by $mg\left[\dfrac{R}{R+y}\right]^2$, where g is the acceleration due to gravity at the earth's surface and R is the "average" radius of earth (about 6370 km). The magnitude of the aerodynamic drag force that resists the motion of the spacecraft is approximated by $k\dot{y}^2 e^{-y/r}$ where k and r are positive and constant parameters and $\dot{y} = \dfrac{dy}{dt}$. In this expression, the exponential term represents the lowering of the air density at higher elevations.

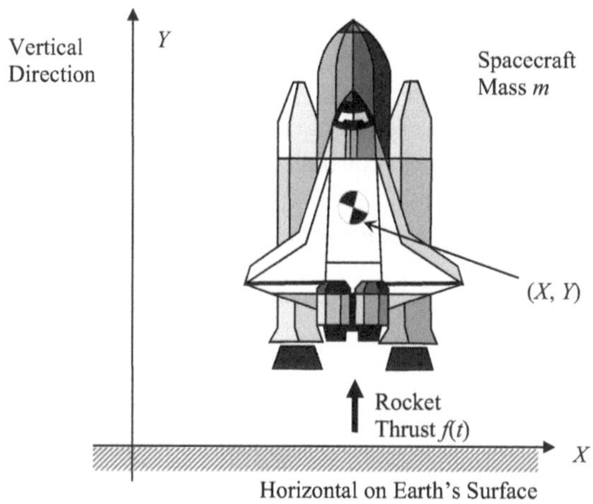

FIGURE P4.9 Coordinate system for the spacecraft problem.

a. Treating f as the input and y as the output, derive the input–output differential equation for the system.
b. The spacecraft accelerates to a height of y_o and then maintains at constant speed v_o, still moving in the same vertical (Y) direction. Determine an expression for the rocket force that is needed for this constant-speed motion. Express your answer in terms of y_o, v_o, time t, and the system parameters m, g, R, r, and k. Show that this force decreases as the spacecraft ascends.
c. Linearize the input–output model (Part (a)) about the steady operating condition (part (b)), for small variations $\hat{y}$ and $\hat{\dot{y}}$ in the position and speed of the spacecraft, due to a force disturbance $\hat{f}(t)$.
d. Treating y and $\dot{y}$ as the state variables and y as the output, derive a complete (nonlinear) state-space model for the vertical dynamics of the spacecraft.
e. Linearize the state-space model in (d) about the steady conditions in (b) for small variations $\hat{y}$ and $\hat{\dot{y}}$ in the position and the speed of the spacecraft, due to force disturbance $\hat{f}(t)$.
f. From the linear state model (Part (e)) derive the linear input–output model and show that the result is identical to what you obtained in Part (c).
g. Solve this problem by using the unified and systematic approach, where the across-variables of the independent A-type elements and the through-variables of the independent T-type variables are used as the state variables, leading to a unique result.

4.10 Characteristic curves of an armature-controlled dc motor are as shown in Figure P4.10. These are torque versus speed curves, measured at a constant armature voltage, at steady state.

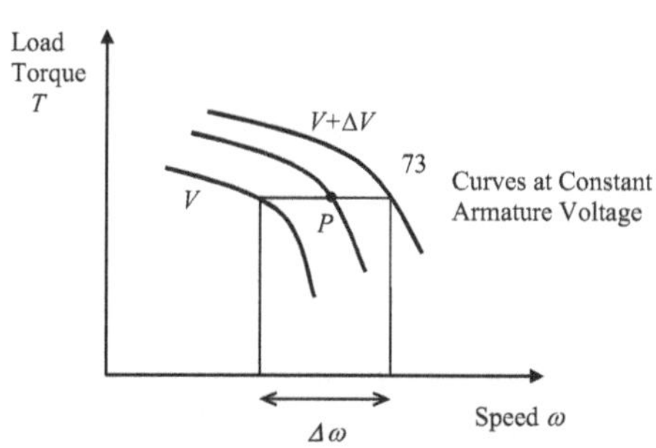

FIGURE P4.10 Characteristic curves of an armature-controlled dc motor.

For the neighborhood of point P, a linear model of the form $\hat{\omega} = k_1\hat{v} + k_2\hat{T}$ needs to be determined, for use in motor control. The following information is given:

The slope of the curve at $P = -a$

The voltage change in the two adjacent curves at point $P = \Delta V$

Corresponding speed change (at constant load torque through P) $= \Delta\omega$.

Estimate the parameters k_1 and k_2.

4.11 An air circulation fan system of a building is shown in Figure P4.11a.

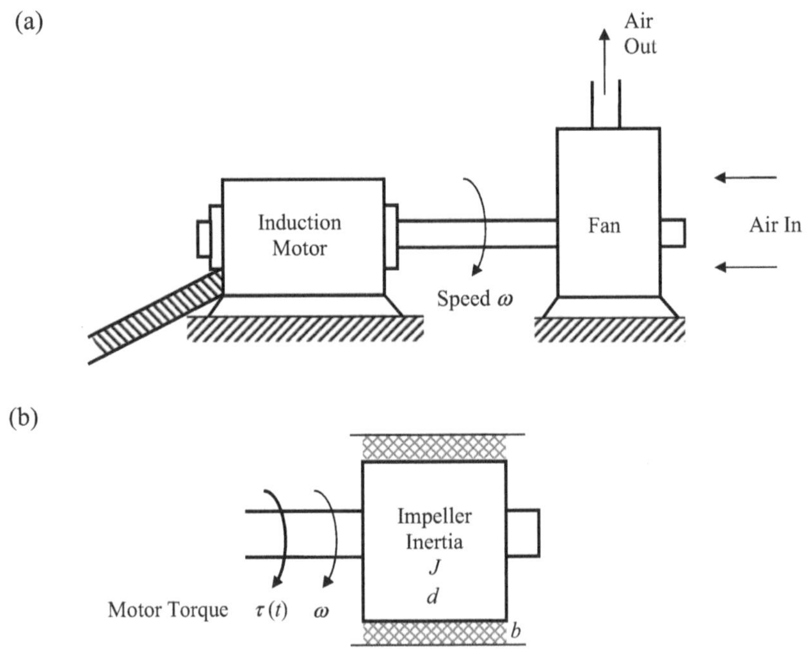

FIGURE P4.11 (a) A motor-fan combination of a building ventilation system; (b) a simplified model of the ventilation fan.

A simplified model of the system may be developed, as represented in Figure P4.11b. The induction motor is represented as a torque source $\tau(t)$. *Note*: In this manner, motor rotor inertia and motor bearing friction do not have to be considered explicitly. The speed ω of the fan, which determines the volume flow rate of air, is of interest. The moment of inertia of the fan impeller is J. The energy dissipation in the fan is modeled as a linear viscous damping component (of the bearings with damping constant b) and a quadratic aerodynamic damping component (of coefficient d).

a. Show that the system equation may be given by $J\dot{\omega} + b\omega + d|\omega|\omega = \tau(t)$.
b. Suppose that the motor torque is given by $\tau(t) = \bar{\tau} + \hat{\tau}_a \sin \Omega t$, in which $\bar{\tau}$ is the steady torque and $\hat{\tau}_a$ is a very small amplitude (compared to $\bar{\tau}$) of the torque fluctuations at frequency Ω. Determine the steady-state operating speed $\bar{\omega}$ (which is assumed positive) of the fan.
c. Linearize the model about the steady-state operating conditions and express it in terms of the speed fluctuations $\hat{\omega}$. From this result, estimate the amplitude of the speed fluctuations.

4.12 Consider a double pendulum (or a two-link robot arm with revolute joints) having link lengths l_1 and l_2, and the end masses m_1 and m_2, as shown in Figure P4.12.

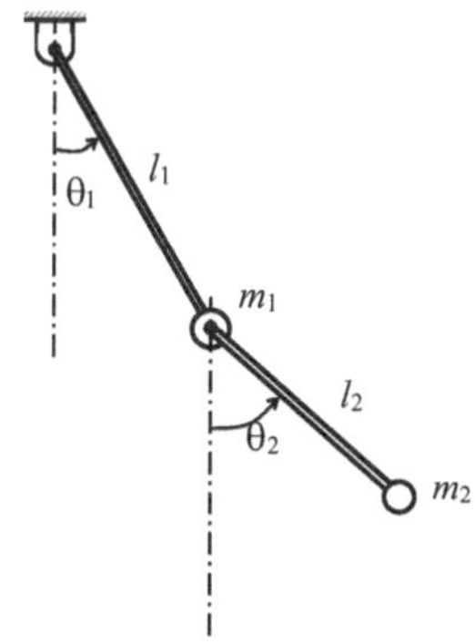

FIGURE P4.12 A double pendulum (robot arm with revolute joints).

Obtain the equations of motion for this nonlinear mechanical system in terms of the absolute angles of swing θ_1 and θ_2 about the vertical equilibrium configuration. Linearize the equations for small motions of θ_1, θ_1, θ_2 and θ_2.

Note: This is an example of geometric nonlinearity. Also, the nonlinear model itself can be used in feedback linearization (an "active" control method of linearization) of this system.

4.13

a. Linearized models of nonlinear systems are commonly used in many applications, including system analysis, design, and model-based control. What primary assumption is made in using a linearized model (not a specific method of linearization) to represent a nonlinear system?

b. A three-phase induction motor is used to drive a centrifugal pump for incompressible fluids. To reduce the shaft misalignment and associated problems such as vibration, noise, and wear, a flexible coupling is used for connecting the motor shaft to the pump shaft. A schematic representation of the system is shown in Figure P4.13.

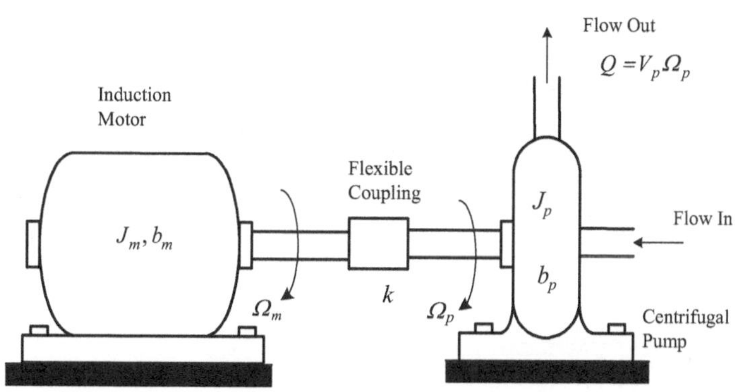

FIGURE P4.13 A centrifugal pump driven by an inductor motor.

Assume that the motor is a "torque source" of torque T_m, which is being applied to the motor of rotor inertia J_m. Also, the following variables and parameters are defined for the system:

J_p = moment of inertia of the pump impeller assembly
Ω_m = angular speed of the motor rotor and shaft
Ω_p = angular speed of the pump impeller and shaft
k = torsional stiffness of the flexible coupling

T_f = torque transmitted through the flexible coupling
Q = volume flow rate of the pump
b_m = equivalent viscous damping constant of the motor rotor

Also, assume that the net torque required at the ump shaft, to pump fluid steadily at a volume flow rate Q, is given by $b_p\Omega_p$, where $Q = V_p\Omega_p$ and V_p = volumetric parameter of the pump (assumed constant).

Using T_m as the input and Q as the output of the system, and through our unified and systematic approach that leads to a unique result, develop a complete state-space model for the system. Identify the model matrices **A**, **B**, **C**, and **D** in the usual notation, in this model. What is the order of the system?

c. In Part (a) suppose that the motor torque is given by $T_m = \dfrac{aSV_f^2}{\left[1+\left(S/S_b\right)^2\right]}$

where the fractional slip S of the motor is defined as $S = 1 - \dfrac{\Omega_m}{\Omega_s}$

Note: a and S_b are constant parameters of the motor. Also,
Ω_s = no-load (i.e., synchronous) speed of the motor
V_f = amplitude of the voltage applied to each phase winding (field) of the motor.

In *voltage control*, V_f is the input, and in *frequency control*, Ω_s is the input. For combined voltage control and frequency control, derive a linearized state-space model, using the incremental variables $\hat{V}_f$ and $\hat{\Omega}_s$, about the operating values $\bar{V}_f$ and $\bar{\Omega}_s$, as the inputs to the system and the incremental flow $\hat{Q}$ as the output.

4.14 A system that is used to pump an incompressible fluid from a reservoir into an open overhead tank is schematically shown in Figure P4.14. The tank has a uniform across section of area A.

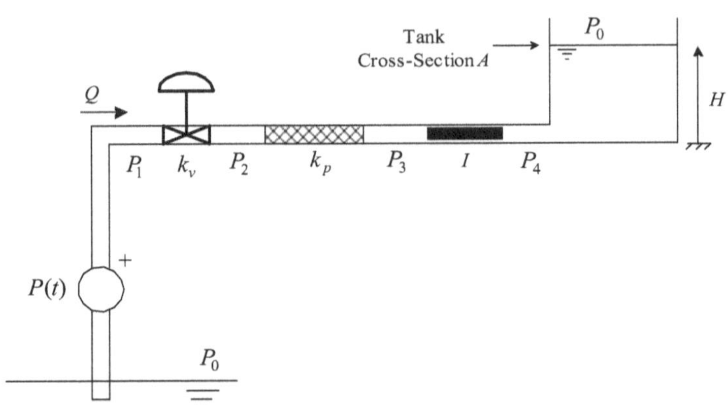

FIGURE P4.14 A pumping system for an overhead tank.

The pump is considered as a pressure source of gauge pressure (pressure difference wrt atmosphere) $P(t)$. A valve of constant k_v is placed near the pump in the long pipe segment, which leads to the overhead tank. The valve equation is $Q = k_v\sqrt{P_1 - P_2}$ in which Q is the volume flow rate of the fluid. The resistance to the fluid flow in the pipe may be modeled as $Q = k_p\sqrt{P_2 - P_3}$ in which k_p is a pipe flow constant. The effect of the accelerating fluid is represented by the linear equation $I\dfrac{dQ}{dt} = P_3 - P_4$ in which I denotes the fluid inertance. Pressures P_1, P_2, P_3, and P_4 are as marked along the pipe length, in Figure P4.14. Also, P_0 denotes the ambient pressure.

i. Using Q and P_{40} as the state variables, the pump pressure $P(t)$ as the input variable, and the fluid level H in the tank as the output variable, obtain a complete (nonlinear) state-space model for the system, using our unified and systematic approach, which leads to a unique model. *Note*: $P_{40} = P_4 - P_0$. Mass density of the fluid$=\rho$.
ii. Linearize the state equations about an operating point given by the flow rate $\bar{Q}$. Determine the model matrices $\boldsymbol{A}$, $\boldsymbol{B}$, $\boldsymbol{C}$, and $\boldsymbol{D}$ for the linear model.
iii. What is the combined linear resistance of the valve and the piping?

4.15

a. A local water storage tank with circular cross-section, with internal radius R at its base, is shown in Figure P4.15a. The inclination of the tank wall (with respect to the base) is θ. The outlet valve is closed and the inlet valve provides an inflow at the volumetric rate Q. At a given instant, the water level in the tank is h. Obtain an expression for the fluid capacitance C_f at that instant, in terms of h, R, θ, mass density ρ of water, and acceleration due to gravity (g).

Note: C_f varies with h in this case.

Hint: Gage pressure (w.r.t the atmosphere pressure) of water at the tank base $P = \rho gh$

Let r=radius of the water (cylinder) cross-section (at height h).

If a water volume (incremental) δV enters the tank, the water level will rise by δh. Then, $\delta V = \pi r^2 \cdot \delta h$

Divide throughout by the incremental time δt

$$\rightarrow \frac{\delta V}{\delta t} = \pi r^2 \frac{\delta h}{\delta t} \rightarrow \frac{dV}{dt} = \pi r^2 \frac{dh}{dt}$$

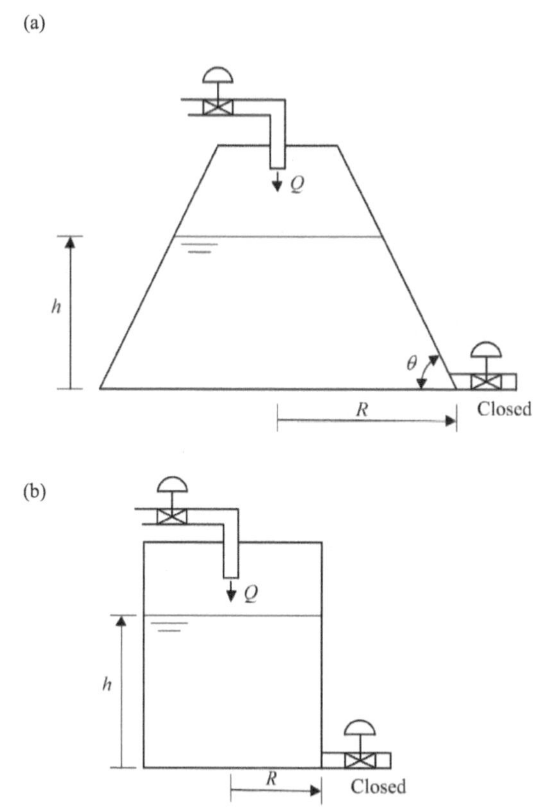

FIGURE P4.15 (a) Conical water tank with water level h; (b) cylindrical water tank with water level h.

b. From your result, show that for a cylindrical tank (i.e., $\theta = \frac{\pi}{2}$) of constant X-sectional radius R (see Figure P4.15b), we have the standard result, $C_f = \frac{\pi R^2}{\rho g} = \frac{A}{\rho g}$ = constant

Note: A=constant area of X-section of the tank.

4.16 Suppose that the volume V of a fluid is a nonlinear function $V(P)$ of its pressure P. Show that its fluid capacitance C_f is given by $C_f = \frac{dV}{dP}$ which itself is a function of P.

Using this expression, verify the result obtained in Problem 4.15.

4.17 Figure P4.17a shows a liquid pump driven by a dc motor through a flexible shaft. The moment of inertia of the motor rotor is J and the torsional stiffness of the flexible shaft is K. The torque T generated by the motor is a function of its speed Ω and the input voltage V. The steady-state characteristics of the motor (which neglects or somehow includes the

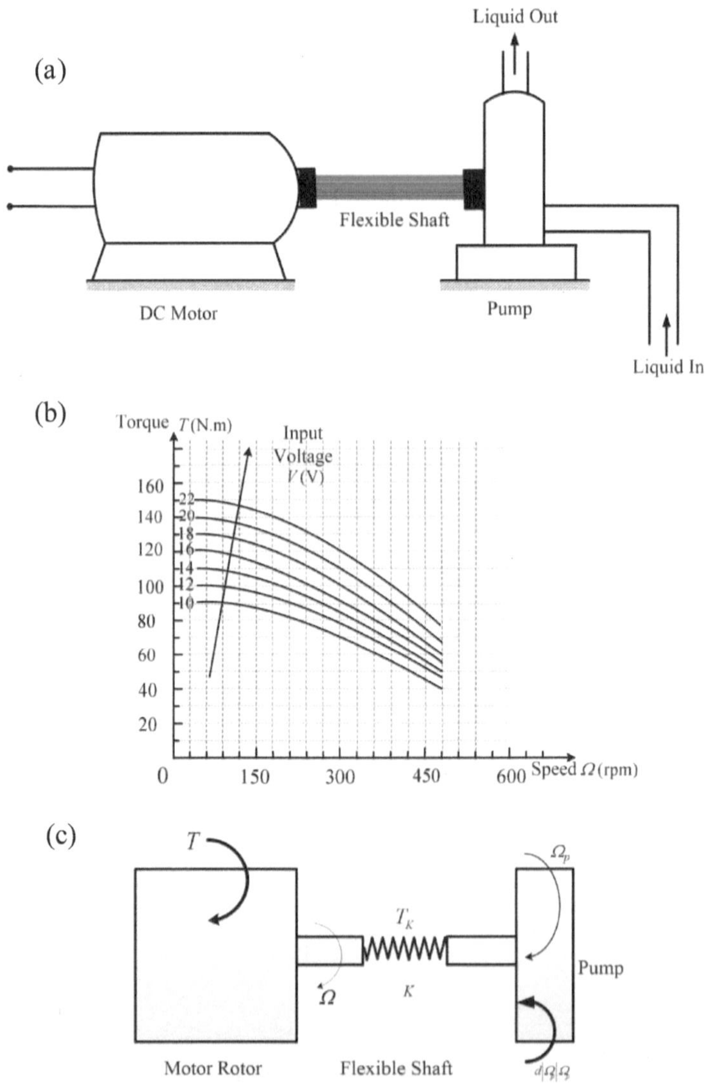

FIGURE P4.17 (a) Schematic diagram of a dc motor-driven pump; (b) steady-state torque versus speed characteristics of the motor; (c) loading on the system.

friction in the motor bearings) given as curves of $T(\Omega, V)$ with respect to Ω for different constant values of V are shown in Figure P4.17b. The speed of the pump is Ω_p. The load torque of the pump varies quadratically with the speed and is given by $d|\Omega_p|\Omega_p$, where d is a pump constant (*Note*: This formulation neglects the inertia of the pump impeller). The loading on the system is shown in Figure P4.17c.

Note: Take Ω_p as the output of the system.

a. Show that the constitutive equations of the system are

$$\text{Motor Rotor}: J_M \frac{d\Omega}{dt} = T(\Omega, V) - T_K$$

$$\text{Flexible Shaft}: \frac{dT_K}{dt} = K(\Omega - \Omega_p)$$

$$\text{Pump}: T_p = d|\Omega_p|\Omega_p$$

and the node equation at the shaft-pump joint is: $T_K - T_p = 0$
where T_K = torque in the flexible shaft; T_p = drive torque of the pump.

b. Linearize the system about the steady-state operating point: $\Omega_o = 300$ rpm and $V_o = 18$ V, and determine the elements of the corresponding model matrices $\boldsymbol{A}$, $\boldsymbol{B}$, $\boldsymbol{C}$, and $\boldsymbol{D}$.

Note: First express the matrix elements in terms of the given parameters (i.e., their symbols) and then compute their numerical values using the parameter values.
Given: $J = 0.005\ \text{kg}\cdot\text{m}^2$, $K = 10.0\ \text{N}\cdot\text{m/rad}$, $d = 5.0\ \text{N}\cdot\text{m/rad}^2/\text{s}^2$.
Denote the incremental variables, for the linear model, as:
$\delta\Omega = \omega$, $\delta T_K = \tau_K$, $\delta\Omega = \omega$, $\delta T_K = \tau_K$, $\delta V = v$, $\delta\Omega_p = \omega_p$
State vector: $\boldsymbol{x} = [\omega, \tau_K]^T$; input vector: $\boldsymbol{u} = [v]$, which is a scalar; and
Output vector: $\boldsymbol{u} = [\omega_p]$, which is also a scalar.

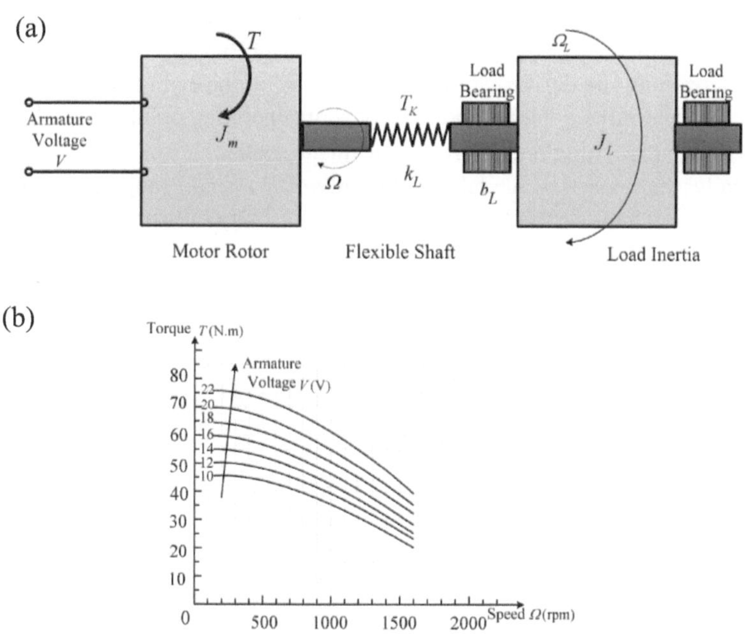

FIGURE P4.18 (a) Schematic diagram of a dc motor-driven load; (b) steady-state torque versus speed characteristics of the motor.

4.18 An electromechanical motion system is sketched in Figure P4.18a. It consists of an armature-controlled dc motor, which drives a load of moment of inertia J_L through a flexible shaft of torsional stiffness k_L. The shaft that carries the load has a set of bearings, which provides damping, assumed to be linear viscous with the overall damping constant b_L. The moment of inertia of the motor rotor is J_M. The torque T generated by the motor is a function of its speed Ω and the input armature voltage V. The steady-state characteristics of the motor, measured as curves of $T(\Omega, V)$ versus Ω for different constant values of V, are shown in Figure P4.18b.

Note: The mechanical friction in the motor bearings is also accounted for in T.

The speed of the load is Ω_L.

a. Determine a state-space model for the system.
b. Linearize the system about the steady-state operating point: $\Omega_o = 1000$ rpm and $V_o = 16$ V and express the corresponding matrices: $\boldsymbol{A}$ and $\boldsymbol{B}$. Give the numerical values of the elements of these matrices.

Given: $J_M = 0.005$ kg·m^2, $J_L = 0.010$ kg·m^2, $k_L = 10.0$ N·m/rad, $b_L = 1.0$ N·m/rad/s

Incremental variables in the linear model are: $\delta\Omega = \omega$, $\delta\Omega_L = \omega_L$, $\delta T_K = \tau_K$, $\delta V = v$

State vector: $\boldsymbol{x} = \left[\omega, \omega_L, \tau_K\right]^T$

Input vector: $\boldsymbol{u} = [v]$, which is a scalar.

4.19 A thermistor (a semiconductor-based temperature sensor) is modeled as

$$R = R_o \exp\left[\beta\left(\frac{1}{T} - \frac{1}{T_o}\right)\right]$$

where R=resistance of the thermistor at temperature T; and T_o, R_o, and β are model parameters, whose values are known.

Typically the resistance R is measured, and from it, the corresponding temperature T is computed using the model (or read using a calibration curve).

a. What is the input and what is the output of this model? Justify your answer.
b. Is this a static model or a dynamic model? Justify your answer.
c. What do the model parameters T_o and R_o represent? Explain.
d. By determining the derivative of the expression on the right-hand side of the model equation, determine a linear model about some operating point $(\bar{T}, \bar{R})$. Use the incremental variables $\hat{T}$ and $\hat{R}$ to express your linear model.
e. Typically, the linearized model is valid only over a small range about the operating point. Suggest another method to linearize the given thermistor model so that it will be accurate for any value of T and R.

Hint: $\dfrac{d\left(e^{ax}\right)}{dx} = ae^{ax}$

4.20 List several response characteristics of nonlinear systems that are not exhibited by linear systems in general.

Determine the response y of the nonlinear system $\left[\dfrac{dy}{dt}\right]^{1/3} = u(t)$ when excited by the input $u(t) = a_1 \sin\omega_1 t + a_2 \sin\omega_2 t$.

What characteristic of a nonlinear system does this result illustrate?

4.21 A mechanical component, whose response is x, is governed by the relationship

$$f = f\left(x, \dot{x}\right)$$

where f denotes the applied force (input) and $\dot{x}$denotes the velocity response (output). Consider the following four special cases for this component:

a. Spring: $f = kx$
b. Damper: $f = b\dot{x}$
c. Spring with a damper: $f = kx + b\dot{x}$
d. Spring with Coulomb friction: $f = kx + f_c\,\mathrm{sgn}(\dot{x})$

Suppose that a harmonic excitation of the form $f = f_o \sin\omega t$ is applied in each case. Sketch the force–displacement curves for the four cases, at steady state. Which components exhibit hysteresis? Which components are nonlinear? Discuss your answers.

4.22 Consider a Coulomb-friction damper model given by the constitutive relation $f = f_c\,\mathrm{sgn}(\dot{x})$ where,

f = damping force
x = relative displacement of the damper
$\dot{x}$ = relative velocity of the damper
f_c = magnitude (constant) of the damping force.

Suppose that a cyclic force is applied to the damper. At steady state, take the harmonic component (dominant component) of the displacement x as

$$x = x_0 \sin(\omega t + \phi)$$

a. Determine the energy dissipation of the Coulomb damper in one cycle of motion (i.e., in the period $T = 2\pi / \omega$). Express your result in terms of f_c, x_0 and ω (as necessary).
b. Now consider the linear viscous damper model given by $f = b\dot{x}$ where b = viscous damping coefficient. By following the same procedure as for the Coulomb damper, determine the energy dissipation in one cycle of motion of the linear damper. Express your result in terms of b, x_0 and ω.
c. From the two results, give an equivalent linear viscous damping coefficient b_{eq} that results in the same energy dissipation per cycle as for the nonlinear damper. Express your result in terms of f_c, x_0 and ω.
d. Comment on the limitations in the use of this method of linearization.

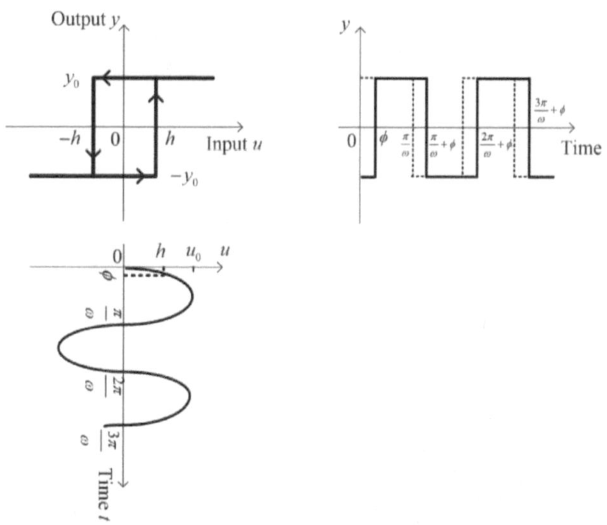

FIGURE P4.23 The output of a relay with hysteresis for a sine input.

4.23 The input–output behavior of a relay (two-state switching function) with hysteresis is shown in Figure P4.23. Specifically, consider a sinusoidal input, given by $u = u_0 \sin \omega t$.

u_0 = amplitude of the input
ω = frequency of the input
y_0 = amplitude of the output
h = hysteresis threshold at the input.

Determine a describing function for this nonlinearity.

4.24 Discuss why feedback linearization could be very useful in the "active" control of complex mechanical systems with nonlinear and coupled dynamics. What are the shortcomings of linearizing feedback?

Consider the two-link manipulator that carries a point load (weight W) at the end effector, as shown in Figure P4.24. Its dynamics can be expressed as: $\boldsymbol{I}\ddot{\boldsymbol{q}} + \boldsymbol{b} = \boldsymbol{\tau}$

where

$\boldsymbol{q}$ = vector of (relative rotations) q_1 and q_2
$\boldsymbol{\tau}$ = vector of drive torques τ_1 and τ_2 at the two joints, corresponding to the coordinates q_1 and q_2.

$\boldsymbol{I}$ = second-order inertia matrix = $\begin{bmatrix} I_{11} & I_{12} \\ I_{21} & I_{22} \end{bmatrix}$

$\boldsymbol{b}$ = vector of joint frictional, gravitational, centrifugal and Coriolis torques (components are b_1 and b_2)

Neglecting joint friction, and with zero payload (W=0), we can write the following expressions for the model parameters:

$$I_{11} = m_1 d_1^2 + I_1 + I_2 + m_2\left(\ell_1^2 + d_2^2 + d\ell_1 d_2 \cos q_2\right)$$

$$I_{12} = I_{21} + I_2 + m_2 d_2^2 + m_2 \ell d_2 \cos q_2$$

$$I_{22} = I_2 + m_2 d_2^2$$

$$b_1 = m_1 g d_1 \cos q_1 + m_2 g\left[d_1 \cos q_1 + d_2 \cos\left(q_1 + q_2\right)\right] - m_2 \ell_1 d_2 \dot{q}_2^2 \sin q_2 - 2 m_2 \ell_1 d_2 \dot{q}_1 \dot{q}_2 \sin q_2$$

$$b_2 = m_2 g d_2 \cos\left(q_1 + q_2\right) + m_2 \ell_1 d_2 \dot{q}_1^2 \sin q_2$$

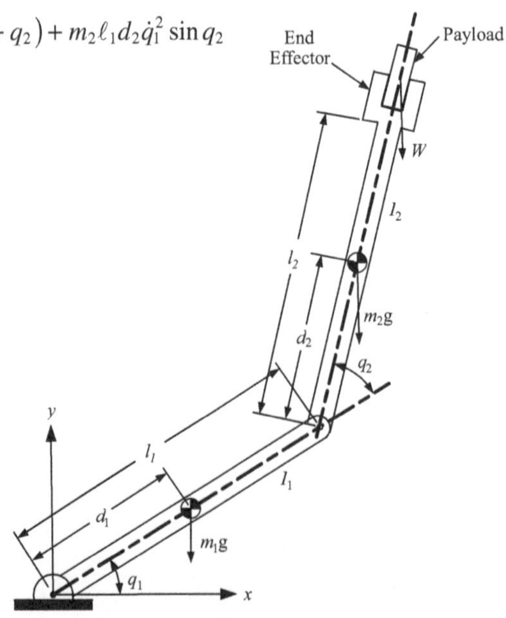

FIGURE P4.24 A two-link robotic manipulator.

where I_1, I_2 = moments of inertia of the links about their centroids

m_1, m_2 = masses of the links

The geometric parameters ℓ_1, ℓ_2, d_1, and d_2 are as defined in Figure P4.24.

What variables have to be measured for linearizing feedback control? Noting that the elements of $\boldsymbol{b}$ are more complex (even after neglecting backlash, and payload) than the elements of $\boldsymbol{I}$, justify using linearizing feedback control for this system, which uses both a nonlinear model and online measurements.

4.25 A sketch of a household stationary exercise bicycle is shown in Figure P4.25a. A photo of a corresponding unit is shown in Figure P4.24b. A model of its resistance mechanism is

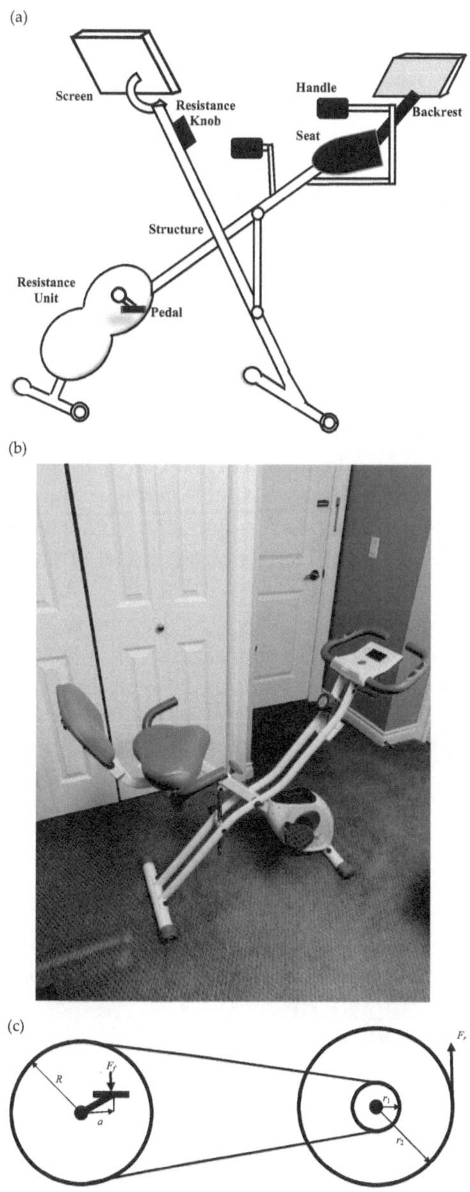

FIGURE P4.25 (a) A household stationary exercise bicycle; (b) A photo of a commercial unit; (c) A model of the resistance mechanism.

shown in Figure P4.25c. The resistance force (F_r) on the flywheel is provided by a magnetic strip, which is pressed on to the flywheel (radius r_2). This force can be adjusted manually by using the adjustment knob. The flywheel has an integral smaller pulley (radius r_1), which is driven by a belt loop linked to the pedal wheel (radius R). To the pedal pair, an equivalent normal force F_f is applied by the feet of the user, at speed v. The corresponding lever arm distance is a. The pedals rotate the pedal wheel, and the required energy for it is provided by the operator. The foot speed, the operator power, and the energy burned by the operator (calories, which is the time integral of power) are displayed on the screen of the bicycle.

a. Derive an expression for the power p provided by the operator to the bicycle, in terms of n, a, r_2, F_r and v. Here, $n = \frac{R}{r_1} > 1$ is the angular speed ratio (increase) achieved through the belt drive. Show that, for a given resistance force (F_r), the power p increases linearly with the pulley radius (speed) ratio n, which is a benefit.

   Note: Neglect the bearing friction and inertia (or, assume static conditions).
b. Linearize the derived expression for power.
c. Now consider the dynamic conditions of the system. Specifically, approximate the bearing friction by equivalent viscous damping, with rotatory damping constants b_f and b_r at the pedal wheel and the resistance flywheel, respectively. Also, let the moments of inertia of the pedal wheel and the resistance flywheel (including its belt pulley) be J_f and j_r, respectively. Derive an expression for the power provided by the operator to the bicycle, under these conditions.
d. Comment on: (i) benefit of the dynamic model; (ii) benefit of the speed ratio n, (iii) need for linearization, and (iv) accuracy of the numbers displayed on the screen.

4.26 An automated wood cutting system contains a cutting unit, which consists of a dc motor and a cutting blade, linked by a flexible shaft and a coupling. The purpose of the flexible shaft is to position the blade unit at any desirable configuration, away from the motor itself. The coupling unit helps with the shaft alignment (compensates for possible misalignment). A simplified, lumped-parameter, dynamic model of the cutting device is shown in Figure P4.26.

The following parameters and variables are shown in the figure:

J_m=axial moment of inertia of the motor rotor
b_m=equivalent viscous damping constant of the motor bearings
k=torsional stiffness of the flexible shaft
J_c=axial moment of inertia of the cutter blade

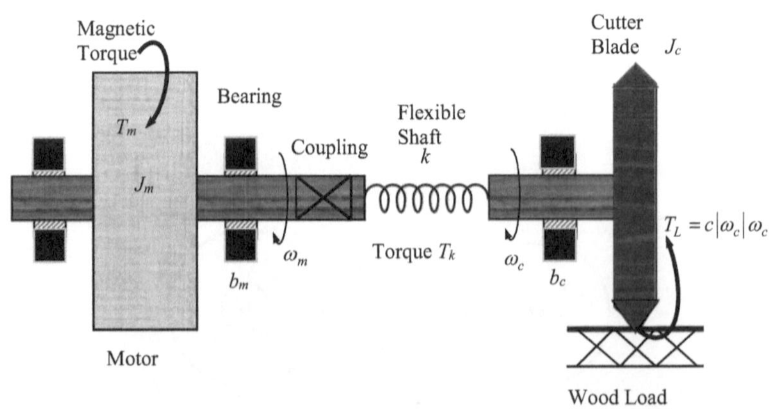

FIGURE P4.26 A wood cutting machine.

b_c = equivalent viscous damping constant of the cutter bearings
T_m = magnetic torque of the motor
ω_m = motor speed
T_k = torque transmitted through the flexible shaft
ω_c = cutter speed
T_L = load torque on the cutter from the workpiece (wood).

In comparison with the flexible shaft, the coupling unit is assumed rigid and is also assumed light. The cutting load is given by

$$T_L = c|\omega_c|\omega_c$$

The parameter c, which depends on such factors as the depth of cut and the material properties of the workpiece, is assumed constant in the present analysis.

a. Using T_m as the input, T_L as the output, and $\begin{bmatrix}\omega_m & T_k & \omega_c\end{bmatrix}^T$ as the state vector, develop a complete (nonlinear) state-space model for the system shown in Figure P4.26. What is the order of the system?
b. Using the state model derived in Part (a), obtain a single input–output differential equation for the system, with T_m as the input and ω_c as the output.
c. Consider steady operating conditions where $T_m = \bar{T}_m$, $\omega_m = \bar{\omega}_m$, $T_k = \bar{T}_k$, $\omega_c = \bar{\omega}_c$, $T_L = \bar{T}_L$ are all constants. Express the operating point values $\bar{\omega}_m$, $\bar{T}_k$, $\bar{\omega}_c$, and $\bar{T}_L$ in terms of the steady input $\bar{T}_m$ and the model parameters only. You must consider both cases $\bar{T}_m > 0$ and $\bar{T}_m < 0$.
d. Now consider an incremental change $\hat{T}_m$ in the motor torque and the corresponding changes $\hat{\omega}_m$, $\hat{T}_k$, $\hat{\omega}_c$, and $\hat{T}_L$ in the system variables. Determine a linear state model (**A**, **B**, **C**, **D**) for the incremental dynamics about the operating point of the system, using $\boldsymbol{x}=\begin{bmatrix}\hat{\omega}_m & \hat{T}_k & \hat{\omega}_c\end{bmatrix}^T$ as the state vector, $\boldsymbol{u}=\begin{bmatrix}\hat{T}_m\end{bmatrix}$ as the input, and $\boldsymbol{y}=\begin{bmatrix}\hat{T}_L\end{bmatrix}$ as the output.
e. In the incremental model (see Part (d)), if the angle of twist of the flexible shaft (i.e., $\theta_m - \theta_c$) is used as the output, what will be a suitable state model? What is the system order then?
f. In the incremental model, if the angular position θ_c of the cutter blade is used as the output variable, explain how the state model obtained in Part (d) should be modified. What is the system order in this case?

4.27 An object of mass m is propelled vertically upward by a rocket with force $F(t)$. When it is at height (elevation) Y from the earth's surface, the gravitational pull F_g is given by $F_g = mg\left(\frac{R}{R+Y}\right)^2$, where R is the average radius of the earth (approximately 6370 km). See Figure P4.27a.

Note: When the force is a function of the position of the object on which it is applied, it may be interpreted as a "generalized spring force." In the present example, it is a gravitational spring, which is nonlinear. Also, in this example, we neglect the aerodynamic forces such as drag. This amounts to considering the behavior of the object at very high elevations only, where the aerodynamic forces are negligible.

a. Taking the velocity v and the gravitational spring force F_g as the state variables (*Note*: This is the systematic and unified approach where we choose the across-variable v of the A-type element—mass, and the through-variable F_g of the T-type element—spring, as the state variables) and $F(t)$ as the input variable, develop a state-space model for the mass dynamics.
b. From the nonlinear state-space model, determine the nonlinear input–output (I-O) model, with v as the output.

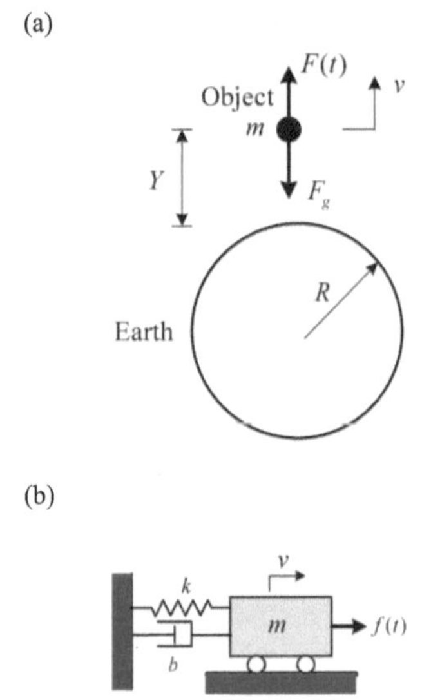

FIGURE P4.27 (a) Propelled object subjected to gravitational force; (b) Forced mass-spring-damper unit.

c. For the condition of constant upward velocity v_o of the object, determine the corresponding propelling force $F(t)$ and the gravitational force F_{go}.
d. Linearize the state-space model obtained in Part (a), about the operating condition established in Part (c).
e. Linearize the I-O model obtained in (b) about the operating condition determined in Part (c). Also, obtain the linear I-O model by using the linear state-space model of Part (d). Compare the two results.
f. Compare the linearized I-O model of Part (e) with a simple (linear) mass-spring-damper system subjected to an external force, as shown in Figure P4.27b.

4.28 A simplified representation of the vertical dynamics of a hovercraft is shown in Figure P4.28.

A flow-controlled pump produces air flow at the volume rate $Q_s(t)$. This air enters the cylindrical space between the hovercraft and the ground, at gauge pressure P_f, and exits to the atmosphere (zero gauge pressure).

Note: Gauge pressure = Absolute pressure – Atmospheric pressure

The following parameters are known:

M = mass of the hovercraft

A = cross-sectional area of the cylindrical space underneath the hovercraft

h = height of the interior wall of the cylindrical space

The following analytical model may be used to study the vertical dynamics of the hovercraft:

$$M\dot{v} = AP_f - Mg \tag{i}$$

$$C_f \frac{dP_f}{dt} = Q_f \tag{ii}$$

$$P_f = k_r Q_e^{\,2} \tag{iii}$$

$$Q_s(t) - Q_e - Av - Q_f = 0 \tag{iv}$$

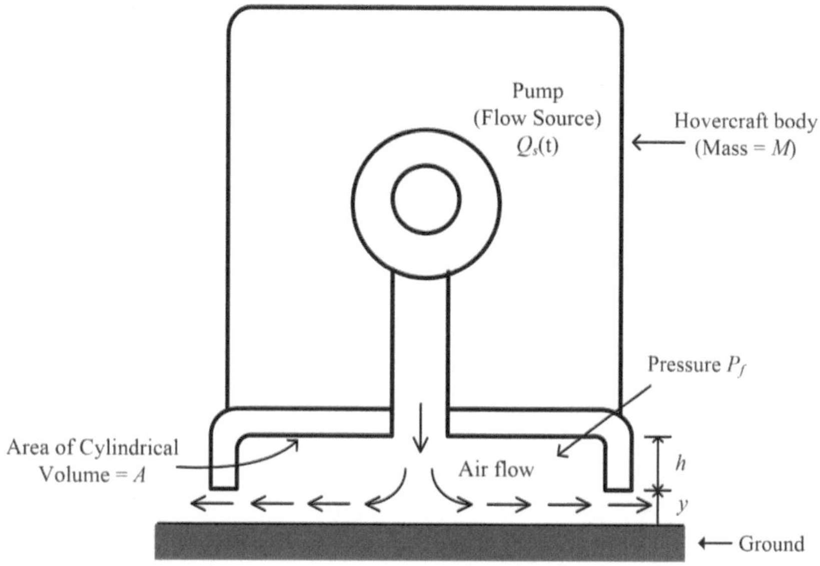

FIGURE P4.28 A simplified model for vertical dynamics of a hovercraft.

Here,

Q_e = volume flow rate of air exiting from the bottom edge of the hovercraft into the atmosphere

y = height of the bottom edge of the hovercraft from the ground

v = vertical upward velocity of the hovercraft = $\frac{dy}{dt}$

k_r = a known flow parameter

a. Among the equations (i) through (iv), which ones represent the following: state-space shell, remaining constitutive equation, node equation?

b. In equation (ii), what do Q_f and C_f represent? If the air in the cylindrical space below the hovercraft obeys the gas law $P_f V_f^{\,k}$ = constant c, where k is the adiabatic constant, give an expression for C_f (without derivation) in terms of V_f, k, and P_f.

Note: V_f = volume of air in the cylindrical space underneath the hovercraft

c. Give a linearized state-space model for the system for small vertical motions about the steady state. The steady sate is when $Q_s(t) = Q_0$ = constant, $y = y_0$ = constant, and $\frac{dP_f}{dt} = 0$

Note: Express the model in terms of the parameters A, h, y_0, k, M, g, and k_r.

4.29 A model of a mechanical load driven by an actuator and supported by an air spring is shown in Figure P4.29. The following parameters and variables of the system are defined:

m = mass of the mechanical load

$f(t)$ = actuator force

P_1, P_2 = air pressures on the two sides of the piston

V_1, V_2 = volumes of the two compartment of the cylinder, separated by the piston

x, v = displacement and velocity of the mechanical load

A = area of the piston (or cylinder cross-section)

L = length of each compartment of the cylinder under the static conditions: $P_1 = P_2 = P_0$;; $V_1 = V_2$; $f(t) = 0$; $x = 0$; $\gamma = c_p/c_v$ = ratio of the specific heats of air at constant pressure and constant volume

Note: There is sufficient thermal insulation around the cylinder.

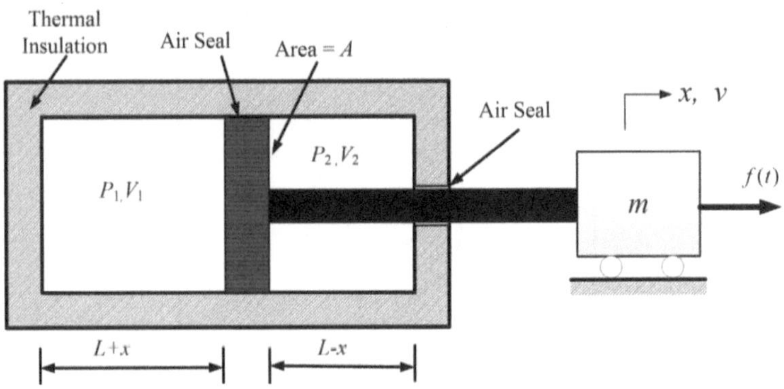

FIGURE P4.29 A mechanical system with an air spring.

a. What assumptions may be in dynamic modeling of the system?
b. Derive a complete (nonlinear) state-space model for the system. Plot the characteristic curve of the nonlinear air spring (force f_s versus displacement x) in the non-dimensional form: $\frac{f_s}{AP_0}$ versus $\frac{x}{L}$. Show that it is a one-to-one relationship (i.e., there are no multiple values for the other variable corresponding to a value of one variable).
c. Linearize the state-space model. Discuss how you can adjust the stiffness of the linearized spring.

4.30 The input and output axes (x and y axes) of a hysteresis curve and also the direction (indicated by an arrow) of the curve are important considerations. For example, suppose that the input is a force and the output is a displacement. Then the work done is represented by the area between the curve and the y-axis (the displacement axis). We see that,

Net work done in a cycle=area of hysteresis loop=energy dissipation in a damped system.

Consider the case of linear viscous damping where the damping force is proportional to the relative speed of the damper. Specifically, damping force $f = b\frac{dx}{dt}$, where b=damping constant, x=displacement.

Show that even though the device is linear, it exhibits a hysteresis loop.

5 Linear Graphs

HIGHLIGHTS

- Benefits of Linear Graphs (LGs)
- Sign Convention
- Use of Through- and Across-Variables
- Action and Reference Points of a Branch
- LG Elements (*A*-Type, *T*-Type, and *D*-Type)
- Sources (*A*-Source and *T*-Source)
- Multi-port Elements (Transformer, Gyrator)
- LG Topological Relation
- LG Solvability
- State Model Development (Constitutive, Node, Loop Equations)
- Examples in Mechanical, Electrical, Fluid, Thermal, Mixed Domains (Multi-physics)

5.1 INTRODUCTION

In the previous chapters, we presented a unified, integrated, and systematic approach for developing an analytical model, particularly a state-space model, of an engineering dynamic system, which leads to a single (unique) model. A starting step of that approach involves sketching a *structural diagram* (e.g., circuit diagram, schematic diagram, free-body diagrams) of the system to show how the elements or components of the system are interconnected. Such a graphical representation can immensely facilitate the process of model formulation. We assert that the particular graphical approach should be unified, integrated, and systematic as well.

There are several graphical approaches for representing the structure of a lumped-parameter analytical model of an engineering dynamic system. Leading among them are:

1. Linear graphs (LGs)
2. Bond graphs.

This chapter presents the approach of LGs. The focus of Chapters 5–7 is LGs in view of the advantages of the approach, as highlighted in this book. More advanced concepts and material on LGs are found under *graph trees*, in Appendix A. A toolbox for using LGs is presented in Appendix B.

State-space models of lumped-parameter dynamic systems, regardless of whether they are mechanical, electrical, fluid, thermal, or multi-physics (multi-domain or mixed), can be conveniently developed by LGs. In other words, LGs support the multi-domain, "unified" (or analogous) and "integrated" (or concurrent) approach of model development, as presented in Chapter 3. Interconnected line segments (called *branches*) connected at *nodes* are used in an LG to represent a dynamic model. In this regard, LGs have the particular advantage that the model structure is analogous across physical domains. For example, two electrical components (say, an inductor and a resistor) connected in parallel and two analogous mechanical components (a spring and a damper) connected in parallel are represented by similar LGs (i.e., two branches in parallel). The term "linear graph" stems from this use of "line" segments and does not mean that the system itself has to be linear. The term "line graph" is also used in some literature synonymously with "linear graph."

Note: Nonlinear models can be represented by LGs.

DOI: 10.1201/9781003124474-5

Particular characteristics and advantages of LGs in model development and representation include the following:

- Applicable for lumped-parameter engineering dynamic systems
- Line segments (branches) are used to represent model elements
- By interconnecting branches at "nodes," an LG provides a graphical representation of a model that is true to the physical structure of the modeled system (it allows visualization of the system structure prior to model formulation)
- LGs facilitate an integrated (i.e., concurrent → all physical domains can be represented in a single LG and analyzed together) methodology for multi-physics systems
- LGs facilitate a unified (i.e., analogous methodology is used in multiple domains) modeling approach
- The model structure is retained across domain (i.e., interconnected components in one domain and similarly interconnected analogous elements in another domain have the same LG structure)
- LGs help identify similarities (in the physical domain, structure, behavior, etc.) in systems
- LGs facilitate the development of computer-based modeling tools and software (systematic, unified, integrated, graphical; see Appendix B)
- A different treatment is not needed when modeling multi-functional devices (e.g., a piezoelectric device, which can function as both a sensor and an actuator, can be represented simply as a reversible source).

This chapter presents the use of LGs in the development of analytical models (particularly, state-space models) for mechanical, electrical, fluid, thermal, and multi-physics systems.

5.2 VARIABLES AND SIGN CONVENTION

LGs systematically use through-variables and across-variables in providing a unified approach for the modeling of dynamic systems in multiple physical domains (mechanical, electrical, fluid, thermal, and any mixt of them). In accomplishing this objective, it is important to adhere to standard and uniform conventions so that there will not be ambiguities in a given LG representation. In particular, a standard sign convention must be established. These issues have to be addressed in presenting a "systematic" approach of modeling. They are discussed now.

5.2.1 Through-Variables and Across-Variables

Each branch (a line segment) in an LG model has one *through-variable* (denoted by f) and one *across-variable* (denoted by v) associated with it. Typically, their product is the power variable. For instance, in a hydraulic or pneumatic system, a pressure "across" an element causes some change of fluid flow "through" the element. The across-variable is the pressure and the through-variable is the flow. The ordered variable pair (f, v) of the branch should be marked on one side of the branch. Table 5.1 lists the through- and across-variable pairs for the four domains that are considered in the present treatment.

TABLE 5.1
Through- and Across-Variable Pairs in Several Physical Domains

System Type (Domain)	Through-Variable	Across-Variable
Hydraulic/pneumatic	Flow rate	Pressure
Electrical	Current	Voltage
Mechanical	Force/torque	Velocity/angular velocity
Thermal	Heat transfer rate	Temperature

5.2.1.1 Sign Convention

- **Reference Point and Action Point**: Consider Figure 5.1 where a general basic element (strictly, a single-port element, as will be discussed later) of a "mechanical" dynamic system is shown. In the LG representation, as shown in Figure 5.1b, the element is drawn as a branch (i.e., a line segment). One end of the branch is selected as the *point of reference* and the other end automatically becomes the *point of action* (see Figure 5.1a and c). The choice is somewhat arbitrary in many situations. However, it may reflect how other elements are connected to this particular element and the physics of the actual system (*Note*: For an inertia element, the point of reference is always the ground—the inertial reference, with zero velocity).
- **Oriented Branch**: An *oriented* branch is one to which a direction is assigned, using an arrowhead, as in Figure 5.1b. The arrow head denotes the positive direction of power flow at each end of the element. By convention, the positive direction of power is taken as "into" the element at the point of action and "out of" the element at the point of reference. According to this convention, the arrowhead of a branch is pointed from the point of action toward the point of reference. Then, the arrowhead also represents the direction of the "drop of the across-variable." In this manner, the point of reference and the point of action are easily identified. (*Note*: There is an exception—the *T*-source, where the arrow head is from the point of reference to the point of action, as will be discussed later).
- Figure 5.1 specifically concerns a mechanical element for the following reason. The through-variable "force" at the point of action is applied in the direction of the "velocity drop" (across-variable) which is direction of the arrowhead, so that their product, power, is going into the element at that point. At the point of reference, the force on the element is in fact the "reaction" from the elements to which this element is connected there. Hence, at the point of reference, the force is in the direction opposite to the velocity so that the power (positive) is coming out of the element at that point. For an electrical element, the through-variable is the current. At the point of action, the current flows into the element, and at the point of reference, the current flows out of the element. Hence, current, unlike force (which is a reaction at the reference), has to be marked in the same direction at the point of reference, so that the power (the product of current and voltage) is coming out at the point of reference (positive) just like in a mechanical element.
- **Through-Variables and Across-Variables**: The through-variable f and the across-variable v are indicated as an ordered pair (f, v) on one side of the branch, as in Figure 5.1b. The across-variable of a branch is always given relative to the point of reference. The relationship between f and v (the *constitutive relation* or physical relation, as discussed in Chapters 2 and 3) can be linear or nonlinear. The parameter of the element (e.g., mass, capacitance) is shown on the other side of the branch. It should be noted that the direction of a branch does not represent the positive

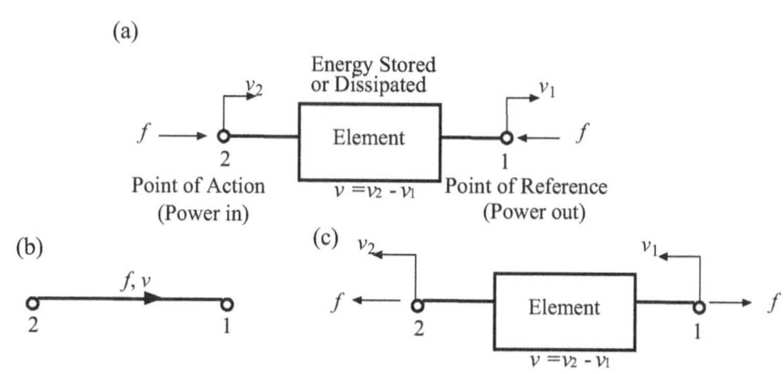

FIGURE 5.1 Sign convention for a linear graph: (a) A basic (mechanical) element and positive directions of its variables; (b) linear graph branch of the element; (c) an alternative sign convention.

direction of f or v. For example, when the positive directions of both f and v are reversed, as in Figure 5.1c, compared to Figure 5.1a, the LG remains unchanged, as in Figure 5.1b, because the positive direction of power flow is unchanged. In a given problem, the positive direction of any one of the two variables f and v should be pre-established for each branch. Then the corresponding positive direction of the other variable is automatically determined by the convention used to orient LGs. It is customary to assign the same positive direction for f (and v) and the power flow in at the point of action (i.e., the convention shown in Figure 5.1a, not Figure 5.1c, is customary). Based on that, the positive directions of the variables at the point of reference are automatically established (because the power flows out there).

Note: In a branch (line segment), the through-variable (f) is transmitted through the element without changing its value. The absolute value of the across-variable, however, changes across the element (from v_2 to v_1, in Figure 5.1a). It is this change ($v = v_2 - v_1$) across the element (i.e., the value at the point of action with respect to that at the point of reference, or the "variable drop") that is called the across-variable. For example, v_2 and v_1 may represent the electric potentials at the two ends of an electric element (e.g., a resistor) and then v represents the voltage across the element (or the voltage "drop" from the action to the reference). In other words, the across-variable is measured relative to the point of reference of the particular element. A special case is an inertia (mass) element, whose point of reference is always the inertial reference (where $v_1 = 0$).

Note: Since the absolute value of the across-variable drops from the point of action to the point of reference, it should be clear that the arrow also indicates the direction of "drop" in the across-variable. There is an exception to this convention, in the case of *T*-source, as discussed later.

According to the sign convention shown in Figure 5.1, the work done (by an external device) on the element at the point of action is positive (i.e., power flows in there), and the work done (on an external load or environment) by the element at the point of reference is positive (i.e., power flows out there). The difference of the work done on the element and the work done by the element (i.e., the difference in the energy flow at the point of action and the point of reference) is either stored as energy (e.g., kinetic energy of a mass; potential energy of a spring; electrostatic energy of a capacitor; electromagnetic energy of an inductor—see Chapter 2), and this stored energy has the capacity to do additional work or dissipated (e.g., mechanical damper; electrical resistor) through various mechanisms that are manifested as heat transfer, noise, wear, and other phenomena. In summary:

1. An element (a single-port element) is represented by a line segment (branch). One end is the point of action and the other end is the point of reference
2. The through-variable f is the same at the point of action and at the point of reference of an element; the across-variable differs, and it is this difference (value at the point of action relative to the point of reference) that is called the across-variable v
3. The variable pair (f, v) of the element is shown on one side of the branch. Their relationship (constitutive relation) can be linear or nonlinear. The parameter of the element is shown on the other side of the branch
4. Typically, the power flow p in a branch is the product of the through-variable and the across-variable. By convention, at the point of action, f and p (power flow in) are taken to be positive in the same direction; and at the point of reference, f that is coming out of the element (which is provided to the connected element) is positive in the direction corresponding to power flowing out of the element
5. The positive direction of power flow p (or energy or work) is into the element at the point of action and out of the element at the point of reference. This direction is shown by an arrow on the LG branch (an oriented branch). *Note*: An exception is a "source" element (e.g., voltage source, force source)
6. The difference in the energy flows at the two ends of the element is either stored (with capacity to do further work) or dissipated, depending on the element type.

LG representation is particularly useful in understanding the rates of energy transfer (power) associated with various phenomena. In particular, dynamic interactions in a physical system (mechanical, electrical, fluid, etc.) can be interpreted in terms of power transfer. Power is the product of a through-variable (a generalized force or current) and the corresponding across-variable (a generalized velocity or voltage). For example, consider a mechanical system. The total work done on the system is, in part, used as stored energy (kinetic and potential) and the remainder is dissipated. Stored energy can be completely recovered when the system is brought back to its original state (i.e., when the cycle is completed). Such a process is said to be *reversible.* On the other hand, dissipation corresponds to irreversible energy transfer, which cannot be recovered by returning the system to its initial state. (A fraction of the mechanical energy that is lost in dissipation could be recovered, in principle, by operating a heat engine, but we shall not go into these thermodynamic details which are beyond the present scope). Energy dissipation may appear in many forms including temperature rise (a molecular phenomenon), noise (an acoustic phenomenon), or work used up in wear mechanisms (also, a molecular or mechanical phenomenon).

5.3 LINEAR GRAPH ELEMENTS

Many types of basic elements exist, which can be used to develop an LG for a dynamic system. In this section, we discuss two types of basic elements under the categories of single-port elements and two-port elements. Analogous elements in these categories exist across the physical domains (mechanical, electrical, fluid, and thermal) for the most part. *Note*: General multi-port elements can be identified as well. However, for typical modeling situations, single-port and two-port elements are adequate. They can be combined, if necessary, to form multi-port modules.

5.3.1 Single-Port Elements

Single-port (or, *single energy port)* elements are those that can be represented by a single branch (single line segment) of LG. These elements possess only one power (or energy) variable (the product of the through-variable and the across-variable, which is the difference between the input power and the output power). This power is "ported" through a single branch; hence, the name "single-port." They have two terminals (the point of action and the point of reference) as noted before. The general form of a single-port LG element is shown in Figure 5.1b.

5.3.1.1 Mechanical Elements

In modeling a mechanical system, we require three passive single-port elements, as shown in Figure 5.2. These lumped-parameter mechanical elements are mass (or inertia), spring, and damper (dashpot). Although only the *translatory* mechanical elements are presented in Figure 5.2, the corresponding *rotary* elements have the same LG form, and the corresponding parameters and variables are easy to establish. In particular, in the latter case, f denotes the torque through the element and v is the relative angular velocity in the same direction at the point of action.

Note: The LG of an inertia element has a broken-line segment. This is because in an inertia element there is no direct physical link between the point of action (lumped inertia element) and the point of reference (ground reference). The "inertia force" does not directly travel from the point of action to the point of reference. However, this force in the inertia element is transmitted indirectly ("felt") at the point of reference, as a "reaction" at the ground, for the source that applies a force to the inertia element. For example, the "source" that applies the force to the inertia element, which generates the "inertia force," will have its other end at the ground. Hence, there will be an equal force transmitted to the ground. Imagine the situation where you are pushing a mass, causing it to accelerate (and creating an inertial force equal to the product of mass and acceleration). Then your feet on the ground will transmit an equal force to the ground. This issue will be further discussed in Example 5.1.

Analogous single-port electrical elements may be represented in a similar manner. These are shown in Figure 5.3.

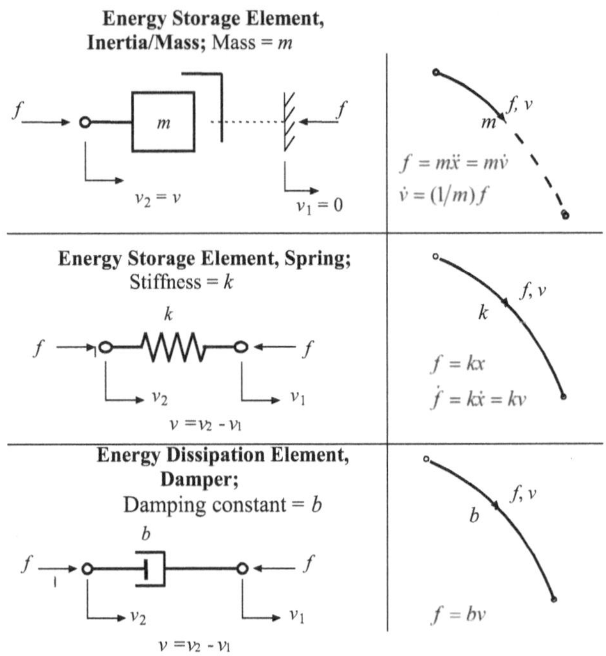

FIGURE 5.2 Single-port mechanical elements and their linear graph representations.

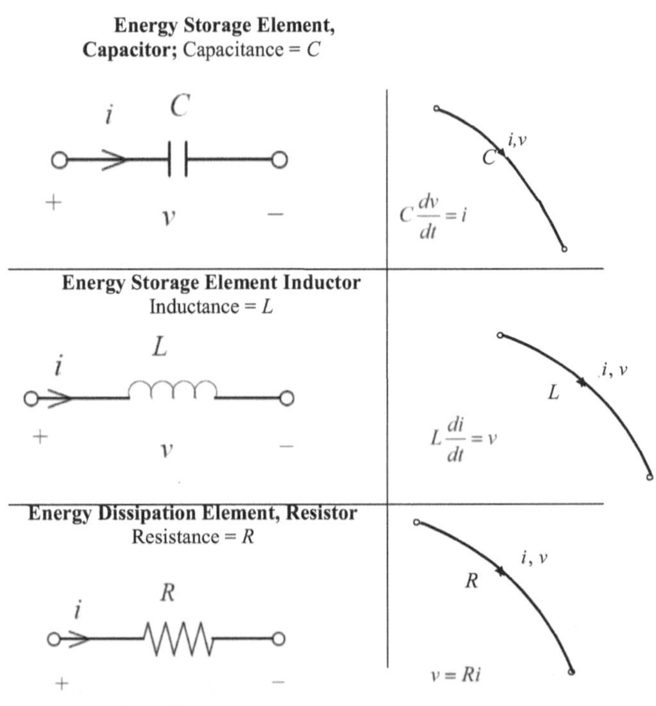

FIGURE 5.3 Single-port electrical system elements and their linear graph representations.

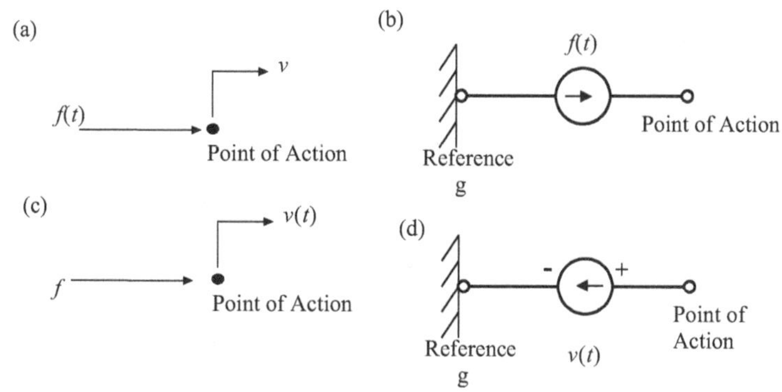

FIGURE 5.4 (a) *T*-source (through-variable source); (b) linear graph representation of a *T*-source; (c) *A*-source; (d) linear graph representation of an *A*-source.

5.3.2 Source Elements

In LG models, the *inputs* into a system are represented by *source elements.* There are two types of sources, as shown in Figure 5.4.

- ***T*-Type Source (e.g., Force Source, Current Source)**: For a *T*-source, the independent variable (i.e., the source output, which is the system input) is the through-variable *f*. The arrowhead indicates the positive direction of *f*.
- *Note* 1: For a *T*-source, the sign convention that the arrow gives the positive direction of *f* still holds. However, the sign convention that the arrow is from the point of action to the point of reference (or the direction of the drop in the across-variable) does not hold.
- *Note* 2: The independent variable of a source is the variable that is not affected by the other elements that are connected to the source. The other variable is the "dependent variable," which is affected, in general.
- ***A*-Type Source (e.g., Velocity Source, Voltage Source)**: For an *A*-source, the independent variable is the across-variable *v*. The arrowhead, if shown on the element, will indicate the direction of the "drop" in *v*. The + and – signs are always indicated for an *A*-source, where the drop in *v* occurs from the + terminal to the – terminal.
- *Note*: For an *A*-source, the sign convention that the arrow is from the point of action to the point of reference (or the direction of the drop in the across-variable) holds. However, the sign convention that the arrow gives the positive direction of *f* does not hold. Often, an arrowhead is not shown for an A-source.

In an ideal force source (a through-variable source), the force variable is the independent variable, and it is not affected by interactions of the source with the rest of the system. The corresponding relative velocity across the force source, however, will vary as determined by the dynamics of the overall system. It should be clear that the direction of *f*(*t*) as shown in Figure 5.4a is the applied force. The reaction on the source would be in the opposite direction. An ideal velocity source (across-variable source) supplies a velocity without being affected by the dynamics of the system to which it is applied. Hence, velocity of the source is the independent variable. The corresponding force is the dependent variable, and it is determined by the system dynamics. Similar facts are applicable to source elements in other physical domains.

5.3.2.1 Interaction Inhibition by Source Elements

As we have noted, the source variable (or input variable) is the independent variable of a source that is unaffected by the dynamics of the system to which the source is connected. But the co-variable (dependent variable) will change. A related property of a source element is identified now. Source

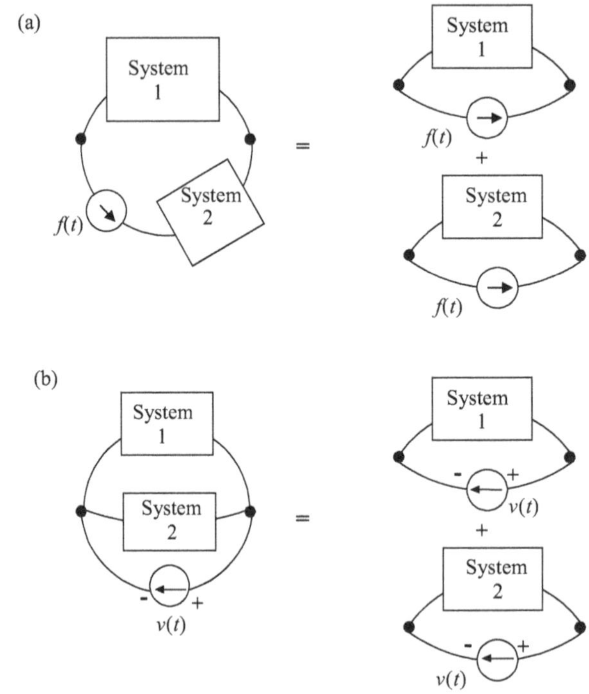

FIGURE 5.5 (a) Two systems connected in series with a *T*-source; (b) two systems connected in parallel with an *A*-source.

elements can serve as means to inhibit dynamic interactions between systems. Specifically, it follows from the definition of an ideal source that the dynamic behavior of a system is not affected by connecting a system in series with a *T*-source (e.g., force source or current source) or in parallel with an *A*-source (e.g., velocity source or voltage source), because the variable that is common to the two connecting systems is the independent variable of the source, which is not affected.

5.3.2.1.1 Corollary

Components can be connected in series with a *T*-source, or a series-connected component can be removed from a *T*-type source without affecting the dynamics of the resulting system. Similarly, components can be connected in parallel with an *A*-source, or a parallel-connected component can be removed from an *A*-type source without affecting the dynamics of the resulting system. Examples of these two situations are given in Figure 5.5.

Another interpretation of these situations is that a source can uncouple (i.e., decouple) subsystems in a system. Specifically, two systems in each case of Figure 5.5 are uncoupled. In other words, System 1 and System 2 in Figure 5.5 are two uncoupled subsystems driven by the same input source. In this sense, the order of the overall system is the sum of the orders of the individual (uncoupled) subsystems.

Note: In general, linking (networking) a subsystem will change the order of the overall system (because new dynamic interactions are introduced due to the linking).

5.3.3 Two-Port Elements

A two-port element has two points of action (with two corresponding energy ports) and two coupled branches corresponding to them. A two-port element can be interpreted as a pair of single-port elements, with a common point of reference, whose net power is zero (i.e., it does not store or dissipate energy). A transformer (mechanical, electrical, fluid, etc.) is a two-port element. Also, a mechanical gyrator is a two-port element. An example of a translatory mechanical transformer is a

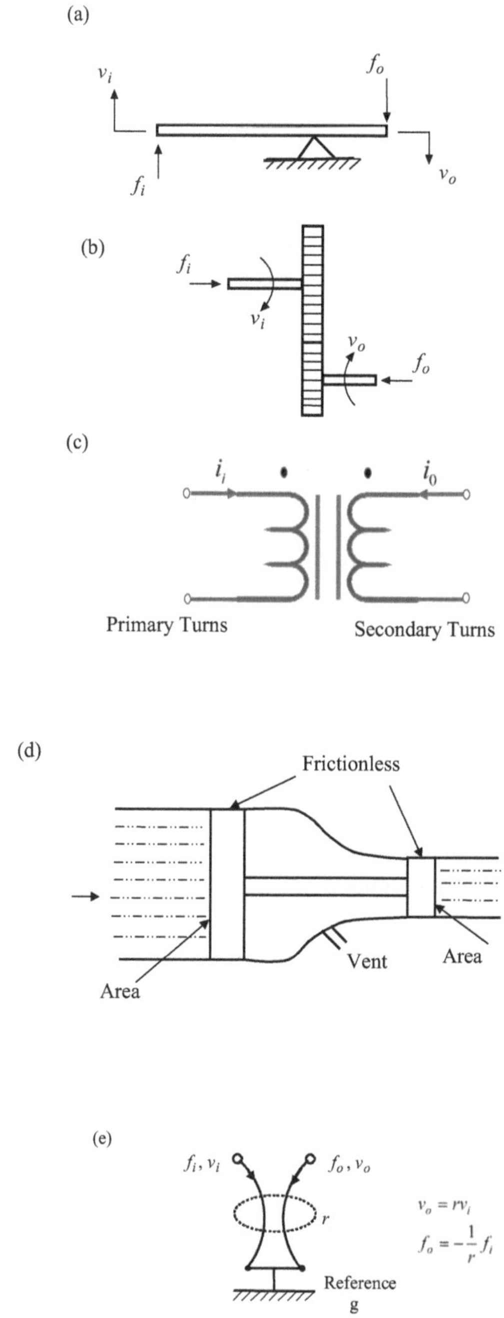

FIGURE 5.6 Transformer. (a) Lever; (b) meshed gear wheels; (c) electrical transformer; (d) fluid transformer; (e) linear graph representation.

lever (Figure 5.6a). An example of rotatory mechanical transformer is a meshed pair of gear wheels (Figure 5.6b). A "mechanical" gyrator is typically an element that possesses gyroscopic properties. We shall consider only the linear cases; that is, ideal transformer and ideal gyrator. The extension to the nonlinear case should be clear.

Note: There are multi-physics (multi-domain) two-port elements as well, examples of which are given in the sequel.

5.3.3.1 Transformer

In an ideal transformer, the across-variables in the two ports (branches) are changed according to some relationship (parameter) without dissipating or storing energy during the process. Hence, the through-variables in the two ports will also change accordingly. Examples of mechanical, electrical, and fluid transformers are shown in Figure 5.6a–d. The LG representation of a transformer is given in Figure 5.6e, which is the same for any physical domain, thereby facilitating the "unified" approach to modeling.

In Figure 5.6e, as for a single-port passive element, the arrow goes from the point of action to the point of reference, in each of the two branches (line segments). It shows the positive direction of power flow (i.e., when the product of the through-variable and the across-variable for that segment is positive).

Note: Since the convention is for the power to enter the device at both ports, and since an ideal transformer does not store or dissipate energy, one of these two power flows has to be negative. This is the reason for the −ve sign in one of the two constitutive equations of a transformer. Also, note the marked directions for force (or torque) and velocity at the two ends of the device in Figure 5.6a and b, which are consistent with the convention of power flow into the device at both ends.

One of the two ports (branches) may be considered the input port and the other branch automatically becomes the output port. This choice depends on what the input segment and what the output segment of the system is. It has nothing to do with the concept of input variables and output variables. Let

v_i and f_i = across- and through-variables at the input port
v_o and f_o = across- and through-variables at the output port.

The (linear) transformation ratio r of the transformer is a model parameter and is given by

$$v_o = r\,v_i \tag{5.1}$$

Due to the conservation of power, we require

$$f_i v_i + f_o v_o = 0 \tag{5.2}$$

By substituting equation (5.1) into (5.2), we get

$$f_o = -\frac{1}{r} f_i \tag{5.3}$$

Here r is a non-dimensional parameter because the input domain and the output domain are the same. A two-domain (multi-physics) transformer will have a dimensional r. The two constitutive relations of a transformer are given by equations (5.1) and (5.3).

5.3.3.1.1 Electrical Transformer

As shown in Figure 5.6c, an electrical transformer has a primary coil, which is energized by an alternating-current (ac) voltage (v_i), a secondary coil in which an ac voltage (v_o) is induced, and a common core, which helps the linkage of magnetic flux between the two coils. A transformer converts v_i to v_o without making use of an external power source. Hence it is a *passive device*, just like a capacitor, inductor, or resistor. The parameter of the electrical transformer is the turn ratio:

$$r = \frac{\text{number of turns in the secondary coil } (N_o)}{\text{number of turns in the primary coil } (N_i)} \tag{5.4}$$

Note: In Figure 5.6c, the two dots on the top side of the two coils indicate that the two coils are wound in the same direction.

In a pure and ideal transformer, there will be full flux linkage without any dissipation of energy. Then, the flux linkage will be proportional to the number of turns. Hence

$$\lambda_o = r\lambda_i \tag{5.5}$$

where λ denotes the flux linkage in each coil. By differentiating equation (5.5), while noting that the induced voltage in the coil is given by the rate of charge of flux, we get equation (5.1). In an *ideal electrical transformer*, there is no energy dissipation or storage and also the signals will be in phase.

5.3.3.2 Gyrator

A gyrator converts the through-variable at the input port into the across-variable at the output port according to some relationship (and correspondingly, the across-variable at the input port into the through-variable at the output port) without any energy storage or dissipation. Mixed-domain (multi-physics) gyrators where the input and the output are in different physical domains are available. First we consider a single-domain mechanical gyrator where the input and the output both are in the mechanical domain.

5.3.3.2.1 Mechanical Gyrator

A mechanical *gyrator* is an ideal gyroscope (Figure 5.7a). It is simply a spinning top that rotates about its own axis at high angular speed ω (positive in the x-direction), which is assumed to remain unaffected by other small motions that are present. If the moment of inertia about this axis of rotation (x in the shown configuration) is J, the corresponding angular momentum is $h = J\omega$, and this vector is also directed in the positive x-direction, as shown in Figure 5.7b.

- **Method 1: With a Rotation About z-axis**

Suppose that the angular momentum vector h is given an incremental rotation $\delta\theta$ about the positive z-axis, as shown in Figure 5.7b. As a result, the free end of the gyroscope will move in the positive y-direction (i.e., the +ve v_i direction). The resulting change in the angular momentum "vector" is $\delta h = J\omega\delta\theta$ in the positive y-direction, as shown in Figure 5.7b. Hence, the rate of change of angular momentum is

$$\frac{\delta h}{\delta t} = \frac{J\omega\delta\theta}{\delta t} \tag{i}$$

where δt is the time increment of the motion. In the limit (as $\delta t \to 0$), the rate of change of angular momentum is

$$\frac{dh}{dt} = J\omega\frac{d\theta}{dt} \tag{ii}$$

The velocity v_i of the free end of the gyroscope, in the positive y-direction, generates this motion. *Note*: There has to be a corresponding force f_i at the free end, in the positive y-direction. The corresponding angular velocity about the positive z-axis is

$$\frac{d\theta}{dt} = \frac{v_i}{L} \tag{iii}$$

Here L is the length of the gyroscope. Substitute (iii) into (ii). The rate of change of angular momentum is

$$\frac{dh}{dt} = \frac{J\omega v_i}{L} \tag{5.6}$$

about the positive y-direction. By Newton's second law, to sustain this rate of change of angular momentum, a torque equal to $\frac{J\omega v_i}{L}$ is required in the same direction (i.e., the y-axis). That means, a

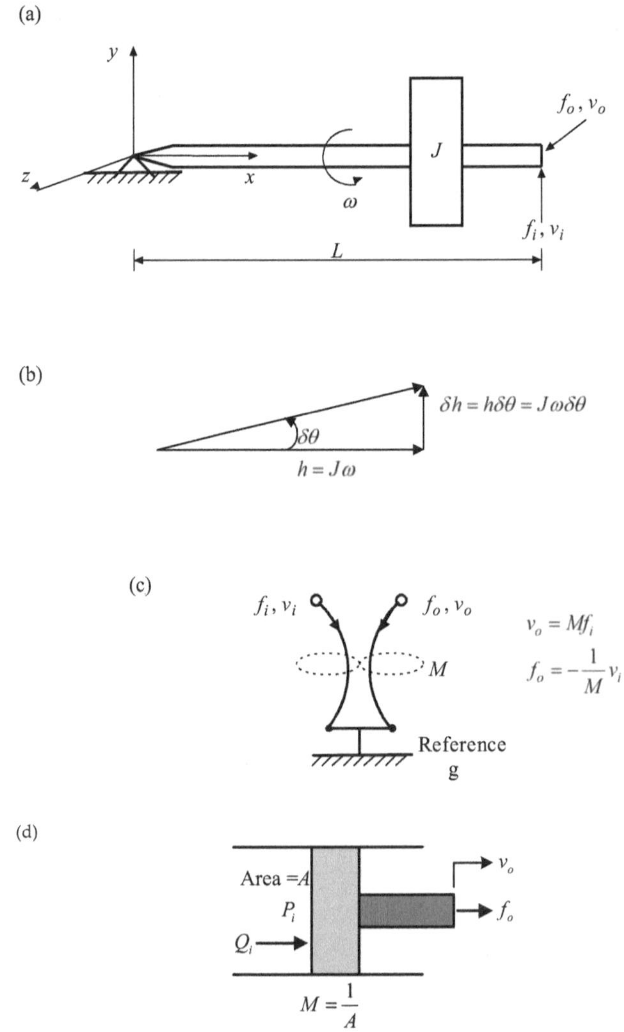

FIGURE 5.7 (a) Mechanical gyrator (gyroscope or spinning top)—a two-port element; (b) derivation of the constitutive equations; (c) linear graph representation; (d) hydraulic-mechanical gyrator.

force is required in the z-direction. If this force at the free end of the gyroscope is f_o in the positive z-direction, the corresponding torque is f_oL acting about the negative y-direction. From Newton's second law, it follows that $-f_oL = \dfrac{J\omega v_i}{L}$. This may be expressed as

$$f_o = -\frac{1}{M}v_i \tag{5.7}$$

This is one constitutive equation for the gyrator.

By the conservation of power (equation (5.2)) for an ideal gyroscope, it follows from equation (5.7) that

$$v_o = Mf_i \tag{5.8}$$

This is the other constitutive equation for the gyrator.

The corresponding gyroscope parameter is

$$M = \frac{L^2}{J\omega} \tag{5.9}$$

Note: For a "mechanical gyrator," M is a "mobility" parameter (velocity/force), as discussed in Chapter 6.

Equations (5.7) and (5.8) are the constitutive equations of a gyrator. The LG representation of a gyrator is shown in Figure 5.7c.

- **Method 2: With a Rotation About y-axis**

The same constitutive relations can be obtained using a different approach (this time getting equation (5.8) first). For that, give the incremental rotation $\delta\theta$ about the −ve y-axis. As a result, the free end of the gyroscope will move in the positive z-direction (i.e., the +ve v_o direction). Now we proceed similarly as before:

1. Incremental change of angular momentum "vector" of magnitude $\delta h = h\delta\theta = J\omega\delta\theta$ is in the +ve z-direction
2. Then $\frac{dh}{dt} = h\frac{d\theta}{dt} = J\omega\frac{d\theta}{dt} = J\omega\frac{v_o}{L}$ whose vector is in the +ve z-direction
3. Corresponding torque is $f_i L$
4. From Newton's second law: $J\omega\frac{v_o}{L} = f_i L \rightarrow v_o = \frac{L^2}{J\omega} f_i \rightarrow v_o = Mf_i$
5. From conservation of energy/power: $f_i v_i + f_o v_o = 0 \Diamond \rightarrow .f_o = -\frac{1}{m}v_i$ with $M = \frac{L^2}{J\omega}$

5.3.3.2.2 *Hydraulic-Mechanical Gyrator*

An example of a mixed-domain gyrator is a hydraulic-mechanical gyrator, which consists of a piston and a cylinder of fluid (see Figure 5.7d). The fluid flow at the volume rate Q_i into the cylinder is accommodated by the movement of the piston (area=A) at velocity v_o. We have

$$Q_i = Av_o \tag{i}$$

Force on the piston due to the fluid pressure P_i is AP_i. The corresponding force on the piston rod is f_o. Then, the force balance of the piston gives

$$AP_i + f_o = 0 \tag{ii}$$

From (i) and (ii), we have the constitutive equations of a hydraulic-mechanical (mixed-domain) gyrator:

$$v_o = MQ_i \tag{5.10}$$

$$f_o = -\frac{1}{M}P_i \tag{5.11}$$

$$\text{with the gyrator parameter } M = \frac{1}{A} \tag{5.12}$$

5.4 LINEAR GRAPH EQUATIONS

As discussed in Chapter 3, in our systematic approach for formulating a state-space model of an engineering dynamic system, three types of equations have to be written: constitutive, loop, and node equations. In using an LG to facilitate this formulation, the same three types of equations are written, using the LG, and further manipulated:

1. Constitutive equations for independent energy storage elements (this forms the state-space shell)
2. Constitutive equations for the remaining elements that are not sources (inputs)
3. Compatibility equations (loop equations), using across-variables, for all the independent closed paths (primary loops) formed by two or more branches
4. Continuity equations (node equations), using through-variables, for all the independent junctions (nodes) of two or more branches (i.e., the total number of nodes in the system – 1)
5. Using equations in items 2–4, eliminate the unnecessary (auxiliary) variables in the state-space shell (item 1).

Note: Auxiliary variables are those other than the state variables and the input variables.

Constitutive equations of elements have been discussed in detail in Chapters 2 and 3 and also in the beginning of the present chapter. In the examples in Chapter 3, not all compatibility equations and continuity equations were stated explicitly because sometimes the system variables were chosen to satisfy these two types of equations implicitly. In the modeling of complex dynamic systems, systematic approaches, which can be computer-automated, will be useful. In that context, approaches are necessary to explicitly write the compatibility equations and continuity equations even when they seem obvious or unnecessary in a manual process of model formulation. The related approaches and issues are discussed next.

5.4.1 Compatibility (Loop) Equations

A loop in an LG is a closed path formed by two or more branches. A loop equation (compatibility equation) is obtained by algebraically (i.e., taking into account the proper sign) summing to zero all the across-variables along the branches of the loop. This is a necessary condition because, at a given point in the LG, there must be a unique absolute value for the across-variable, at a given time. In other words, in a system, a component connected at a point (node) must not be broken off causing discontinuity in the across-variable there. For example, a mass and an end of a spring connected to the same point must remain connected (and hence, must have the same velocity there at a particular time instant). Since this point must be intact (i.e., does not break or snap thereby separating the connected components), the system (loop) is said to remain "compatible."

5.4.1.1 Sign Convention

1. In writing a loop equation, go in the counter-clockwise (ccw) direction of the loop, starting from a convenient node
2. The across-variable drops in the direction of a branch arrow. This direction is taken to be positive (i.e., the associated across-variable is positive). *Note*: The exception is a *T*-source (through-variable source), where the arrow direction indicates the increasing direction of the across-variable.

The arrow in each branch is important in writing a loop equation. Clearly, we cannot always go in the direction of the arrow in a branch that forms a loop. If we go opposite to the arrow, a negative sign is used with the associated across-variable.

5.4.1.2 Number of "Primary" Loops

Primary loops are a "minimal" set of loops from which any other loop in the LG can be formed. Hence, a set of primary loops is a complete and "independent" set. Such a set is not unique (different sets of primary loops can be formed and chosen). Regardless, a primary set of loops will generate all the independent loop equations and also can generate any other loop in the LG.

Note: Loops closed by broken-line branches (i.e., of inertia elements) should be included as well in determining the primary loops.

Not all "primary" loop equations are useful in the development of state equations (i.e., in the elimination of auxiliary variables). Specifically, the equations of loops that include one or more *T*-sources are not useful. This is because the across-variable of a *T*-source is the *dependent variable* of the source and should not remain in the state equations (because it is neither a state variable nor an input variable. It is an auxiliary variable, which should be eliminated).

Example 5.1

Figure 5.8 shows a mass-spring-damper system (a mechanical oscillator) and its LG. Each element in the LG forms a branch. As noted before, always, an inertia element is (virtually) connected to the ground reference point (g) by a partially dotted line (or a partially broken line) through which the "inertia force" transmits. This is because the velocity (across-variable) of a mass is given with respect to the ground reference (which has zero velocity) but it is not directly connected to the ground. The ground is always the reference point of a mass element, and it "indirectly feels" the *inertia force* of the mass. To understand this further, suppose that we push an unconstrained mass upward by our hands, imparting it an acceleration. The pushing force of the hands is equal to the inertia force, which is the product of mass and acceleration. An equal force is transmitted to the ground through our feet. Clearly, the mass itself is not directly connected to the ground, yet the force applied to the mass (or the inertia force) is "felt" at the ground (indirectly transmitted to the ground). Hence the force "appears" to travel directly from the "lumped" mass element to the ground through this "virtual" branch.

Note: If you think that in this example, the human body is a solid link between the mass and the ground, instead imagine applying a magnetic force to a magnetic mass using a "magnetic bearing" (the housing of the magnetic bearing/actuator is fixed to the ground, but it is not directly connected to the pushed mass element). Yet, the inertia force of the mass travels to the ground.

Furthermore, in Figure 5.8, the input force from the "force source" travels to (or "felt at") the ground reference point.

Note: For any mechanical source (i.e., force source or velocity source) assume that the reference point is the ground (and the action point where the input from the source is applied).

In the present example, there are three primary loops. These three loops are not unique and can be chosen in many ways.

Note: A set of primary loops is an "independent set," hence, a primary loop cannot be formed by combining the remaining loops in the primary set. Furthermore, any loop in the LG can be formed by combining the primary loops.

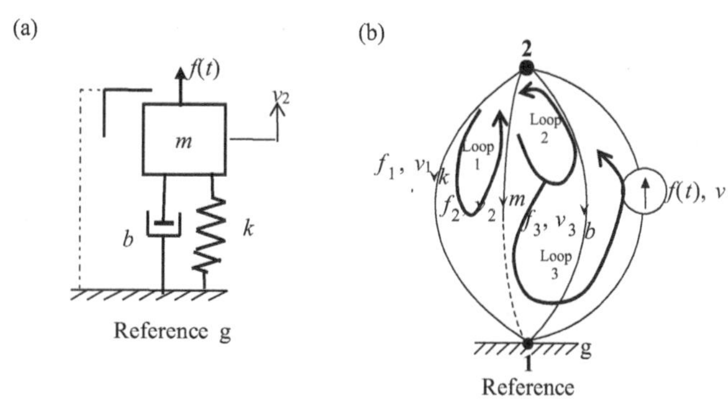

FIGURE 5.8 (a) A mass-spring-damper system; (b) linear graph having two nodes and three primary loops.

In this example, we have chosen the set of three primary loops as marked in Figure 5.8:

Loop 1: (*k, m*)
Loop 2: (*m, b*)
Loop 3: (*m, s*).

Note: "*s*" denotes the source element. The pair within the parentheses indicates the branches that are combined to form the corresponding loop.

Observations Concerning this Choice of Primary Loops:

1. Every loop includes the branch corresponding to the *A*-type element (mass). Hence, every loop equation will contain its across-variable, which is known to be a state variable (which must be retained in the final state equations)
2. The loop that contains the *T*-source (i.e., *m, s* loop) is not a useful loop (even though it is a primary loop) because the across-variable of a *T*-source is the "dependent variable" of the source, which does not have to be known. It is an "auxiliary variable" (not an input variable or a state variable) and should not be retained in the final state equations
3. Many other primary sets of three loops can be chosen. Three other possible sets of primary loops are [(*k, b*), (*m, b*), (*b, s*)], [(*k, b*), (*m, s*), (*m, b*)], and [(*k, b*), (*m, s*), (*k, s*)]. But the corresponding loop equations will require more manipulation in the development of the state equations (i.e., in the elimination of the auxiliary variables) than what is needed with the choice shown in Figure 5.8.

Loop Equations (go ccw):

Loop 1: $v_1 - v_2 = 0$
Loop 2: $v_2 - v_3 = 0$
Loop 3: $v_2 - v = 0$ (not useful because v is an unwanted variable)

Once we select a set of primary loops (three loops in this example), any other loop can be formed by combining this primary set. For example, the loop (*k, b*) can be obtained by combining the loops (*k, m*) and (*m, b*). In particular, its loop equation can be obtained by algebraically adding the loop equations of the latter two loops. Specifically:

$$v_1 - v_3 = (v_1 - v_2) + (v_2 - v_3)$$

LEARNING OBJECTIVES

1. Use of the sign convention (go ccw in a loop; arrow indicates *A*-variable drop, except in a *T*-source)
2. Appropriate choice of primary loops (pick loops that have *A*-type state variables and/or *A*-sources)
3. Identification of primary loops that are not useful (a loop with a *T*-source).

General Observations

- A set of primary loops is a "minimal" and "independent" set: 1. No loop in the primary set can be formed by combining the remaining loops in the primary set; 2. any loop in an LG can be formed by combining the primary loops
- Many choices are available in selecting the primary loop set
- The set of primary loops provide all the independent loop equations

- The best choice of primary loops will include *A*-type energy storage elements (providing *A*-type state variables) and *A*-sources (providing *A*-type input variables), if these elements are present in the system. Such a choice will minimize the necessary mathematical manipulations in the development of the state equations (in the process of eliminating the auxiliary variables)
- Some primary loops (and their loop equations) may not be useful in the development of a state model (i.e., in the elimination of the auxiliary variables). Specifically, 1. ignore the loops that include *T*-type sources (because their across-variables are dependent variables, which should not be in the final state equations); and 2. the loop equations from the remaining primary loops are the set of *useful* loop equations.

5.4.2 Continuity (Node) Equations

A node is a "junction" where two or more branches meet. A node equation (or, continuity equation) is created by equating to zero the algebraic sum of all the through-variables at a node. This equation holds in view of the fact that a node can neither store nor dissipate energy; in effect, it amounts to "what goes in must come out." Hence, a node equation dictates the *continuity* of the through-variables at a node. For this reason, one must use proper signs for the variables when writing node equations.

- **Sign Convention**: A through-variable going "into" the node is positive (and coming out of the node is negative)
- The meaning of a node equation in various domains is indicated below.
- **Mechanical Systems**: Force balance; equilibrium equation; Newton's third law; etc.
- **Electrical Systems**: Current balance; Kirchoff's current law; conservation of charge; etc.
- **Hydraulic Systems**: Conservation of fluid
- **Thermal Systems**: Conservation of thermal energy

5.4.2.1 Primary Node Equations

If an LG has n nodes, the number of primary nodes is $n-1$. This is because the equation for any node can be obtained by algebraically combining the equations for the other $n-1$ nodes.

As in the case of loop equations, not all "primary" node equations are useful in the development of state equations (i.e., in the elimination of auxiliary variables). Specifically, the equation for a node that connects an *A*-sources is not useful. This is because the through-variable of an *A*-source is the *dependent variable* of the source and should not remain in the final state equations (because it is neither a state variable nor an input variable. It is an auxiliary variable).

Example 5.2

Revisit the problem shown in Figure 5.8. The system has two nodes. The corresponding node equations are identical, as given below.

Node 1 Equation: $f_1+f_2+f_3-f=0$
Node 2 Equation: $-f_1-f_2-f_3+f=0$

Note: The equation of Node 1 is obtained simply by reversing the signs in the Node 2 equation. Hence, there is only one independent (primary) node in this example.

In the present example, this node equation is clearly a useful equation. It contains "f," which is an input variable and should be retained in the state equations. Also, it contains f_1, which is a state variable. Hence, this node equation is useful in the elimination of the auxiliary variables f_2 and f_3, when generating the final state equations.

LEARNING OBJECTIVES

1. Application of the sign convention for node equations
2. Required number of node equations = total number of nodes − 1
3. Identification of useful node equations.

General Observations

- For an LG with n nodes, there are only $n-1$ primary (independent) node equations. The remaining node equation is the algebraic sum of the first $n-1$ node equations
- All primary node equations may not be useful in the development of a state model (i.e., in the elimination of the auxiliary variables). Specifically, the equation of a node that connects an A-type source is not useful (because the through-variable of an A-type source is a dependent variable, which should not be present in a state equation. It is an auxiliary variable, which should be eliminated).

Example 5.3

Consider the *L-C-R* electrical circuit (an electrical oscillator) shown in Figure 5.9a. Its LG is drawn in Figure 5.9b. The system has three primary loops; one primary node; and a voltage source (an *A*-source). It should be clear that this system is "not" analogous to the mechanical system of Figure 5.8 because that system has a force source (a *T*-source), not a velocity source.

Loop Equations:
There are many choices for the three primary loops. However, the best choices (which yield loop equations that require minimal manipulation in eliminating auxiliary variables, in the generation of state equations) have the following properties:

1. Loops that contain *A*-sources are desirable (because their across-variables are input variables, which must be retained in the final state equations)
2. Loops that contain *A*-type (energy storage) elements are desirable (because their across-variables are state variables, which must be retained in the final state equations)
3. Loops that contain *T*-sources should be avoided (because their across-variables are the dependent variables of the sources, which must not be present in the final state equations).

According to these criteria, in Figure 5.9b, we have selected the following set of primary loops:

Loop 1: (s, C)
Loop 2: (L, C)
Loop 3: (C, R)

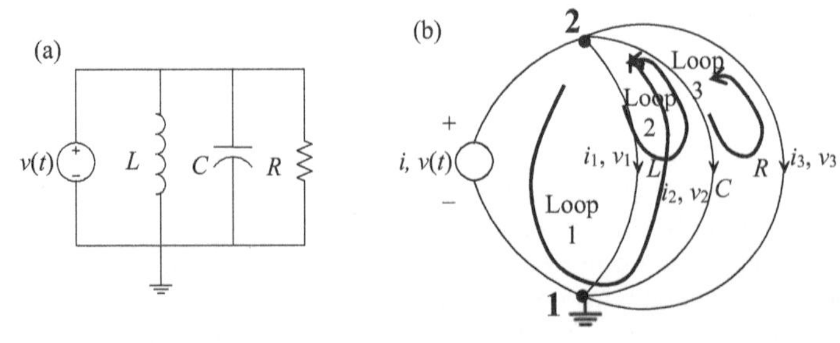

FIGURE 5.9 (a) An *L-C-R* circuit; (b) its linear graph.

Note: All three loops have the *A*-element (capacitor). One loop has the *A*-source (voltage source).

The corresponding loop equations (three) are given below, with the standard sign convention.

Loop 1: $v - v_2 = 0$
Loop 2: $v_1 - v_2 = 0$
Loop 3: $v_2 - v_3 = 0$

We may select many other sets of three loops as primary loops. Some examples are [(*s*, *L*), (*L*, *C*), (*s*, *R*)] or [(*s*, *L*), (*L*, *C*), (*C*, *R*)] or [(*s*, *L*), (*s*, *C*), (*s*, *R*)]. No matter what primary set is chosen, it will include all the branches of the LG, and the resulting loop equations will be complete.

Node Equations:

In this example, there are two nodes. Their equations (according to the standard sign convention) are

Node 1: $-i + i_1 + i_2 + i_3 = 0$
Node 2: $i - i_1 - i_2 - i_3 = 0$

Since Node 2 equation can be obtained simply by reversing the signs of the terms in the Node 1 equation, there is only one independent (primary) node equation.

Note: This node equation has the dependent variable *i* of the source (which is an auxiliary variable, which should not be present in the state equations). Hence, this primary node equation is in fact not a useful equation (in the generation of the final state equations).

LEARNING OBJECTIVES

1. Selection of a proper set of primary loops
2. Selection of the independent (primary) nodes
3. Sign convention in writing loop equations
4. Sign convention in writing node equations
5. Identification of the useful/useless loop equations
6. Identification of the useful/useless node equations.

5.4.3 Series and Parallel Connections

If two or more elements are connected in *series*, their through-variables are the same but the across-variables are not the same (they add algebraically). If two or more elements are connected in *parallel*, their across-variables are the same but the through-variables are not the same (they add algebraically). These facts are given in Table 5.2.

Let us consider two systems with a spring (*k*) and a damper (*b*), and an applied force (*f*(*t*)). In Figure 5.10a, they are connected in parallel, and in Figure 5.10b, they are connected in series. Their LGs are as shown in the corresponding figures. The LG in (a) has two primary loops (two elements in parallel with the force source), whereas in (b) there is only one loop because all the elements are in series with the force source. From Table 5.2, we can appreciate the nature of the node equations and the loop equations in these two cases, without even having to write these equations.

TABLE 5.2
Series-Connected and Parallel-Connected Components

Components in Series	Components in Parallel
Through-variables are the same	Across-variables are the same
Across-variables are not the same (they add algebraically)	Through-variables are not the same (they add algebraically)

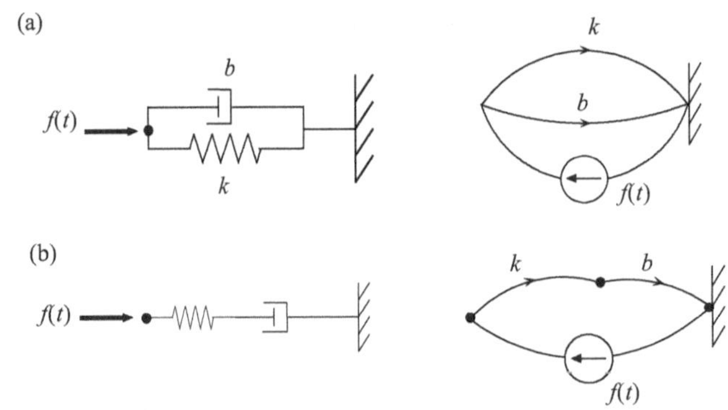

FIGURE 5.10 Spring-damper systems with a force source and their linear graphs. (a) Elements in parallel; (b) elements in series.

Another important consequence of the interconnection structure of components with source elements has been observed before. Specifically,

1. Components connected in series with a T-source have no dynamic interactions (i.e., they are uncoupled and can be treated separately; a series component can be removed without affecting the rest of the system)
2. Components connected in parallel with an A-source have no dynamic interactions (i.e., they are uncoupled and can be treated separately; a parallel component can be removed without affecting the rest of the system).

5.5 STATE MODELS FROM LINEAR GRAPHS

In Chapter 3, we presented a systematic way to develop a unique state-space model of an engineering dynamic system, by incorporating a unified and integrated approach. In that approach, an initial step is to sketch a "structural" diagram of the system showing how the components (or basic elements) are interconnected. A powerful graphic representation that not only provides the model structure but also the nature of the interconnected components, model variables and parameters (in short, the complete dynamic model) is an LG. Another comparable graphic representation is provided by bond graphs; that topic is beyond our focus. The focus of this chapter is LGs. An LG presents:

- The "structure" of the system (how the components are interconnected)
- The nature/type of the components (e.g., A-type and T-type elements and sources, D-type elements, multi-port elements)
- Variables and parameters of the elements (through-variable and across-variable pair for each element or port; characteristic parameters of the model elements, which are needed to formulate the constitutive equations)
- Directions (according to some sign convention) of the variables (through-variables, across-variables, power flow).

Once the LG of a dynamic system is sketched, we use the same systematic procedure as presented in Chapter 3, albeit facilitated by the LG, to obtain a state-space model of the system.

5.5.1 Sketching of a Linear Graph

An LG is indeed a complete model of a dynamic system because it contains all the information about the system, particularly what is needed to formulate a state-space model (an analytical model). In developing an LG, we sketch a branch starting from its point of reference and connect it to other branches from its point of action (except in the process of returning to the original point of reference, when it is the point of reference of the returning branch that is connected to the original reference), according to the specific model structure, until all the branches (elements/components) of the system are included. Next we orient the graph (i.e., insert the arrows for all the branches) according to a sign convention. Finally, we indicate the through-across-variable pair of each branch on one side of the branch and the parameter of the branch on the other side. In summary, a systematic way to sketch an LG is as follows:

1. Identify the energy storage elements, energy dissipation elements, and source elements in the system (these are single-port elements, each represented by one branch)
2. Identify any multi-port elements (e.g., transformers, gyrators), which need more than one branch (*Note*: Two-port elements are adequate for this process because a larger number of ports can be formed by combining single-port and two-port elements)
3. Recognize how the elements (branches) are interconnected (series or parallel and to what elements) and sketch a schematic diagram (e.g., circuit diagram). This is the parent graph from which the LG is generated. *Note*: Often, such a parent graph is not needed. We can sketch the LG directly
4. Where possible or clear, identify the terminals of each element or branch (i.e., the point of action and the point of reference). *Note*: For some elements (e.g., inertia element and source elements), the action and the reference points are pre-established due to their nature
5. Starting from a convenient node point (typically, the ground reference or some other established reference) draw a branch (typically, for a source), link it to another appropriate branch through a node (this automatically determines the point of action of the linked branch), and so on, to form a loop
6. Repeat step 5 until the entire system is completed (i.e., all the elements/branches in the system are included and connected)
7. Orient each branch of the LG (this could have been done in an earlier step)
8. Write the corresponding through-across variable pair on one side of each branch and the corresponding element parameter on the other side (this may have been done in an earlier step). *Note*: For an ideal source, we can only indicate the variable pair, not a parameter, because one variable is independent.

Once the LG is developed in this manner, a state-space model of the system can be formulated using it.

5.5.2 State Models from Linear Graphs

As discussed in Chapters 2 and 3, in the systematic approach of formulating a state-space model (now using LGs), which incorporates a unified and integrated treatment, we use as

a. **State Variables**: Across-variables of independent *A*-type (energy-storage) elements and through-variables of independent *T*-type (energy-storage) elements
b. **Input Variables**: Across-variables of *A*-sources and through-variables of *T*-sources. These are the "independent variables" of the source elements.

Note: When there are several "dependent" energy-storage elements, a single common state variable should be used to represent the dynamic state of all of them. A systematic way to identify dependent elements (or "conflicts") in an LG is provided by the *graph tree* approach, which is presented in Appendix A.

In obtaining an analytical model from the LG of a system, we write three types of equations:

1. Constitutive equations (i.e., "physical" equations) for all the independent energy storage elements. This forms the state-space shell
2. Constitutive equations for the remaining branches, excluding the source (input) elements
3. Compatibility equations (i.e., "loop" equations) for the independent (primary) loops (as noted before, some of these loop equations may not be useful)
4. Continuity equations (i.e., "node" equations) for the independent nodes (as noted before, some of these node equations may not be useful).

This approach is further elaborated in the present section.

5.5.2.1 System Order

As we know, A-type elements and T-type elements are energy storage elements. The *system order* is given by any one of the following, which is equivalent:

1. The number of independent energy storage elements in the system
2. Number state variables
3. The order of the state-space model (the order of the system matrix $\mathbf{A}$ in the linear case, and the number of state equations in general)
4. The number of initial conditions required to solve for the response of the analytical model
5. The order of the input–output differential equation model.

Note: In the Laplace (or frequency) domain (see Chapter 6), the system order is also equal to (for a linear system):

1. The order of the characteristic polynomial (denominator of the transfer function)
2. The number of poles (or eigenvalues) in the system (counting any repeated poles separately)

These are also equivalent.

As noted before, the total number of energy storage elements in a system can be greater than the system order because some of these elements might not be independent. Any *dependent elements* can be identified in an ad hoc manner (see Example 5.6) or using the graph tree approach (see Appendix A). A group of dependent energy storage elements can be represented by a single equivalent element with a corresponding single state variable.

5.5.2.2 Sign Convention

The important first step of developing a state-space model using LGs is indeed to draw an LG for the considered system. A sign convention should be established to accomplish this task in a systematic manner, as discussed before. In summary, the sign convention which we use is as follows:

1. Power flows into the *point of action* and out of the *point of reference* of an element (branch). This direction is shown by the branch arrow (a branch with an arrow is called an *oriented* branch). **Exception**: In a source element, power flows out of the point of action and into the element that is connected to the source
2. Through-variable (f), across-variable (v), and power flow (fv) of a branch are taken to be positive in the same direction at the point of action. *Note*: v is measured at the point of action with respect to the point of reference because it is the across-variable. Once the direction of f is known at the point of action, its direction is also known at the point of reference because it is the through-variable
3. In writing a node equation: Flow into a node is positive

4. In writing a loop equation: (a) Go in the ccw direction of the loop, starting from a convenient node; (b) an across-variable is positive in the direction of the branch arrow (across-variable "drops" from the point of action to the point of reference; i.e., in the direction of the arrow, because it is measured wrt the point of reference). *Exception*: In a *T*-source, the arrow is in the "negative" direction of its across-variable (i.e., the increasing direction) and out of the source at its point of action.

Note: Once the sign convention is established, the actual values of the variables can be positive or negative depending on their actual direction.

5.5.2.3 Steps of Obtaining a State Model

The systematic steps for obtaining state equations (i.e., a state-space model) from an LG are the same as those presented in Chapter 3. The only difference is, now the method is assisted by the LG, which by itself is a complete model. The key steps are:

1. Choose as state variables: across-variables of independent *A*-type elements and through-variables of independent *T*-type elements
2. Write the constitutive equations for the independent energy storage elements. This set of equations is called the *state-space shell. Note*: If the dependent elements cannot be identified at this stage, you may write the constitutive equations for all the energy storage elements, but keep in the state-space shell only those for the independent energy storage elements. A formal method of identifying the dependent elements is the *graph-tree approach* (see Appendix A)
3. Write the constitutive equations for the remaining elements (dependent energy storage elements, dissipation—*D*-type elements, two-port elements)
4. Write compatibility equations for the primary loops. *Note*: If a particular primary loop equation is not useful (e.g., the loop has a *T*-source), you may skip it
5. Write continuity equations for the independent (primary) nodes (total number of nodes – 1). *Note*: If a particular primary node equation is not useful (e.g., the node has an *A*-source), you may skip it
6. In the state-space shell, retain the state variables and the input variables only. Eliminate all other variables (called *auxiliary variables*) using the loop and node equations and the additional constitutive equations.

5.5.3 Characteristics of Linear Graphs

Being a graphical representation (or a network), an LG possesses important topological relations with regard to the

Number of nodes (n)
Number of branches (b)
Number of primary loops (ℓ)

Also, being a representation of an engineering dynamic system, there must be a physical link to such topological characteristics. In particular, for the system of model equations to be *solvable*, it is required that

The number of unknown variables = the number of equations.

In particular, the number of unknown variables must be related to b and

The number of sources (s)

From the formulation of an analytical model (e.g., a state-space model), it is clear that the total number of model equations is equal to the sum of

1. The number of constitutive equations
2. The number of compatibility equations
3. The number continuity equations.

Also, these three numbers of equations must be related to b, s, ℓ, and n.

We now establish the two relations:

1. Relation concerning system dynamics (solvability of the analytical model),
2. Topological relation of an LG (a network containing branches, nodes, and loops).

Then we will show these two relations are identical, thereby confirming what we suspected.

5.5.3.1 LG Variables and Relations

Each source branch has one unknown variable (because one variable is the known input to the system—the *independent variable* of the source, which is known) and all other passive branches have two unknown variables each. Hence, we have

Total number of unknown variables $= 2b - s$

Since a source branch does not provide a constitutive equation, and the remaining branches provide one constitutive equation each, we have

The number of constitutive equations $= b - s$

Also, each primary loop gives a compatibility equation. Hence,

The number of loop (compatibility) equations $= \ell$

Since one of the total n nodes does not provide an extra independent node equation, we have

The number of node (continuity) equations $= n - 1$

Hence, total number of equations $= (b - s) + \ell + (n - 1) = b + \ell + n - s - 1$

To uniquely solve the analytical model, we must have

The number of unknowns = the number of equations $\rightarrow 2b - s = b + \ell + n - s - 1$

Hence, we have the result

$$\ell = b - n + 1 \text{ or } b = \ell + (n - 1) \tag{5.13}$$

Note: It is easy to remember this relation, in words, as the number of branches = the number of independent loop equations + the number of independent node equations.

Even though this relation was obtained by considering the characteristics of a "dynamic system," (i.e., by considering the variables and the equations, specifically, the "solvability" of the analytical model), it is indeed a "topological result" (relating the number of loops, nodes, and branches), which must be satisfied by any LG (network). Hence, we should be able to obtain the relationship (5.13) from purely topological considerations, as we will show next.

5.5.3.2 Topological Result

Now a topological relationship is determined for an LG (a network) in terms of its geometric (topological) characteristics (nodes, loops, branches) only. Consequently, we will observe that the resulting relationship is identical to equation (5.13), which we obtained before by considering the

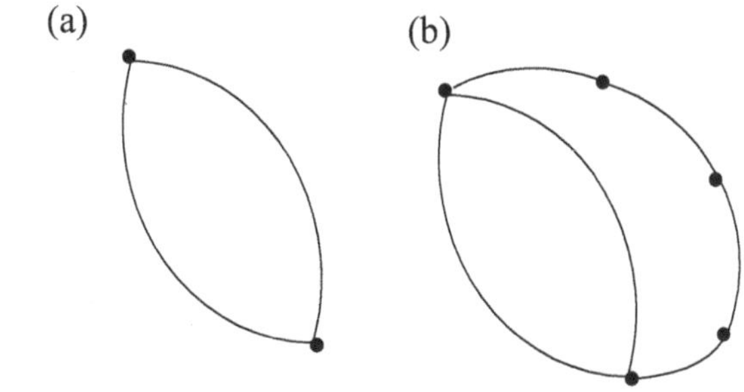

FIGURE 5.11 Proof of the topological result for an LG. (a) Basic LG (single loop with two branches); (b) adding new general branches to create a new loop.

analytical model of a system. We use the well-known "mathematical induction" to carry out the proof.

Consider Figure 5.11. Using the earlier notation, we proceed with the following steps of *mathematical induction*:

Step 1: Start with the basic graph of a single loop, shown in Figure 5.11a: For this graph, $b=2$, $\ell = 1$, and $n=2$. Hence, equation (5.13) — $b = \ell + (n-1)$ is satisfied by this basic graph.

Step 2: To the basic graph in Figure 5.11a, add a new "general" loop using m new nodes connecting $m+1$ new branches in series. This step is shown in Figure 5.11b. For this new graph, we have $b=2+m+1=m+3$, $\ell=2$, and $n=2+m$. Hence, equation (5.13), $b = \ell + (n-1)$, is still satisfied by this new graph.

Note: $m=0$ is a special case (where we connect only one new branch to the existing two nodes, in parallel, to form the new loop, without adding new nodes).

Step 3: Start with a general LG having b ranches, ℓ loops, and n nodes, and assume that it satisfies equation (5.13). This is now the general case of step 1—Figure 5.11a.

As before, add a new "general" loop by using m nodes and $m+1$ branches (as in step 2 above).

Then we have: $b+m+1$ branches, $\ell+1$ loops, and $n+m$ nodes.

Substitute these into equation (5.13). We get

$$(b+m+1) \stackrel{?}{=} (\ell+1)+(n+m-1) \rightarrow b \stackrel{?}{=} \ell+(n-1).$$

Note: We have incorporated a "?" mark into the "=" sign because, to complete the proof, we need to check whether the equation is satisfied.

Since we started by assuming that it is satisfied, the proof is complete by mathematical induction.

In summary, we proceeded as follows: First we showed that the rule was satisfied for a basic network of two branches, one loop, and two nodes. Then, we did the next step of adding one new "general" loop (by using any "general" number of nodes m and the corresponding "general" number of branches $m+1$). We showed that the relation is still satisfied. Finally, we started with the general case of having b ranches, ℓ loops, and n nodes; assumed that it satisfied the required relation; added a new "general" loop (by using m new nodes connecting $m+1$ new branches in series); and showed that it too satisfied the required relation. This means, since the relation is true for the basic network, it must be true for the next incremented network, and hence with the subsequent incremented network, and so on, until all network possibilities are exhausted. In other words, we proved that the required relation was true for all networks, by mathematical induction.

We considered topological characteristics only in this proof. Yet the result is identical to equation (5.13), which was obtained by considering only the analytical modeling of a dynamic system. Hence, equation (5.13) is true from both dynamic modeling and topological contexts.

Now we present several examples of using LGs to develop state-space models, in different physical domains and in "mixed" domains. Since the present LG approach is unified across various physical domains, the same approach is applicable in mechanical, electrical, fluid, thermal, and multi-domain (i.e., mixed) systems, as we will illustrate.

5.6 LINEAR GRAPH EXAMPLES IN MECHANICAL DOMAIN

LGs can be used to develop state-space models in different physical domains. Specifically, since the approach is a unified one across various physical domains, the same approach is applicable in mechanical, electrical, fluid, and thermal domains, and also in multi-domain (i.e., multi-physics or mixed) systems. Several examples in the mechanical domain are given now to illustrate the use of LGs in the systematic development of state-space models, through the unified and integrated approach. Examples in other physical domains and in multiple domains are given in subsequent sections.

Example 5.4

A dynamic absorber is a passive device for vibration suppression (i.e., a "passive controller" for vibration) and is mounted at the vibrating area of a dynamic system. By properly tuning (i.e., selecting the parameters of) the absorber, it is possible to "absorb" most of the power supplied by an unwanted excitation (e.g., support motion, imbalance in rotating parts) by sustaining the absorber motion such that, in steady operation, the vibratory motions of the main system are inhibited. In practice, some damping should be present (and is present) in the absorber to dissipate the energy that flows into the absorber, without generating excessive motions in the absorber mass itself. In the example shown in Figure 5.12a, the main system and the absorber are modeled as simple oscillators with parameters (m_2, k_2, b_2) and (m_1, k_1, b_1), respectively. The LG of this system can be drawn in the usual manner, as shown in Figure 5.12b. The external excitation (system input) is the velocity $u(t)$ of the support (an A-type source).

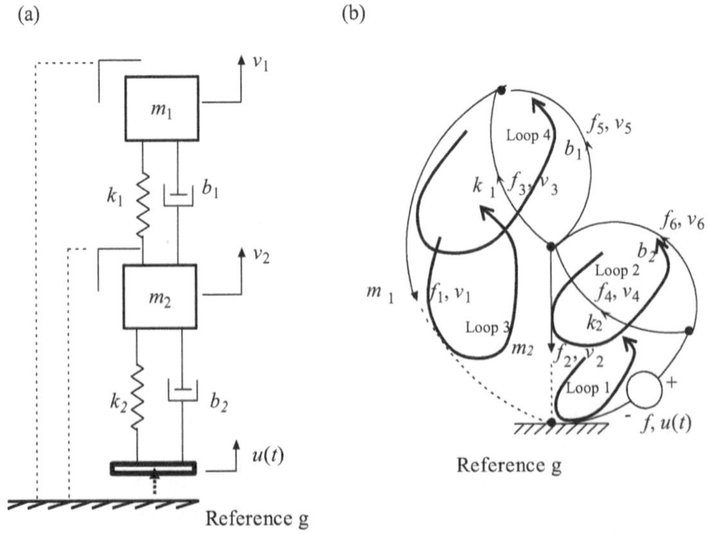

FIGURE 5.12 (a) A mechanical system containing a shock absorber; (b) linear graph of the system.

Note: The LG is oriented, starting from the source element. It acts on the two elements k_2 and b_2, at a particular point. That automatically determines the points of reference of these two elements. The other end of the two elements becomes their points of action, which in turn becomes the point of reference of the element to which these two elements are connected (unless that element closes the loop back to the ground reference, e.g., an inertia element, when that point is also the point of action of the connected element). We proceed in this manner until all the branches are connected and oriented.

From the resulting LG, we observe the following:

Number of branches = b = 7
Number of nodes = n = 4
Number of sources = s = 1
Number of independent loops = l = 4
Number of unknowns = $2b - s$ = 13
Number of constitutive equations = $b - s$ = 6
Number of (primary) node equations = $n - 1$ = 3
Number of (primary) loop equations = 4

The four loop equations are provided by using four independent (primary) loops from the LG.

Check: Number of unknowns = $2b - s$ = 13

Number of equations = $(b - s) + (n - 1) + l = 6 + 3 + 4 = 13$.

Hence the LG satisfies equation (5.13), and accordingly, the analytical model is solvable.

Now we formulate the state-space model of the system, systematically, through the following sequence of steps:

Step 1: Since the system has four independent energy storage elements (m_1, m_2, k_1, k_2), it is a fourth-order system. The state variables are chosen as the across-variables of the two masses (velocities v_1 and v_2) and the through-variables of the two springs (forces f_1 and f_2):

State vector $x = \begin{bmatrix} x_1 & x_2 & x_3 & x_4 \end{bmatrix}^T = \begin{bmatrix} v_1 & v_2 & f_3 & f_4 \end{bmatrix}^T$
Input (A-source velocity) = $[u(t)]$

Step 2: The skeleton state equations (state-space shell):

Newton's second law for mass m_1>: $\dot{v}_1 = \frac{1}{m_1} f_1$
Newton's second law for mass m_2>: $\dot{v}_2 = \frac{1}{m_2} f_2$
Hooke's law for spring k_1>: $\dot{f}_3 = k_1 v_3$
Hooke's law for spring k_2>: $\dot{f}_4 = k_2 v_4$

Step 3: The remaining constitutive equations:

$$\text{For damper } b_1: f_5 = b_1 v_5 \quad \text{(i)}$$

$$\text{For damper } b_2: f_6 = b_2 v_6 \quad \text{(ii)}$$

Step 4: The node equations:
We leave out one node (the ground node) and take the remaining three nodes as the independent (primary) nodes. The corresponding equations are

$$-f_1 + f_3 + f_5 = 0 \quad \text{(iii)}$$

$$-f_3 - f_5 - f_2 + f_4 + f_6 = 0 \quad \text{(iv)}$$

$$-f_4 - f_6 + f = 0 \quad \text{(not useful because of } f) \quad \text{(v)}$$

The loop equations:

The four primary loops are chosen as in Figure 5.12b with the objective of including as many desirable across-variables as possible in a loop while skipping undesirable/unwanted across-variables. The corresponding equations are

$$\text{Loop 1}: v_2 - u + v_4 = 0 \quad \text{(vi)}$$

$$\text{Loop 2}: v_2 - u + v_6 = 0 \quad \text{(vii)}$$

$$\text{Loop 3}: v_1 - v_2 + v_3 = 0 \quad \text{(viii)}$$

$$\text{Loop 4}: v_1 - v_2 + v_5 = 0 \quad \text{(ix)}$$

Note: All four of these loop equations are useful.

Step 5: Eliminate the auxiliary variables in the state-space shell.

$$f_1 = f_3 + f_5 = f_3 + b_1 v_5 = f_3 + b_1(v_2 - v_1) \quad \text{(from (iii), (i), and (ix))}$$

$$f_2 = -f_3 - b_1 v_5 + f_4 + b_2 v_6 = -f_3 - b_1(v_2 - v_1) + f_4 + b_2 v_6$$
$$= -f_3 - b_1(v_2 - v_1) + f_4 - b_2(v_2 - u) \quad \text{(from (iv), (i), (ii), (ix), and (vii))}$$

$$v_3 = -v_1 + v_2 \quad \text{(from (viii))}$$

$$v_4 = -v_2 + u \quad \text{(from (vi)}$$

The following state equations are obtained:

$$\dot{v}_1 = -(b_1/m_1)v_1 + (b_1/m_1)v_2 + (1/m_1)f_3$$

$$\dot{v}_2 = (b_1/m_2)v_1 - [(b_1 + b_2)/m_2]v_2 - (1/m_2)f_3 + (1/m_2)f_4 + (b_2/m_2)u(t)$$

$$\dot{f}_3 = -k_1 v_1 + k_1 v_2$$

$$\dot{f}_4 = -k_2 v_2 + k_2 u(t)$$

The proper output for this system is the velocity of the main mass m_2. The corresponding output equation (algebraic) is $y = v_2$

The corresponding model matrices are

$$\text{System matrix}: \boldsymbol{A} = \begin{bmatrix} -b_1/m_1 & b_1/m_1 & 1/m_1 & 0 \\ b_1/m_2 & -(b_1+b_2)/m_1 & -1/m_2 & 1/m_2 \\ -k_1 & k_1 & 0 & 0 \\ 0 & -k_2 & 0 & 0 \end{bmatrix}$$

$$\text{Input distribution matrix}: \boldsymbol{B} = \begin{bmatrix} 0 \\ b_2/m_2 \\ 0 \\ k_2 \end{bmatrix}$$

$$\text{Output matrix}: \boldsymbol{C} = \begin{bmatrix} 0 & 1 & 0 & 0 \end{bmatrix}^T$$

$$\text{Feedforward gain matrix } \boldsymbol{D} = [0]$$

Note: There is no feedforward property in this system, and hence, the algebraic output equation does not contain the input variable.

LEARNING OBJECTIVES

1. Checking the solvability of a dynamic model (i.e., the number of unknown variables = the number of independent equations)
2. Selection of proper state variables
3. Formulation of the constitutive equations, starting from the state-space shell
4. Selection of "useful" primary loops and identification of primary loops that can be skipped in the formulation
5. Selection of independent (primary) nodes and identification of which of their node equations can be skipped in the formulation
6. Sign conventions for an LG
7. Systematic development of a state-space model for a mechanical system
8. A complete state-space model should include the algebraic output equation as well.

Example 5.5

Commercial motion controllers are digitally controlled (using microcontrollers with software control or digital hardware controllers) high-torque devices that are capable of applying a prescribed motion to a system. Such a controlled actuator may be considered as a velocity source. Consider an application where a rotatory motion controller is used to position an object, which is coupled through a gear transmission. The system is modeled as in Figure 5.13a. Systematically develop a state-space model for this system using the LG approach. The system output is the velocity of m_1.

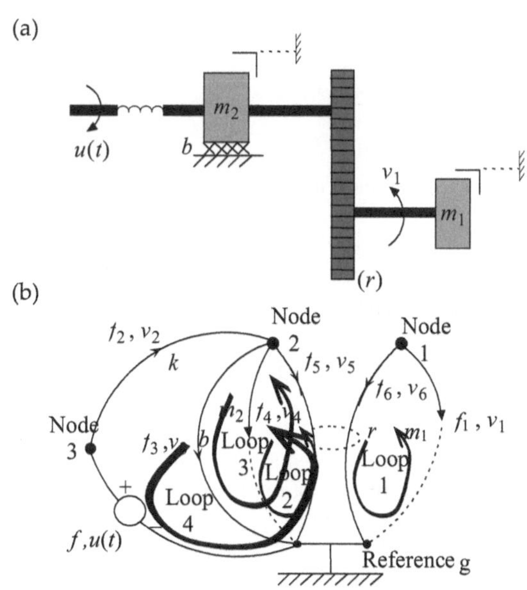

FIGURE 5.13 (a) Rotary-motion system with a gear transmission; (b) linear graph of the system.

Solution

The LG of the system is sketched in Figure 5.13b. The steps of the model development are given now.

Step 1: It is easy to notice that the two inertia elements m_1 and m_2 are not independent (because they are linked through a gear, without any flexible components, and hence their across-variables are directly related). Together they comprise a single (equivalent) storage element. Thus, along with the stiffness element, there are only two independent energy storage elements. Hence, the system is second order.

Note: A formal and systematic way to identify such dependent elements is through the graph-tree approach, which is described in Appendix A.

Let us choose as state variables, v_1—the across-variable of one of the inertia elements (because the other inertia is "dependent") and f_2 —the through-variable of the spring.

With $x_1 = v_1$ and $x_2 = f_2$

Note: We could have chosen m_2 instead of m_1 here, but the present choice is better because, as clear from the system structure, m_1 is the "output" element, and its state variable is a "proper" output variable.

State vector $\mathbf{x} = \begin{bmatrix} x_1 & x_2 \end{bmatrix}^T = \begin{bmatrix} v_1 & f_2 \end{bmatrix}^T$; input vector $\mathbf{u} = [u(t)]$; output vector $\mathbf{y} = [v_1]$

Step 2: The constitutive equations for m_1 and k (state-space shell):

$$\dot{v}_1 = \frac{1}{m} f_1; \quad \dot{f}_2 = k v_2$$

Step 3: The remaining constitutive equations:

$$\text{For damper}: f_3 = b v_3 \text{(i)}$$

$$\text{For the "dependent" inertia } m_2: \dot{v}_4 = \frac{1}{m_2} f_4 \quad \text{(ii)}$$

For the transformer (pair of meshed gear wheels)

$$v_6 = r v_5 \quad \text{(iii)}$$

$$f_6 = -\frac{1}{r} f_5 \quad \text{(iv)}$$

Step 4: Node equations (three independent nodes)

$$\text{Node 1:} \quad -f_6 - f_1 = 0 \quad \text{(v)}$$

$$\text{Node 2:} \quad f_2 - f_3 - f_4 - f_5 = 0 \quad \text{(vi)}$$

$$\text{Node 3:} \quad f - f_2 = 0 \quad \text{(not useful)} \quad \text{(vii)}$$

Note: The node 3 equation is not useful because it has the dependent variable f of the *A*-source (which is an auxiliary variable and should not remain in the final state equations)

Loop equations (for the four primary loops):

$$\text{Loop 1:} \quad v_6 - v_1 = 0 \quad \text{(viii)}$$

$$\text{Loop 2:} \quad v_4 - v_5 = 0 \quad \text{(ix)}$$

$$\text{Loop 3:} \quad v_3 - v_5 = 0 \tag{x}$$

$$\text{Loop 4:} \quad -v_2 + u(t) - v_5 = 0 \tag{xi}$$

Note: Our choice of primary loops seems desirable, as we have avoided the across-variable of the dependent inertia when possible because it is not a state variable and has to be eliminated in the final result.

Step 5: Eliminate the auxiliary variables.

Using equations from steps 3 and 4, the auxiliary variable f_1 can be expressed as:

$$f_1 = -f_6 = \frac{1}{r} f_5 = \frac{1}{r}(f_2 - f_3 - f_4) = \frac{1}{r}(f_2 - bv_3 - m_2\dot{v}_4) = \frac{1}{r}(f_2 - bv_5 - m_2\dot{v}_5)$$

$$= \frac{1}{r}\left[f_2 - \frac{b}{r}v_6 - \frac{m_2}{r}\dot{v}_6\right] = \frac{1}{r}\left[f_2 - \frac{b}{r}v_1 - \frac{m_2}{r}\dot{v}_1\right]$$

(obtained by substituting the two node equations, equations of loops 3, 4, and 1, and the constitutive equations of the dependent inertia and the transformer)

The auxiliary variable v_2 can be expressed as:

$$v_2 = u(t) - v_5 = u(t) - \frac{1}{r}v_6 = -\frac{1}{r}v_1 + u(t)$$

(obtained by substituting node the equations of loops 4 and 1, and a constitutive equation of the transformer)

By substituting these equations into the state-space shell, we obtain the following two state equations:

$$\dot{v}_1 = -\left[\frac{b}{(m_1 r^2 + m_2)}\right]v_1 + \left[\frac{r}{(m_1 r^2 + m_2)}\right]f_2$$

$$\dot{f}_2 = -\frac{k}{r}v_1 + ku(t)$$

Note: The system is second order; only two state equations are present.

$$\text{State model}: \dot{\boldsymbol{x}} = \boldsymbol{Ax} + \boldsymbol{Bu}$$

The output equation (algebraic) is $y = v_1$.

The corresponding system matrix, the input-gain matrix (input distribution matrix), the output matrix, and the feedforward gain matrix are

$$\boldsymbol{A} = \begin{bmatrix} -b/m & r/m \\ -k/r & 0 \end{bmatrix}; \boldsymbol{B} = \begin{bmatrix} 0 \\ k \end{bmatrix}; \boldsymbol{C} = [1 \quad 0]; \boldsymbol{D} = [0]$$

Note: There is no feedforward property in this system, and hence, the algebraic output equation does not contain the input variable.

Here, $m = m_1 r^2 + m_2$=*equivalent inertia* of m_1 and m_2 when determined at the location of the inertia m_2.

Note: It is more desirable to use the equivalent at the output element m_1, which is $m_{eq} = m_1 + \frac{m_2}{r^2}$.

LEARNING OBJECTIVES

1. Dealing with situations where the energy storage elements are dependent (typical approach: When there are several energy storage elements of the same type that are dependent, select the element among them that is at the system output, assign it a state variable, and include it in the state-space shell)
2. Selection of proper state variables
3. Formulation of the constitutive equations, starting from the state-space shell
4. Selection of proper primary loops
5. Identification of primary nodes that can be skipped in the formulation
6. Sign conventions
7. Systematic development of a state-space model for a mechanical system
8. A complete state-space model should include the algebraic output equation as well.

5.7 LINEAR GRAPH EXAMPLES IN ELECTRICAL DOMAIN

In the previous section, we illustrated the use of LGs in the modeling of lumped-parameter mechanical systems—systems with mass/inertia, flexibility, and mechanical energy dissipation. In view of the "unified" nature of the approach, the same procedures may be extended (in an "analogous" manner) to the other three physical domains (electrical, fluid, and thermal). In the present section, we illustrate the use of LGs to formulate the state-space models of an electrical dynamic system. Also, we introduce two practical components: amplifier and dc motor, which are useful in electrical, electromechanical, and other types of multi-physics (multi-domain) systems.

Example 5.6

Consider the circuit shown in Figure 5.14a. It consists of one inductor (L_1), two capacitors (C_2 and C_3), and two resistors (R_4 and R_5). The input to the circuit is the voltage $v_i(t)$ and the output of the circuit is the voltage v_o (across the capacitor C_3).

a. Draw a complete and oriented LG for the circuit. Indicate all the variables and parameters on the LG.
b. Using the LG, systematically derive a complete state-space model for the circuit. Identify the model matrices ***A***, ***B***, ***C***, and ***D***.
c. Sketch a structural diagram (not a LG) for a mechanical system that is completely analogous to the electrical circuit shown in Figure 5.14a.
d. Explain why a completely analogous mechanical system cannot be established for the electrical circuit shown in Figure 5.14b.

Solution

a. The LG of the circuit is shown in Figure 5.15a. The primary loops and the primary nodes are indicated.
b. State vector $\boldsymbol{x} = \begin{bmatrix} i_1 & v_2 & v_3 \end{bmatrix}^T$; input vector $\boldsymbol{u} = [v_i(t)]$; output vector $\boldsymbol{y} = [v_o]$.

Constitutive Equations
State-space shell:

$$L_1: \quad L_1\dot{i}_1 = v_1$$

$$C_2: \quad C_2\dot{v}_2 = i_2$$

$$C_3: \quad C_3\dot{v}_3 = i_3$$

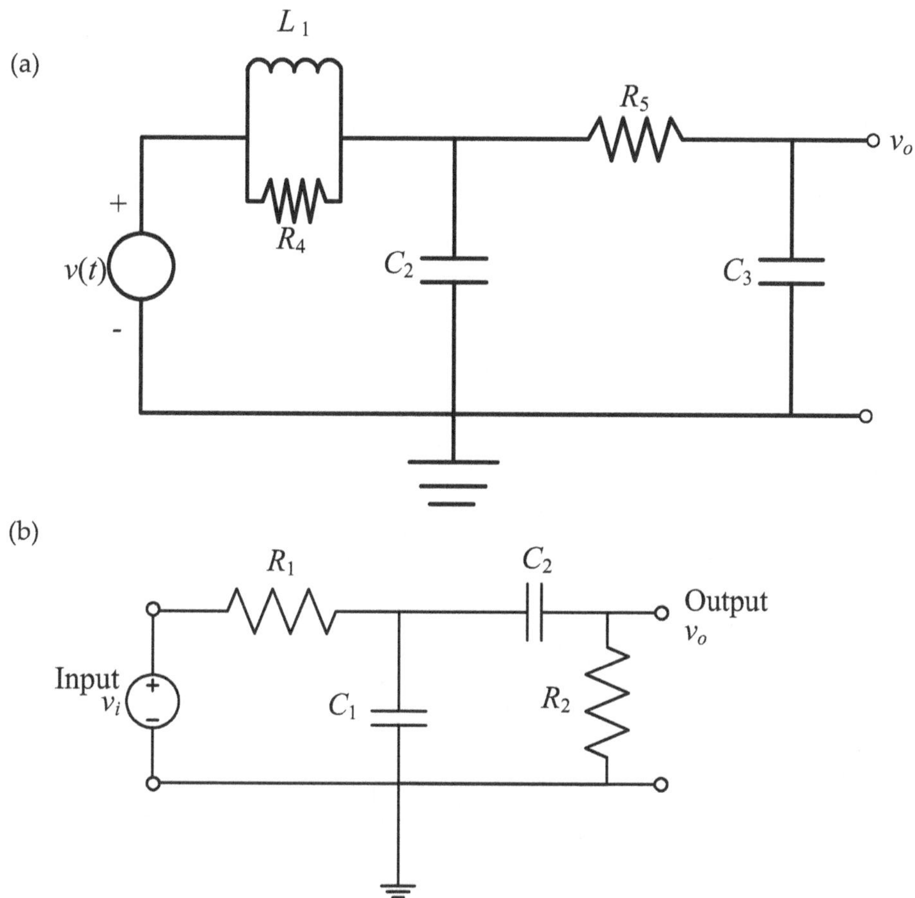

FIGURE 5.14 (a) A passive electrical circuit; (b) the circuit of a passive band-pass filter.

Remaining constitutive equations:

$$R_4: \quad v_4 = R_4 i_4$$

$$R_5: \quad v_5 = R_5 i_5$$

Node Equations

$$\text{Node 1}: \quad i - i_1 - i_4 = 0 \left\{ \text{not useful because } i \text{ is the dependent variable of the source} \right\}$$

$$\text{Node 2}: \quad i_1 + i_4 - i_2 - i_5 = 0$$

$$\text{Node 3}: \quad i_5 - i_3 = 0$$

Loop Equations

$$\text{Loop 1}: \quad -v_2 - v_4 + v_i(t) = 0$$

$$\text{Loop 2}: \quad v_4 - v_1 = 0$$

$$\text{Loop 3}: \quad -v_3 - v_5 + v_2 = 0$$

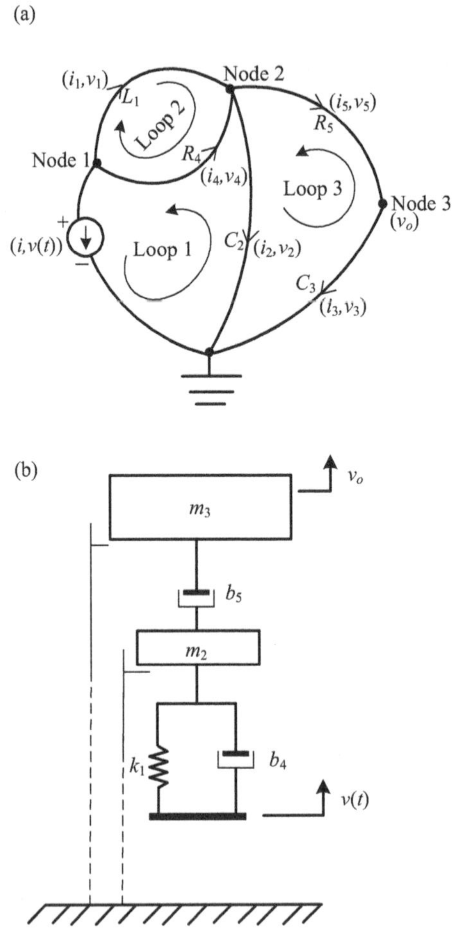

FIGURE 5.15 (a) The linear graph of the circuit; (b) Analogous mechanical system.

Eliminate Auxiliary Variables:

$$v_1 = v_4 = -v_2 + v_i(t)$$

$$i_2 = i_1 + i_4 - i_5 = i_1 + \frac{v_4}{R_4} - \frac{v_5}{R_5} = i_1 + \frac{-v_2 + v_i(t)}{R_4} - \frac{v_2 - v_3}{R_5}$$

$$i_3 = i_5 = \frac{v_5}{R_5} = \frac{v_2 - v_3}{R_5}$$

State-Space Model:

$$L_1 \frac{di_1}{dt} = -v_2 + v_i(t)$$

$$C_2 \frac{dv_2}{dt} = i_1 - \left(\frac{1}{R_4} + \frac{1}{R_5}\right) v_2 + \frac{v_3}{R_5} + \frac{v_i(t)}{R_4}$$

$$C_3 \frac{dv_3}{dt} = \frac{v_2}{R_5} - \frac{v_3}{R_5}$$

Output equation $v_o = v_3$

Model Matrices:

$$A = \begin{bmatrix} 0 & -\dfrac{1}{L_1} & 0 \\ \dfrac{1}{C_2} & -\left(\dfrac{1}{R_4}+\dfrac{1}{R_5}\right)\dfrac{1}{C_2} & \dfrac{1}{R_5C_2} \\ 0 & \dfrac{1}{R_5C_3} & -\dfrac{1}{R_5C_3} \end{bmatrix}; B = \begin{bmatrix} \dfrac{1}{L_1} \\ \dfrac{1}{R_4C_2} \\ 0 \end{bmatrix}; C = \begin{bmatrix} 0 & 0 & 1 \end{bmatrix}; D = [0]$$

Note: There is no feedforward property in this system, and hence, the algebraic output equation does not contain the input variable.

c. Analogous mechanical system is shown in Figure 5.15b.
d. The circuit in Figure 5.14b has a capacitor (C_2) with a terminal that is not grounded. For an inertia element, one of the terminals must be the ground reference. Hence, a mass element cannot be established that would correspond to this electrical capacitor.

5.7.1 Amplifiers

An electrical amplifier is a common component in a practical electrical system. Purely mechanical, fluid, and thermal amplifiers have been either developed or envisaged as well. Common characteristics of an amplifier are:

1. They accomplish tasks of signal amplification
2. They are active devices (i.e., they need external power to operate)
3. They are not affected (ideally) by the load that they drive (i.e., loading effects are negligible)
4. They have a decoupling effect on the systems (this is a desirable effect that reduces the dynamic interactions between components).

The electrical signals voltage, current, and power are amplified using voltage amplifiers, current amplifiers, and power amplifiers, respectively. Operational amplifiers (op-amps) are the basic building block in constructing these amplifiers. An op-amp has a very high-input impedance, low-output impedance, and a very high open-loop gain. But, in the open-loop form, an op-amp is not a practical (stable) device. However, an op-amp with feedback provides the desirable characteristics of

1. Very high-input impedance
2. Low-output impedance
3. Stable operation.

For example, due to its impedance characteristics, the output of a good amplifier is not affected by the device (load) that is connected to it. Furthermore, due to its high-input impedance, an amplifier does not distort the signal that is coming into it from an electrical device. In other words, electrical loading errors can be greatly reduced by an amplifier.

Analogous to electrical amplifiers, a mechanical amplifier can be designed to provide force amplification (a *T*-type amplifier) or a fluid amplifier can be designed to provide pressure amplification (an *A*-type amplifier). In these situations, typically, the device is active and an external power source is needed to operate the amplifier (to drive a combination of motor and a mechanical load, for example).

5.7.1.1 Linear Graph Representation

In its LG representation, an amplifier is considered as a *Dependent Source* or a *Modulated Source.* Specifically, the output of the amplifier depends on (modulated by) the input of the amplifier, and this output is not affected by the dynamics of the devices that are connected to the output of the amplifier (i.e., the load of an amplifier will not "load" the amplifier). This is the ideal case. In practice, some loading error will be present (i.e., the output of the amplifier will be affected by the load that it drives).

The LG representations of an *A*-type amplifier (e.g., voltage amplifier, pressure amplifier) and a *T*-type amplifier (e.g., current amplifier, force amplifier) are shown in Figure 5.16a and b, respectively. The pertinent constitutive equations in the general and linear cases are given as well in the figures.

5.7.2 Power-Information Transformer

An amplifier is a power-to-power transformer. Another transformer that is useful in modeling engineering dynamic systems is a *power-information transformer.* For example, in control and communication of an engineering system, data streams and control sequences are in fact information signals, which are present in sensors, data acquisition systems, controllers and communication networks. On the other hand, the signals in the physical domains (mechanical, electrical, fluid, and thermal) are power signals, with associated through-variables and across-variables. Thus far, we have used only the power signals in our LG models, as these models represent the four specific physical domains. However, to incorporate such aspects as sensing, control, and communication into an LG model, we need means for power-to-information transformation and information-to-power transformation. For this purpose, the concept of modulated source, as in the LG representation of an amplifier, may be extended. Then we have *information sources* and *power sources* with power signals (same as those used thus far in our models) and information signals (those needed in data acquisition, control, and communication purposes) as their respective modulating signals. This aspect is revisited in Problem 5.21, at the end of this chapter.

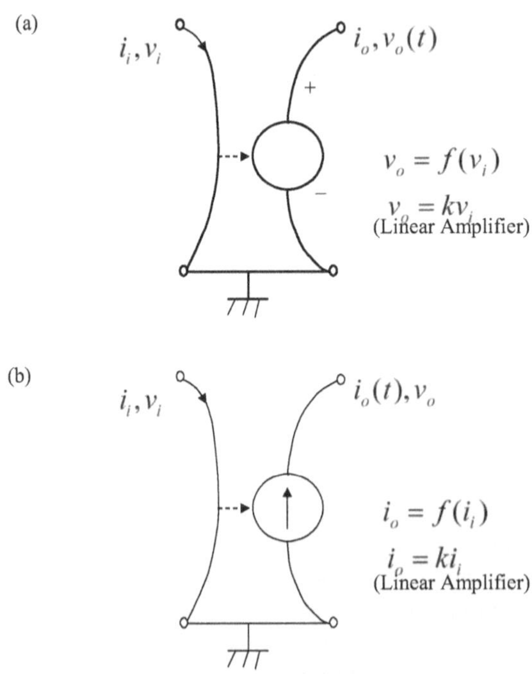

FIGURE 5.16 Linear graph representation of: (a) An *A*-type amplifier; (b) a *T*-type amplifier.

5.7.3 DC Motor

The direct current (dc) motor is a popular electrical actuator. It converts dc electrical energy into mechanical energy. The principle of operation is based on the fact that when a conductor carrying current is placed in a magnetic field, a force is generated (Lorentz's law) in the conductor, which is able to push the conductor. This force results from the interaction of the two magnetic fields (one from the current in the conductor and the other is the magnetic field in which the conductor is placed), and it is available as the magnetic torque at the rotor of the motor, to move its rotor (and any load that is connected to it).

A dc motor has a stator and a rotor (armature) with windings, which provide the two magnetic field for the motor operation. The stator and the rotor are excited by a field voltage v_f and an armature voltage v_a, respectively (in the case of separately excited dc motor). The equivalent circuit of a dc motor is shown in Figure 5.17a, where the field circuit and the armature circuit are shown separately, with the corresponding supply voltages. This is the *separately excited* case. If the stator filed is provided by a permanent magnet, then the stator circuit that is shown in Figure 5.17a is simply an equivalent circuit, where the stator current i_f can be assumed constant. Similarly, if the rotor is a permanent magnet, what is shown in Figure 5.17a is an equivalent circuit where the armature current i_a can be assumed constant. The magnetic torque of the motor is generated by the interaction of the stator field (proportional to i_f) and the rotor field (proportional to i_a) and is given by

$$T_m = ki_f i_a \tag{5.14}$$

A back-electromotive force (*back emf*) is generated in the rotor (armature) windings to oppose its rotation when these windings rotate in the magnetic field of the stator (lenz's law). This voltage is given by

$$v_b = k' i_f \omega_m \tag{5.15}$$

where i_f = field current; i_a = armature current; and ω_m = angular speed of the motor (rotor).

For perfect conversion of electrical energy into mechanical energy in the rotor, we need

$$T_m \omega_m = i_a v_b \tag{5.16}$$

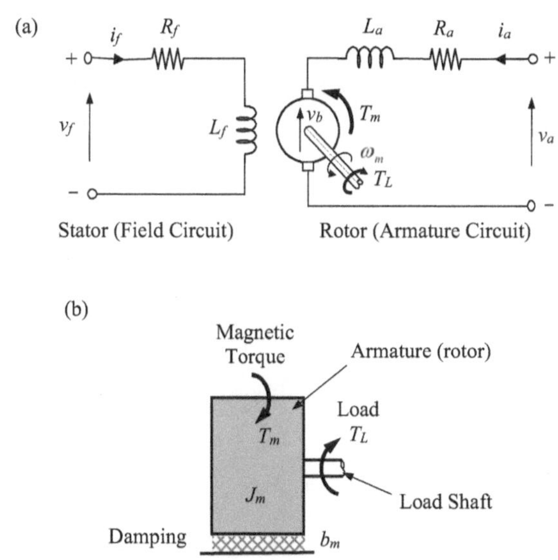

FIGURE 5.17 (a) Equivalent circuit of a dc motor (separately excited); (b) mechanical loading on the armature.

This corresponds to an ideal *electromechanical transformer.*

$$\text{Field Circuit Equation}: v_f = R_f i_f + L_f \frac{di_f}{dt} \tag{5.17}$$

where v_f = supply voltage to stator; R_f = resistance of the field windings; L_f = inductance of the field windings.

$$\text{Armature (Rotor) Circuit Equation}: v_a = R_a i_a + L_a \frac{di_a}{dt} + v_b \tag{5.18}$$

where v_a = armature supply voltage; R_a = resistance of the armature windings; L_a = leakage inductance of the armature.

Suppose that the motor drives a load whose equivalent torque is T_L. Then from Figure 5.17b,

$$\text{Mechanical (Load) Equation}: J_m \frac{d\omega_m}{dt} = T_m - T_L - b_m \omega_m \tag{5.19}$$

where J_m = moment of inertia of the rotor; b_m = equivalent (mechanical) damping constant for the rotor; T_L = load torque.

Note: By using T_L to denote the torque applied to the load, we allow for any "general" load (whose nature does not have to be defined yet).

In the field control of the motor, the armature supply voltage v_a is kept constant and the field voltage v_f is controlled. In the armature control of the motor, the field supply voltage v_f is kept constant and the armature voltage v_a is controlled.

5.8 LINEAR GRAPH EXAMPLES IN FLUID DOMAIN

The LG approach is unified across various physical domains, and hence, the same approach is applicable in mechanical, electrical, fluid, and thermal domains. Now an example in the fluid domain is given to illustrate the use of LGs in the systematic development of state-space models in that domain.

Example 5.7

A pump supplies water to a tank through a long narrow pipe. The tank is connected to a second tank through a short pipe with a valve. There is an outflow valve in the second tank. A schematic diagram of the system is shown in Figure 5.18. The pump may be considered as a pressure source of absolute pressure $P_s(t)$.

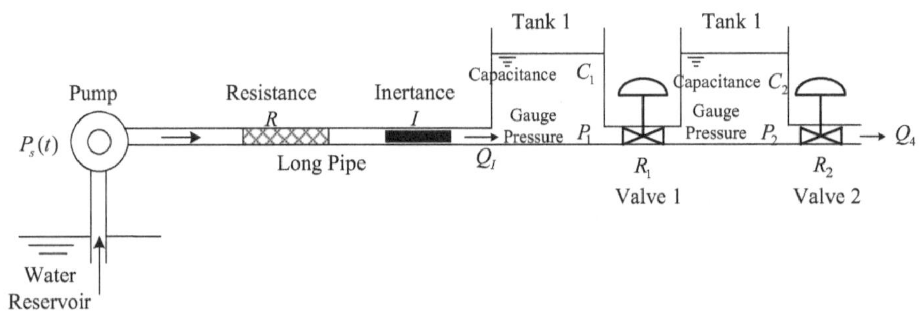

FIGURE 5.18 Two-tank cascade supplied by a single water pump.

The following parameters are given:

R = fluid resistance of the long pipe
L = fluid inertance of the long pipe
C_1 = fluid capacitance of the first tank
C_2 = fluid capacitance of the second tank
R_1 = fluid resistance of the first valve
R_2 = fluid resistance of the second valve

a. Draw an LG for the system.
b. Determine a complete state-space model for the system using the state variables:
Q_I = volume flow rate through the fluid inertor (I)
P_1 = gauge pressure at the bottom of Tank 1
P_2 = gauge pressure at the bottom of Tank 2

Input variable: $P_s(t)$ = outflow pressure of the pump
Output variable: Q_4 = volume flow rate through the exit valve.

Solution

(a)

The system has a single A-type source ($P_s(t)$); three independent energy storage elements (two A-type elements: fluid capacitors C_1 and C_2; a T-type element: fluid inertor I); and three D-type elements (fluid resistors R, R_1 and R_2).

The LG of the system is shown in Figure 5.19.

(b)

Constitutive Equations

State-Space Shell:

Fluid Intertor: $I\dot{Q}_I = P_I$
First Tank: $C_1\dot{P}_1 = Q_1$
Second Tank: $C_2\dot{P}_2 = Q_2$

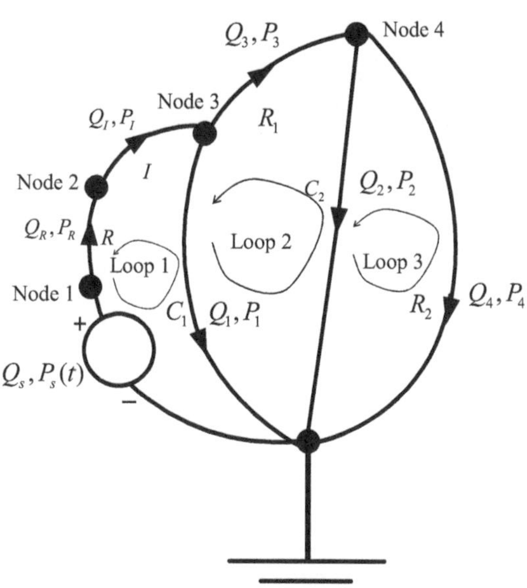

FIGURE 5.19 Linear graph of the system.

Remaining Constitutive Equations:

$$\text{Fluid Resistance in Long Pipe}: P_R = RQ_R \quad \text{(i)}$$

$$\text{First Valve}: P_3 = R_1 Q_3 \quad \text{(ii)}$$

$$\text{Second Valve}: P_4 = R_2 Q_4 \quad \text{(iii)}$$

Node Equations (five nodes → four independent nodes, as marked in the figure)

$$\text{Node 1}: Q_s - Q_R = 0 \left\{ \text{Not useful because } Q_s \text{ is the dependent variable of the source} \right.$$

$$\text{Node 2}: Q_R - Q_I = 0 \quad \text{(iv)}$$

$$\text{Node 3}: Q_I - Q_1 - Q_3 = 0 \quad \text{(v)}$$

$$\text{Node 4}: Q_3 - Q_2 - Q_4 = 0 \quad \text{(vi)}$$

Loop Equations

There are three primary loops. They are chosen as shown in Figure 5.19 so as to include the A-source and the two A-type elements.

$$\text{Loop 1}: -P_1 - P_I - P_R + P_s(t) = 0 \quad \text{(vii)}$$

$$\text{Loop 2}: -P_2 - P_3 + P_1 = 0 \quad \text{(viii)}$$

$$\text{Loop 3}: -P_4 + P_2 = 0 \quad \text{(ix)}$$

To obtain the state equations, now we eliminate the auxiliary variables in the state-shell equations.

$$P_I = P_s(t) - P_1 - P_R = P_s(t) - P_1 - RQ_R = P_s(t) - P_1 - RQ_I \quad \text{(from (vii), (i), and (iv))}$$

$$Q_1 = Q_I - Q_3 = Q_I - \frac{P_3}{R_1} = Q_I - \frac{P_1 - P_2}{R_1} \quad \text{(from (v), (ii), and (viii))}$$

$$Q_2 = Q_3 - Q_4 = \frac{P_3}{R_1} - \frac{P_4}{R_2} = \frac{P_1 - P_2}{R_1} - \frac{P_2}{R_2} \quad \text{(from (vi), (ii), (iii), (viii), and (ix))}$$

Hence, we have the final state equations:

$$I\dot{Q}_I = -RQ_I - P_1 + P_s(t)$$

$$C_1\dot{P}_1 = Q_I - \frac{P_1}{R_1} + \frac{P_2}{R_1}$$

$$C_2\dot{P}_2 = \frac{P_1}{R_1} - \left(\frac{1}{R_1} + \frac{1}{R_1}\right)P_2$$

Output equation (algebraic):

$$Q_4 = \frac{P_4}{R_2} = \frac{P_2}{R_2}$$

In the matrix-vector form,

$$\dot{x} = Ax + Bu$$

$$y = Cx + Du$$

where

$$A = \begin{bmatrix} -\frac{R}{I} & -\frac{1}{I} & 0 \\ \frac{1}{C_1} & -\frac{1}{C_1R_1} & \frac{1}{C_1R_1} \\ 0 & \frac{1}{C_2R_1} & \frac{1}{C_2}\left(\frac{1}{R_1}+\frac{1}{R_1}\right) \end{bmatrix}; B = \begin{bmatrix} \frac{1}{I} \\ 0 \\ 0 \end{bmatrix}; C = \begin{bmatrix} 0 & 0 & \frac{1}{R_2} \end{bmatrix}; D = [0]$$

Note: There is no feedforward property in this system, and hence, the algebraic output equation does not contain the input variable.

LEARNING OBJECTIVES

1. Identification of the element types in a fluid system
2. Selection of proper state variables
3. Formulation of the constitutive equations, starting from the state-space shell
4. Selection of primary loops
5. Identification of primary nodes that can be skipped in the formulation
6. Sign conventions
7. Systematic development of a state-space model for a fluid system

5.9 LINEAR GRAPH EXAMPLES IN THERMAL DOMAIN

Thermal systems have temperature (T) as the across-variable, which is measured with respect to a reference point of the element (i.e., as the temperature difference across the element) and heat transfer (flow) rate (Q) as the through-variable. Heat source and temperature source are the two types of source elements of a thermal system. The former is more common. The latter may correspond to a large reservoir whose temperature is practically not affected by heat transfer into or out of it. There is only one type of energy (thermal energy) in a thermal system. Hence, there is only one type of energy storage element, an A-type element or thermal capacitance, with the associated state variable temperature (T). There is no T-type energy storage element in a thermal system. These issues have been discussed in Chapter 2. Some modeling examples of thermal systems are presented in Chapter 3. In this section, we present an example where an LG is used to develop the state-space model formulation of a thermal system.

5.9.1 Model Equations

In the state model formulation of a thermal system, we follow the systematic procedure as for any other system. Specifically we write:

1. Constitutive equations (for thermal capacitance and resistance elements)
2. Node equations (the algebraic sum of heat transfer rate at a node is zero)
3. Loop equations (the algebraic sum of the temperature drop around a closed thermal path is zero).

Finally, we obtain the state-space model by eliminating the auxiliary variables, which should not be present in the state equations.

Example 5.8

Consider the simplified model of a household hot-water heating system, as shown in Figure 5.20. The furnace supplies heat (thermal energy) to the water in the tank at the rate $Q_s(t)$.This is the deliberate input to the system (it is a *T*-type source). The ambient room temperature T_a is variable and it affects the response (output) of the system. This is an unintentional input or a disturbance input (it is an *A*-type source). The water inside the hot water heater has mass M and specific heat c. This water is assumed to be fully mixed (using a stirrer) and the water temperature is assumed uniform at T_h. It is a thermal capacitor (*A*-type element of capacitance C_h). The radiator, which provides heat to the room, is a thermal resistor (*D*-type element with thermal resistance R_r). Hot water enters the radiator at temperature T_h, transfers heat to the room through a temperature difference $T_h - T_a$ and at a rate Q_r, and leaves the radiator at temperature T_o.

a. Draw an LG for the system.
b. Taking T_h as the state variable; $Q_s(t)$ and $T_a(t)$ as the inputs; and Q_r and T_o as the outputs, determine a complete state-space model for the system.

Given:

M = mass of the water in the heater
c = specific heat of water
$\dot{m}$ = rate of mass flow of the water into the radiator
R_r = thermal resistance of the radiator.

Note: The water heater and the piping are fully insulated.

Solution

(a)

The system has a *T*-type source ($Q_s(t)$) and an *A*-type source ($T_a(t)$); an *A*-type element (thermal capacitor C_h); and a *D*-type element (thermal resistor R_r).

The LG of the system is sketched in Figure 5.21a.

Note: The reference temperature is the absolute zero temperature. In this problem, T_a cannot be taken as the reference temperature because it is variable externally (unintentionally).

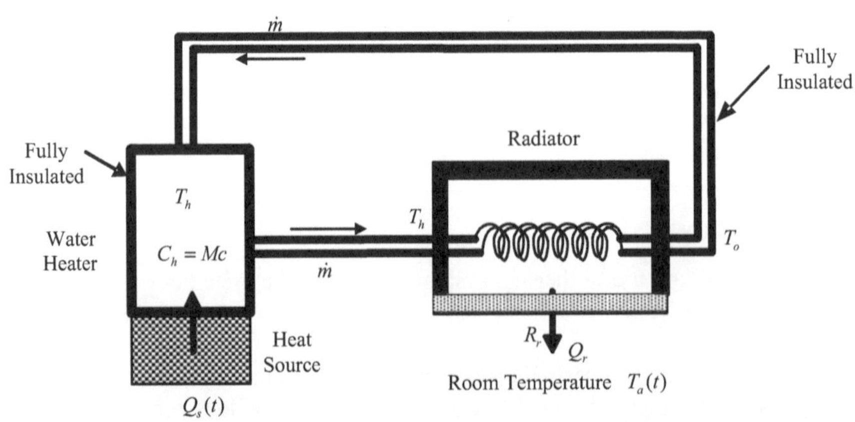

FIGURE 5.20 A simplified hot-water heating system.

(a)

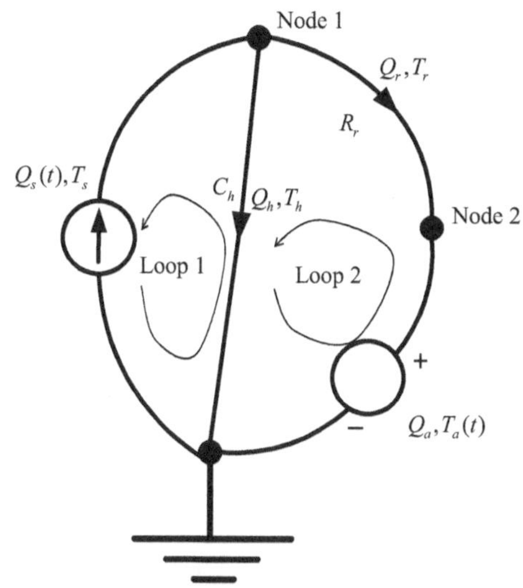

(b)

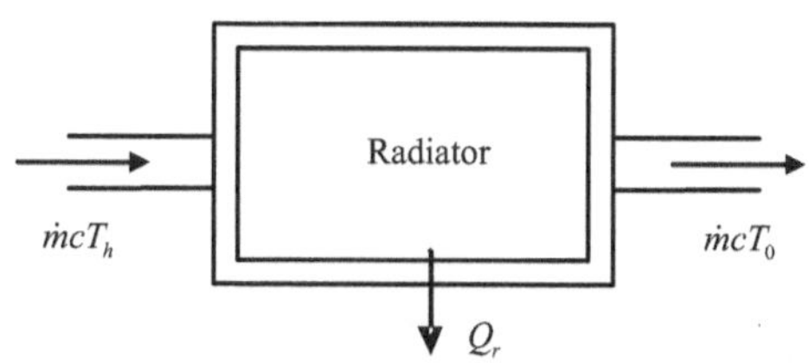

FIGURE 5.21 (a) Linear graph of the hot-water heating system; (b) heat flow rate continuity of the radiator.

(b)

Constitutive Equations

Thermal capacitor (water heater): $C_h \frac{dT_h}{d_t} = Q_h$ with $C_h = Mc$. In fact, this is the state-space shell.

Thermal resistor (Radiator): $T_r = R_r Q_r$

Node Equations

There are two primary nodes, as marked in Figure 5.21a.

$$\text{Node 1}: Q_s(t) - Q_h - Q_r = 0$$

$$\text{Node 2}: Q_r + Q_a = 0 \left\{ \text{Not useful because } Q_a \text{ is the dependent variable of a source} \right.$$

Loop Equations

There are two primary loops in the LG. They are chosen as shown in Figure 5.21a.

$$\text{Loop 1}: T_h + T_s = 0 \left\{ \text{Not useful because } T_s \text{ is the dependent variable of a source} \right.$$

$$\text{Loop 2}: T_r + T_h = 0$$

To determine the final state equation, eliminate the auxiliary variable Q_h in the state-shell equation:
$Q_h = Q_s(t) - Q_r = Q_s(t) - \dfrac{T_r}{R_r} = Q_s(t) - \dfrac{T_h - T_a(t)}{R_r}$

State equation: $C_h \dfrac{dT_h}{dt} = -\dfrac{T_h}{R_r} + Q_s(t) + \dfrac{T_a(t)}{R_r}$

The output equations are obtained as follows:

First output equation: $Q_r = \dfrac{T_r}{R_r} = \dfrac{T_h - T_a(t)}{R_r}$

The second output equation is obtained from the heat flow balance of the radiator (see Figure 5.21b): $\dot{m}cT_n - \dot{m}cT_o - Q_r = 0$

$$T_o = T_h - \frac{Q_r}{\dot{m}c} = T_h - \frac{T_h - T_a(t)}{R_r\dot{m}c} = \left(1 - \frac{1}{R_r\dot{m}c}\right)T_h + \frac{T_a(t)}{R_r\dot{m}c}$$

Now, with

$$\boldsymbol{x} = [T_h];\ \boldsymbol{u} = [Q_s(t), T_a(t)]^T;\ \boldsymbol{y} = [Q_r, T_o]^T$$

We have

$$\dot{\boldsymbol{x}} = \boldsymbol{A}\boldsymbol{x} + \boldsymbol{B}\boldsymbol{u}$$

$$\boldsymbol{y} = \boldsymbol{C}\boldsymbol{x} + \boldsymbol{D}\boldsymbol{u}$$

The model matrices are

$$\boldsymbol{A} = \left[-\frac{1}{C_h R_r}\right];\ \boldsymbol{B} = \begin{bmatrix} \dfrac{1}{C_h} & \dfrac{1}{C_h R_r} \end{bmatrix};\ \boldsymbol{C} = \begin{bmatrix} \dfrac{1}{R_r} \\ 1 - \dfrac{1}{R_r\dot{m}c} \end{bmatrix};\ \boldsymbol{D} = \begin{bmatrix} 0 & -\dfrac{1}{R_r} \\ 0 & \dfrac{1}{R_r\dot{m}c} \end{bmatrix}$$

Note: This system has the feedforward character because the algebraic output equations contain an input variable ($T_a(t)$).

LEARNING OBJECTIVES

1. Identification of the element types in a thermal system
2. Selection of proper state variables
3. Formulation of the constitutive equations
4. Identification of primary loops that can be skipped in the formulation
5. Identification of independent nodes that can be skipped in the formulation
6. Systematic development of a state-space model for a fluid system
7. Understanding the presence of the feedforward character in a system.

5.10 LINEAR GRAPH EXAMPLES IN MIXED DOMAINS

In the previous sections, we illustrated the use of LG in the state model formulation of lumped-parameter mechanical, electrical, fluid, and thermal systems, using analogous (unified) and systematic procedure. Specifically, LGs facilitate a unified approach for developing dynamic models in these various physical domains. LGs facilitate as well an "integrated procedure" to model multi-domain systems (i.e., multi-physics or mixed systems) that use a combination of two or more types of physical components (mechanical, electrical, fluid, and thermal) by considering (and model) them together (i.e., concurrently, in an integrated manner). In other words, a single LG can be developed for a multi-physics system, while using analogous methodology across its physical domains. Mixed-domain examples include electromechanical or mechatronic systems. In this section, we present two examples of the use of LGs in the state model formulation of multi-physics systems.

Example 5.9

A classic problem in robotics is the case of a robotic gripping and turning a doorknob to open a door. The mechanism is schematically shown in Figure 5.22a. Suppose that the actuator of the robotic hand is an armature-controlled dc motor. Revisit the nomenclature and the details of such a dc motor, as presented in Section 5.7.3. The associated circuit is shown in Figure 5.22b. The input signal to the robotic hand is the armature $v_a(t)$, as shown.

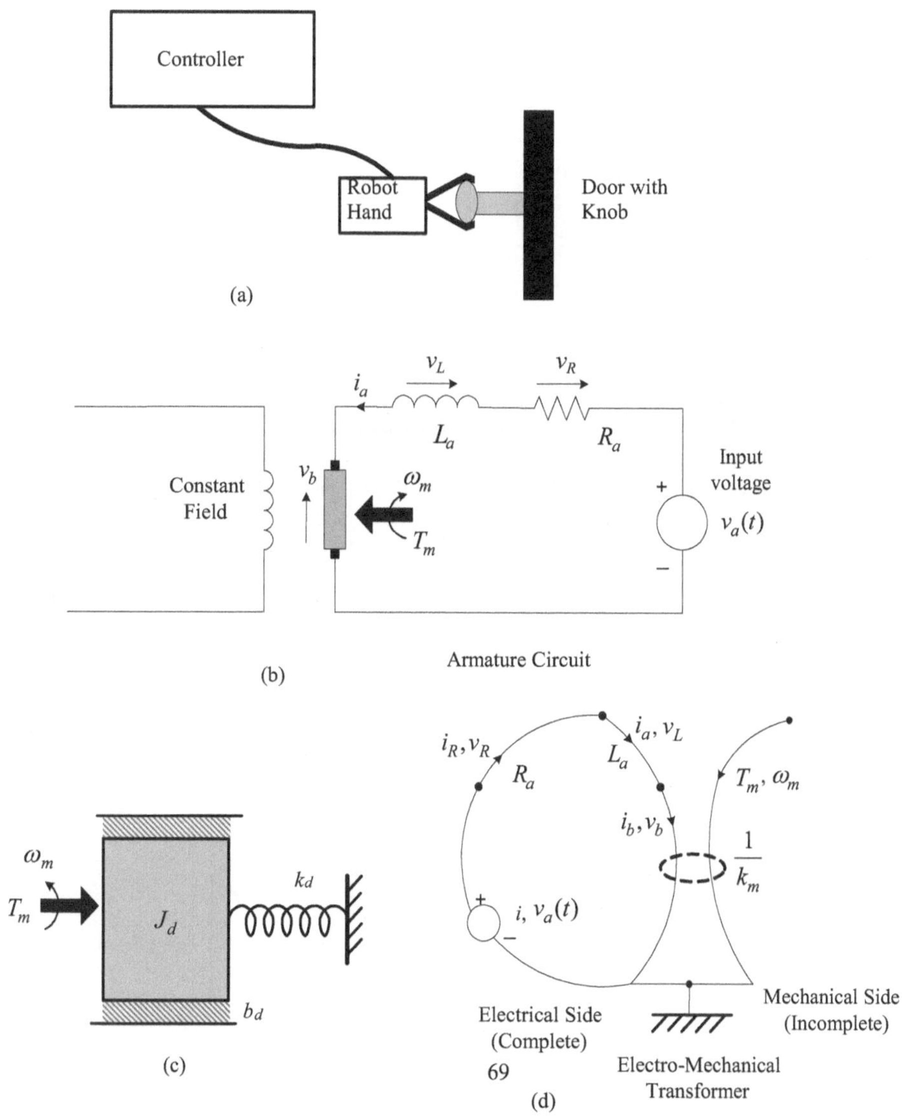

FIGURE 5.22 (a) Robotic hand turning a doorknob; (b) armature-controlled dc motor of the robotic hand; (c) mechanical model of the hand-doorknob system; (d) incomplete linear graph.

Note: In the LG representation, the generated magnetic torque T_m is negative by convention, as marked in Figure 5.22b, because power is considered to be positive when going in at either port of the electromechanical transformer.

This *magnetic torque* is available to turn the doorknob and is resisted by the inertia force (with moment of inertia J_d), the friction (modeled as a linear viscous damper of rotational damping constant b_d), and the flexibility (of torsional stiffness k_d) of the hand-knob-lock combination. A mechanical model is shown in Figure 5.22c. The dc motor may be considered as an ideal electromechanical transducer, which is represented by a transformer in the LG. The associated equations are (see Section 5.7.3):

$$\omega_m = \frac{1}{k_m} v_b \tag{5.20}$$

$$T_m = -k_m i_b \tag{5.21}$$

Note: As mentioned before, the negative sign in equation (5.21) arises due to the specific sign convention in LGs. Specifically, power is assumed to go "into" the electromechanical transformer at either port. In reality, however, a motor takes in electrical power and supplies mechanical power.

The LG may be easily drawn, as shown in Figure 5.22d, for the electrical side of the system. Answer the following questions:

a. Complete the LG by including the mechanical side of the system.
b. Give the number of branches (*b*), nodes (*n*), and the independent loops (*l*) in the complete LG.
c. Take the current through the inductor (i_a), the speed of rotation of the door knob (ω_d), and the resisting torque of the torsional spring within the door lock (T_k) as the state variables, the armature voltage $v_a(t)$ as the input variable, and ω_d and T_k as the output variables. Write the constitutive equations, independent node equations, and independent loop equations for the completed LG. Clearly show the state-space shell. Also verify that the number of unknown variables is equal to the number of equations obtained in this manner (i.e., the problem is solvable).
d. Eliminate the auxiliary variables and obtain a complete state-space model for the system, using the equations written in Part (c) above.

Solution

a. The complete LG is shown in Figure 5.23.
b. $b=8$, $n=5$, $l=4$ for this LG. It satisfies the topological relationship $l=b-n+1$

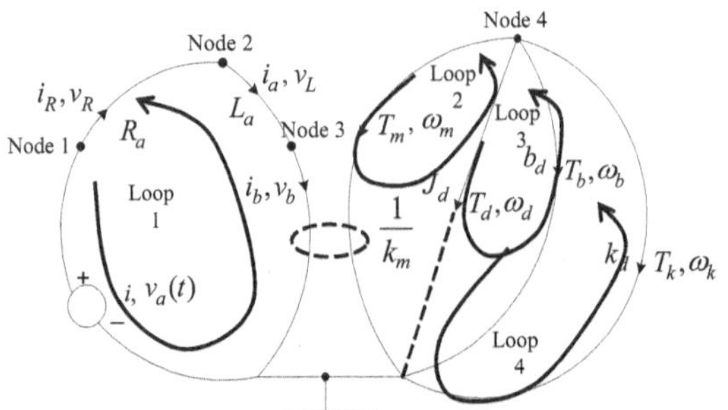

FIGURE 5.23 The complete linear graph of the system.

c.

Constitutive Equations:

$$\left.\begin{aligned} L_a \frac{di_a}{dt} &= v_L \\ J_d \frac{d\omega_d}{dt} &= T_d \\ \frac{dT_k}{dt} &= k_d \omega_k \end{aligned}\right\}\text{State-space shell}$$

Other Constitutive Equations:

$$\text{Resistor:} \quad v_R = R_a i_R \tag{i}$$

$$\text{Mechanical Damper:} \quad T_b = b_d \omega_b \tag{ii}$$

Electromechanical transformer:

$$\omega_m = \frac{1}{k_m} v_b \tag{iii}$$

$$T_m = -k_m i_b \tag{iv}$$

Node Equations:

There are five nodes. The four primary nodes are shown in Figure 5.23, which provide four independent node equations:

$$\text{Node 1:} \quad i - i_R = 0 \quad \text{(not useful)} \tag{v}$$

$$\text{Node 2:} \quad i_R - i_a = 0 \tag{vi}$$

$$\text{Node 3:} \quad i_a - i_b = 0 \tag{vii}$$

$$\text{Node 4:} \quad -T_m - T_d - T_b - T_k = 0 \tag{viii}$$

Note: The node 1 equation is not useful because it has the dependent variable i of the A-source (which is an auxiliary variable and should not remain in the final state equations).

Loop Equations:

There are four primary loops, generating the independent loop equations. They are chosen, as shown in Figure 5.23, to include an A-element in the final three loop equations:

$$\text{Loop 1:} \quad v_a(t) - v_R - v_L - v_b = 0 \tag{ix}$$

$$\text{Loop 2:} \quad \omega_m - \omega_d = 0 \tag{x}$$

$$\text{Loop 3:} \quad \omega_d - \omega_b = 0 \tag{xi}$$

$$\text{Loop 4:} \quad \omega_d - \omega_k = 0 \tag{xii}$$

Note: There are 15 unknown variables (i, i_R, i_a, i_b, T_m, T_d, T_b, T_k, v_R, v_L, v_b, ω_m, ω_d, ω_b, and ω_k) and 15 equations (including the three shell equations). Specifically,

Number of unknown variables $=2b-s=2\times 8-1=15$
Number of independent node equations $=n-1=5-1=4$
Number of independent loop equations $=l=4$
Number of constitutive equations $=b-s=8-1=7$
Check: $15=4+4+7$

d. Eliminate the auxiliary variables from the sate-space shell, through substitution:

$$v_L = v_a(t) - v_R - v_b = v_a(t) - R_a i_a - k_m\omega_m = v_a(t) - R_a i_a - k_m\omega_d \quad \text{(from (ix), (i), (iii), and (x))}$$

$$T_d = -T_k - T_m - T_b = -T_k + k_m i_b - b_d\omega_b = k_m i_a - b_d\omega_d - T_k \quad \text{(from (iv), (ii), (vii), and (xi))}$$

$$\omega_k = \omega_d \quad \text{(from (xii))}$$

Hence, we have the state-space equations:

$$L_a \frac{di_a}{dt} = -R_a i_a - k_m\omega_d + v_a(t)$$

$$J_d \frac{d\omega_d}{dt} = k_m i_a - b_d\omega_d - T_k$$

$$\frac{dT_k}{dt} = k_d\omega_d$$

State vector $\boldsymbol{x} = \begin{bmatrix} i_a & \omega_b & T_k \end{bmatrix}^T$
Input vector $\boldsymbol{u} = [v_a(t)]$
Output vector $\boldsymbol{y} = \begin{bmatrix} \omega_d & T_k \end{bmatrix}^T$
The complete state-space model is

$$\dot{\boldsymbol{x}} = \boldsymbol{Ax} + \boldsymbol{Bu}$$

$$\boldsymbol{y} = \boldsymbol{Cx} + \boldsymbol{Du}$$

The model matrices are

$$\boldsymbol{A} = \begin{bmatrix} -R_a/L_a & -k_m/L_a & 0 \\ k_m/J_d & -b_d/J_d & -1/J_d \\ 0 & k_d & 0 \end{bmatrix}; \boldsymbol{B} = \begin{bmatrix} 1/L_a \\ 0 \\ 0 \end{bmatrix}; \boldsymbol{C} = \begin{bmatrix} 0 & 1 & 0 \\ 0 & 0 & 1 \end{bmatrix}; \boldsymbol{D} = [0]$$

Observations:

- This is a multi-physics (electromechanical) model
- Multi-functional devices (e.g., a piezoelectric device that serves as both actuator and sensor) may be modeled similarly, using an electromechanical transformer (or, through the use of the "reciprocity principle")
- There is no feedforward property in this system, and hence, the algebraic output equation does not contain the input variable.

LEARNING OBJECTIVES

1. Checking the solvability of a dynamic model
2. Selection of proper state variables
3. Formulation of the constitutive equations, starting from the state-space shell
4. Proper selection of primary loops

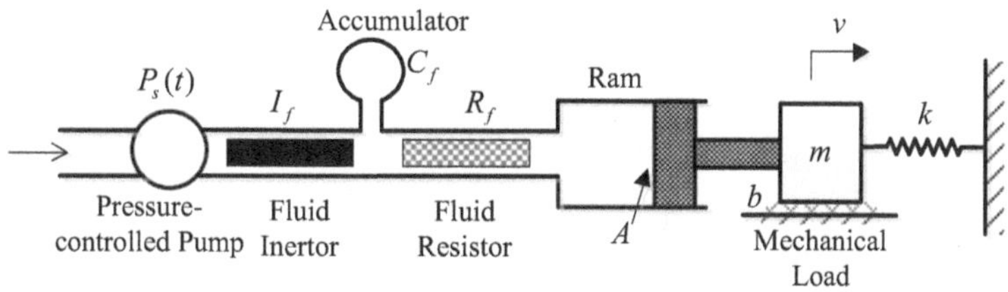

FIGURE 5.24 A mechanical load driven by a pump through a ram.

5. Identification of primary nodes that can be skipped in the formulation
6. Sign conventions
7. Systematic development of a state-space model for an electromechanical (two-domain) system
8. Checking whether there is a feedforward character in the system.

Example 5.10

A pressure-controlled pump drives a mechanical load using a piston-cylinder hydraulic actuator (ram), which is connected to the pump through a long pipe (see Figure 5.24). A hydraulic accumulator of capacitance C_f is used to smoothen any pressure spikes in the input to the ram. The input to the overall system is the output pressure $P_s(t)$ of the pump. The output of the system is the velocity v of the mechanical load. The following parameters are given:

I_f = fluid inertance of the long pipe
R_f = fluid resistance at the inlet of the ram
C_l = fluid capacitance of the ram
m = mass of the load
k = resisting stiffness on the load
b = damping constant of the load motion
A = area of the piston of the ram

a. Sketch a complete LG for the system.
b. Show that the topological relation $b = l + (n-1)$ is satisfied in the LG. Also, show that the number of unknown variables is equal to the number of independent equations that can be written for the LG (i.e., the problem is solvable).
c. Write the constitutive equations (state-space shell equations and the remaining constitutive equations), loop equations, and node equations. By eliminating the auxiliary variables determine a complete state-space model for the system.

Solution

a. The LG for the system is shown in Figure 5.25.
b. For the LG,

$$l = 6, b = 10, \text{ and } n = 5$$

Note: The reference node is common to both ports of the gyrator.
Hence, $b = 10 = l + (n-1) = 6 + (5-1) = 10$ is satisfied.
The number of sources, $s = 1$
The number of unknown variables $= 2b - s = 20 - 1 = 19$

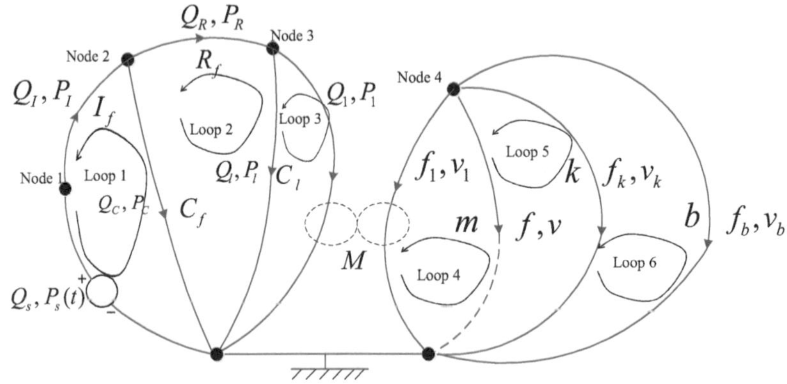

FIGURE 5.25 Linear graph of the system.

The number of constitutive equations $= b - s = 10 - 1 = 9$
The number of loop equations $= l = 6$
The number of node equations $= n - 1 = 4$
$\Rightarrow$ Total number of equations $= 9 + 6 + 4 = 19$
$\Rightarrow$ The number of unknown variables = the number of equations
Hence, the problem is solvable.

c. There are five independent energy storage elements I_f, C_f, C_l, m, and k.
The state vector, $\boldsymbol{x} = \left[Q_I, P_C, P_l, v, f_k\right]^T \Rightarrow$ fifth-order system
The input vector $\boldsymbol{u} = [P_s(t)]$
The output vector $\boldsymbol{y} = [v]$
Constitutive Equations:
State-Space Shell:

$$I_f \dot{Q}_I = P_I$$

$$C_f \dot{P}_C = Q_C$$

$$C_l \dot{P}_l = Q_l$$

$$m\dot{v} = f$$

$$\dot{f}_k = kv_k$$

Remaining Constitutive Equations:

$$P_R = R_f Q_R$$

$$f_b = bv_b$$

$$v_1 = MQ_1$$

$$f_1 = -\frac{1}{M}P$$

$$\text{with } M = \frac{1}{A}$$

Loop Equations

We have picked primary loops that contain the independent A-type elements and the A-source, as shown in Figure 5.25 (which generally provides variables that should be retained in the state-space model).

$$-P_C - P_I + P_s(t) = 0$$

$$-P_l - P_R + P_C = 0$$

$$-P_1 + P_l = 0$$

$$-v + v_1 = 0$$

$$-v_k + v = 0$$

$$-v_b + v = 0$$

Node Equations

The primary nodes are shown in Figure 5.25.

$$Q_S - Q_I = 0 \quad \text{(not a useful equation because it contains } Q_S\text{)}$$

$$Q_I - Q_R - Q_C = 0$$

$$Q_R - Q_l - Q_1 = 0$$

$$-f_1 - f - f_k - f_b = 0$$

Eliminate Auxiliary Variables:

$$P_I = -P_C + P_s(t)$$

$$Q_C = Q_I - Q_R = Q_I - \frac{P_R}{R_f} = Q_I - \frac{(P_C - P_l)}{R_f}$$

$$Q_l = Q_R - Q_1 = \frac{P_R}{R_f} - \frac{1}{M}v_1 = \frac{(P_c - P_l)}{R_f} - \frac{1}{M}v_1$$

$$f = -f_1 - f_k - f_b = \frac{1}{M}P_1 - f_k - bv_b$$

$$= \frac{1}{M}P_l - f_k - bv$$

$$v_k = v$$

State Equations:

$$I_f\dot{Q}_I = -P_c + P_s(t)$$

$$C_f\dot{P}_c = Q_I - \frac{P_c}{R_f} + \frac{P_l}{R_f}$$

$$C_l\dot{P}_l = \frac{P_c}{R_f} - \frac{P_l}{R_f} - \frac{v}{M}$$

$$m\dot{v} = \frac{P_l}{M} - bv - f_k$$

$$\dot{f}_k = kv$$

Output equation: $y = v$

The vector-matrix form of the linear sate-space model:

$$\dot{x} = Ax + Bu$$

$$y = Cx + Du$$

Model Matrices:

$$A = \begin{bmatrix} 0 & -\frac{1}{I_f} & 0 & 0 & 0 \\ \frac{1}{C_f} & -\frac{1}{R_f C_f} & \frac{1}{R_f C_f} & 0 & 0 \\ 0 & \frac{1}{R_f C_l} & -\frac{1}{R_f C_l} & -\frac{1}{MC_l} & 0 \\ 0 & 0 & \frac{1}{mM} & -\frac{b}{m} & -\frac{1}{m} \\ 0 & 0 & 0 & k & 0 \end{bmatrix}; B = \begin{bmatrix} \frac{1}{I_f} \\ 0 \\ 0 \\ 0 \\ 0 \end{bmatrix}; C = \begin{bmatrix} 0 & 0 & 0 & 1 & 0 \end{bmatrix}; D = [0]$$

There is no feedforward property in this system, and hence, the algebraic output equation does not contain the input variable.

LEARNING OBJECTIVES

1. Identification of the element types in a system
2. Checking the solvability of a dynamic model
3. Selection of proper state variables
4. Formulation of the constitutive equations, starting from the state-space shell
5. Selection of convenient primary loops
6. Identification of primary nodes that can be skipped in the formulation
7. Sign conventions
8. Systematic development of a state-space model for a multi-physics (mixed-domain) system

5.11 SUMMARY SHEET

- **Characteristics and Advantages of LGs**: 1. Applicable for lumped-parameter engineering dynamic systems (extension to distributed-parameter systems is taking place); 2. line segments (branches) represent model elements; 3. branches are connected at "nodes" → model structure (graphical representation) → structure visualization before model formulation; 4. loop: closed path formed by two or more branches; 5. LGs facilitate a unified (i.e., analogous methodology is used in multiple domains) modeling approach; 6. LGs facilitate an integrated (i.e., concurrent → all physical domains can be represented in a single LG and analyzed together) methodology for multi-physics (multi-domain) systems; 7. the model structure is retained across domain → interconnected components in one domain and similarly interconnected analogous elements in another domain have the same LG (e.g., parallel connections remain parallel and series connections remain series across domain); 8. identify similarities (in domain, structure, behavior, etc.); 9. facilitate the development of computer-based modeling tools and software (unified, integrated, and systematic, graphical tool); 9. multi-functional devices are modeled conveniently (e.g., a piezoelectric device, which can function as both a sensor and an actuator, can be represented simply by a reversible source)

- **LG Conventions**: 1. Single port → one line segment (branch): one end is *point of action* and the other end is *point of reference*; 2. through-variable f is the same at action point and reference point; across-variable differs (difference = value relative to reference point = across-variable v); 3. pair (f, v) is shown on one side of the branch. Their relationship (constitutive relation) can be linear or nonlinear; element parameter is shown on the other side of the branch; 4. power flow p = [through-variable] × [across-variable]. At action point, f and p (power flow in) are positive in the same direction; at reference point, power flows out (direction of f at reference is determined by this); positive direction of power flow p is shown by arrow on LG branch (oriented branch); 5. arrow also gives the direction of the across-variable drop (i.e., action to reference), except for a T-source (where the arrow is from reference to action); 6. difference in the energy flows at action and reference is either stored (A-type and T-type, with capacity to do work) or dissipated (D-type)
- **Decoupling**: Components in series with T-source or parallel with A-source have no dynamic interaction
- **Two-Port Elements**: Transformer: $v_o = r v_i$, $f_o = -\frac{1}{r} f_i$; gyrator: $v_o = M f_i$, $f_o = -\frac{1}{M} v_i$. They transfer energy from one side of the element to the other side. They can be in two physical domains
- **Constitutive Equations**: Physical equations for elements. May be written for all elements except for source elements
- **Compatibility (Loop) Equations**: Algebraic sum of across-variables around a loop = 0 → no variable discontinuity (incompatibility) at any point in a loop. **Convention**: 1. Go in ccw direction of loop; 2. across-variable drops in branch arrow direction of a branch arrow → across-variable is positive. Exception: T-source, whose arrow direction is increasing direction of across-variable (still, it is the +ve direction of power flow); 3. primary loop set: a "minimal" and "independent" set; (a) any other loop can be formed by combining the loops in this set; (b) no loop in the set can be formed by combining the remaining loops in the set; 4. primary loops provide all the independent loop equations; 5. best choice of primary loops: (a) will contain A-type energy storage elements (providing A-type state variables) and A-sources (providing A-type input variables), (b) minimizes the mathematical manipulations in state equation generation (i.e., in eliminating auxiliary variables; 6. all primary loops (loop equations) may not be useful in the development of a state model (e.g., ignore loop with T-source because its across-variable is an auxiliary variable, which should not be in state equations)
- **Continuity (Node) Equations**: Algebraic sum of through-variables at a junction = 0; **In Mechanical Systems**: Force balance, equilibrium equation, Newton's third law, etc.; **In Electrical Systems**: Current balance, Kirchoff's current law, conservation of charge, etc.; **In Hydraulic Systems**: Conservation of fluid; **In Thermal Systems**: Conservation of thermal energy. *Note*: 1. LG with n nodes → $n-1$ primary (independent) node equations (remaining node equation = algebraic sum of first $n-1$ node equations); 2. all primary node equations may not be useful in the development of a state model (e.g., node equation with an A-source, because its through-variable is an auxiliary variable, which should not be in state equations)
- **Sketching of an** LG: 1. Identify energy storage elements, energy dissipation elements, and source elements in system (single-port elements, each represented by one branch); 2. identify multi-port elements (e.g., transformers, gyrators); 3. recognize how elements (branches) are interconnected (series or parallel and to what elements); 4. starting from a convenient node point (typically, the ground reference) draw a branch (typically, for a source), link it to another appropriate branch through a node (action point of incoming branch → reference point of the next branch, unless the next branch closes the loop at ground reference); 5. repeat step 4 until the entire system is completed (i.e., all system elements are included and connected); 6. orient (with an arrow) each LG branch (this may have been done in an

earlier step); 7. write through-across variable pair on one side of each branch and element parameter on the other side (no element parameter for a source element)

- **Steps for State Model from LG**: 1. Choose as state variables: across-variables for independent *A*-type elements and through-variables for independent *T*-type elements; 2. write constitutive equations for independent energy storage elements → **state-space shell**. *Note*: If a group of dependent elements is present, pick the element at the output for inclusion in the state-space shell (The graph-tree approach in Appendix A gives a formal way to identify dependent elements); 3. write constitutive equations for remaining elements (dependent energy storage elements, dissipation—*D*-type elements, two-port elements, etc.); 4. write compatibility equations for primary loops. *Note*: If a particular primary loop equation is not useful, you may skip it; 5. Write continuity equations for primary (independent) nodes (= total number of nodes – 1). *Note*: If a particular primary node equation is not useful, you may skip it; 6. in the state-space shell, retain state variables and input variables only. Eliminate all other variables (*auxiliary variables*) using loop and node equations and other constitutive equations
- **LG Topological Relation (Also consistent with physical model)**: $b = \ell + (n-1)$; number of branches $= b$; number of primary loops $= \ell$; number of nodes $= n$
- **Model Solvability**: Number of sources $= s$; number of constitutive equations $= b - s$; number of loop (compatibility) equations $= \ell$; number of node (continuity) equations $= n-1$ → total number of equations $= b - s + \ell + n - 1$.

  Number of variables $= 2b - s$. For the analytical model to be solvable, need $b - s + \ell + n - 1 = 2b - s \rightarrow b = \ell + (n-1)$ → identical to the topological fact.

PROBLEMS

5.1 Select the correct answer for each of the following multiple-choice questions:

i. A through-variable is characterized by
   a. being the same at both ends of the element
   b. being listed first in the pair written on an LG branch
   c. requiring no reference value
   d. all the above.

ii. An across-variable is characterized by
   a. the value difference between action and reference points
   b. being listed second in the pair written on an LG branch
   c. requiring a reference point
   d. all the above.

iii. Which of the following could be a through-variable?
   a. pressure
   b. voltage
   c. force
   d. all the above.

iv. Which of the following could be an across-variable?
   a. motion (velocity)
   b. fluid flow
   c. current
   d. all the above.

v. If angular velocity is selected as an element's across-variable, the accompanying through-variable is
   a. force
   b. flow
   c. torque
   d. distance.

vi. The equation written for through-variables at a node is called
   a. a continuity equation
   b. a constitutive equation
   c. a compatibility equation
   d. all the above.

vii. The functional relation between a through-variable and its across-variable of an LG branch is called
   a. a continuity equation
   b. a constitutive equation
   c. a compatibility equation
   d. a node equation.

viii. The equation that equates the algebraic sum of across-variables in a loop to zero is known as
   a. a continuity equation
   b. a constitutive equation
   c. a compatibility equation
   d. a node equation.

ix. A node equation is also known as
   a. an equilibrium equation
   b. a continuity equation
   c. the balance of through-variables at the node
   d. all the above.

x. A loop equation represents
   a. compatibility of across-variables
   b. continuity of through-variables
   c. a constitutive relationship
   d. all the above.

5.2 An LG has ten branches, two sources, and six nodes.
   i. How many unknown variables are there?
   ii. What is the number of independent loops?
   iii. How many inputs are present in the system?
   iv. How many constitutive equations could be written?
   v. How many independent continuity equations could be written?
   vi. How many independent compatibility equations could be written?
   vii. Do a quick check on your answers.

5.3 The circuit shown in Figure P5.3 has an inductor L, a capacitor C, a resistor R, and a voltage source $v(t)$. The output of the circuit is the voltage across the capacitor C.
   i. Sketch the LG denoting the currents through and the voltages across the elements L, C, and R by (i_1, v_1), (i_2, v_2), and (i_3, v_3), respectively. What are the state, input, and output variables? What is the order of the system?
   ii. Follow the systematic steps to obtain the state-space model for this system, what are the system matrix $\boldsymbol{A}$, the input distribution matrix $\boldsymbol{B}$, the measurement gain matrix $\boldsymbol{C}$, and the feedforward gain matrix $\boldsymbol{D}$ for the model?

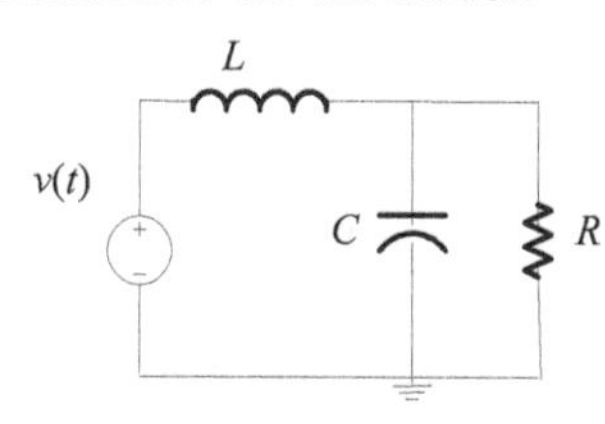

FIGURE P5.3 An electrical circuit.

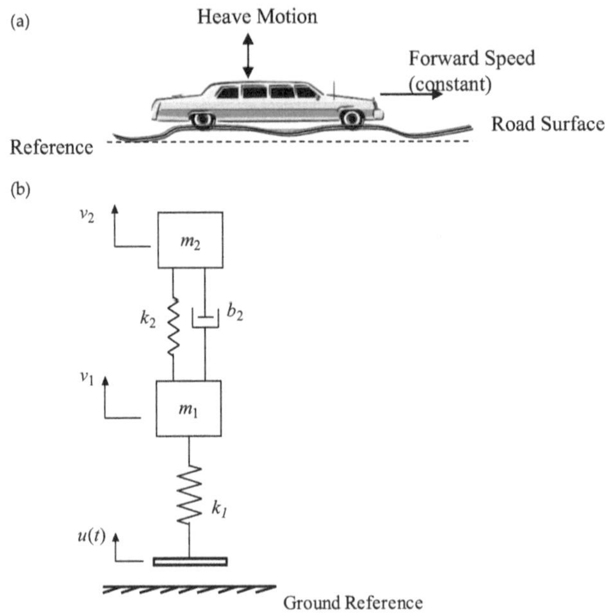

FIGURE P5.4 (a) An automobile traveling at constant speed; (b) a quarter model of the automobile for the heave motion analysis.

iii. Briefly explain what happens if the voltage source $v(t)$ is replaced by a current source $i(t)$.

Note: The fact that L is analogous to a spring, C is analogous to an inertia, R is analogous to a viscous damper, and a voltage source is analogous to a velocity source may be used to solve the analogous mechanical-domain problems.

5.4 Consider an automobile traveling at a constant speed on a rough road, as sketched in Figure P5.4a. The disturbance input in the vertical direction, due to road irregularities, can be considered as a velocity source $u(t)$ at the tires. An approximate one-dimensional model (one-quarter model) shown in Figure P5.4b may be used to study the "heave" (up and down) motion of the automobile. Note that v_1 and v_2 are the velocities of the lumped masses m_1 and m_2, respectively.

a. Briefly state what physical components of the automobile are represented by the model parameters k_1, m_1, k_2, m_2 and b_2. Also, discuss the validity of the assumptions that are made in arriving at this model.

b. Draw an LG for this model, orient it (i.e., mark the directions of the branches), and completely indicate the system variables and parameters. What is the order of the system?

c. By following the step-by-step (systematic) procedure of writing constitutive equations, node equations, and loop equations, develop a complete state-space model for this system. The outputs are v_1 and v_2. Instead of the velocity source $u(t)$, a force source $f(t)$ is applied at the same location and is considered as the system input. Draw an LG for this modified model. Obtain the state equations for this modified model. What is the order of the system now?

Note: In this problem, assume that the gravitational effects are completely balanced by the initial compression of the springs. All the motions in the model are defined from this static equilibrium condition.

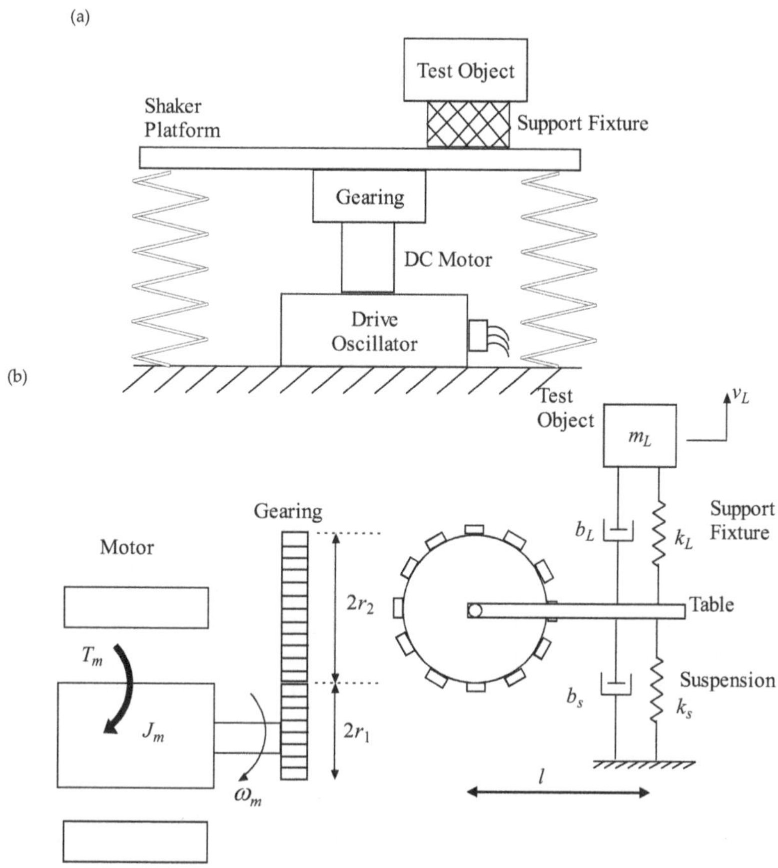

FIGURE P5.5 (a) A dynamic-test system; (b) a model of the system.

5.5

a. List several advantages of using LGs in the development of a state-space model of a dynamic system.

b. Electrodynamic shakers are commonly used in the dynamic testing of products. One possible configuration of a shaker/test-object system is shown in Figure P5.5a. A simple, linear, lumped-parameter model of the mechanical system is shown in Figure P5.5b. The driving motor (actuator) is represented by a torque source T_m.
Also, the following parameters are indicated:
J_m = equivalent moment of inertia of motor rotor, shaft, coupling, gears, and the shaker platform
r_1 = pitch circle radius of the gear wheel attached to the motor shaft
r_2 = pitch circle radius of the gear wheel rocking the shaker platform
l = lever arm from the center of the rocking gear to the support location of the test object
m_L = equivalent mass of the test object and its support fixture
k_L = stiffness of the support fixture
b_L = equivalent viscous damping constant of the support fixture
k_s = stiffness of the suspension system of the shaker table
b_s = equivalent viscous damping constant of the suspension system.

Since the inertia effects are lumped into equivalent elements, assume that the shafts, gearing, platform and the support fixtures are light. The following variables are of interest:

ω_m = angular speed of the drive motor
v_L = vertical speed of motion of the test object
f_L = equivalent dynamic force of the support fixture (force in spring k_L)
f_s = equivalent dynamic force of the suspension system (force in spring k_s)

i. Obtain an expression for the motion parameter:

$$r = \frac{\text{vertical movement of the shaker table at the test object support location}}{\text{angular movement of the drive motor shaft}}$$

ii. Draw an LG to represent the dynamic model.

iii. Using $\boldsymbol{x} = \begin{bmatrix} \omega_m, & f_s, & f_L, & v_L \end{bmatrix}^T$ as the state vector, $\boldsymbol{u} = [T_m]$ as the input, and $\boldsymbol{y} = \begin{bmatrix} v_L & f_L \end{bmatrix}^T$ as the output vector, systematically obtain a complete state-space model for the system. For this purpose, you must use the LG drawn in Part (ii).

5.6 A robotic sewing system consists of a conventional sewing head. During operation, a panel of garment is fed through by a robotic hand into the sewing head. The sensing and control system of the robotic hand ensures that the seam is accurate and the cloth tension is correct, in order to guarantee the quality of the stitch. The sewing head has a frictional feeding mechanism, which pulls the fabric in a cyclic manner away from the robotic hand, using a toothed feeding element. When there is slip between the feeding element and the garment, the feeder functions as a *force source* and the applied force is assumed cyclic with a constant amplitude. When there is no slip, however, the feeder functions as a *velocity source*, which is the case during normal operation. The robot hand has inertia. There is some flexibility at the mounting location of the hand, in the robot. The links of the robot are assumed rigid, and some of its joints can be locked to reduce the number of degrees of freedom, when desired.

Consider the simplified case of a single-degree-of-freedom robot. The corresponding robotic sewing system is modeled as in Figure P5.6. Here the robot is modeled as a single moment of inertia J_r, which is linked to the hand with a light rack-and-pinion device whose speed transmission parameter is given by:

$$\frac{\text{Tanslatory movement of the rack}}{\text{Rotatory movement of the pinion}} = r$$

The drive torque of the robot is T_r and the associated rotatory speed is ω_r. Under conditions of slip, the feeder input to the cloth panel is force f_f. With no slip, the input is the velocity v_f. Various energy dissipation mechanisms are modeled as linear viscous dampers

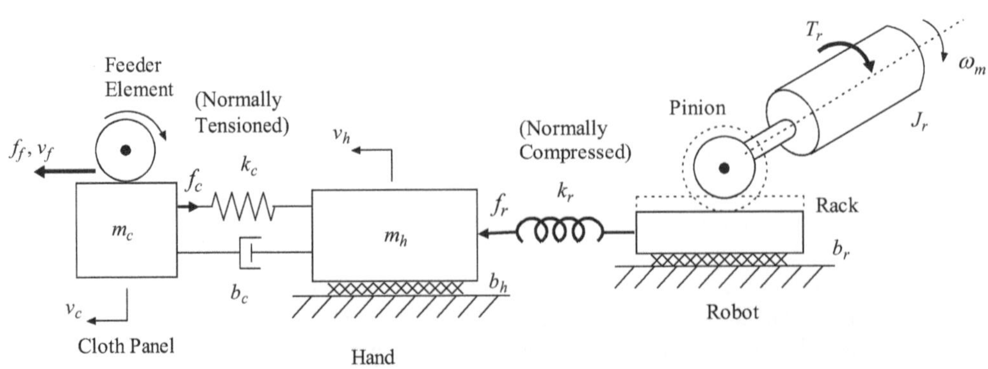

FIGURE P5.6 A robotic sewing system.

of damping constants b (with corresponding subscripts). The flexibility of various system elements is modeled by linear springs with stiffness k (with corresponding subscripts). The inertia effects of the cloth panel and the robotic hand are denoted by the lumped masses m_c and m_h, respectively, having the corresponding velocities v_c and v_h, as shown in Figure P5.6.

Note: The cloth panel is normally in tension with tensile force f_c. In order to push the panel, the robotic wrist is normally in compression with compressive force f_r.

First consider the case of the feeding element with slip:

a. Draw an LG for the model shown in Figure P5.6, orient the graph, and mark all the element parameters, through-variables, and across-variables on the graph.
b. Write all the constitutive equations (element physical equations), independent node equations (continuity), and independent loop equations (compatibility). What is the order of the model?
c. Develop a complete state-space model for the system. The outputs are taken as the cloth tension f_c and the robot speed ω_m, which represent the two variables that have to be measured to control the system. Obtain the model matrices $\boldsymbol{A}$, $\boldsymbol{B}$, $\boldsymbol{C}$, and $\boldsymbol{D}$.

Now consider the case where there is no slip at the feeder element:

d. What is the order of the system now? What is the modified LG of the model, for this situation? Accordingly, modify the state-space model obtained earlier to represent the present situation (no slip) and from that obtain the new model matrices $\boldsymbol{A}$, $\boldsymbol{B}$, $\boldsymbol{C}$, and $\boldsymbol{D}$.
e. Comment on the validity of the assumptions made in obtaining the model shown in Figure P5.6 for a robotic sewing system, in general.

5.7 Systematically develop a state-space model for the system shown in Figure P5.7, using its LG. The output is the velocity v_2 of the mass. How many equations and how many unknowns are present in the initial formulation? Is the model solvable?

5.8 An approximate model of a motor-compressor combination that is used in a process control application is shown n Figure P5.8.

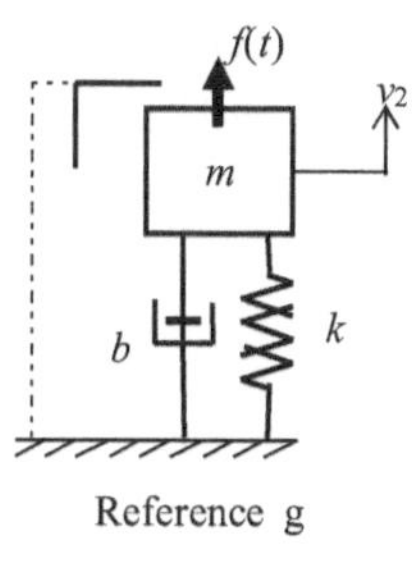

FIGURE P5.7 A mass-spring-damper system.

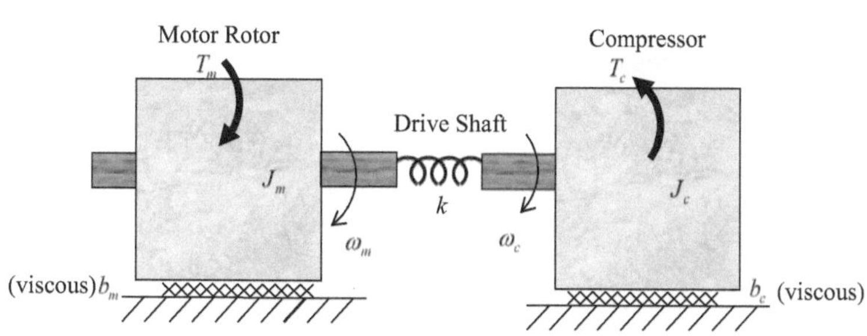

FIGURE P5.8 A model of a motor-compressor unit.

In the figure, T, J, k, b, and ω denote torque, moment of inertia, torsional stiffness, angular viscous damping constant, and angular speed, respectively, and the subscripts m and c denote the motor rotor and the compressor impeller, respectively.

Note: T_c is the torque from the compressor load, which may be considered a "negative actuator" (a negative "torque source" or an input) in the present model. The viscous damping represents the friction at the bearings of the shafts.

a. Sketch a translatory mechanical model that is analogous to this rotatory mechanical model.
b. Draw an LG for the given model, orient it, and indicate all necessary variables and parameters on the graph.
c. By following a systematic procedure and using the LG, obtain a complete state-space representation of the given model. The outputs of the system are the compressor speed ω_c and the torque T transmitted through the drive shaft.

5.9 A model for a single joint of a robotic manipulator is shown in Figure P5.9. The usual notation is used. The gear inertia is neglected and the gear reduction ratio is taken as $1{:}r$ (*Note*: $r<1$).

a. Draw an LG for the model, assuming that no external (load) torque is present at the robot arm.
b. Using the LG derive a state model for the given system. The input is the motor magnetic torque T_m and the output is the angular speed ω_r of the robot arm. What is the order of the system? If the shaft is not flexible, what is the order of the resulting system?
c. Discuss the validity of various assumptions made in arriving at this simplified model for a commercial robotic manipulator.

5.10 Consider the rotatory electromechanical system with feedback control that is schematically shown in Figure P5.10a. The motor rotor has inertia J, moves at angular speed ω_m, and experiences equivalent damping with viscous rotatory damping constant B, from the bearings, as shown. The armature circuit for the dc fixed-field motor is shown in Figure P5.10b. The load is considered as a torque source (negative actuator) that is connected to the motor through a flexible shaft and moves at angular speed ω_l.

For a dc motor, the following relations hold (see Figure P5.10 for the nomenclature):

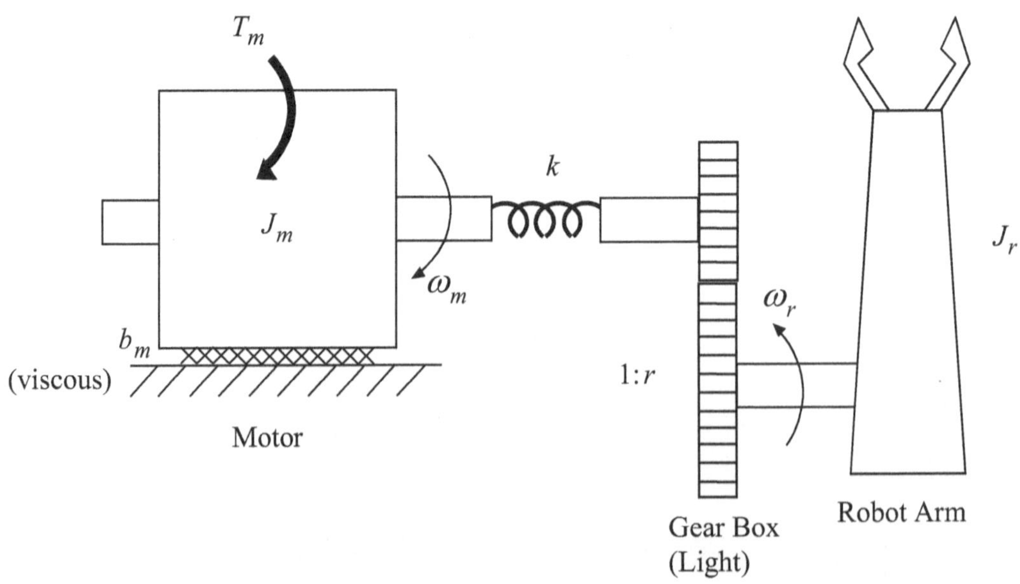

FIGURE P5.9 A model of a single-degree-of-freedom robot.

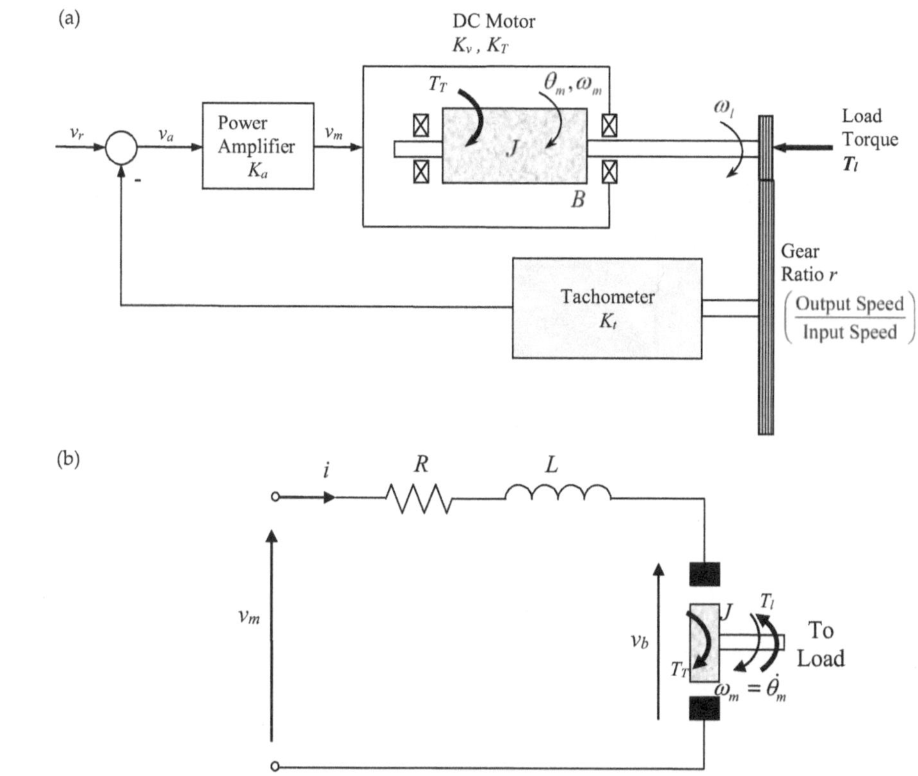

FIGURE P5.10 (a) A rotatory electromechanical system; (b) the armature circuit.

$$\text{The back e.m.f. } v_b = K_V \omega_m$$

$$\text{The motor magnetic torque } - T_T = K_T i$$

Note: The –ve sign in the second equation originates from the LG convention for a transformer, where energy is considered to flow into the transformer at either port (branch), whereas in the actual case, energy is conserved; electrical energy flows in at the input port and mechanical energy flows out at the output port.

For ideal electromechanical energy conversion of a dc motor, with consistent units, we have $K_V = K_T$

a. Identify the system inputs.
b. Give an equivalent translatory model for the system.
c. Sketch the LG of the electromechanical system. Do not include the sensing and control signals, which are "information signals" that can be accommodated separately, as discussed in this book, and not power signals.
d. Write the linear system equations (constitutive, node, and loop equations) using the LG, and systematically obtain a state model for the electromechanical system.

5.11

a. What is the main physical reason for natural oscillatory behavior in a purely fluid system?

   Why do purely fluid systems with large tanks connected by small-diameter pipes rarely exhibit an oscillatory response?

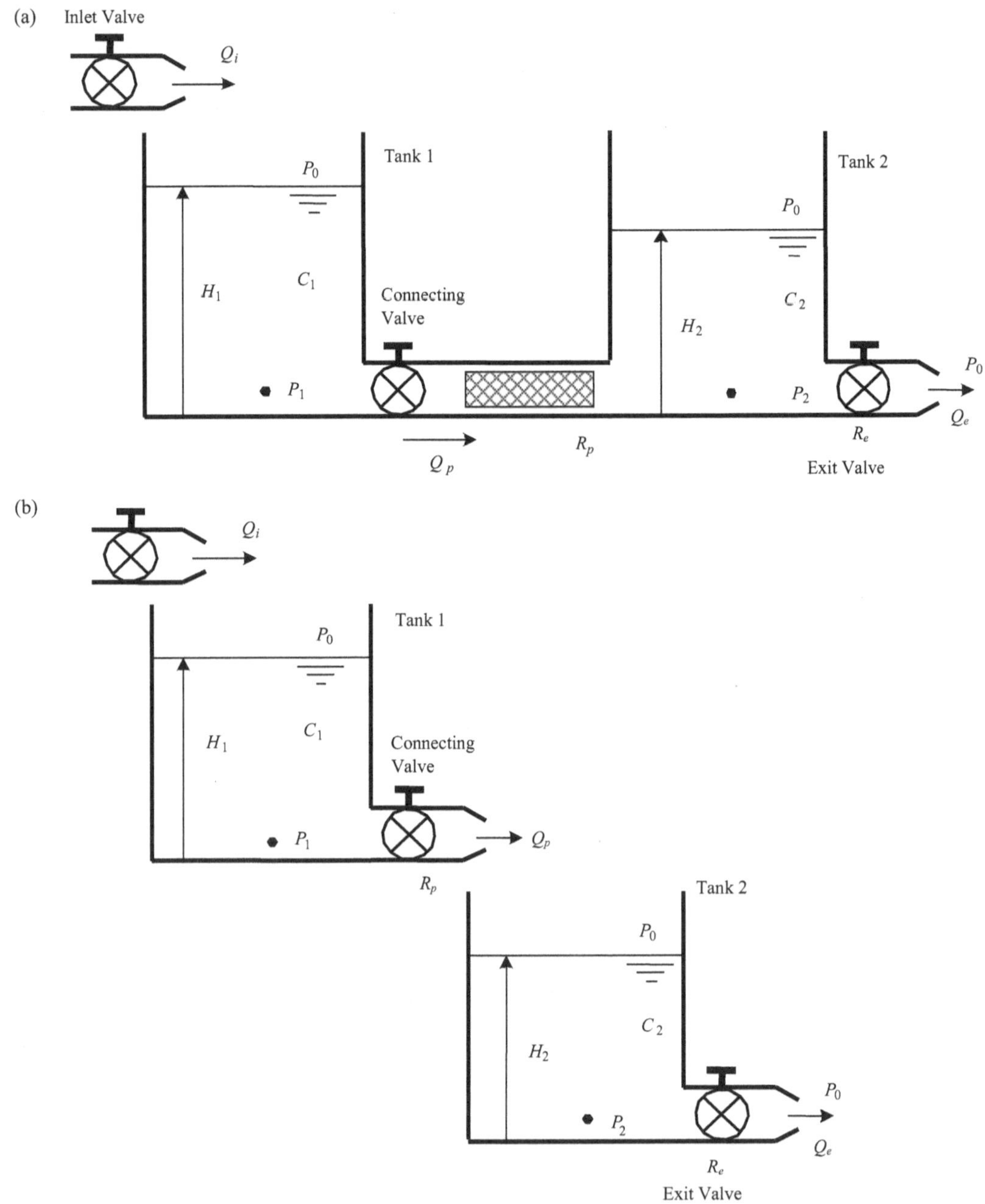

FIGURE P5.11 (a) An interacting two-tank fluid system; (b) a non-interacting two-tank fluid system.

b. Two large tanks whose bottoms are connected by a thin horizontal pipe are shown in Figure P5.11a. Tank 1 receives an inflow of liquid at the volume rate Q_i when the inlet valve is open. Tank 2 has an exit valve, which has a fluid flow resistance R_e and a flow rate Q_e when opened. The connecting pipe also has a valve, and when opened, the combined fluid flow resistance of the valve and the thin pipe is R_p. The following parameters and variables are defined:
C_1, C_2 = fluid (gravity head) capacitances of tanks 1 and 2
ρ = mass density of the fluid

g = acceleration due to gravity

P_1, P_2 = pressures at the bottom of tanks 1 and 2

P_0 = ambient pressure.

Sketch an LG for this system. Using $P_{10} = P_1 - P_0$ and $P_{20} = P_2 - P_0$ as the state variables, and the liquid levels H_1 and H_2 in the two tanks as the output variables, systematically derive a complete, linear, state-space model for the system (by using the LG).

c. Suppose that the two tanks are as in Figure P5.11b. Here Tank 1 has an exit valve at its bottom whose resistance is R_e and the volume flow rate is Q_e when open. This flow directly enters Tank 2, without a connecting pipe. The remaining tank characteristics and the parameters are the same as in Part (b).Sketch a suitable LG for this modified system. Using it, derive a state-space model for the system in terms of the same variables as in Part (b).

5.12 Give reasons for the common experience that in the flushing tank of a household toilet, some effort is needed to move the handle for the flushing action, but virtually no effort is needed to release the handle at the end of the flush.

A simple model for the valve movement mechanism of a household flushing tank is shown in Figure P5.12. The overflow tube on which the handle lever is hinged is assumed rigid. Also, the handle rocker is assumed light and rigid, and the rocker hinge is assumed frictionless.The following parameters are indicated in the figure:

$r = \dfrac{l_v}{l_h}$ = the lever arm ratio of the handle rocker

m = equivalent lumped mass of the valve flapper and the lift rod

k = stiffness of the spring action on the valve flapper.

The damping force f_{NLD} on the valve is assumed quadratic and is given by

$$f_{\mathrm{NLD}} = a\left|v_{\mathrm{NLD}}\right| v_{\mathrm{NLD}}$$

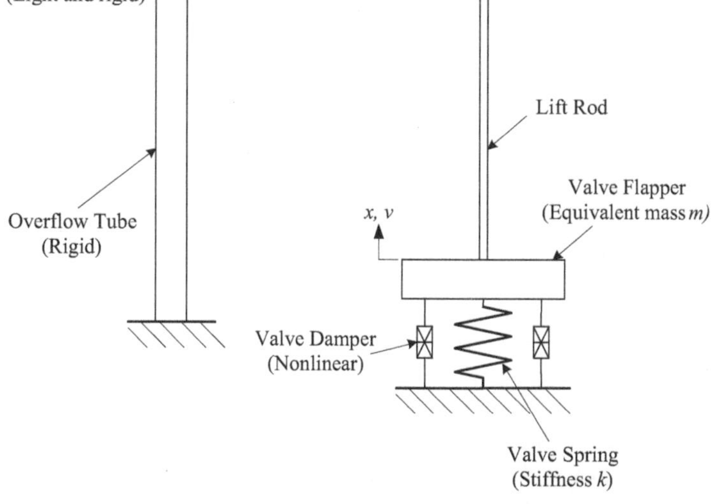

FIGURE P5.12 Simplified model of a toilet-flushing mechanism.

where the positive parameter a is defined as follows:

$a = a_u$ for upward motion of the flapper ($v_{NLD} \geq 0$)

$= a_d$ for downward motion of the flapper ($v_{NLD} < 0$)

with $a_u >> a_d$. The force applied at the handle is $f(t)$, as shown.

We are interested in studying the dynamic response of the flapper valve. Especially, the valve displacement x and the valve speed v are considered outputs, as shown in Figure P5.12. *Note*: x is measured from the static equilibrium point of the spring where the weight mg is balanced by the initial spring force.

a. By defining appropriate through-variables and across-variables, draw an LG for the system shown in Figure P5.12. Clearly indicatc thc power flow arrows.
b. Using the valve speed and the spring force as the state variables, systematically develop a (nonlinear) state-space model for the system, with the aid of the LG. Specifically, start with all the constitutive, continuity, and compatibility equations, and eliminate the auxiliary variables, to obtain the state-space model.
c. Linearize the state-space model about an operating point where the valve speed is $\bar{v}$. For the linearized model, obtain the model matrices $\boldsymbol{A}$, $\boldsymbol{B}$, $\boldsymbol{C}$, and $\boldsymbol{D}$, in the usual notation. The incremental variables $\hat{x}$ and $\hat{v}$ are the outputs in the linear model and the incremental variable $\hat{f}(t)$ is the input.
d. From the linearized state-space model, derive the input–output model (differential equation) relating $\hat{f}(t)$ and $\hat{x}$.

5.13 A common application of dc motors is in accurate positioning of a mechanical load. A schematic diagram of a possible arrangement is shown in Figure P5.13. The actuator of the system is an armature-controlled dc motor. The moment of inertia of its rotor is J_r and the angular speed is ω_r. The mechanical damping of the motor (including that of its bearings) is neglected in comparison to that of the load.

The armature circuit is also shown in Figure P5.13, which indicates a back emf v_b (due to the motor coil rotation in the stator field), a leakage inductance L_a, and a resistance R_a. The current through the leakage inductor is i_L. The input signal is the armature voltage $v_a(t)$ as shown. The interaction of the rotor magnetic field and the stator magnetic field (*Note*: The rotor field rotates at an angular speed ω_m) generates a "magnetic" torque T_m, which is exerted on the motor rotor.

The stator provides a constant magnetic field to the motor and is not important in the present problem. The dc motor may be considered as an ideal electromechanical transducer, which may be represented by an LG transformer. The associated constitutive equations are

$$\omega_m = \frac{1}{k_m} v_b$$

$$T_m = -k_m i_b$$

where k_m is the torque constant of the motor.

Note: The negative sign in the second equation arises due to the specific sign convention used for a transformer, in the conventional LG representation. Specifically, power is taken as positive going in at either port of the transformer, whereas in the actual operation, electrical power goes in at the input port while mechanical power comes out at the output port.

The motor is connected to a rotatory load of moment of inertia J_l using a long flexible shaft of torsional stiffness k_l. The torque transmitted through this shaft is denoted by T_k. The load rotates at angular speed ω_l and experiences mechanical dissipation, which is modeled by a linear viscous damper of damping constant b_l.

Answer the following questions:

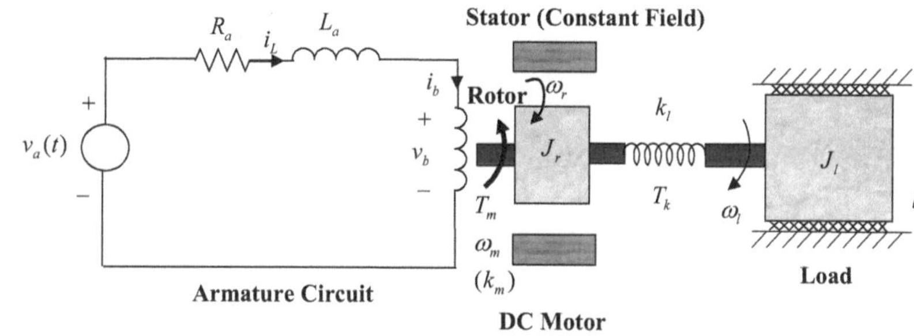

FIGURE P5.13 An electromechanical model of a rotatory positioning system.

a. Draw a suitable LG for the entire system shown in Figure P5.13, mark the variables and parameters (you may introduce new, auxiliary variables but not new parameters), and orient the graph.
b. Give the number of branches (b), nodes (n), and independent loops (l) in the complete LG. What relationship do these three parameters satisfy? How many independent node equations, loop equations, and constitutive equations can be written for the system? Verify the sufficiency of these equations to solve the problem.
c. Take the current through the inductor (i_L), speed of rotation of the motor rotor (ω_r), torque transmitted through the load shaft (T_k), and the speed of rotation of the load (ω_l) as the four state variables; the armature supply voltage $v_a(t)$ as the input variable; and the shaft torque T_k and the load speed ω_l as the output variables. Write the constitutive equations, independent node equations, and the independent loop equations for the complete LG. Clearly show the state-space shell.
d. Eliminate the auxiliary variables and obtain a complete state-space model for the system, using the equations written in Part (c). Express the matrices $\boldsymbol{A}$, $\boldsymbol{B}$, $\boldsymbol{C}$, and $\boldsymbol{D}$ of the state-space model in terms of the system parameters $R_a, L_a, k_m, J_r, k_l, b_l$, and J_l.

5.14 An armature-controlled dc motor is used for motion control of a payload, which is assumed to be a concentrated mass. A lead-screw-and-nut device is used to convert the rotatory motion of the motor into the translatory motion of the nut, which is integral with the carriage on which the payload is firmly mounted (Figure P5.14).

The following system parameters are given:

J=moment of inertia of the motor rotor ($\text{kg} \cdot \text{m}^2$)
B=equivalent rotary damping constant at the motor ($\text{N} \cdot \text{m} \cdot \text{rad}^{-1} \cdot \text{s}$)
m=total mass of the carriage and the payload (kg)
b=equivalent translatory damping constant at the carriage ($\text{N} \cdot \text{m}^{-1} \cdot \text{s}$)
k_m=torque constant of the motor ($\text{N} \cdot \text{m} \cdot \text{A}^{-1}$)
R_a=resistance of the armature windings (Ω)
L_a=leakage inductance of the armature (H)
p=pitch of the lead-screw (i.e., distance between adjacent threads) (m).

Also,

$v_a(t)$=armature voltage, which is the input to the system (V)
v_p=velocity of the payload, which is the output of the system (m/s).

Note: Pitch p of the lead-screw is equal to the motion of the nut per revolution, which assumes that a single-threaded lead-screw is used.

a. Assume that the torsional stiffness K of the lead-screw, from the motor to the nut, is a constant.
Note: Actually K varies because the position of the nut along the lead-screw is not fixed. Draw an LG for the system. Using it and giving all necessary steps, systematically

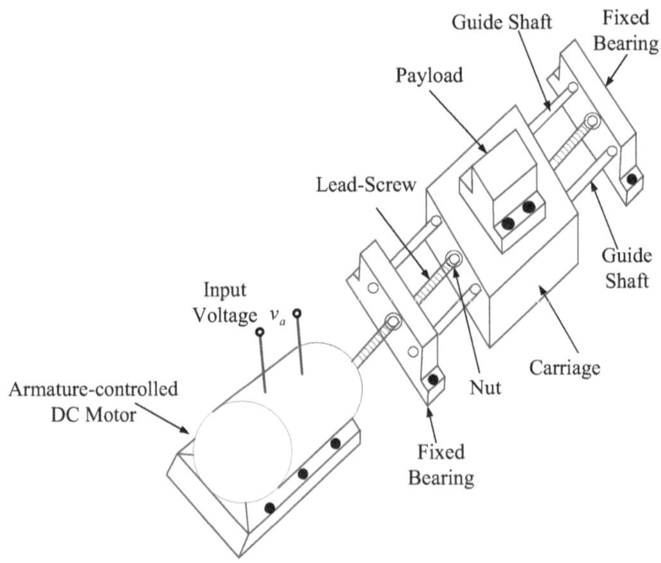

FIGURE P5.14 Motion control device with a dc motor and a lead-screw-nut mechanism.

derive a complete state-space model for the system. Use the following as the state variables:

i_a = armature current (A)
ω_m = motor speed (rad/s)
T_l = torque in the lead-screw (N · m)
v_p = velocity of the payload (m/s)

b. Suppose that the stiffness K of the lead-screw from the motor to the nut is variable. In particular, suppose that the angle of twist of a unit length (1 m) of the lead-screw, when a unit torque (1 N · m) is applied to it = a (rad · m^{-1} · N^{-1} · m^{-1}).
Also,
x_o = initial distance of the nut from the motor
Note: During motion, the distance of the nut from the motor varies.
Derive a state-space model for the system, using the same state variables as before.
Note: Now you must use the constant parameter a instead of K in your equations. You must give all necessary steps and details of your derivation.

5.15 A passive shock-absorber unit has a piston-cylinder mechanism, as schematically shown in Figure P5.15. The cylinder is fixed and rigid and is filled with an incompressible hydraulic fluid (on both sides of the piston). The piston mass is m and its area is A. It has a small opening through which the hydraulic fluid can flow from one side of the cylinder to the other side as the piston moves. The fluid resistance to this flow may be represented by a linear hydraulic resistance R_f.
A spring of stiffness k resists the movement of the piston. Suppose that the input force applied to the shock absorber (piston) is $f(t)$.a.
Sketch an LG for this system.
Hint: It may contain a force source, a spring branch, a mass branch, a mechanical-fluid gyrator (two domains), and a hydraulic resistance. Some of the associated branch variables are:

v = velocity of the piston
Q = fluid volume flow rate through the piston opening (taken positive from left to right)
$P = P_2 - P_1$ = pressure difference between the two sides of the piston
f_k = spring force.

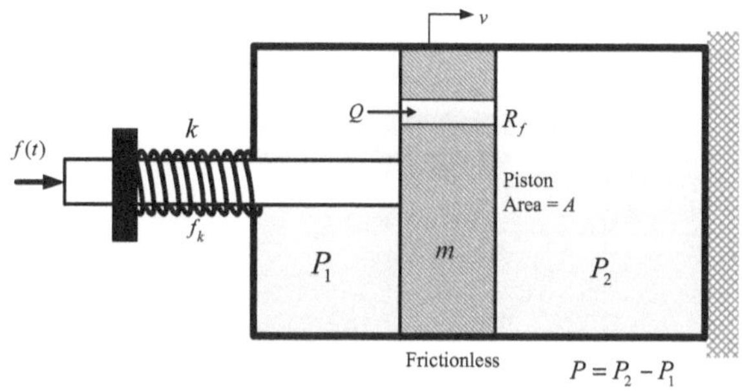

FIGURE P5.15 A passive shock absorber unit.

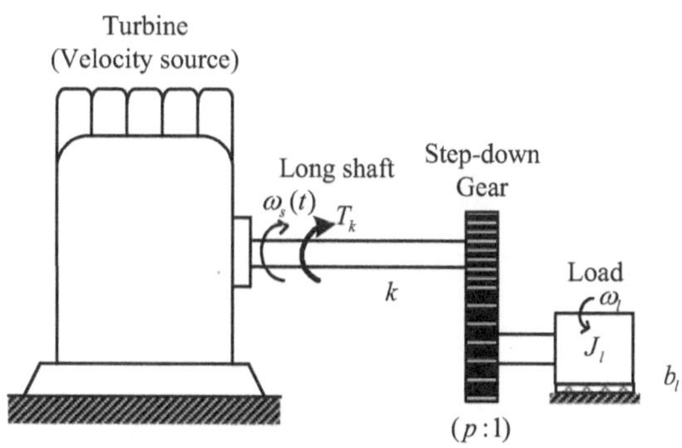

FIGURE P5.16 A load driven by a turbine through a long shaft and step-down gear.

b. Using the LG, systematically develop a state-space model for the system (obtain the model matrices **A** and **B** only) with:
State vector: $\boldsymbol{x} = [v, f_k]^T$
Input vector: $\boldsymbol{u} = [f(t)]$
Express the model parameters in terms of m, k, A, and R_f.
Notes: Assume that,
1. There is no friction between the piston and the cylinder
2. The area of the two sides of the piston is the same (approximately) at value A. That is, neglect the area of the piston rod.

5.16 A turbine is used to drive a rotational device through a long (hence, flexible) shaft and a speed-reduction gear unit, as indicated in Figure P5.16.
The turbine may be considered as a velocity source $\omega_s(t)$. Also,
T_k = torque in the long shaft
p = speed-reduction ratio (p:1 with $p > 1$)
k = torsional stiffness of the shaft
J_l = moment of inertia of the rotational load
b_l = rotational damping constant at the load
ω_l = angular speed of the load.

a. Sketch an oriented LG for the system and mark all the parameters and variables on it.
b. Write the state-space shell equations, remaining constitutive equations, node equations, and the loop equations.
c. Taking ω_l as the output variable, derive a complete state-space model for the system.

Note: In particular, obtain the model matrices **A**, **B**, **C**, and **D** for this "linear" system.

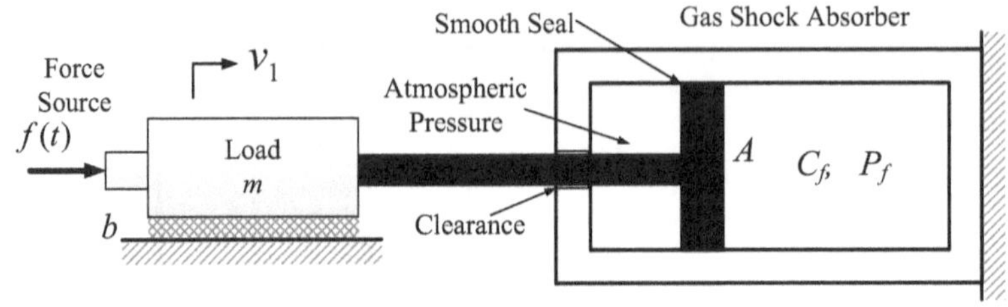

FIGURE P5.17 Mechanical load with a shock absorber.

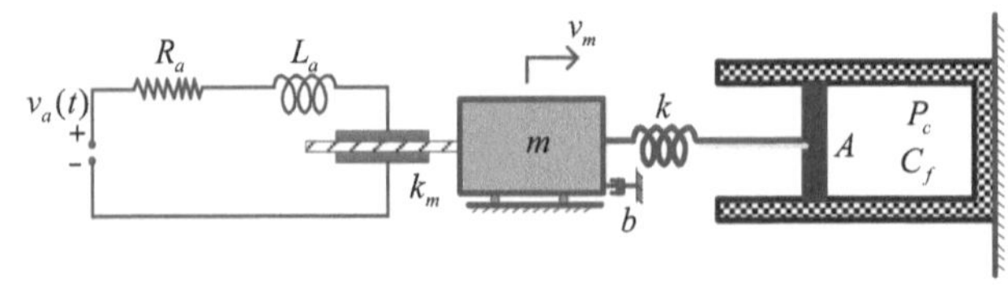

FIGURE P5.18 A linear motor-driven mechanical load.

5.17 A gyrator is an element that is used in the modeling a dynamic system. Briefly describe its function.

A mechanical load of mass m is driven by a force source $f(t)$ and is buffered using a gas shock absorber, as shown in Figure P5.17. The gas cylinder is properly sealed (i.e., gas does not leak out of the cylinder compartment). The piston moves against the cylinder without experiencing appreciable friction (or, this friction component may be incorporated into the damping of the motion of the mass, assuming that the shaft is rigid). The following parameters are given:

m = mass of the moving load
b = equivalent viscous damping constant of the motion of the load
C_f = capacitance of the gas in the shock absorber
A = area of the piston.

a. Sketch a complete LG for the system.
b. Using the state variables: v_1 = velocity of the mass, P_f = gauge pressure in the gas; the input variable $f(t)$ = force of the source element (linear motor); and the output variable v_1 = velocity of the mass, systematically determine a complete state-space model for the system.
c. Express the undamped natural frequency and the damping ratio of the system in terms of b, m, A, and C_f.

Note: Neglect the mass of the piston (or incorporate it into the mass of the load) and assume that the piston rod is rigid.

5.18 Consider a positioning mechanism, which has an armature-controlled linear dc motor that drives a massive load having viscous damping. The load is connected through a flexible rod of stiffness k to a gas cylinder (which serves as a shock absorber) of pressure P_c and fluid (gas) capacitance C_f. A lumped-parameter linear model of the system is shown in Figure P5.18.

The following model parameters are given:

R_a = resistance of the armature coil
L_a = leakage inductance of the armature circuit
k_m = force constant of the linear dc motor
m = overall mass of the load and the armature

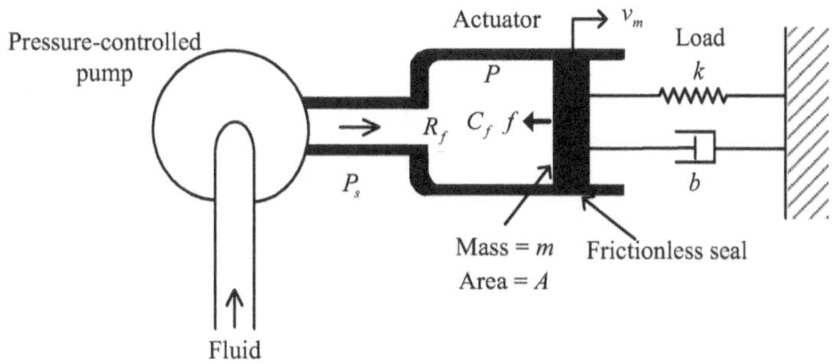

FIGURE P5.19 A fluid actuator moving a mechanical load.

k = stiffness of the rod that connects the load to the shock absorber
b = viscous damping constant of the load
C_f = fluid (gas) capacitance of the shock absorber
A = area of the shock-absorber piston.

Note: The piston is light (neglect its mass) and frictionless.

a. Sketch a complete LG for the system.
Orient the branches of the LG and indicate all the system parameters and branch variables (through- and across-pairs) on them.

b. Systematically derive a complete state-space model using the LG and the following variables:
Input = armature input voltage = $v_a(t)$
Output = velocity of the load = v_m
Give the model matrices ***A***, ***B***, ***C***, and ***D***.

c. Discuss an important characteristic of the model.

5.19 Consider a system where a pressure-controlled fluid actuator moves a mechanical load, as illustrated in Figure P5.19. The pump provides hydraulic fluid at pressure P_S, which enters the actuator chamber through a valve of fluid flow resistance R_f. The fluid capacitance of the actuator chamber is C_f (due to fluid compressibility) and the chamber pressure is P. This pressure moves the load of mass m and area of cross-section A, against a mechanical resistance of stiffness k and damping constant b. The compressive spring force associated with k is f_k. The mass is moved inside the actuator chamber at speed v_m (the output).

Note: Even though the piston seal is assumed frictionless, alternatively the friction of the seal can be included in the damping constant b.

a. Draw an LG for the system.

b. Using $\boldsymbol{x} = [P, f_k, v_m]^T$ as the state vector, $\boldsymbol{u} = [P_s(t)]$ as the input, and $\boldsymbol{y} = [x]$ as the output, systematically determine a complete state-space model with the aid of the LG in Part (a).
Note: f_k = spring force; x = position of the load (of velocity v_m)

c. Determine the input–output differential equation model of the system.

5.20 A flow-controlled pump is used to operate a ram (piston-cylinder actuator), which positions a mechanical load (see Figure P5.20). The piston and the cylinder wall of the ram are rigid. The capacitance C_f results from the flexibility of the fluid only. The mechanical load consists of a mass, spring, and a damper, as shown in the figure.

The following parameters are given:

m_p = mass of the piston
A = effective area of the piston
R_p = fluid resistance at the piston due to the flow through the damping cavities (and also possible leakage flow between the piston and the cylinder)

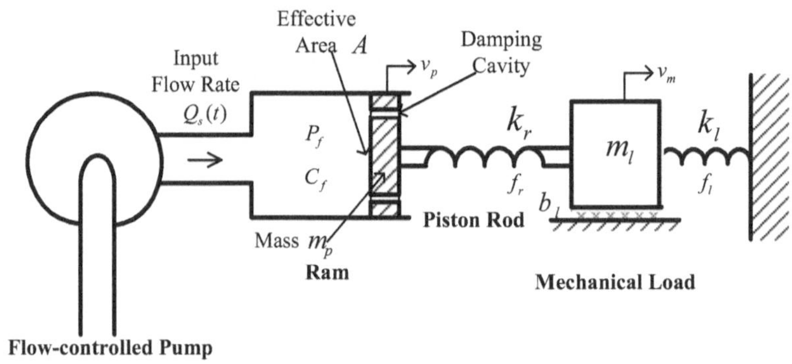

FIGURE P5.20 Mechanical load positioned by a ram using a flow-controlled pump.

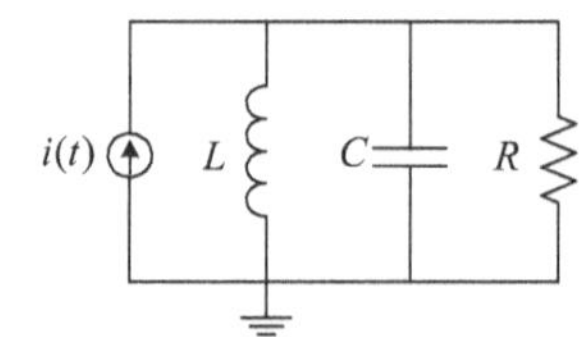

FIGURE P5.21 An electrical circuit with a current source.

k_r = stiffness of the piston rod, which is connected to the mechanical load
m_l = mass of the mechanical load
k_l = resisting stiffness of the mechanical load
b_l = linear viscous damping constant at the mechanical load.

Also, the following state variables are defined:

P_f = fluid pressure in the cylinder
v_p = velocity of the piston
f_r = spring force in the piston rod
v_m = velocity of the mechanical load
f_l = resisting spring force of the mechanical load.
The system input is the volume flow rate $Q_s(t)$ of the pump.
The system output is the velocity v_m of the mechanical load.

a. Sketch a complete LG for the system. Indicate the parameters, through-variables, and across-variables of all the branches.
b. By writing the state-space shell equations, remaining constitutive equations, loop equations, and the node equations, systematically determine a complete state-space model for the system. The state vector is $\boldsymbol{x} = [P_f, v_p, f_r, v_m, f_l]^T$.
c. Determine the matrices $\boldsymbol{A}$, $\boldsymbol{B}$, $\boldsymbol{C}$, and $\boldsymbol{D}$ of the linear state-space model.

5.21 Figure P5.21 shows an electrical circuit with a current source and the passive elements L, C, and R.

a. Sketch a complete LG for this circuit, and indicate the variables, parameters and the branch orientation.
b. Show that the topological relation $b = l + (n - 1)$ is satisfied in this LG. Also, show that the number of unknown variables is equal to the number of independent equations that can be written for the LG.
c. Write the constitutive equations (state-space shell equations and the remaining constitutive equations), node equations, and loop equations. By eliminating the auxiliary variables, systematically determine a state-space model for the system.

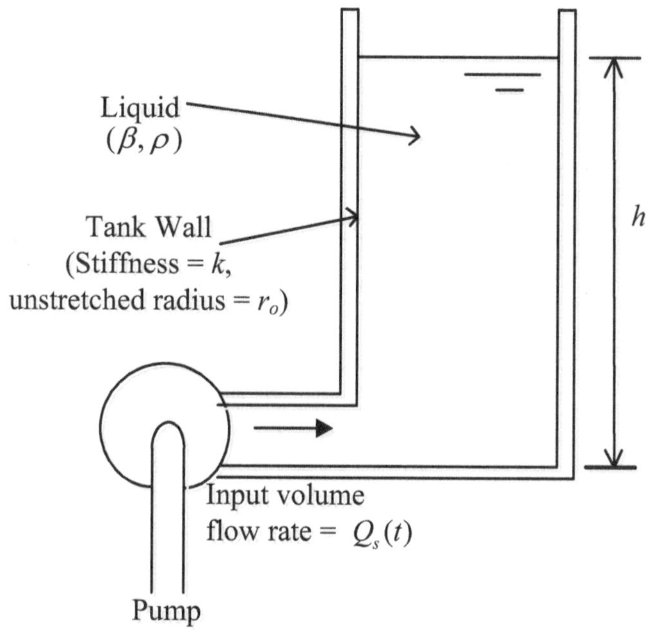

FIGURE P5.22 Compressible liquid pumped into a flexible tank against gravity.

5.22 A liquid input of volume flow rate $Q_s(t)$ is supplied to a storage tank (see Figure P5.22). The tank, when empty, is uniform and cylindrical with radius r_o. The tank wall is flexible. Especially, when a gauge pressure P is exerted at a particular interior location of the tank wall, the tank radius at that location stretches from r_o to $r_o + r$ such that $P = kr$, where k is a known stiffness parameter.

In addition, the following parameters are known:

β=bulk modulus of the liquid in the tank

ρ=mass density of the liquid

Note: The gauge pressure of the liquid in the tank is not uniform. It varies linearly from the highest value ($\rho g h$) at the bottom of the tank to zero at the liquid surface; g=acceleration due to gravity.

It is required to model this dynamic system in order to study its response (liquid level h) as a result of the input $Q_s(t)$.

With sufficient detail develop a lumped-parameter dynamic model (state-space model) for the system for this purpose.

Note 1: In fact this is a distributed-parameter system. Since the liquid pressure varies from the bottom to the top, we cannot use a single lumped capacitor to represent the entire liquid compressibility. Also, we will not be able to use a single lumped capacitor to represent the entire flexibility of the tank. We will have to use several lumped elements for different liquid layers in the tank, in modeling the capacitance. Also, the gravity effect (gravity head capacitance) has to be included in the model.

Note 2: You may appropriately discretize the system (e.g., different liquid layers in the tank) and also introduce other parameters if necessary (in addition to the given parameters γ, ρ, k and r_o).

5.23 A traditional Asian pudding is made by blending roughly equal portions by volume of treacle (a palm honey similar to maple syrup), coconut milk, and eggs; flavoring with cashew nuts, cloves and cardamoms; and baking in a special oven for about 1 hour. The traditional oven uses charcoal fire in an earthen pit that is well insulated, as the heat source.

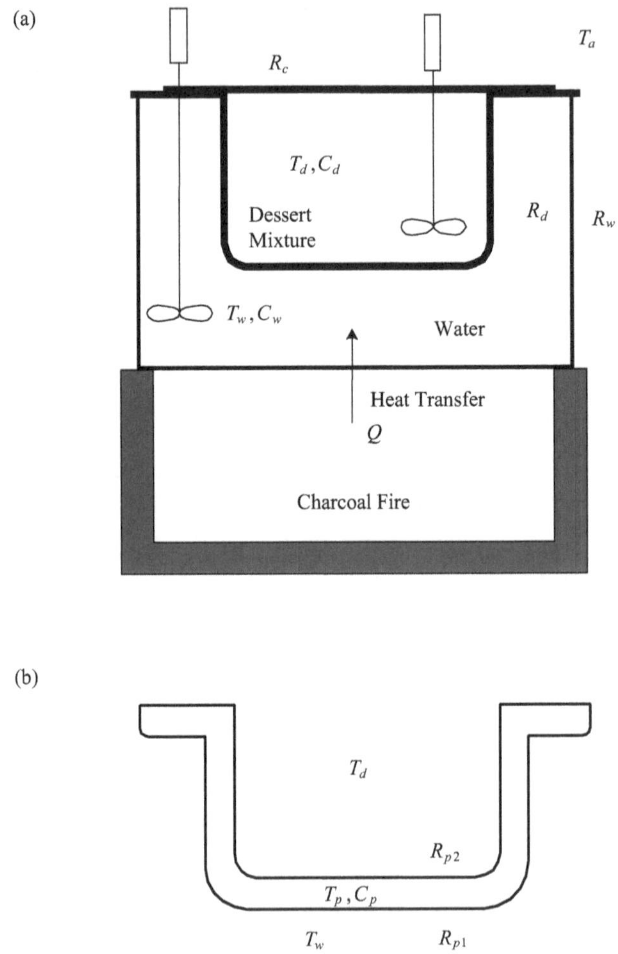

FIGURE P5.23 (a) A simplified model of an Asian dessert oven; (b) an improved model of the dessert pot.

An aluminum container half filled with water is placed on the fire. A smaller stainless steel pot containing the dessert mixture is placed inside the water bath and covered fully with a metal lid. Both the water and the dessert mixture are well stirred and assumed to have uniform temperatures. A simplified model of the oven is shown in Figure P5.23a.

Assume that the thermal capacitances of the aluminum water container, dessert pot, and the lid are negligible. The following equivalent (linear) parameters and variables are defined:

C_w = thermal capacitance of the water bath
C_d = thermal capacitance of the dessert mixture
R_w = thermal resistance between the water bath and the ambient air
R_d = thermal resistance between the water bath and the dessert mixture
R_c = thermal resistance between the dessert mixture and the ambient air, through the covering lid
T_w = temperature of the water bath
T_d = temperature of the dessert mixture
T_a = ambient temperature
Q = heat flow rate from the charcoal fire into the water bath.

a. Assuming that T_d is the output of the system, develop a complete state-space model for the system. What are the system inputs?
b. In Part (a), suppose that the thermal capacitance of the dessert pot is not negligible and is given by C_p. Also, as shown in Figure P5.23b, thermal resistances R_{p1} and R_{p2} are defined for the two interfaces of the pot. Assuming that the pot temperature is maintained uniform at T_p show how the state-space model of Part (a) should be modified to include this improvement. What parameters do R_{p1} and R_{p2} depend on?
c. Draw the LGs for the systems in (a) and (b). Indicate in the LG only the system parameters, input variables, and the state variables.

5.24 Consider a positioning mechanism, which has an armature-controlled linear dc motor, driving a load by using a flexible rod and buffered with a gas cylinder (shock absorber) as shown in Figure P5.24a. A lumped-parameter linear model of the system is shown in Figure P5.24b.

The following model parameters are given:
R_a=resistance of the armature coil
L_a=leakage inductance of the armature circuit
k_m=force constant of the linear dc motor
m_m=mass of the linear-moving mass of the armature
k=stiffness of the driving rod of the load
b_m=viscous damping constant of the moving part of the motor
m_p=mass of the positioned load and the connected piston of the shock absorber
C_f=fluid capacitance of the shock absorber
A=area of the shock-absorber piston.
Neglect the friction between the piston and the cylinder.

a. Sketch a complete LG for the system model.
Note: Orient the branches and indicate all the system parameters and branch variables (through- and across-variable pairs).
b. Derive a complete state-space model, systematically, using the LG and the following variables:
Input=armature input voltage=$v_a(t)$
Output=velocity of the load=v_p

State variables:
i_L=armature current

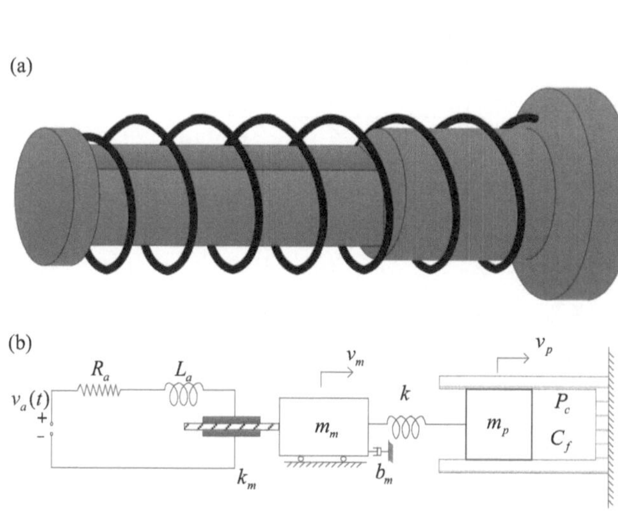

FIGURE P5.24 (a) A positioning mechanism driven by a linear motor; (b) lumped-parameter model.

v_m = motor speed
f_k = force in the rod that moves the load
v_p = speed of the load/piston
P_c = pressure in the shock absorber

Give the model matrices $\boldsymbol{A}, \boldsymbol{B}, \boldsymbol{C}$, and $\boldsymbol{D}$.

5.25 Consider the mechanical system shown by the structural diagram (mechanical circuit) in Figure P5.25. It consists of two masses (m_1 and m_3), one spring (k_2), and one linear viscous damper (b_4). The input of the system is the velocity $v_i(t)$ of the moving platform and the output of the system is the velocity v_o (of mass m_3).

a. Draw a complete LG for the system. Indicate all the variables and parameters on the LG.
b. Using the LG, systematically derive a complete state-space model for the system. Identify the model matrices **A**, **B**, **C**, and **D**.

5.26 The circuit of a passive band-pass filter is shown in Figure P5.26. It consists entirely of two capacitor elements and two resistor elements. The filter input is v_i and the filter output is .

a. Draw a complete LG for the circuit. Indicate all the variables and parameters on the LG.
b. Using the LG, systematically derive a complete state-space model for the circuit. Identify the matrices **A**, **B**, **C**, and **D**.
c. From the state equations derive an input–output differential equation for the filter circuit.
d. Explain why a completely analogous mechanical system cannot be determined for this electrical circuit.

5.27 Consider the circuit shown in Figure P5.27. It consists of two capacitors (C_1 and C_3), one inductor (L_2), and one resistor (R_4). The circuit input is the voltage $v_t(t)$ and the circuit output is the voltage v_o (across the capacitor C_3).

a. Draw a complete LG for the circuit. Indicate all the variables and parameters on the LG.

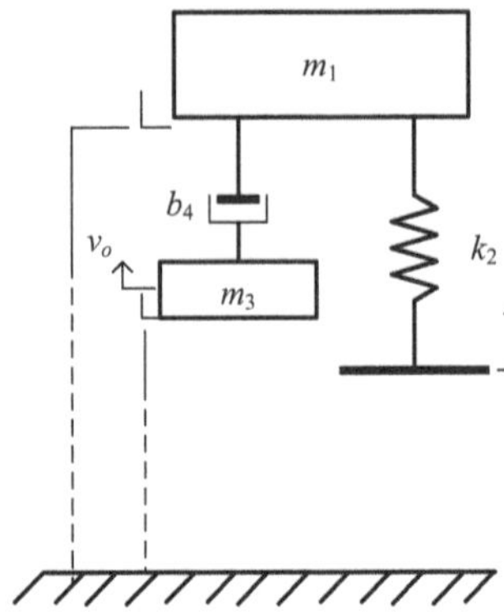

FIGURE P5.25 A mechanical system (structural circuit).

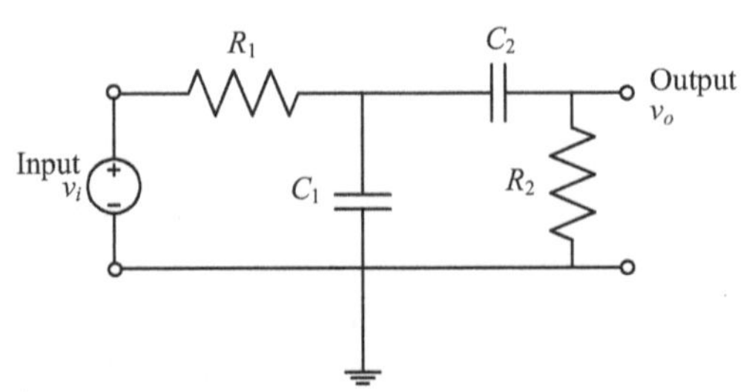

FIGURE P5.26 The circuit of a passive band-pass filter.

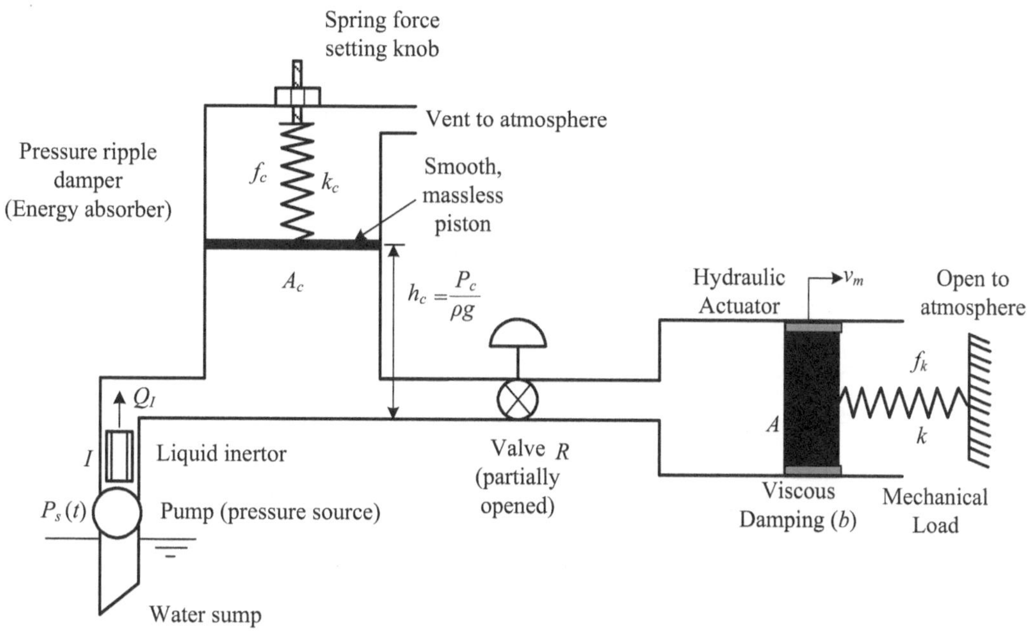

FIGURE P5.27 A passive electrical circuit.

b. Using the LG, systematically derive a complete state-space model for the circuit. Identify the model matrices $\boldsymbol{A}$, $\boldsymbol{B}$, $\boldsymbol{C}$, and $\boldsymbol{D}$.

5.28 Consider the multi-domains (mixed) system consisting of both mechanical components and fluid components, as shown in Figure P5.28. A pump of pressure $P_s(t)$, which is a pressure source, pumps water into a uniform horizontal cylinder of area of cross-section A, which serves as the hydraulic actuator that drives a mechanical load. The combined mass of the actuator piston and the mechanical load is m, the resisting stiffness of the mechanical load is k, and the combined viscous damping constant of the actuator piston and the mechanical load is b. The water is pumped through a short pipe of circular cross-section. *Note*: Assume that the water is incompressible.

The pressure ripples in the water flow of the pipe are reduced before entering the actuator, by means of an energy absorber (hydraulic capacitor) consisting of a small fluid tank of area of cross-section A_c and a spring-loaded, light (massless) and smooth (no energy dissipation) piston with the resisting stiffness k_c. *Note*: This stiffness is adjustable using a nut, as shown, but in this question, assume it to be a constant.

The water flow into the actuator cylinder can be adjusted by means of a valve, as shown. It offers a fluid resistance R. Even though this is also adjustable, assume that it is a constant in the present question. This is the only notable fluid resistance that is present in the entire system (i.e., neglect any other hydraulic resistances).

Also given,

I = fluid inertance in the pipe from the pump up to the energy absorber (passive pressure controller). Neglect any other fluid inertances.

ρ = mass density of the water

g = acceleration due to gravity.

i. Draw a complete LG for the system. Orient all the branches. Mark all the variables and parameters of the LG branches.

ii. Using the LG, systematically develop a complete (and linear) state-space model for the system. Use the following state variables:

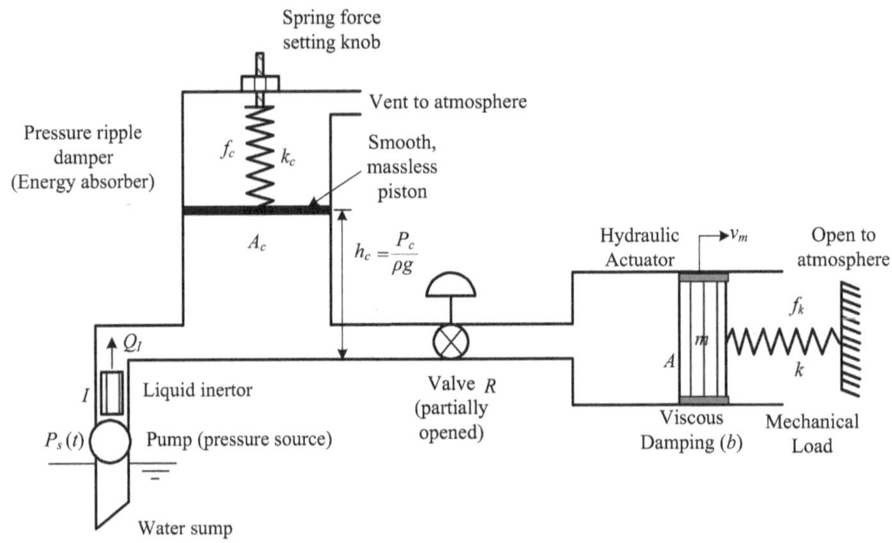

FIGURE P5.28 Water pumping system for a hydraulic actuator.

v_m = velocity of the mechanical load (and also of the actuator piston)
f_k = spring force of the mechanical load (attached to the actuator piston)
f_c = compressive force of the spring of the energy absorber
Q_I = volume flow rate of the water in the pipe before reaching the energy absorber
P_c = pressure difference of the water column (of height h_c) in the energy absorber tank (not the pressure at the bottom of this water column).
Note: $P_c = \rho g h_c$.
System output = v_m = velocity of the load (also, of the actuator piston).
Note: Neglect the bulk modulus of water and the flexibility of the pipes, actuator cylinder, and the absorber tank.
Give all the vectors and matrices of the state-space model.

iii. From the state-space model, determine the input–output differential equation (input = $P_s(t)$, output = v_m). From that equation, write the system transfer function. *Note*: Details on transfer functions are found in Chapter 6. For the purpose of the current question, a transfer function is output/input in the Laplace domain, where the Laplace variable s represents time-derivative.

5.29 Consider the electrical system shown in Figure P5.29, which has a voltage source v_t, inductor L, capacitors C_1 and C_2, resistors R_1 and R_2, and a transformer with the turn ratio r. The output v_o of the system is the voltage across the capacitor C_2.

a. Sketch an analogous system (physical diagram, not an LG) in the *mechanical domain* using realizable physical components (e.g., velocity source, spring, inertias, dampers, and gear pair). Indicate the corresponding component parameters (e.g., spring stiffness), input variable, and output variable.
b. Sketch an analogous system (physical diagram, not an LG) in the *fluid domain* using physical components (e.g., pressure source, fluid inertor, fluid capacitors, fluid resistors, and fluid transformer). Indicate the corresponding component parameters, input variable, and output variable.
c. Is there a thermal system that is analogous to the system shown in Figure P5.29? Explain.
d. Sketch an LG for the system in Figure P5.29. Orient the LG and indicate all the variables and parameters on the LG branches. Using the LG, systematically develop a complete state-space model for the system. Give the matrices ***A***, ***B***, ***C***, and ***D***.

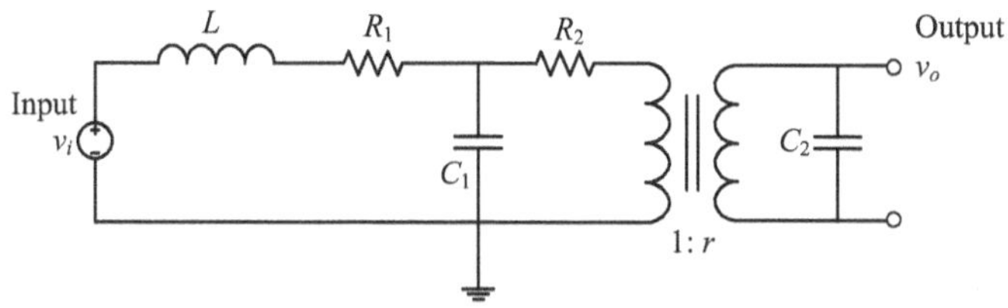

FIGURE P5.29 An electrical system.

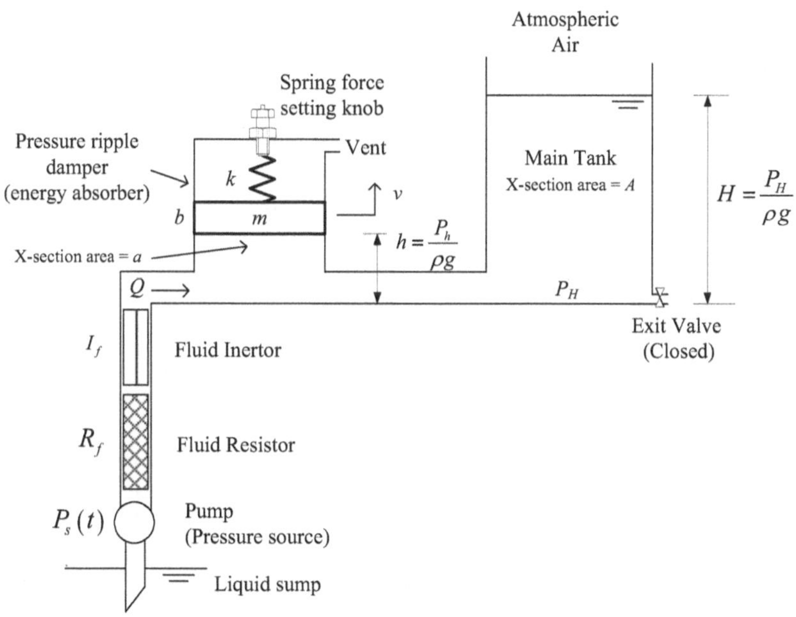

FIGURE P5.30 Liquid pumping system.

5.30

a. If two components are connected in series, it is known that their through-variables are common (i.e., equal) and the across-variables add. Using sketches describe a physically realizable fluid system that has two fluid capacitors connected in series. *Note*: It must use realistic physical components, not components that are not physically realizable. Use sketches of physical components, not circuits or LGs.

b. Consider the multi-domains (mixed) system consisting of both mechanical components and fluid components, as shown in Figure P5.30. A pump of pressure $P_s(t)$, which is a pressure source, pumps liquid into a uniform tank of area of cross-section A, through a pipe of circular cross-section. The pressure ripples in the liquid flow of the pipe are reduced before entering the tank, by means of an energy absorber consisting of a small fluid tank of area of cross-section a and a spring-loaded piston of mass m, stiffness k, and viscous damping constant b.

Also given,

I_f = fluid inertance in the pipe up to the energy absorber (passive pressure controller)

R_f = fluid resistance in the pipe up to the energy absorber

ρ = mass density of the liquid

Draw an oriented LG for the system. Mark all the variables and parameters of the LG branches.

Using the LG, systematically develop a complete state-space model for the system. Use the following state variables:

Q = volume flow rate of the liquid in the pipe before reaching the energy absorber

P_h = pressure difference of the liquid column in the energy absorber tank (not the pressure at the bottom of the liquid column)

v = upward velocity of the piston

f_k = compressive force of the resisting spring attached to the piston

P_H = pressure at the bottom of the main tank

Also, output = H = liquid level in the main tank.

Note: Neglect the bulk modulus of the liquid and the flexibility of the pipes and the tanks.

5.31 Consider a multi-domain engineering system that you are familiar with (in your projects, research, engineering practice, informed imagination, through the literature that you have read, etc.). It should include the mechanical structural domain (i.e., with inertia, flexibility, and damping) and at least one other domain (e.g., electrical, fluid, thermal).

a. Using sketches, describe the system, by giving at least the following information:
   i. The practical purpose and functions of the system
   ii. Typical operation/behavior of the system
   iii. System boundary
   iv. Inputs and outputs
   v. Characteristics of the main components of the system
b. Sketch a lumped-parameter model of the system, by approximating any significant distributed effects using appropriate lumped elements and showing how the lumped-parameter elements (including sources) are interconnected (i.e., show the system structure). You must justify your choice of elements and approximation decisions. Also, you must retain any significant nonlinearities in the original system.
c. Develop an analytical model of the system by writing the necessary constitutive equations, continuity equations, and compatibility equations. The model should be at least fifth order but not greater than 10th order.

Note: Draw an LG for the system (assuming that you plan to use the LG approach to obtain the analytical model).

d. Approximate the nonlinear elements by suitable linear elements.
e. Identify suitable state variables for the linear system and develop a complete state-space model (i.e., matrices $\boldsymbol{A}$, $\boldsymbol{B}$, $\boldsymbol{C}$, and $\boldsymbol{D}$) for the system.

6 Frequency-Domain Models

HIGHLIGHTS

- Transfer Function (TF) Models
- Laplace Transform and Fourier Transform
- Differentiation and Integration Operators
- TF Matrix (MIMO)
- Frequency Domain Models (FTF or FRF)
- Bode and Nyquist Diagrams of FTF
- Harmonic Response (Magnitude and Phase)
- Mechanical TFs (Mechanical Impedance and Mobility)
- Component Interconnection Laws (Series and Parallel)
- Force Transmissibility and Motion Transmissibility
- Vibration Isolation and Suspension/Mount Design
- Maxwell's Reciprocity Property (and Generalization)

6.1 INTRODUCTION

In the previous chapters, we primarily studied time-domain models of engineering dynamic systems. In their analytical form, these models are represented by differential equations with respect to time (where the independent variable is time t). In particular, we paid much attention to state-space models, which are presented by a set of first-order ordinary differential equations in time. This chapter formally studies another popular type of *input–output model* called transfer functions (TFs). In the general case, the *Laplace transform* is used to convert a linear and time invariant time-domain model into a *Laplace TF* model. A special, yet very practical, version of Laplace TF is the *frequency transfer function* (FTF) or *frequency response function* (FRF). In its analytical form, an FTF model is represented as a ratio of polynomials in frequency (where the independent variable is frequency ω). Analytically in particular, the *Fourier transform* is used to convert a linear and time-invariant time-domain model into an FTF model.

This chapter specifically studies several mechanical TFs that are very practical and illustrates how they are used in various practical applications including model formulation, response analysis, and design. The main focus is on mechanical systems. However, in view of the existing analogies (as extensively covered in the previous chapters), the methods covered in this chapter for mechanical systems can be easily extended to other domains (electrical, fluid, and thermal). In fact, many procedures in the present chapter use the concept of *mechanical circuits*, which are analogous to electrical circuits. Hence, some of the methods that are applied to mechanical systems in this chapter are widely used in electrical systems and electrical circuit analysis.

6.1.1 TRANSFER FUNCTION MODELS

A TF model is an *input–output model* (*I-O model*) and is equivalent to a linear input–output differential equation in the time domain (see Chapters 3 and 5). Simplistically, a TF model provides an "algebraic" (not differential-equation) representation of $\frac{\text{System output}}{\text{System input}}$. Since it is an algebraic representation, its analytical manipulation is much easier than for differential equations.

DOI: 10.1201/9781003124474-6

A TF model (strictly, a *Laplace TF*) is based on the Laplace transform and is a versatile means of representing a linear system that has constant (time-invariant) parameters. Strictly, it is a dynamic model in the *Laplace domain*. A frequency-domain model (or an FTF) is an equivalent model of a Laplace domain model, and it is based on the *Fourier transform*. In particular, these two types of models are interchangeable—it is a trivial exercise to convert a Laplace-domain model into the corresponding frequency-domain model and vice versa. Similarly, it is a simple and straightforward exercise to convert a *linear*, constant-coefficient (*time-invariant*) time-domain model (e.g., input–output differential equation or a state-space model) into a TF model and vice versa. It follows that all these types of models are entirely equivalent (and no information is lost through a model conversion, at least analytically).

A system with just one input (excitation) and one output (response) can be represented uniquely by one TF. For example, in a mechanical dynamic system, the response characteristics at a given location and direction (say, along a particular coordinate or in a given degree of freedom) of the system to a forcing input at the same or a different location and direction can be modeled using a single FTF. When a system has two or more inputs (i.e., an *input vector*) and/or two or more outputs (i.e., an *output vector*), its TF representation needs several TFs (i.e., a *TF matrix* is needed).

TF models have been widely used in early studies of dynamic systems, when digital computers were not commonly available because they are algebraic functions rather than differential equations and are easier to analyze. In view of the simpler algebraic operations that are involved in TF approaches, a substantial amount of information regarding the dynamic behavior of a system can be obtained with minimal computational effort. This is the primary reason for the popularity enjoyed by the TF methods prior to the advent of the digital computer. One might think that the abundance of high-speed and low-cost digital processing would lead to a dominance of time-domain methods, over TF methods. But there is evidence to the contrary in many areas, particularly in dynamic systems and control, due to the analytical simplicity and the intuitive appeal of TF techniques. Only minimal knowledge of the theory of Laplace transform and Fourier transform is needed to use TF methods in system *modeling*, *analysis*, *simulation*, *design*, and *control*. Furthermore, just as a time-domain model presents the behavior of a dynamic system in its true physical domain as the time changes, a frequency-domain model is also a very realistic and practical model, which shows the behavior of a dynamic system as the excitation frequency changes.

In its core, a frequency-domain model represents how a dynamic system responds to a sinusoidal (i.e., *harmonic*) excitation. Since, according to the theory of Fourier analysis, any signal can be represented by a collection of harmonic components, a frequency-domain model can represent as well the true physical behavior of a dynamic system to any type of input. Indeed, a frequency-domain model is completely equivalent to a time-domain model and to a Laplace-domain model. Techniques of TF models, both in the Laplace domain and the special frequency (Fourier) domain, are studied in this chapter. In particular, we will see that *mobility*, *mechanical impedance*, and *transmissibility* are convenient TF representations of mechanical dynamic systems, in the frequency domain. We will study how such a model can be formulated with the knowledge of the element (component) TFs and how they are interconnected (series, parallel) in the system. The complementary (dual) characteristic of force transmissibility and motion transmissibility is recognized. Also, the application of Maxwell's reciprocity property in mechanical dynamic systems and its extension to other physical domains are examined. As practical examples, transmissibility is applicable in vibration isolation of machines and suspension systems in vehicles; and mechanical impedance is useful in such tasks as cutting, joining, and assembly that employ machine tools and robots.

6.2 LAPLACE AND FOURIER TRANSFORMS

In mathematics, we encounter various transforms. Typically, a transform is used to covert a mathematical problem into a different analytical form in order to take advantage of possible analytical convenience of the transformed problem. For example, the *logarithm* is a transform, which converts the multiplication operation into an addition and the division operation into a subtraction, thereby making the analysis much simpler. Furthermore, it enables the capability to represent information on a much wider scale (as multiples rather than linearly—proportionally) while paying more attention to the lower-magnitude information. In a similar manner, the Laplace transform converts "differentiation" into a "multiplication by the Laplace variable s" and "integration" into a "division by s," thereby providing significant analytical convenience. Even though Fourier transform is sometimes considered as a special case of the Laplace transform, a Fourier transform of a function contains all the information of the original function and also of the Laplace-transformed result. Hence, they are fully reversible. The Fourier result corresponding to a Laplace result is obtained simply by setting $s = j\omega$ in the Laplace result, where ω is the frequency variable. Even though this conversion itself is trivial, the underlying analytical basis is quite sophisticated and comprehensive, which is beyond the scope of the present need.

6.2.1 Laplace Transform

The Laplace transform involves the mathematical transformation of a function ($y(t)$) in the time domain into an equivalent function ($Y(s)$) in the Laplace domain (also termed the *s-domain* or the *complex frequency domain*) according to

$$Y(s) = \int_0^\infty y(t)\exp(-st)\,dt \text{ or } Y(s) = \mathcal{L}y(t) \tag{6.1}$$

Here, the Laplace transform operator is denoted by $\mathcal{L}$, and the Laplace variable is $s = \sigma + j\omega$, which is complex, since $j = \sqrt{-1}$.

The real positive value σ is chosen sufficiently large so that the transform integral (6.1) is finite (i.e., $\int e^{-\sigma t} y(t)\,dt$ is finite) even when $\int y(t)\,dt$ is not finite.

Note: Mathematically, $y(t)$ can be complex in general. However, in a practical dynamic system, it is a real function of time t (representing a time response of the dynamic system). Even in this practical case, $Y(s)$ will be complex, because s is complex, as clear from equation (6.1).

The inverse Laplace transform is given by

$$y(t) = \frac{1}{2\pi j}\int_{\sigma-j\infty}^{\sigma+j\infty} Y(s)\exp(st)\,ds \text{ or } y(t) = \mathcal{L}^{-1}Y(s) \tag{6.2}$$

This is obtained simply through mathematical manipulation (multiply both sides by the appropriate exponential and integrate with respect to s) of the forward transform (6.1).

Note: For a given time function, its Laplace transform is unique, and the Laplace operation is completely reversible. Specifically, the time function can be fully recovered from its Laplace transform.

6.2.1.1 Laplace Transform of a Derivative

Using (6.1), the Laplace transform of the time derivative $\dot{y} = \frac{dy}{dt}$ may be determined as

$$\mathcal{L}\dot{y} = \int_0^\infty e^{-st}\frac{dy}{dt}dt = sY(s) - y(0) \tag{6.3}$$

Note: Integration by parts: $\int u\,dv = uv - \int v\,du$ is used in obtaining the result (6.3). Also $y(0)$ is the initial condition (IC) of $y(t)$ at $t=0$.

By repeatedly applying (6.3), we can get the Laplace transform of the higher derivatives; specifically, $\mathcal{L}\ddot{y}(t) = s\mathcal{L}[\dot{y}(t)] - \dot{y}(0) = s[sY(s) - y(0)] - \dot{y}(0)$

This gives the result

$$\mathcal{L}\ddot{y}(t) = s^2\mathcal{L}[y(t)] - sy(0) - \dot{y}(0) \tag{6.4}$$

Similarly, we can obtain

$$\mathcal{L}\dddot{y} = s^3Y(s) - s^2y(0) - s\dot{y}(0) - \ddot{y}(0) \tag{6.5}$$

Proceeding in this manner, we get the general result

$$\mathcal{L}\frac{d^n y(t)}{dt^n} = s^nY(s) - s^{n-1}y(0) - s^{n-2}\dot{y}(0) - \cdots - \frac{d^{n-1}y}{dt^{n-1}}(0) \tag{6.6}$$

Note: With zero ICs, we have:

$$\mathcal{L}\frac{d^n y(t)}{dt^n} = s^nY(s) \tag{6.7}$$

This means the time derivative corresponds to multiplication by s in the Laplace domain. As a result, *differential equations* (in time domain models) become *algebraic equations* (in TFs), which require easier mathematics. We will explore this issue further in the next section. From result (6.7), it is clear that the Laplace variable s can be interpreted as the *derivative operator* in the context of a dynamic system.

Note: ICs can be added separately to a Laplace model after using (6.7) to transform the derivatives of a time-domain dynamic model into polynomials in s. The polynomials in a TF represent the model and do not depend on the ICs. Hence, in the transformation of a time-domain model (ordinary differential equation) into a Laplace-domain model (TF), first, the ICs are assumed zero.

6.2.1.2 Laplace Transform of an Integral

The Laplace transform of the time integral $\int_0^t y(\tau)d\tau$ is obtained by the direct application of (6.1) as

$$\mathcal{L}\int_0^t y(\tau)d\tau = \int_0^\infty e^{-st}\int_0^t y(\tau)d\tau\,dt = \int_0^\infty\left(-\frac{1}{s}\right)\frac{d}{dt}(e^{-st})\int_0^t y(\tau)d\tau\,dt$$

Integrate by parts using $\int u\,dv = uv - \int v\,du$ as

$$\mathcal{L}\int_0^t y(\tau)d\tau = \left(-\frac{1}{s}\right)e^{-st}\int_0^t y(\tau)d\tau\,|_0^\infty - \int_0^\infty\left(-\frac{1}{s}\right)e^{-st}y(t)dt = 0 - 0 + \int_0^\infty\left(\frac{1}{s}\right)e^{-st}y(t)dt$$

We get

$$\mathcal{L}\int_0^t y(\tau)d\tau = \frac{1}{s}Y(s) \tag{6.8}$$

It follows that integration in the time domain becomes multiplication by $1/s$ in the Laplace domain. In particular, $1/s$ can be interpreted as the *integration operator*, in the context of a dynamic system.

In using techniques of Laplace transform in the analysis of dynamic systems, the general approach is to first convert the time-domain problem into an equivalent s-domain problem (conveniently, Laplace transform tables are available as well, to facilitate this task), perform the necessary analysis (algebra rather than calculus) in the s-domain and finally convert the results back into the time domain (again, Laplace transform tables may be used).

6.2.2 Fourier Transform

The Fourier transform involves the mathematical transformation from the time domain into the frequency domain) according to

$$Y(j\omega) = \int_{-\infty}^{\infty} y(t)\exp(-j\omega t)\,dt \text{ or } Y(j\omega) = \mathcal{F}\ y(t) \tag{6.9}$$

where the *cyclic frequency* variable is f (in Hz) and the *angular frequency* variable is $\omega = 2\pi f$ (in rad/s). Also, the Fourier transform operator is denoted by $\mathcal{F}$.

Note: Mathematically, $y(t)$ can be complex (i.e., not real) even though, in practice, it is a real function of time t (representing a time response of a dynamic system). Even when $y(t)$ is real, $Y(j\omega)$ will be complex in general, because $\exp(-j\omega t) = \cos\omega t - j\sin\omega t$ is complex, as clear from equation (6.9).

The inverse Fourier transform is given by

$$y(t) = \frac{1}{2\pi}\int_{-\infty}^{\infty} Y(j\omega)\exp(j\omega t)\,d\omega \text{ or } y(t) = \mathcal{F}^{-1}\ Y(j\omega) \tag{6.10}$$

This is obtained simply through mathematical manipulation (multiply both sides by the appropriate exponential and integrate with respect to ω) of the forward transform (6.9).

By examining the transforms (6.1) and (6.9), it is clear that the conversion from the Laplace domain into the Fourier (frequency) domain may be done simply by setting $s = j\omega$. Strictly, the *one-sided Fourier transform* is used for this purpose (where the lower limit of integration in (6.9) is set to $t=0$) because it is then that (6.1) becomes identical to (6.9) with $s = j\omega$.

6.2.2.1 Summary of Results

We summarize below the main results and observations.

- **Laplace Transform:**
  - Time domain $\leftrightarrow$ Laplace (complex frequency) domain
  - Time derivative $\leftrightarrow$ Multiplication by Laplace variable s
  - Differential equations $\leftrightarrow$ Algebraic equations (ratios of two polynomials in s)
  - Time integration $\leftrightarrow$ Multiplication by $1/s$

- **Fourier Transform:**
  - Time domain $\leftrightarrow$ Frequency domain
  - Laplace result $\underset{\leftarrow j\omega = s}{\overset{\rightarrow s = j\omega}{}}$ Fourier result (one-sided)

It should be clear that, in modeling and analysis of dynamic systems (that are linear and time-invariant, in particular), the Laplace approach and the Fourier approach are completely equivalent and interchangeable. The conversion $s = j\omega$ is used to get the Fourier result from the equivalent Laplace result and vice versa. In other words, for our purposes, when we have a Laplace result, there is an equivalent Fourier result (and vice versa). Mathematically, however, a Fourier result may not exist even when a Laplace result exists. This is because, in the Laplace transform (see equation (6.1)), there is the multiplier $e^{-\sigma t}$ with $\sigma > 0$ in the integrand, which decays to zero very rapidly, whereas in the Fourier transform (see equation (6.9)), there is no multiplier $e^{-\sigma t}$ in the integrand. For $\sigma > 0$, the rapid (exponential) decay property of $e^{-\sigma t}$ makes the Laplace transform much more "convergent" than the Fourier transform.

6.3 TRANSFER FUNCTION

The TF is a dynamic model that is represented in the Laplace domain. Specifically, the TF $G(s)$ of a linear, time-invariant, single-input single-output (SISO) system is given by:

$G(s) = \dfrac{\text{Laplace-transformed output}}{\text{Laplace-transformed input}}$, assuming zero ICs. This is a unique function, which represents the system (model); it does not depend on the specific input, the output, or the initial conditions.

A linear, constant-parameter system possesses a unique TF even if the Laplace transform of a particular input to the system (and of the corresponding output) may not exist. For example, suppose that the Laplace transform of a particular input $u(t)$ is infinite. Then the Laplace transform of the corresponding output $y(t)$ will also be infinite. But the TF itself will be finite and will represent the actual system (irrespective of the input or the output).

Consider the nth-order linear, constant-parameter system given by the ordinary differential equation (ODE) in time t:

$$a_n \frac{d^n y}{dt^n} + a_{n-1} \frac{d^{n-1} y}{dt^{n-1}} + \cdots + a_0 y = b_0 u + b_1 \frac{du}{dt} + \cdots + b_m \frac{d^m u}{dt^m} \tag{6.11}$$

where input $= u(t)$ and output $= y(t)$.

For systems that possess dynamic delay (i.e., systems whose response does not feel the excitation (input) either instantly or ahead of time (i.e., systems whose excitation and/or its derivatives are not directly fed forward to the output), we have $m < n$. These are the systems that concern us most in real applications.

Note: We will assume that $m < n$, or at worst $m \le n$. The requirement $m \le n$ is satisfied by a *physically realizable* system. In the extreme case of $m = n$, the input $u(t)$ is instantly felt at the output $y(t)$. In other words, the input is directly fed forward into the output. Then we say that the system possesses the "feedforward character."

Use result (6.7) in (6.11), assuming zero ICs. We obtain the system TF:

$$\frac{Y(s)}{U(s)} = G(s) = \frac{b_0 + b_1 s + \cdots + b_m s^m}{a_0 + a_1 s + \cdots + a_n s^n} \tag{6.12}$$

It should be clear from (6.11) and (6.12) that the TF that corresponds to the differential equation of a dynamic system can be written simply by inspection, without requiring any mathematical manipulation or knowledge of Laplace-transform theory. Conversely, once the TF is given, the corresponding time-domain (differential) equation should be immediately obvious as well.

Note: The dominator polynomial of a TF is called the *characteristic polynomial*, and the corresponding equation (obtained by setting the denominator polynomial to zero) is called the *characteristic equation*: $a_0 + a_1 s + \cdots + a_n s^n = 0$. Its roots are the *poles* (or *eigenvalues*) of the system, and they determine such characteristics as the natural response and the stability of the system.

TFs are simple algebraic expressions. Specifically, differential equations (time-domain models) are transformed into algebraic relations through the Laplace transform. This is a major advantage of the TF approach. Once the analysis is performed using TFs, the inverse Laplace transform can convert the Laplace results into the corresponding time-domain results. This can be accomplished with the aid of Laplace transform tables.

Example 6.1

Consider the mechanical oscillator (mass-spring-damper system) shown in Figure 6.1a. Its dynamic equation is obtained in a straightforward manner (by applying Newton's second law to the lumped mass) as $m\ddot{y} = -b\dot{y} - ky + f(t)$ or

$$m\ddot{y} + b\dot{y} + ky = ku(t) \tag{6.13a}$$

where the response (output) y of the mass is measured from its static equilibrium position (so, the gravitational force is balanced by the initial static force in the spring). Also, the input $u(t)$ is a "scaled" version of the force applied to the mass, according to

$$f(t) = ku(t) \tag{6.14}$$

Note: Alternatively, $u(t)$ may be considered as the displacement "applied" to the base of the spring only (not the damper, whose base is still fixed to the ground). See Figure 6.1b and also Problem 6.3. Here, to move the platform, a "displacement actuator" is needed. Then, there will be a "dependent force" in this actuator (see the broken line in Figure 6.1b), which will be reacted on the ground.

Take the Laplace transform of the system equation (6.13) with zero ICs

$$\left(ms^2 + bs + k\right)Y(s) = kU(s)$$

The corresponding TF is

$$G(s) = \frac{Y(s)}{U(s)} = \frac{k}{\left(ms^2 + bs + k\right)} \tag{6.15a}$$

Define $\omega_n^2 = k/m$ and $2\zeta\omega_n = b/m$ where

It is known that undamped natural frequency$=\omega_n$ (the frequency at which the "undamped" system naturally oscillates—for example, in response to an IC, without a sustaining input), damped natural frequency$=\omega_d = \sqrt{1-\zeta^2}\,\omega_n$ (the frequency at which the "damped" system naturally oscillates), and damping ratio$=\zeta$.

Then, we can write the TF as

$$G(s) = \frac{\omega_n^2}{\left(s^2 + 2\zeta\omega_n s + \omega_n^2\right)} \tag{6.15b}$$

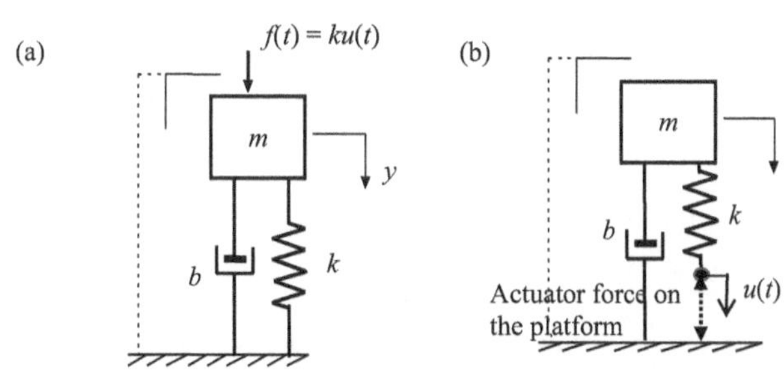

FIGURE 6.1 (a) A damped mechanical oscillator, (b) an equivalent system.

This is the TF corresponding to the *displacement* output.

If the output of the oscillator is taken as the *velocity* of the mass, we have

$$\frac{sY(s)}{U(s)} = sG(s) = \frac{s\omega_n^2}{\left(s^2 + 2\zeta\omega_n s + \omega_n^2\right)} \tag{6.16}$$

Similarly, if the output is the *acceleration* of the mass, the corresponding TF is

$$\frac{s^2Y(s)}{U(s)} = s^2G(s) = \frac{s^2\omega_n^2}{\left(s^2 + 2\zeta\omega_n s + \omega_n^2\right)} \tag{6.17}$$

Note: The results (6.16) and (6.17) are obtained from the fact that the Laplace transform of the velocity is $sY(s)$, and the Laplace transform of the acceleration is $s^2Y(s)$. See equation (6.7).

In equation (6.17), the numerator order is equal to the denominator order (i.e., $m=n=2$). This means, the input (applied force) is instantly felt in the acceleration of the mass, which is intuitive as clear from the Newton's second law and also may be verified experimentally by using an accelerometer (acceleration sensor) mounted on the mass. The TF (6.17) corresponds to feeding forward the input directly into the output (as noted before, with regard to the case of $m=n$) with zero dynamic delay. For example, this is the primary mechanism through which road disturbances are felt inside a vehicle compartment that has hard suspensions.

The *characteristic equation* of the oscillator is

$$\Delta(s) = s^2 + 2\zeta\omega_n s + \omega_n^2 = 0 \tag{6.18}$$

LEARNING OBJECTIVES

1. Converting an input–output differential equation of a system into a TF
2. Natural frequency and damping ratio of a damped mechanical oscillator
3. Characteristic polynomial of a system
4. Obtaining the TF corresponding to an input force and an output displacement of a mechanical dynamic system
5. Derivation of the TFs corresponding to related different outputs (displacement → velocity → acceleration)
6. A situation where the input is directly present in the output (feedforward situation).

6.3.1 Transfer-Function Matrix

Consider the state variable representation (state-space model) of a linear, time-invariant system (see Chapters 3 and 5):

$$\dot{\boldsymbol{x}} = \boldsymbol{A}\boldsymbol{x} + \boldsymbol{B}\boldsymbol{u} \tag{6.19a}$$

$$\boldsymbol{y} = \boldsymbol{C}\boldsymbol{x} + \boldsymbol{D}\boldsymbol{u} \tag{6.20a}$$

where $\boldsymbol{x}(t)$ is the nth-order state vector; $\boldsymbol{u}$ is the rth-order input (excitation); and $\boldsymbol{y}$ is the mth-order output (response) vector. This is a multi-input multi-output (MIMO) system. The corresponding TF model relates the output vector $\boldsymbol{y}$ to the input vector $\boldsymbol{u}$. We will need $m\times n$ TFs, or a *transfer-function matrix*, to represent this MIMO system.

To obtain an expression for the TF matrix, we first apply Laplace transform to equations (6.19) and (6.20), with zero ICs for $\boldsymbol{x}$. We get

$$s\boldsymbol{X}(s) = \boldsymbol{A}\boldsymbol{X}(s) + \boldsymbol{B}\boldsymbol{U}(s) \tag{6.19b}$$

$$\boldsymbol{Y}(s) = \boldsymbol{C}\boldsymbol{X}(s) + \boldsymbol{D}\boldsymbol{U}(s) \tag{6.20b}$$

From equation (6.19b), we have

$$\boldsymbol{X}(s) = (s\boldsymbol{I} - \boldsymbol{A})^{-1}\boldsymbol{B}\boldsymbol{U}(s) \tag{6.19c}$$

in which $\boldsymbol{I}$ is the nth-order "identity matrix" (a matrix with 1s as its diagonal elements and 0s for all other elements). By substituting equation (6.19c) into (6.20b), we get the TF relation

$$\boldsymbol{Y}(s) = \left[\boldsymbol{C}\left((s\boldsymbol{I} - \boldsymbol{A})^{-1}\boldsymbol{B}\right) + \boldsymbol{D}\right]\boldsymbol{U}(s) \tag{6.21a}$$

or

$$\boldsymbol{Y}(s) = \boldsymbol{G}(s)\boldsymbol{U}(s) \tag{6.21b}$$

The TF matrix $\boldsymbol{G}(s)$ is an $m \times n$ matrix given by

$$\boldsymbol{G}(s) = \boldsymbol{C}\left((s\boldsymbol{I} - \boldsymbol{A})^{-1}\boldsymbol{B}\right) + \boldsymbol{D} \tag{6.22a}$$

In most practical systems with dynamic delay, the excitation $\boldsymbol{u}(t)$ is not directly fed forward to the response $\boldsymbol{y}$; and as a result, it is not instantaneously felt in the response $\boldsymbol{y}$. Then we have $\boldsymbol{D}=\boldsymbol{0}$, and equation (6.22a) becomes

$$\boldsymbol{G}(s) = \boldsymbol{C}(s\boldsymbol{I} - \boldsymbol{A})^{-1}\boldsymbol{B} \tag{6.22b}$$

The characteristic equation of the system is

$$\mathrm{Det}(s\boldsymbol{I} - \boldsymbol{A})^{-1} = 0 \tag{6.23}$$

Examples are presented now to illustrate the conversion of a time-domain (differential-equation) model into a transfer-function (matrix) model.

Example 6.2

Let us consider again the damped oscillator of Example 6.1, given by

$$\ddot{y} + 2\zeta\omega_n\dot{y} + \omega_n^2 y = \omega_n^2 u(t) \tag{6.13b}$$

Define the state variables as

$$\boldsymbol{x} = \begin{bmatrix} x_1 & x_2 \end{bmatrix}^T = \begin{bmatrix} y & \dot{y} \end{bmatrix}^T$$

where $\boldsymbol{y}$=position and $\dot{y}$=velocity. Then, a state model for the system may be expressed as

$$\dot{\boldsymbol{x}} = \begin{bmatrix} 0 & 1 \\ -\omega_n^2 & -2\zeta\omega_n \end{bmatrix}\boldsymbol{x} + \begin{bmatrix} 0 \\ \omega_n^2 \end{bmatrix}u(t)$$

If we consider both displacement and velocity as outputs, we have

$$\boldsymbol{y} = \boldsymbol{x}$$

Note: In this example, the output gain matrix (measurement matrix) is the *identity matrix*: $\boldsymbol{C}=\boldsymbol{I}$. Also, there is no direct feedforward of the input to the output. Hence, $\boldsymbol{D}=\boldsymbol{0}$. From equation (6.21a), we get

$$X(s) = Y(s) = \begin{bmatrix} s & -1 \\ \omega_n^2 & s+2\zeta\omega_n \end{bmatrix}^{-1} \begin{bmatrix} 0 \\ \omega_n^2 \end{bmatrix} U(s)$$

$$= \frac{1}{\left(s^2+2\zeta\omega_n s+\omega_n^2\right)} \begin{bmatrix} s+2\zeta\omega_n & 1 \\ -\omega_n^2 & s \end{bmatrix} \begin{bmatrix} 0 \\ \omega_n^2 \end{bmatrix} U(s)$$

$$= \frac{1}{\left(s^2+2\zeta\omega_n s+\omega_n^2\right)} \begin{bmatrix} \omega_n^2 \\ s\omega_n^2 \end{bmatrix} U(s) \qquad \text{(i)}$$

We observe that the TF matrix of the system is

$$G(s) = \begin{bmatrix} \omega_n^2/\Delta(s) \\ s\omega_n^2/\Delta(s) \end{bmatrix}$$

The *characteristic polynomial* of the system is $\Delta(s) = s^2 + 2\zeta\omega_n s + \omega_n^2$ and the *characteristic equation* is $\Delta(s) = s^2 + 2\zeta\omega_n s + \omega_n^2 = 0$.

Note: This result can be further verified by using equation (6.23).

The TF matrix in the present example is in fact a column vector. The first element in **G**(*s*) is the displacement-output TF, and the second element is the velocity-output TF. These results agree with the expressions obtained in Example 6.1.

Now, let us consider the acceleration $\ddot{y}$ as an output and denote it by y_3. It is clear from the system equation (6.13b) that

$$y_3 = \ddot{y} = -2\zeta\omega_n\dot{y} - \omega_n^2 y + \omega_n^2 u(t)$$

or in terms of the state variables

$$y_3 = -2\zeta\omega_n x_2 - \omega_n^2 x_1 + \omega_n^2 u(t)$$

This output explicitly contains the input term. This is a direct "feedforward" situation, which implies that the matrix ***D*** becomes non-zero when acceleration $\ddot{y}$ is chosen as an output. In this case, by substituting (i), we get

$$Y_3(s) = -2\zeta\omega_n X_2(s) - \omega_n^2 X_1(s) + \omega_n^2 U(s) = -2\zeta\omega_n \frac{s\omega_n^2}{\Delta(s)} U(s) - \omega_n^2 \frac{\omega_n^2}{\Delta(s)} U(s) + \omega_n^2 U(s)$$

This simplifies to

$$Y_3(s) = -2\zeta\omega_n X_2(s) - \omega_n^2 X_1(s) + \omega_n^2 U(s) = \frac{s\omega_n^2}{\Delta(s)} U(s)$$

This agrees with the acceleration-output TF obtained in Example 6.1.

LEARNING OBJECTIVES

1. The vector-matrix approach to converting an input–output differential equation of a system into a state-space model and then into a TF
2. Characteristic polynomial and characteristic equation of a system
3. Derivation of the TFs of related outputs (displacement → velocity → acceleration)
4. A situation where the input is directly present in the output (feedforward situation).

6.4 FREQUENCY DOMAIN MODELS

The TF of a device is given by output/input of device. As presented previously, if the output and the input are expressed in the Laplace domain (i.e., by the Laplace transforms of the corresponding time-domain signals), we have the *Laplace* TF. Alternatively, if the output and the input are expressed in the frequency domain (i.e., by the Fourier transforms of the corresponding time-domain signals), we have the FTF. Specifically, in the frequency domain

$$\text{Frequency transfer function} = \frac{\text{Fourier transform of output}}{\text{Fourier transform of input}}$$

FRF is another name for FTF. Laplace TF and Fourier (frequency) TF are completely equivalent and reversible through the change of variables $s = j\omega$.

Frequency-domain representations are particularly useful in the analysis, simulation, design, control, and testing of electro-mechanical systems, and generally, multi-physics systems. The signal waveforms encountered in such systems can be interpreted and represented as a combination of sinusoidal components (i.e., a *Fourier spectrum*). In particular, any periodic signal can be represented as a summation of sinusoidal (harmonic) components, which forms the *Fourier series expansion*. Such periodic excitations are used, for example, in dynamic testing (vibration testing, shaker testing) of products and equipment. Usually, by testing, it is easier to determine the frequency-domain models than the associated time-domain models.

6.4.1 Frequency Transfer Function (Frequency Response Function)

Consider the time-domain system (6.11) whose TF (in the Laplace domain) is given by equation (6.12). The corresponding FTF is obtained by substituting $s = j\omega$ in (6.12). This fact can be easily shown, as presented next.

6.4.1.1 Response to a Harmonic Input

Suppose that the input to a system is harmonic (sinusoidal). This input can be expressed in the complex form

$$u = u_o e^{j\omega t} = u_o(\cos\omega t + j\sin\omega t) \tag{6.24}$$

Note: Mathematically, we can always use a "complex" input. Then, the response of the actual (real) system is obtained from either the real part or the imaginary part of the analytical response, which corresponds to the real part and the imaginary part of the "complex" input. A complex signal has the exponential form (6.24), which is easier to analyze than either the cosine signal or the sine signal (because the derivative of a cosine is a sine, the derivative of a sine is a cosine, and the derivative of an exponential is an exponential of the same form).

On applying the input, eventually, the response of the system will settle down (i.e., steady state will be reached). Then, the output (response) of the system will also be harmonic, at the same frequency (ω), and given by

$$y = y_o e^{j\omega t} = y_o(\cos\omega t + j\sin\omega t) \tag{6.25}$$

By substituting equations (6.24) and (6.25) into (6.11) and canceling the common term $e^{j\omega t}$, we get

$$y_o = \left[\frac{b_m(j\omega)^m + b_{m-1}(j\omega)^{m-1} + \cdots + b_0}{a_n(j\omega)^n + a_{n-1}(j\omega)^{n-1} + \cdots + a_0}\right]u_o \tag{6.26a}$$

or

$$y_o = G(j\omega)u_o \tag{6.26b}$$

(*Note*: $\frac{de^{j\omega t}}{dt} = j\omega e^{j\omega t}$)

Hence, it is seen that $G(j\omega)$ in equation (6.26) is obtained precisely by substituting $s = j\omega$ in (6.12). Specifically, the FTF or the FRF is given by

$$G(j\omega) = G(s)\,|_{s=j\omega} = \frac{b_0 + b_1(j\omega) + \cdots + b_m(j\omega)^m}{a_0 + a_1(j\omega) + \cdots + a_n(j\omega)^n} \tag{6.27a}$$

Note: Angular frequency variable (rad/s) is $\omega = 2\pi f$, where f=cyclic frequency (Hz).

The meaning of the FRF is clear from equation (6.26a). When a harmonic input of frequency ω and amplitude u_o is applied to a system (linear, time-invariant), its output at steady state will also be harmonic at frequency ω. However, the amplitude will be magnified by the magnitude of $G(j\omega)$ and the phase angle will change by the phase angle of $G(j\omega)$.

Further interpretation of the FRF $G(j\omega)$ can be given because Fourier transform and Laplace transform are directly related (as discussed before). Hence, Fourier results can be obtained directly from the Laplace-domain results, simply by substituting $s = j\omega$. Accordingly, the Laplace result (6.12) has the frequency-domain (Fourier) result

$$G(j\omega) = \frac{Y(j\omega)}{U(j\omega)} = G(j\omega) = G(s)\,|_{s=j\omega} = \frac{b_0 + b_1(j\omega) + \cdots + b_m(j\omega)^m}{a_0 + a_1(j\omega) + \cdots + a_n(j\omega)^n} \tag{6.27b}$$

where $Y(j\omega) = \mathcal{F}\ y(t)$ and $U(j\omega) = \mathcal{F}\ u(t)$ with $\mathcal{F}$ denoting the Fourier transform operator. In other words, FRF is obtained by dividing the Fourier spectrum of the output by the Fourier spectrum of the input.

6.4.1.2 Magnitude (Gain) and Phase

Let us denote the magnitude of $G(j\omega)$ by

$$|G(j\omega)| = M \tag{6.28a}$$

and the phase angle of $G(j\omega)$ by

$$\angle G(j\omega) = \phi \tag{6.28b}$$

Then we can write

$$G(j\omega) = M\cos\phi + jM\sin\phi = Me^{j\phi} \tag{6.27c}$$

Now by combining (6.24) and (6.25) and (6.27c), we have

$$y = u_o M e^{j(\omega t + \phi)} \tag{6.29}$$

We summarize the following observations on the harmonic response of a system.

OBSERVATIONS

1. The FRF is given by

$$G(j\omega) = \frac{Y(j\omega)}{U(j\omega)} = \frac{\text{Fourier spectrum of the output}}{\text{Fourier spectrum of the input}}$$

2. The FTF is obtained by substituting $s = j\omega$ into the Laplace TF

$$G(j\omega) = G(s)\,|_{s=j\omega} = \frac{b_0 + b_1(j\omega) + \cdots + b_m(j\omega)^m}{a_0 + a_1(j\omega) + \cdots + a_n(j\omega)^n}$$

When a harmonic input of frequency ω is applied to the system, at steady state

1. The output is magnified by $M = |G(j\omega)|$
2. The output has a *phase lead* wrt the input of $\phi = \angle G(j\omega)$.

Note: For practical systems, typically, the phase lead $\angle G(j\omega)$ is negative; that is, the output typically lags the input.

These observations further confirm that $G(j\omega)$ constitutes a complete model for a linear, constant-parameter system (as does $G(s)$).

6.4.2 Bode Diagram (Bode Plot)

We have established that the FTF $G(j\omega)$ is a complete model of a (linear time-invariant) system. In general, FRF is a complex function of frequency ω, which is a real variable. Hence, the FRF has a magnitude and a phase angle, or a real part and an imaginary part, expressed as a function of frequency.

Experimental determination of $G(j\omega)$ can be done in several ways. One straightforward method is indicated by the result (6.28):

Step 1: Decide on the frequency range of interest $[\omega_s, \omega_e]$. Set $\omega = \omega_s$

Step 2: Apply a harmonic (i.e., sinusoidal) excitation of known amplitude to the system, at frequency ω and measure the amplitude and the phase change of the output (response), at steady state

Step 3: Increment the excitation frequency by a small step ($\Delta\omega$) according to $\omega \to \omega + \Delta\omega$. If $\omega > \omega_e$, go to Step 4. Otherwise, go to Step 2

Step 4: For each frequency, compute (a) gain $|G(j\omega)|$=[output amplitude]/[input amplitude]; (b) phase lead $\angle G(f)$=[phase angle of output]−[phase angle of input]

Either a *sine-sweep* or a *sine-dwell* excitation may be used in this test. The frequency of excitation is varied continuously in a sine sweep and in steps in a sine dwell. The sweep rate should be sufficiently slow, or the dwell times should be sufficiently long, to achieve the steady-state response in each measurement, in these methods.

An alternative method of determining $G(j\omega)$ is by using the Fourier transform according to equation (6.27b). The associated steps are as follows:

Step 1: Apply a transient excitation $u(t)$ that has all the frequency components of interest, to the system, and measure the output (response) $y(t)$

Step 2: Compute the Fourier spectrum $Y(j\omega)$ of $y(t)$ and the Fourier spectrum $U(j\omega)$ of $u(t)$

Step 3: Compute the FRF according to $G(j\omega) = \dfrac{Y(j\omega)}{U(j\omega)}$.

The analytical FRF or the dynamic test results are usually presented as the pair of curves

$$|G(j\omega)| \text{ versus } \omega$$

$$\angle G(j\omega) \text{ versus } \omega$$

with the log scales for both the magnitude axis (e.g., in decibels or $20\log_{10}()$) and the frequency axis (e.g., in *decades*, which are multiples of 10; *octaves*, which are multiples of 2; *one-third octaves*, which are multiples of $2^{1/3}$). This pair of curves is called the *Bode plot* or the *Bode diagram*.

Note: A linear scale is used for the phase angle axis.

Example 6.3

The TF of a dynamic system is given by $G(s) = \dfrac{(s+3)}{(s^2+4s+16)}$

a. Tabulate the values of the magnitude $|G(j\omega)|$ and the phase angle $\angle G(j\omega)$ for about 6 points of frequency in the range $\omega = 0$ to $\omega = 5$
b. Plot the Bode diagram using MATLAB®.
c. If the system (G) is given the sinusoidal input $u = 2\cos 2t$, what is the corresponding output at steady state?

Solution

By setting $s = j\omega$, we get the FTF

$$G(j\omega) = \frac{j\omega + 3}{16 - \omega^2 + 4j\omega} \tag{i}$$

Hence,

$$|G(j\omega)| = \sqrt{\frac{3^2 + \omega^2}{(16-\omega^2)^2 + 16\omega^2}} \tag{ii}$$

and

$$\begin{aligned} \angle G(j\omega) &= \tan^{-1}\frac{\omega}{3} - \tan^{-1}\frac{4\omega}{16-\omega^2} \quad \text{for } \omega < 4 \\ &= \tan^{-1}\frac{\omega}{3} - \pi + \tan^{-1}\frac{4\omega}{\omega^2 - 16} \quad \text{for } \omega > 4 \end{aligned} \tag{iii}$$

Note: When $\omega > 4$, the real part of the denominator of the FTF (i) is negative (and the imaginary part is positive). Hence, the denominator term is in quadrant 2 of the complex plane. The denominator phase angle $= \pi -$ [phase angle obtained by using +ve real part] $= \pi - \tan^{-1}\dfrac{4\omega}{\omega^2 - 16}$. This has to be subtracted from the numerator phase angle $\left(\tan^{-1}\dfrac{\omega}{3}\right)$, to get the overall phase angle of $G(j\omega)$. This gives the second part of equation (iii).

(a)

Frequency ω	**0**	**1**	**2**	**3**	**4**	**5**	**∞**
Magnitude $\lvert G(j\omega)\rvert$	3/16	0.204	0.25	0.305	0.3125	0.266	0
(dB)	(−14.5)	(−13.8)	(−12)	(−10.3)	(−10.1)	(−11.5)	(−∞)
Phase $\angle G(j\omega)$ (degrees)	0	3.5	0	−14.7	−36.8	−55.2	−90

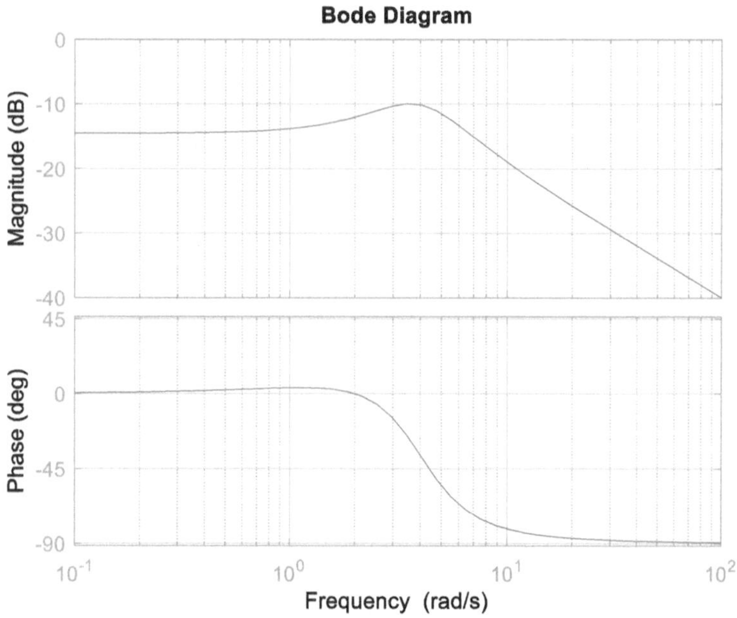

FIGURE 6.2 Bode plot using MATLAB®.

(b)

The Bode curves may be obtained using the following MATLAB commands:

```
>> num=[1 3];
>> den=[1 4 16];
>> G=tf(num, den);
>> bode(G), grid
```

The resulting curves are shown in Figure 6.2.

(c)

FTF of the system is $G(j\omega) = \dfrac{j\omega + 3}{\left(16 - \omega^2 + 4j\omega\right)}$

At ω=2 rad/s, $|G(j\omega)| = 0.25$ and $\angle G(j\omega) = 0° = 0$ radians
Hence, the steady-state response for an input of $u = 2\cos 2t$ is $y = 2 \times 0.25\cos(2t + 0)$
Or, $y = 0.5\cos 2t$

LEARNING OBJECTIVES

1. Manual computation of the Bode plots of a TF
2. Use of MATLAB to get the Bode curves
3. Determination of the harmonic response of a system using its FTF.

6.4.3 Bode Diagram Using Asymptotes

The denominator (characteristic) polynomial of a TF may be factorized into first-order terms of the form $(s + a)$ and second-order oscillatory terms $\left(s^2 + 2\zeta\omega_n s + \omega_n^2\right)$, $0 \le \zeta < 0$. The numerator polynomial also can be similarly factorized. The Bode plot of each factor (component) can be easily determined. Then, the Bode plot of the entire TF can be constructed (by using only additions and subtractions of the component plots, in view of their log scales). The rationale for this is the following:

1. In a product of complex numbers, the magnitudes multiply and the phase angles add
2. In a quotient of complex numbers, the magnitudes divide and the phase angles subtract
3. Since a log scale (dB) is used for the magnitudes, the multiplications and divisions of the magnitudes are "transformed" into additions and subtractions.

An advantage of the log scale for magnitude is that the Bode diagram for a product of several TFs can be obtained by simply adding the Bode plots for the individual TFs. In this manner, the Bode plot of a complex system can be conveniently obtained with the knowledge of the Bode plots of its components.

Note: In a Bode plot, a linear scale is used for the phase angle. Hence, the component phase angles also add (or subtract).

When a log scale is used, it emphasizes the lower values in the range. The x-axis (frequency axis) of the Bode plot is marked in units of frequency, which may be incremented by factors of 2 (*octaves*) or factors of 10 (*decades*). Typically in a Bode plot, the frequency axis is scaled in *decades*. This is a linear $\log_{10}$ scale. The amplitude axis is given in decibels (dB), which is also a $\log_{10}$ scale, specifically $20\log_{10}(\ )$ in decibels (dB).

Note: $20\log_{10}(\ ) = 10\log_{10}(\ )^2$. Since power and energy are represented by the square of a signal such as voltage, current, velocity, and force, it is clear that 10 dB corresponds to a power (or energy) increase by a factor of 10 or a signal increase by a factor of $\sqrt{10}$. Similarly, 20 dB corresponds to a signal increase by a factor of 10 or a power increase by a factor of 100.

The exercise of sketching a Bode diagram may be further simplified by first sketching the asymptotes of the elementary terms $(s+a)$ and $\left(s^2 + 2\zeta\omega_n s + \omega_n^2\right)$, $0 \le \zeta < 0$ and then approximating the actual curves to approach the asymptotes in the limit. This approach is illustrated now using examples.

Example 6.4

Sketch the Bode plot of the TF [output speed/input voltage] of an armature-controlled dc motor, given by $G(s) = \dfrac{K}{(\tau s + 1)}$

where

K = gain parameter (depends on motor constants, armature resistance, and damping)
τ = time constant (depends on motor inertia, motor constants, armature resistance, and damping).

Solution

This is a first-order system. The FTF corresponding to the given TF is

$$G(j\omega) = \frac{K}{(\tau j\omega + 1)} \text{ or } G(j2\pi f) = \frac{K}{(\tau j2\pi f + 1)} \tag{i}$$

Here ω is the *angular frequency* (in rad/s) and f is the *cyclic frequency* (in cycles/s or Hz).

Note: The complex functions $G(j\omega)$ and $G(j2\pi f)$ may be denoted by $G(\omega)$ and $G(f)$, respectively, for notational convenience (even though contrary to strict mathematical meanings).

The numerator term in the TF is a constant. The asymptotes for the numerator and the denominator of the TF are determined now. First, we define the critical frequency (break frequency) where the real part and the imaginary part of the considered factor are equal:

$$f_b = \frac{1}{2\pi\tau} \tag{ii}$$

$$\text{When } f << f_b : G(f) \approx K \tag{iii}$$

The corresponding magnitude is K (or $20\log_{10} K$ dB). This asymptote is a horizontal line as shown in Figure 6.4. The phase angle of this asymptote is zero.

$$\text{When } f >> f_b : G(f) \approx \frac{K}{\tau j2\pi f} \tag{iv}$$

The magnitude of this function is $K/(\tau 2\pi f)$. It monotonically decreases with frequency. If decibel scale (i.e., $20\log_{10}(\ )$ dB) is used for the magnitude axis and decade scale (i.e., multiples of 10) for the frequency axis, the slope of this asymptote is –20 dB/decade. The phase angle of this asymptote is 90°.

The two asymptotes intersect at $f = f_b$. This frequency is the *break frequency* (or the *corner frequency*).

Note: Since a significant magnitude attenuation takes place for input signal frequencies greater than f_b and in view of the fast decay of the natural response for large f_b, it is appropriate to consider f_b, given by equation (ii), as the limiting frequency for the frequency response, a measure of the *bandwidth*, for a dc motor.

The asymptotes are drawn and the approximate Bode plots are sketched based on them (so as to approach them in the limit) as shown in Figure 6.3.

Suppose that a sinusoidal signal is used as the input test signal to the dc motor. As the input frequency is raised, the output amplitude decreases and the phase lag increases, as confirmed by the Bode plot.

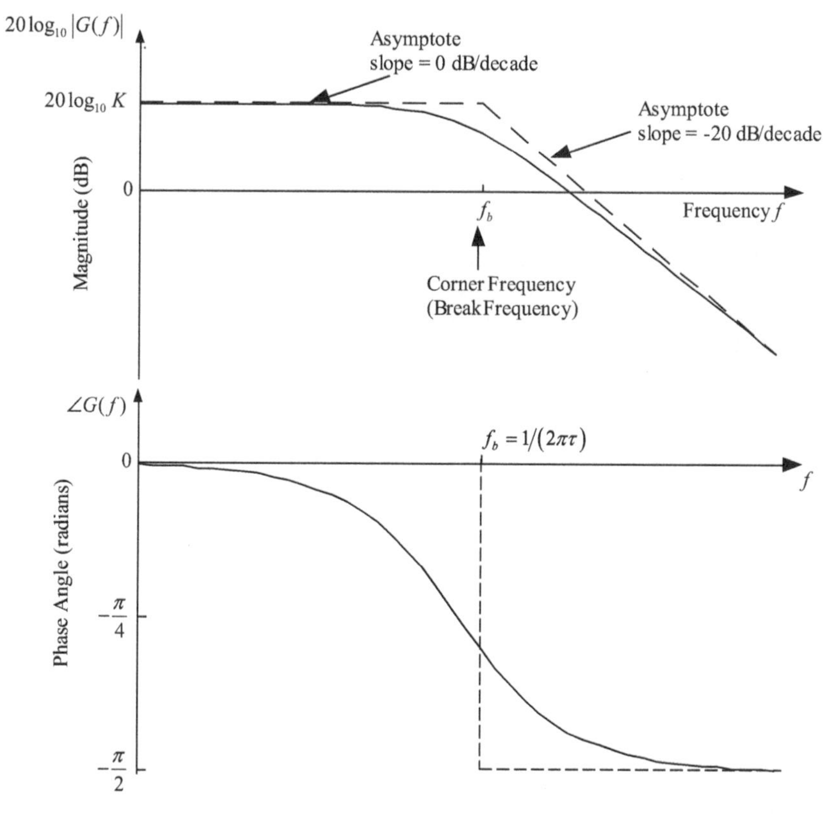

FIGURE 6.3 Bode diagram of a dc motor transfer function.

Note 1: Similarly, it can be shown that for a TF "numerator" component of the form $G(s) = K(\tau s + 1)$, the second asymptote (beyond the break point of $\omega_b = 1/\tau$) of the magnitude (gain) will have a positive slope of +20 dB/decade, and the asymptote of the corresponding phase angle will be a constant at +90°.

Note 2: The advantages of using a log scale for frequency are the fact that a wide range of frequencies can be accommodated in a limited plotting area, and that asymptotes to the magnitude curve become straight lines with slopes differing by fixed increments (by ±20 dB/decade if the decibel scale is used for magnitude and the decade scale is used for frequency).

LEARNING OBJECTIVES

1. The use of asymptotes to sketch a Bode plot
2. The use of log scales in a Bode plot
3. Break frequency, corner frequency, and bandwidth of a simple TF
4. The asymptote slope of a first-order factor in the denominator is –20 dB/decade
5. The asymptote slope of a first-order factor in the numerator is +20 dB/decade.

Example 6.5

Consider a damped oscillator, which has the FTF:

$$G(j\omega) = \frac{K}{\left(\omega_n^2 - \omega^2 + 2j\zeta\omega_n\omega\right)} \quad 0 < \zeta < 1 \tag{i}$$

Note: This is an underdamped system, with its damping ratio ζ less than 1. For it to have a resonance, we need $0 < \zeta < 1/\sqrt{2}$.

The break point for the asymptotes is the undamped natural frequency ω_n.

For $\omega << \omega_n$, the FTF (i) can be approximated by the static gain (i.e., the zero-frequency magnitude)

$$G(j\omega) \approx \frac{K}{\omega_n^2} \tag{ii}$$

This is a real TF. Its magnitude is a constant, and hence, the slope of its Bode plot is zero. The phase angle is zero as well, in this region (because the FTF is real and +ve). The corresponding gain and phase asymptote pair (for $\omega = 0$ to ω_n) are shown in Figure 6.4a.

For $\omega >> \omega_n$, the FTF (i) can be approximated by

$$G(j\omega) \approx -\frac{K}{\omega^2} \tag{iii}$$

In this region, the magnitude in decibels is $20\log_{10}\left(\frac{K}{K_o}\right) - 40\log_{10}\left(\frac{\omega}{\omega_o}\right)$ dB

Note: K and ω are non-dimensionalized because, mathematically, it is not correct to obtain the logarithm of a dimensional quantity. An important observation, however, is that when the frequency changes by 1 decade (i.e., when $\omega/\omega_o = 10$), the magnitude of this expression changes by –40dB.

Hence, the slope of this asymptote is –40 dB/decade. Since (iii) represents a negative real quantity, its phase angle is –180°. The corresponding gain and phase asymptote pair (for $\omega = \omega_n$ to ∞) are shown in Figure 6.4a.

For the sake of completion, the Bode plot of the damped oscillator, as obtained using MATLAB, is shown in Figure 6.4b.

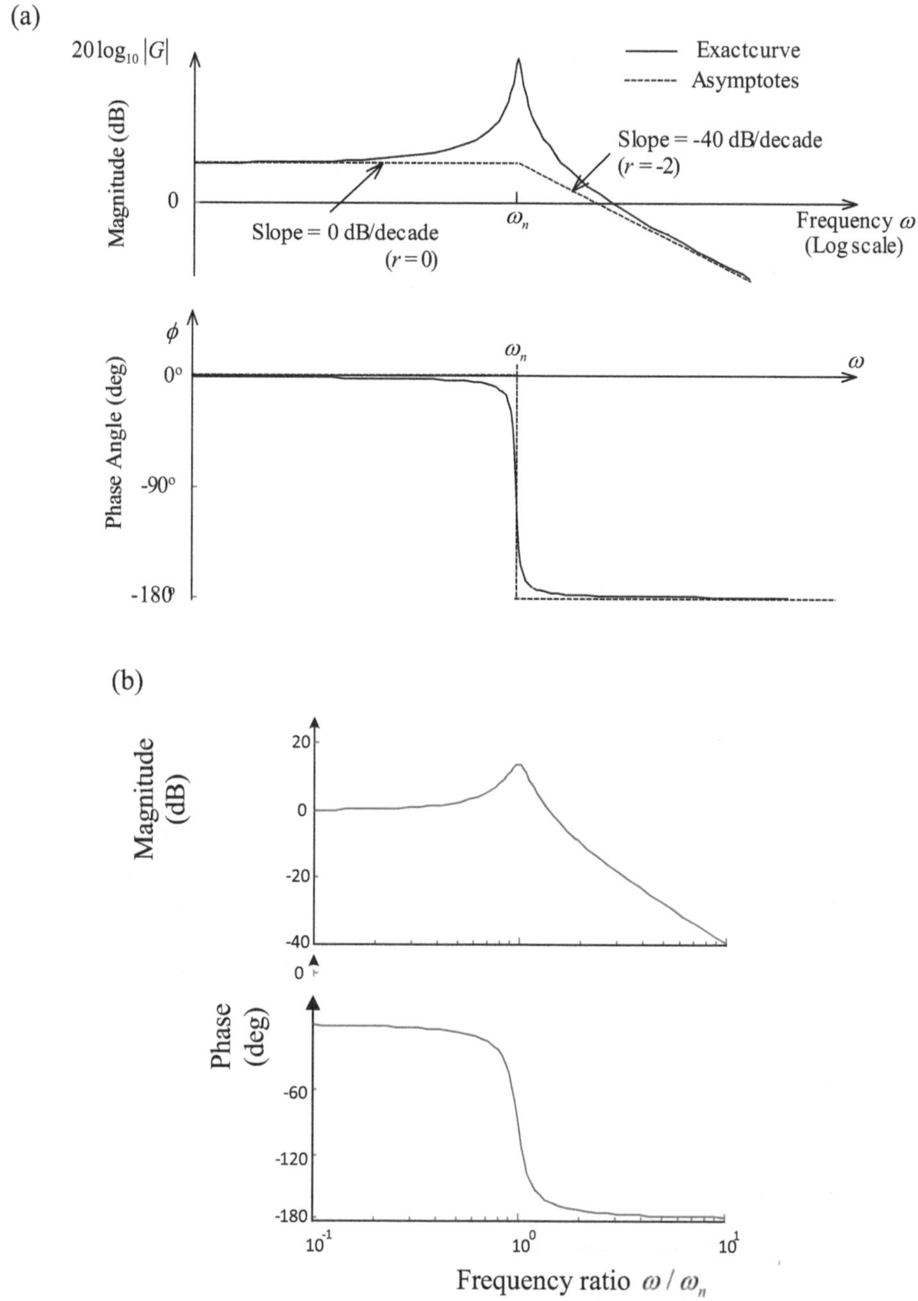

FIGURE 6.4 Damped oscillator: (a) asymptotes of Bode plot; (b) Bode plot using MATLAB.

LEARNING OBJECTIVES

1. Use of asymptotes to sketch a Bode plot
2. The asymptote slope of a second-order factor in the denominator is −40 dB/decade
3. Special consideration in Bode sketching near a resonance (in an underdamped system)
4. Further understanding of decibels (dB) and decades (frequency multiples of 10).

6.5 MECHANICAL IMPEDANCE AND MOBILITY

Impedance is a TF, which is useful in both mechanical and electrical systems. However, mechanical impedance is not directly analogous to electrical impedance. It is the mobility, which is the inverse of mechanical impedance that is analogous to electrical impedance. Mechanical impedance is analogous to electrical admittance. Specifically, electrical impedance and mechanical mobility are *A-type* TFs (across-variable/through-variable) or, *generalized impedances*. Mechanical impedance and electrical admittance are *T-type* TFs (through-variable/across-variable). In view of the existing analogies (particularly, the force-current analogy, as used in linear graphs), which have been studied in Chapter 5, similar treatments are possible concerning analogous TFs in mechanical and electrical systems. Several relevant topics are addressed next. Of course, in view of the "unified approach" that is used in this book, the same concepts can be directly extended to other physical domains (fluid and thermal, in particular) as well.

6.5.1 Transfer Functions in Mechanical Systems

The use of the FTF as a dynamic model has been addressed in the previous sections. Its significance can be illustrated by considering again the damped mechanical oscillator (i.e., a single degree-of-freedom mass-spring-damper system) shown in Figure 6.1. Its force-displacement TF, in the frequency domain, is

$$G(j\omega) = \frac{1}{ms^2 + bs + k} \text{ with } s = j\omega \tag{6.30}$$

in which m, b, and k denote mass, damping constant, and stiffness, respectively. When the excitation frequency ω is small in comparison to the system's undamped natural frequency $\omega_n = \sqrt{k/m}$, the terms ms^2 and bs can be neglected with respect to k, and the system behaves as a simple spring. When the excitation frequency ω is much larger than the system natural frequency, the terms bs and k can be neglected in comparison to ms^2. In this case, the system behaves like a simple lumped mass. When the excitation frequency ω is very close to the natural frequency (i.e., $s = j\omega \approx j\omega_n = j\sqrt{k/m}$), it is seen from (6.30) that the term $ms^2 + k$ in the denominator of the TF (i.e., the characteristic polynomial) becomes almost zero and can be neglected. Then the TF can be approximated by $G(j\omega) = \dfrac{1}{bj\omega}$.

In summary:

1. When the excitation frequency approaches the resonant frequency or natural frequency (i.e., for intermediate values of excitation frequencies), system damping becomes the most important parameter. Then the system behaves like a damper, and the largest response occurs at the resonant frequency. Specifically, we have As $\omega \to \omega_n,\ G(j\omega) \to \dfrac{1}{bj\omega}$
2. At low-excitation frequencies, the system stiffness is the most significant parameter. Then the system behaves like a spring, giving a somewhat "static" response. Specifically, we have As $\omega \to 0,\ G(j\omega) \to \dfrac{1}{k}$
3. At high-excitation frequencies, the mass is the most significant parameter. Then the system behaves like a mass, and its response tends to zero (i.e., it is very difficult for a mass to move at very high frequencies). Specifically, we have As $\omega \to \infty,\ G(j\omega) \to -\dfrac{1}{m\omega^2} \to 0$

Note: In these observations, instead of the physical parameters m, k, and b, we could use natural frequency $\omega_n = \sqrt{k/m}$ and the damping ratio $\zeta = b/\left(2\sqrt{mk}\right)$ as the system parameters. Then the number of the system parameters reduces to two, which is an advantage in parametric and sensitivity studies.

6.5.1.1 Mechanical Transfer Functions

Any force variable or motion variable of a system may be used as the input and the output in defining an "analytical" TF in a mechanical system. However, some such "analytical" TFs may not be *physically realizable* (Then, for practical purposes, the input and the output must be reversed, and the inverted TF should be used). We can define several versions of FTFs that may be useful in the modeling and analysis of mechanical systems. Some relatively common ones are given in Table 6.1.

Furthermore, we have established the following facts in the frequency domain:

$$\text{Acceleration} = (j\omega) \times (\text{Velocity})$$

$$\text{Displacement} = (\text{Velocity}) / (j\omega)$$

In view of these relations, it can be shown that many of the alternative types of TFs defined in Table 6.1 are related to the mechanical impedance and mobility through the factor $j\omega$. Specifically,

$$\text{Dynamic Stiffness} = \text{Force} / \text{Displacement} = \text{Mechanical Impedance} \times j\omega$$

$$\text{Receptance} = \text{Displacement} / \text{Force} = \text{Mobility} / (j\omega)$$

$$\text{Dynamic Inertia} = \text{Force} / \text{Acceleration} = \text{Impedance} / (j\omega)$$

$$\text{Accelerance} = \text{Acceleration} / \text{Force} = \text{Mobility} \times j\omega$$

In these definitions, the variables force, acceleration, and displacement should be interpreted as the corresponding Fourier spectra and not functions of time.

6.5.1.2 Mechanical Impedance and Mobility

In studies of mechanical systems, three types of FTFs are particularly useful. They are *Mechanical Impedance, Mobility*, and *Transmissibility*, as presented in Table 6.1. In a mechanical impedance function, velocity is considered the input variable and the force is the output variable, whereas in a mobility function, the converse applies. It is clear that mobility is the inverse of mechanical impedance. Either TF may be used in the "analysis" of a given problem, for convenience of analysis, as will be clear from the examples presented in this chapter. However, in the context of practical application, some TFs may not be physically realizable (even though they have analytical expressions).

TABLE 6.1
Definitions of Useful Mechanical Transfer Functions

Transfer Function	Definition (in Laplace or Frequency Domain)
Dynamic stiffness	Force/displacement
Receptance (dynamic flexibility or compliance)	Displacement/force
Mechanical impedance (Z)	Force/velocity
Mobility (M)	Velocity/force
Dynamic inertia	Force/acceleration
Accelerance	Acceleration/force
Force transmissibility (T_f)	Transmitted force/applied force
Motion transmissibility (T_m)	Transmitted velocity/applied velocity

6.5.2 Interconnection Laws

Once the TFs of the basic components (elements) of a system are known, the interconnection laws may be used to determine the overall TF of the system. Two types of interconnection are possible for two components:

1. Series connection
2. Parallel connection.

Determination of the interconnection laws is straightforward in view of the fact that:

1. For series-connected elements, the through-variable is common and the across-variables add
2. For parallel-connected elements, the across-variable is common and the through-variables add.

6.5.2.1 Interconnection Laws for Mechanical Impedance and Mobility

Since mobility is given by an across-variable (velocity) divided by a through-variable (force), it is clear (on dividing throughout by the common through-variable) that for series-connected elements, the mobilities add (or, the inverse of mechanical impedance is additive) or generalized impedances add.

Since mechanical impedance is given by a through-variable (force) divided by an across-variable (velocity), it is clear (on dividing throughout by the common across-variable) that for parallel-connected elements, the mechanical impedances add (or, the inverse of mobility is additive).

These interconnection laws are presented in Table 6.2.

6.5.2.2 Interconnection Laws for Electrical Impedance and Admittance

Since electrical impedance is given by

$$\text{Electrical Impedance} = \frac{\text{Across-variable (voltage)}}{\text{Through-variable (current)}},$$

TABLE 6.2
Interconnection Laws for Mechanical Impedance (Z) and Mobility (M)

Series Connection

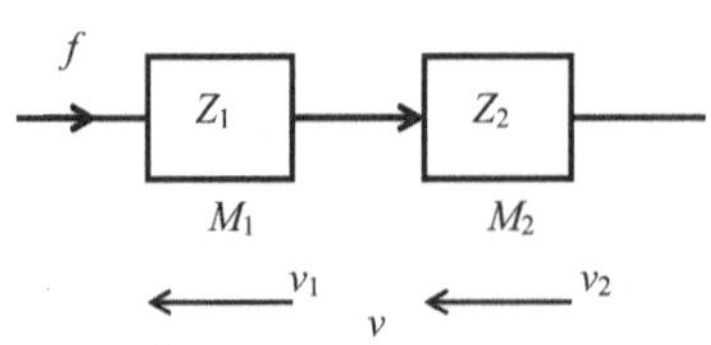

$v = v_1 + v_2$

$\frac{v}{f} = \frac{v_1}{f} + \frac{v_2}{f}$

$M = M_1 + M_2$

$\frac{1}{Z} = \frac{1}{Z_1} + \frac{1}{Z_2}$ or $Z = \frac{Z_1 Z_2}{Z_1 + Z_2}$

Parallel Connection

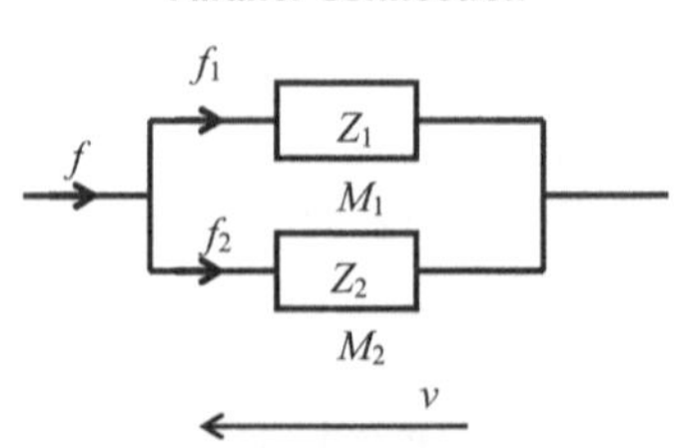

$f = f_1 + f_2$

$\frac{f}{v} = \frac{f_1}{v} = \frac{f_2}{v}$

$Z = Z_1 + Z_2$

$\frac{1}{M} = \frac{1}{M_1} + \frac{1}{M_2}$ or $M = \frac{M_1 M_2}{M_1 + M_2}$

TABLE 6.3
Interconnection Laws for Electrical Impedance (*Z*) and Admittance (*W*)

Series Connections	Parallel Connections
$v = v_1 + v_2$	$i = i_1 + i_2$
$\frac{v}{i} = \frac{v_1}{i} + \frac{v_2}{i}$	$\frac{i}{v} = \frac{i_1}{v} = \frac{i_2}{v}$
$Z = Z_1 + Z_2$	$W = W_1 + W_2$
$\frac{1}{W} = \frac{1}{W_1} + \frac{1}{W_2}$ or $W = \frac{W_1 W_2}{W_1 + W_2}$	$\frac{1}{Z} = \frac{1}{Z_1} + \frac{1}{Z_2}$ or $Z = \frac{Z_1 Z_2}{Z_1 + Z_2}$

it is clear (on dividing throughout by the common through-variable) that for series-connected elements, the electrical impedances add (or, the inverse of admittance is additive).

Since admittance is given by

$$\text{Electrical Admittance} = \frac{\text{Through-variable (current)}}{\text{Across-variable (voltage)}},$$

it is clear (on dividing throughout by the common across-variable) that for parallel-connected elements, the admittances add (or, the inverse of electrical impedance is additive). These interconnection laws for electrical systems are presented in Table 6.3.

6.5.2.3 *A*-Type Transfer Functions and *T*-Type Transfer Functions

Electrical impedance and mechanical mobility are "*A*-Type TFs" because

$$A\text{-type Transfer Function} = \frac{[\text{Across-variable}]}{[\text{Through-variable}]}$$

They follow the same interconnection laws (compare Tables 6.2 and 6.3).

Electrical admittance and mechanical impedance are "*T*-Type TFs" because

$$T\text{-type Transfer Function} = \frac{[\text{Through-variable}]}{[\text{Across-variable}]}$$

They follow the same interconnection laws (compare Tables 6.2 and 6.3).

6.5.3 Transfer Functions of Basic Elements

Since a complex system can be formed through series and parallel interconnections of basic elements, it is possible to systematically generate the TF of a complex system by using the TFs of the basic elements and the appropriate interconnection laws.

In Chapter 2, the linear constitutive relations for the mass, spring, and the damper elements are presented as time-domain relations. The corresponding TFs are obtained by simply replacing the derivative operator $\frac{d}{dt}$ by the Laplace operator s. The FTFs are obtained by substituting $j\omega$ or $j2\pi f$ for s. In this manner, the TFs of the basic (linear) mechanical elements, mass, spring, and damper are obtained, as given in Table 6.4.

Similarly, in Chapter 2, the linear constitutive relations for the electrical capacitor, inductor, and resistor elements are presented as time-domain relations. The corresponding TFs are obtained by

TABLE 6.4
Mechanical Impedance and Mobility of Basic Mechanical Elements

Element	Time-Domain Model	Mechanical Impedance	Mobility (Generalized Impedance)
Mass m	$m\frac{dv}{dt} = f$	$Z_m = ms$	$M_m = \frac{1}{ms}$
Spring k	$\frac{df}{dt} = kv$	$Z_k = \frac{k}{s}$	$M_k = \frac{s}{k}$
Damper b	$f = bv$	$Z_b = b$	$M_b = \frac{1}{b}$

TABLE 6.5
Impedance and Admittance of Basic Electrical Elements

Element	Time-Domain Model	Impedance (Z)	Admittance (W)
Capacitor C	$C\frac{dv}{dt} = i$	$Z_C = \frac{1}{Cs}$	$W_C = Cs$
Inductor L	$L\frac{di}{dt} = v$	$Z_L = Ls$	$W_L = \frac{1}{Ls}$
Resistor R	$Ri = v$	$Z_R = R$	$W_R = \frac{1}{R}$

replacing the derivative operator $\frac{d}{dt}$ by the Laplace operator s. In this manner, the TFs of the basic (linear) electrical elements are obtained, as given in Table 6.5.

Examples are given next to demonstrate the use of mechanical impedance and mobility methods in the development of frequency-domain models. In particular, the interconnection of the TFs (mechanical impedance and mobility) of the basic mechanical elements (mass—an A-type element, spring—a T-type element, damper—a D-type element), along with the input elements (force source, which is a T-source, and velocity source, which is an A-source), is demonstrated. We do not primarily consider examples in the electrical domain in the present section mainly because the present approach is common knowledge in the field of electrical engineering. In fact, the mechanical circuits that we use here are analogous to electrical circuits, and they also are the parent schematic diagrams used in the process of generating linear graphs (see Chapters 5 and 7). Similarly, the present approach can be easily extended to the fluid domain and the thermal domain (*Note*: There is no T-type element in the thermal domain). That extension is quite straightforward and is not focused as well, in this chapter. However, the use of TFs (particularly, transfer function linear graphs, or TFLGs) in all four physical domains and for multi-physics (multi-domain) systems is exclusively addressed in Chapter 7.

Example 6.6: Ground-Based Mechanical Oscillator

Consider the damped mechanical oscillator shown in Figure 6.5a. This may represent a simple machine that is mounted on a rigid floor. Its mechanical impedance circuit is given in Figure 6.5b. This circuit will clearly show the "structure" of the system (i.e., whether the elements are connected in series or parallel).

Note: In the mechanical impedance circuit, we have indicated the two ends (terminals) of each element. In particular, a mass element always has the ground as the reference terminal (with a

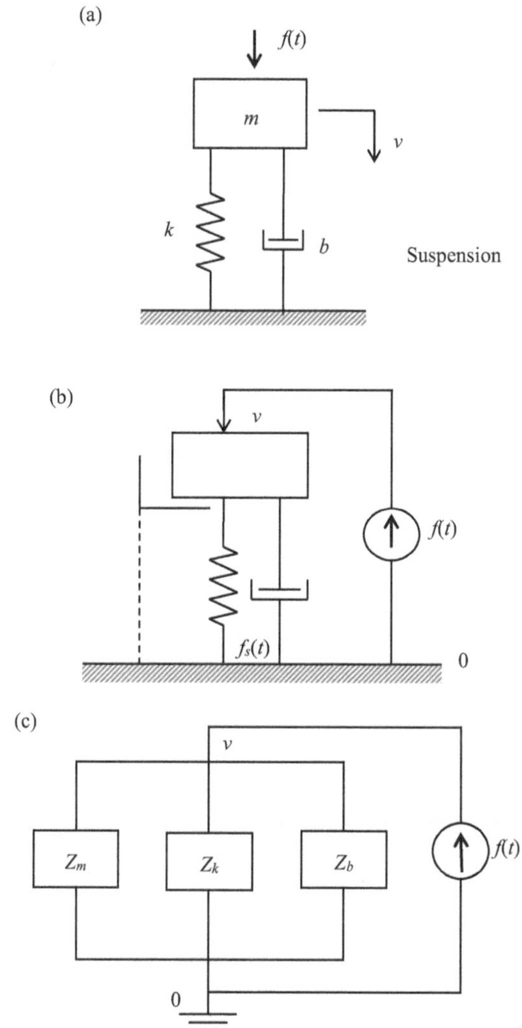

FIGURE 6.5 (a) Ground-based mechanical oscillator; (b) schematic mechanical circuit; (c) mechanical impedance circuit.

virtual connection, indicated by a broken line—see Chapter 5). Also, a source element has the ground as its reference terminal.

Since the input in this example is the force $f(t)$, the source element is a force source (a through-variable source or *T*-source).

The output (response) of the system is the velocity v. The corresponding TF $\frac{V(j\omega)}{F(j\omega)}$ is a *mobility function*.

From Figure 6.5b, it is clear that the four elements (mass, spring, damper, and source) in the system are connected in *parallel*. This structure is clear as well from the mechanical impedance circuit shown in Figure 6.5c.

Note: Analytically, the mechanical impedance representation (rather than the mobility representation) is more convenient when combining parallel elements because mechanical impedance is a *T*-type TF: across-variable/through-variable), and they simply add in parallel connection. Of course, it is okay to indicate mobilities in Figure 6.5c and call it a "mobility circuit." Whether a particular TF is physically realizable is a different issue, which needs to be addressed in the final result and not in the intermediate analytical steps.

The overall impedance function of the system is

$$Z(j\omega) = \frac{F(j\omega)}{V(j\omega)} = Z_m + Z_k + Z_b = ms + \frac{k}{s} + b\Big|_{s=j\omega} = \frac{ms^2 + bs + k}{s}\Big|_{s=j\omega} \tag{6.31}$$

The mobility function is the inverse of $Z(j\omega)$

$$M(j\omega) = \frac{V(j\omega)}{F(j\omega)} = \frac{s}{ms^2 + bs + k}\Big|_{s=j\omega} \tag{6.32}$$

It is known that this TF is *physically realizable* because the numerator order (1) is not greater than the denominator order (2). This is further confirmed by the physical fact that the input to the system is the force and the output is the velocity, and the corresponding *mobility function* governs the system behavior. In this TF, the *characteristic polynomial* is $s^2 + bs + k$, which corresponds to a damped oscillator (a second-order system). Accordingly, the behavior (particularly, the free or natural response) of the system is governed by this characteristic polynomial.

Another useful TF in the present example is that corresponding to the suspension force f_s. The suspension system has two parallel elements of mechanical impedance, $Z_k = \frac{k}{s}$ and $Z_b = b$. Their combined mechanical impedance is $\frac{F_s}{V} = Z_s = Z_k + Z_b = \frac{k}{s} + b = \frac{bs + k}{s}$. Combining this with (6.32), we get

$$\frac{F_s(j\omega)}{F(j\omega)} = \frac{bs + k}{ms^2 + bs + k}\Big|_{s=j\omega} \tag{6.33}$$

In fact, this is the *force transmissibility* function, which we will further discuss in the next section.

Note: Uppercase letters are used for variables, to denote their Fourier spectra, while lowercase letters are used to denote time functions. But, often, lowercase letters are used to denote the Fourier spectra of time functions as well, for analytical convenience, even though the two are quite different.

Degenerate Situation: Suppose that in this example, the force source (*T*-source) is replaced by a velocity source (*A*-source). Then, as we have discussed in Chapter 5, the three elements (mass spring and damper) become completely decoupled. Each element can be analyzed separately where the given velocity source is the input to each element. In fact, with a velocity input, the mass element becomes physically non-realizable (TF $Z_m = ms$) and the damping element becomes "algebraic" having a constant TF $Z_b = b$. In this case, the spring element has the TF $Z_k = \frac{k}{s}$ and is physically realizable. In essence, then, with a velocity input, the system in Figure 6.5 becomes degenerate and practically useless.

OBSERVATION

Suppose that using a force source, a known forcing function is applied to this system (with zero ICs) and the velocity response is measured. Next, using a velocity source if we move the mass exactly according to this predetermined velocity, the force generated at the source (the dependent variable of the velocity source) will be identical to the originally applied force. This is because mobility is the reciprocal (inverse) of impedance. This reciprocity should be intuitively clear because we are dealing with the same system and same ICs. Due to this property, for analytical convenience, we may use either the impedance representation or the mobility representation, depending on whether the elements are connected in parallel or in series, irrespective of whether the input is a force or a velocity or whether the TFs are physically realizable. This will be a matter of analytical convenience rather than practical importance.

LEARNING OBJECTIVES

1. Use of the concepts of mechanical impedance and mobility in frequency-domain modeling

2. System structure representation using a schematic mechanical circuit or a mechanical impedance/mobility circuit
3. Determination of the system TF using the element TFs and the system structure (series or parallel connection)
4. Physical realizability of a TF.

Example 6.7: Oscillator with Support Motion

Consider the system shown in Figure 6.6a, where a mass-spring-damper system is supported on a movable platform. This may represent a simple (1-D) model of a vehicle or an elevator. In this example, the motion of the mass m is not associated with an external force. The support (platform) velocity $v(t)$ is provided by a "velocity actuator," the input (an A-source), with the associated force f, which is the *dependent variable* of the actuator (see the broken line). This "dependent force" will be reacted on the ground.

A schematic mechanical circuit for the system is shown in Figure 6.6b and the corresponding impedance circuit is shown in Figure 6.6c. These circuits clearly indicate that the spring and the damper are connected in parallel, and the mass is connected in series with this pair. By impedance addition for parallel elements, and mobility addition for series elements, it is seen that the overall mobility function of the system is

$$\frac{V(j\omega)}{F(j\omega)} = M_m + \frac{1}{(Z_k + Z_b)} = \frac{1}{ms} + \frac{1}{(k/s + b)}\bigg|_{s=j\omega} = \frac{ms^2 + bs + k}{ms(bs + k)}\bigg|_{s=j\omega} \tag{6.34}$$

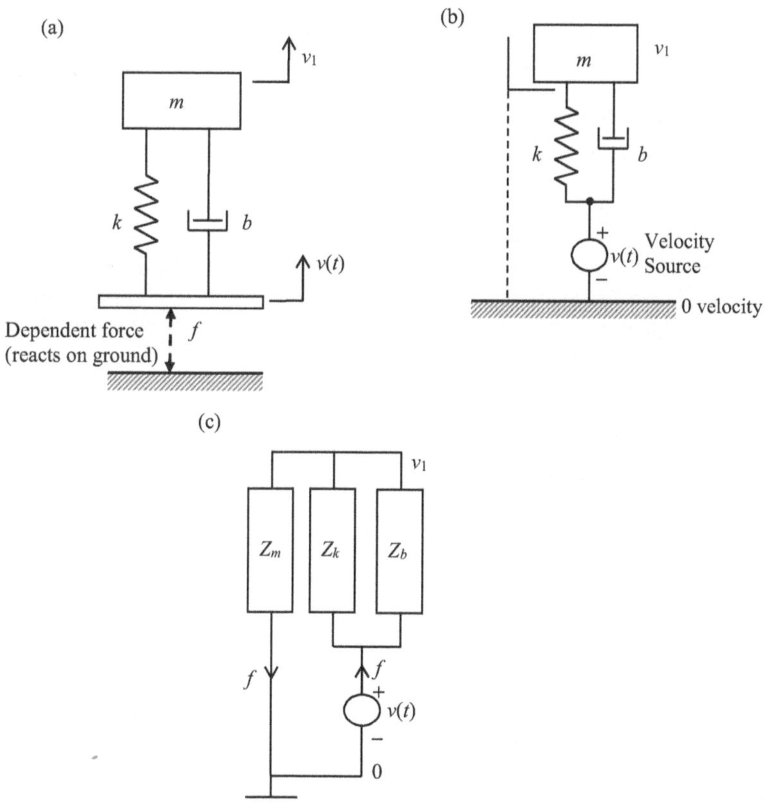

FIGURE 6.6 (a) A mechanical oscillator with support motion; (b) schematic mechanical circuit; (c) mechanical impedance circuit.

However, since we know that the input is the velocity $v(t)$, the proper TF is the mechanical impedance function

$$\frac{F(j\omega)}{V(j\omega)} = \left.\frac{ms(bs+k)}{ms^2+bs+k}\right|_{s=j\omega} \tag{6.35}$$

Furthermore, since the same force goes through to the mass element (see the circuit diagrams in Figure 6.6), we have

$$\frac{V_1(j\omega)}{F(j\omega)} = \left.\frac{1}{ms}\right|_{s=j\omega} \tag{6.36}$$

A practically more useful TF is the one that uses the velocity of the mass (v_1) as the output. The corresponding TF is obtained by combining (6.36) with (6.34). We get

$$\frac{V_1(j\omega)}{V(j\omega)} = \left.\frac{(bs+k)}{ms^2+bs+k}\right|_{s=j\omega} \tag{6.37}$$

Clearly, this TF is physically realizable. In fact, this is the *motion transmissibility* function, which we will further discuss in the next section. It is interesting to note that this TF is identical to that given by (6.33) (the expressions are the same but the meanings are different).

Note: If the force at the support (platform) is the input (a force source, which is a *T*-source), it is clear that the same input force (f) goes through the combined spring-damper unit and the mass element. Then, the spring-damper unit becomes completely uncoupled from the mass element, and they may be treated as two independent modules with input f applied separately. This situation is problematic for several reasons, as listed below:

1. Consider the mass element as an uncoupled unit. The corresponding TF is the mobility function $\frac{V_1(j\omega)}{F(j\omega)}$. This is a simple integrator, as given by (6.36). Physically, when a force f is applied to the mass, it will accelerate in proportion, regardless of the behavior of the other components (spring and damper) in the system. In particular, when f is constant, a constant acceleration is produced at the mass, causing its velocity to increase linearly (a behavior of simple "integration"). This is an "unstable" device (such as a solid rocket)
2. If the support (platform) velocity is considered as the output, the system TF is given by the inverse of (6.35): $\frac{V(j\omega)}{F(j\omega)} = \left.\frac{ms^2+bs+k}{ms(bs+k)}\right|_{s=j\omega}$. In theory, this is physically realizable (because the numerator order is not greater than the denominator order). However, the characteristic polynomial of the system is $ms(bs+k)$, which is known to correspond to an inherently unstable system, due to the presence of the free integrator, and has a non-oscillatory transient response.

Note: Uppercase letters are used here for the variables to denote their Fourier spectra.

LEARNING OBJECTIVES

1. Use of the concepts of mechanical impedance and mobility in frequency-domain modeling
2. System structure representation using a schematic mechanical circuit or a mechanical impedance/mobility circuit
3. Determination of the system TF using the element TFs and the system structure (series or parallel connection)
4. Proper inputs, proper TFs, and physical realizability.

Example 6.8

a. Give four ways how impedance information of the associated devices can be used in procedures of instrumentation of an engineering system. Explain why it is the *mobility* of a mechanical device that is analogous to the electrical impedance of an electrical device.

b. Human eyeball is manipulated by eye muscles (see Figure 6.7a). This ability of eye movement is affected by various health conditions and can be used to diagnose those conditions. A pulley model is commonly used to represent the associated dynamics. In this model, a muscle is modeled by a spring and a damper, with an actuator to move the muscle, and the eyeball is modeled as a rigid rotatory inertia.

   Consider only a single degree of freedom of eye movement, modeled as in Figure 6.7b. In this model, the passive component and the active component of an eye muscle are separately represented. The following model parameters are given:

   k=muscle stiffness (i.e., force per unit rectilinear deflection)

   b=muscle damping constant (i.e., force per unit rectilinear velocity)

   J=moment of inertia about the central axis of rotation of the eyeball (pulley)

   r=eyeball radius.

   The following variables are defined:

   ω_e=angular speed of the pulley (eyeball), which is the model output

   $r\omega_a$=rectilinear speed of the active muscle movement.

   Note: ω_a is the model input, which has the same units as ω_e.

   i. Taking ω_a as the model input and ω_e as the model output, sketch a mobility circuit for the model in Figure 6.7b, while representing various elements in the model as mobility blocks. Give expressions for the element mobilities (indicated within the blocks) in terms of the given model parameters.

   ii. Determine the input mobility and the output mobility of the circuit in terms of the given parameters and the Laplace variable *s*.

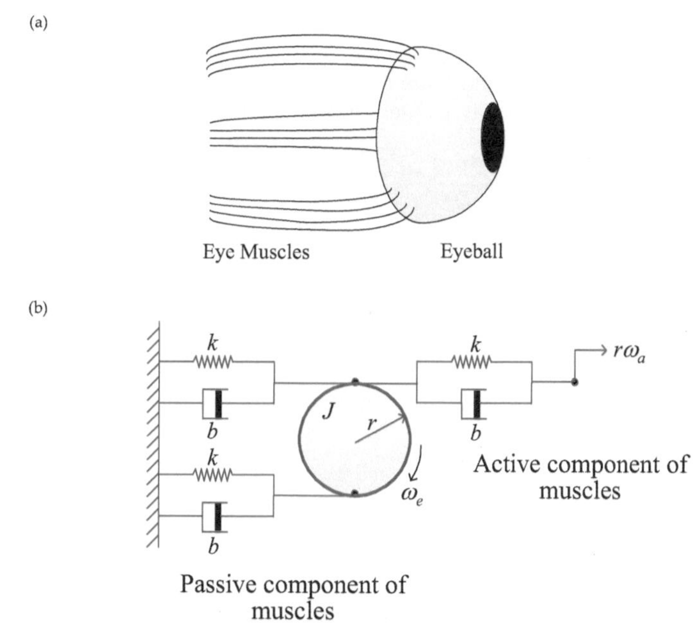

FIGURE 6.7 (a) Human eyeball and eye muscles; (b) a single-degree-of-freedom pulley model of eyeball and eye muscles.

Solution

(a)

The impedances of the interconnected components will determine: 1. the amount of power transferred to a load; 2. the efficiency of power transfer; 3. how much of a signal is transmitted through an impedance discontinuity, due to signal reflection; and 4. the level of loading from one component on to another connected component. Proper instrumentation procedures are able to adjust (match) component impedances so as to 1. maximize the amount of power transfer; 2. maximize the efficiency of power transfer; 3. reduce the loading, by making the input impedance of the connected second component larger than the output impedance of the first component; and 4. avoid the signal reflection at a junction of two components (an impedance discontinuity), by making the component impedances equal.

$$\text{Mobility} = \frac{\text{Generalized Velocity}}{\text{Generalized Force}} = \frac{\text{Across-variable}}{\text{Through-variable}}$$

$$\text{Electrical Impedance} = \frac{\text{Voltage}}{\text{Current}} = \frac{\text{Across-variable}}{\text{Through-variable}}$$

Hence, the two are analogous, and they will provide analogous circuit structures (i.e., parallel mechanical-component connections ←→ parallel electrical-component connections; series mechanical-component connections ←→ series electrical-component connections).

(b)

(i)

First, we observe from Figure 6.7b that all the passive elements are in parallel because the velocity (the across-variable) is common (the same) for these elements (*Note*: An inertia element always has an inertial/ground reference). The two active elements are also parallel on their own (separately), for the same reason (i.e., their velocity is common).

In the mobility circuit, we must show the proper input (ω_a) and proper output (ω_e). The corresponding mobility circuit is shown in Figure 6.8a.

Note: Actually there are two passive springs in parallel, which we have combined (with an equivalent stiffness $2k$) and there are two passive dampers as well in parallel, which we have combined (with an equivalent damping constant $2b$).

Note: Since angular velocity is the across-variable in this circuit, torque is the through-variable (just like using translatory velocity and force in the translatory case). So, we need to know the torque corresponding to each mobility element in the circuit, in order to determine the corresponding mobility expression.

Inertial torque of J is $Js\omega_e \rightarrow M_J = \dfrac{1}{Js}$

Note: $s\omega_e$ is angular acceleration, in the Laplace domain.

Torque from damper $2b$ is $2br\omega_e \times r = 2br^2\omega_e \rightarrow M_b = \dfrac{1}{2br^2}$

Torque from spring $2k$ is $2kr\dfrac{\omega_e}{s} \times r = \dfrac{2kr^2}{s}\omega_e \rightarrow M_k = \dfrac{s}{2kr^2}$

Note: $\dfrac{\omega_e}{s}$ is angular displacement (rotation), in the Laplace domain.

Similarly,

$$M_{ba} = \frac{1}{br^2}$$

$$M_{ka} = \frac{s}{kr^2}$$

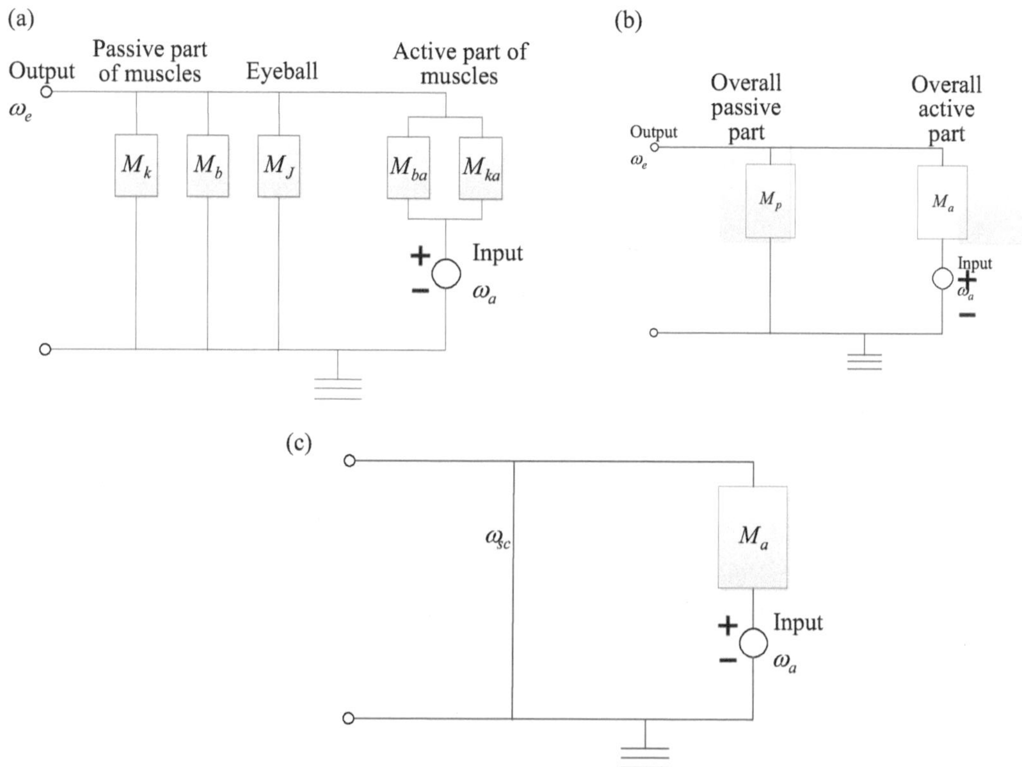

FIGURE 6.8 (a) The mobility circuit of the system. (b) The reduced mobility circuit. (c) Mobility circuit with output short-circuited.

(ii)

The parallel mobilities add in inverse (just like parallel electrical impedances). The mobility circuit can be reduced by using this fact, as shown in Figure 6.8b.

The corresponding mobilities are governed by the following relations:

$$\frac{1}{M_p} = \frac{1}{M_J} + \frac{1}{M_b} + \frac{1}{M_k} = Js + 2br^2 + \frac{2kr^2}{s} \tag{i}$$

$$\frac{1}{M_a} = \frac{1}{M_{ba}} + \frac{1}{M_{ka}} = br^2 + \frac{kr^2}{s} \tag{ii}$$

Input Mobility:

Maintain the output of the system in open circuit, as shown in in Figure 6.8b.

Angular velocity at the input $= \omega_a$

Torque at the input $\tau_a = \dfrac{\omega_a}{M_p + M_a}$

Input mobility $M_i = \dfrac{\omega_a}{\tau_a} = M_p + M_a$

Substitute (i) and (ii):

$$M_i = \frac{1}{Js + 2br^2 + \dfrac{2kr^2}{s}} + \frac{1}{br^2 + \dfrac{kr^2}{s}} = \frac{br^2 + \dfrac{kr^2}{s} + Js + 2br^2 + \dfrac{2kr^2}{s}}{\left(Js + 2br^2 + \dfrac{2kr^2}{s}\right)\left(br^2 + \dfrac{kr^2}{s}\right)}$$

Or,

$$M_i = \frac{\left(Js^2 + 3br^2 s + 3kr^2\right)s}{r^2\left(Js^2 + 2br^2 s + 2kr^2\right)(bs + k)}$$

Output Mobility:

Angular velocity at the output, in open circuit (see Figure 6.8a; note the velocity divider)

$$\omega_{oc} = \omega_e = \frac{M_p}{M_p + M_a}\omega_a \qquad \text{(iii)}$$

Maintain the output of the system in short circuit, as shown in Figure 6.8c.

$$\text{Short-circuit torque at the output, } \tau_{sc} = \frac{\omega_a}{M_a} \qquad \text{(iv)}$$

Substitute (iii) and (iv):

$$\text{Output mobility } M_o = \frac{\omega_{oc}}{\tau_{sc}} = \frac{M_p}{M_p + M_a}\omega_a \times \frac{M_a}{\omega_a} = \frac{M_p M_a}{M_p + M_a} = \frac{1}{1/M_p + 1/M_a}$$

Substitute (i) and (ii):

$$M_o = \frac{1}{br^2 + \dfrac{kr^2}{s} + Js + 2br^2 + \dfrac{2kr^2}{s}} = \frac{s}{Js^2 + 3br^2 s + 3kr^2}$$

Example 6.9

a. Figure 6.9a shows a circuit with a resistor of resistance R and a capacitor of capacitance C.
   v_i = input voltage
   v_o = output voltage
   Sketch an impedance circuit diagram for it. Using the impedances in the circuit elements, obtain the TF of the circuit.
b. Figure 6.9b shows the circuit obtained by cascading two circuit modules of that given in Figure 6.9a. Again,
   v_i = input voltage
   v_o = output voltage
   Sketch an impedance circuit diagram for this cascaded circuit. Using the impedances in the circuit elements, obtain the TF of the cascaded circuit.
   Comment on the result.

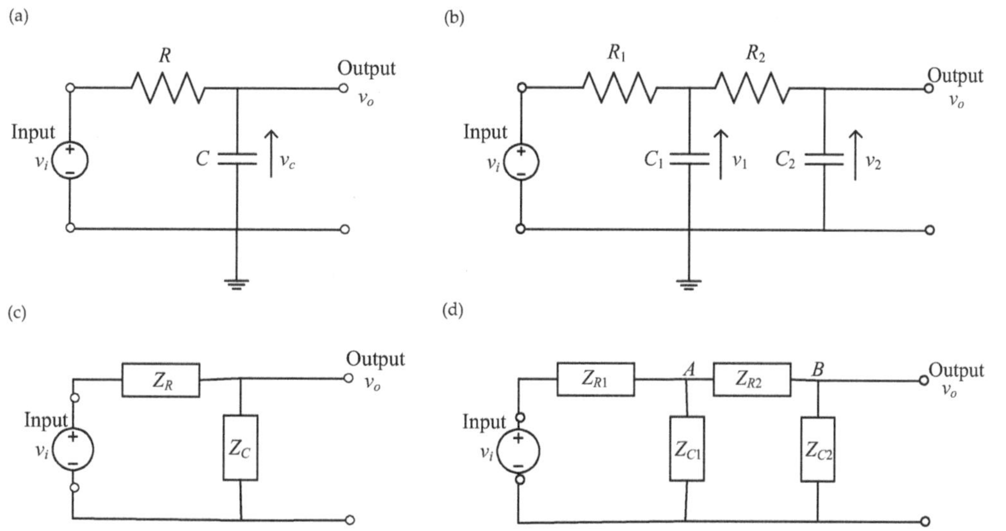

FIGURE 6.9 (a) A circuit module; (b) cascade circuit. (c) Impedance circuit of the original circuit; (d) impedance circuit of the cascade circuit.

Solution

a. Figure 6.9c shows the impedance circuit of the original circuit. The impedances are

$$\text{For } R: Z_R = R \tag{i}$$

$$\text{For } C: Z_c = \frac{1}{Cs} \tag{ii}$$

By voltage division in a series branch, we directly get $v_o = \frac{Z_C}{Z_R + Z_C} v_i$. Hence, the TF is

$$G_a = \frac{Z_C}{Z_R + Z_C} = \frac{\frac{1}{Cs}}{R + \frac{1}{Cs}} = \frac{1}{RCs + 1} \tag{iii}$$

b. Figure 6.9d shows the impedance circuit of the cascade circuit. The impedances in it are

$$\text{For } R_1 : Z_{R1} = R_1 \tag{iv}$$

$$\text{For } R_2 : Z_{R2} = R_2 \tag{v}$$

$$\text{For } C_1 : Z_{C1} = \frac{1}{C_1 s} \tag{vi}$$

$$\text{For } C_2 : Z_{C2} = \frac{1}{C_2 s} \tag{vii}$$

The combined impedance of the circuit modules beyond Node A (a series connection then a parallel connection)

$$Z_A = \frac{Z_{C1} \times (Z_{R2} + Z_{C2})}{Z_{C1} + Z_{R2} + Z_{C2}} \text{ or } \frac{1}{Z_A} = \frac{1}{Z_{C1}} + \frac{1}{Z_{R2} + Z_{C2}} = \frac{1}{Z_{C1}} + \frac{1/Z_{C2}}{Z_{R2}/Z_{C2} + 1}$$

$$\text{Or, } \frac{1}{Z_A} = C_1 s + \frac{C_2 s}{R_2 C_2 s + 1} \tag{viii}$$

By voltage division in a series branch, we get the voltage at Node A

$$v_A = \frac{Z_A}{Z_{R1} + Z_A} v_i \tag{ix}$$

Again by voltage division in a series branch, and substitution of (ix), we get the voltage at Node B (which is v_o),

$$v_o = \frac{Z_{C2}}{Z_{R2} + Z_{C2}} v_A = \frac{Z_{C2}}{(Z_{R2} + Z_{C2})} \frac{Z_A}{(Z_{R1} + Z_A)} v_i = \frac{1}{(Z_{R2}/Z_{C2} + 1)} \frac{1}{(Z_{R1}/Z_A + 1)} v_i$$

Substitute (viii)

$$v_o = \frac{1}{(R_2 C_2 s + 1)} \frac{1}{\left(R_1 \left(C_1 s + \frac{C_2 s}{R_2 C_2 s + 1}\right) + 1\right)} v_i$$

Hence, the TF is

$$G_b = \frac{1}{(R_2 C_2 s + 1)} \frac{1}{\left(R_1 \left(C_1 s + \frac{C_2 s}{R_2 C_2 s + 1}\right) + 1\right)} = \frac{1}{(R_2 C_2 s + 1)} \frac{1}{\left(R_1 C_1 s + \frac{R_1 C_2 s}{R_2 C_2 s + 1} + 1\right)}$$

Certainly, this is not equal to $\frac{1}{(R_1 C_1 s + 1)} \frac{1}{(R_2 C_2 s + 1)}$, which would have been the case if the two cascaded modules were independent. The reason is as follows. In the original circuit module, the output is in open circuit, and the output current is zero. When the second circuit module is connected to the first circuit module, the latter is no longer in open circuit, and an output current is generated. This is called a *loading effect*. The "coupling" term $\frac{R_1 C_2 s}{R_2 C_2 s + 1}$ was caused by this.

6.6 TRANSMISSIBILITY FUNCTION

Transmissibility is another TF that is quite versatile in mechanical systems. Transmissibility functions are TFs that are particularly useful in the design and analysis of fixtures, suspensions of vehicles, and mounts and support structures for machinery and other engineering systems (with moving parts and inertia, flexibility, and damping characteristics). They are directly applicable in the studies of vibration isolation (particularly, in relation to engine mounts) and vehicle suspension design. Two types of transmissibility functions—force transmissibility and motion transmissibility—can be defined. Due to a reciprocity (or dual or complementary) characteristic of linear systems, it can be shown that these two TFs are identical and, consequently, it is sufficient to analyze only one of them. First we consider both types of transmissibility functions and show their equivalence.

6.6.1 Force Transmissibility

Consider a mechanical system supported on a rigid foundation through a suspension system (see Figure 6.10a). If a forcing excitation is applied to the system (this can be a force generated within the machine as well, during its operation), it is not directly transmitted to the foundation. The suspension system acts as an "isolation" device. The force transmissibility determines the fraction of the forcing excitation that is directly transmitted to the foundation through the suspension system, as a function of the excitation (force) frequency, and is defined as

$$\text{Force Transmissibility } T_f = \frac{\text{Suspension Force } F_s}{\text{Applied Force } F} \tag{6.38}$$

Note: This function is defined in the frequency domain, and accordingly, F_s and F should be interpreted as the Fourier spectra of the corresponding forces. This is a non-dimensional TF.

A schematic diagram of a force transmissibility mechanism is shown in Figure 6.10a. The reason that the suspension force F_s is not equal to the applied force F is attributed to the inertia force path (broken line in Figure 6.10a) that is present in the mechanical system.

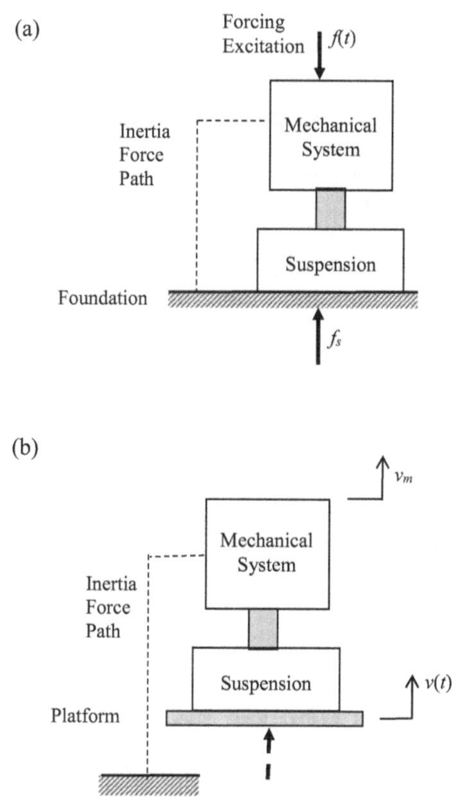

FIGURE 6.10 (a) Force transmissibility mechanism; (b) motion transmissibility mechanism.

6.6.2 Motion Transmissibility

Consider a mechanical system supported through a suspension mechanism on a movable platform or structure, which may be subjected to deliberate or undesirable motions (e.g., elevator movement, seismic disturbances, road irregularities effecting the tires of a vehicle, machinery disturbances). Motion transmissibility determines the fraction of the support motion that is transmitted to the system through its suspension, as a function of the excitation (platform velocity) frequency. It is defined as

$$\text{Motion Transmissibility } T_m = \frac{\text{System Motion } V_m}{\text{Support Motion } V} \tag{6.39}$$

Note: The velocities V_m and V are expressed in the frequency domain, as Fourier spectra. This is also a non-dimensional TF.

A schematic representation of the motion transmissibility mechanism is shown in Figure 6.10b. The platform force, indicated by a broken-line arrow, represents the dependent force of the velocity source. Also, the broken line from the mechanical system to the foundation (ground) is the path of the inertia force. Typically, the representative motion of the system is the velocity of one of its critical masses. Different transmissibility functions are obtained when different mass points (or degrees of freedom) of the system are considered.

Next, two single-degree-of-freedom examples are given to show the reciprocity property, which makes the force transmissibility and the motion transmissibility functions equivalent.

Example 6.10

Consider the single-degree-of-freedom systems shown in Figure 6.11. In these examples, the system is represented by a lumped mass *m*, and the suspension system is modeled as a spring of stiffness *k* and a viscous damper of damping constant *b*. The model shown in Figure 6.11a is used to study force transmissibility. Its (mechanical) impedance circuit is shown in Figure 6.12a. The model shown in Figure 6.11b is used in determining the motion transmissibility. Its mechanical impedance (or, mobility) circuit is shown in Figure 6.12b.

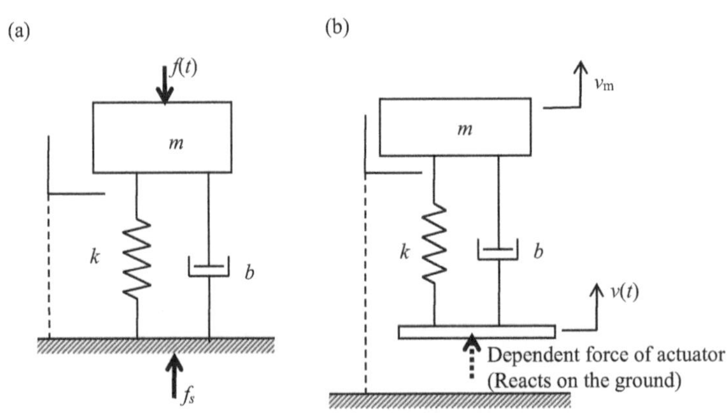

FIGURE 6.11 Single-degree-of-freedom systems. (a) Fixed on ground; (b) with support (platform) motion.

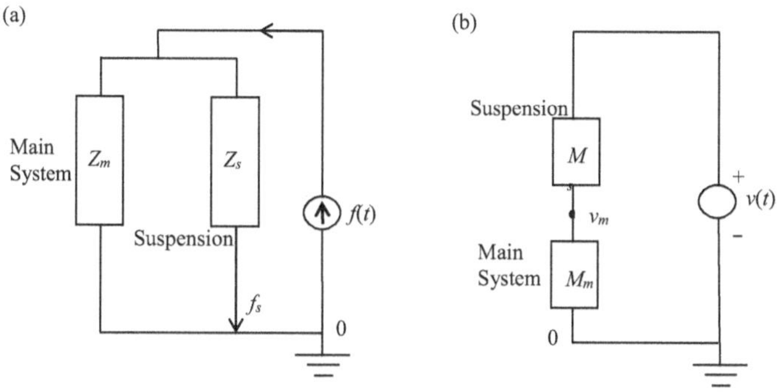

FIGURE 6.12 Mechanical impedance circuit of: (a) System in Figure 6.9(a); (b) system in Figure 6.9(b).

Note: Mechanical impedance representation is suitable for force transmissibility studies, and mobility representation is suitable for motion transmissibility studies (even though either representation is acceptable in analysis).

By the definition of mechanical impedance (force/velocity), a force is divided among parallel branches in proportion to their impedances. Hence, in Figure 6.112a, we have

$$T_f = \frac{F_s}{F} = \frac{Z_s}{Z_m + Z_s} \tag{6.40}$$

By the definition of mobility (velocity/force), a velocity is divided among series elements in proportion to their mobilities. Hence, in Figure 6.12b, we have

$$T_m = \frac{V_m}{V} = \frac{M_m}{M_m + M_s} \tag{6.41}$$

But $\dfrac{M_m}{M_m + M_s} = \dfrac{1/Z_m}{1/Z_m + 1/Z_s} = \dfrac{Z_s}{Z_m + Z_s}$

Consequently, we have

$$T_f = T_m \tag{6.42}$$

In view of this result, a distinction between the two types of transmissibility is not particularly necessary. Hence, let us denote them by a common *transmissibility function T*. It can be concluded as well that Figure 6.11a and b are *complementary systems* or *duals*, for transmissibility.

Since $Z_m = ms$ and $Z_s = \dfrac{k}{s} + b$, it follows that

$$T = \left[\frac{bs + k}{ms^2 + bs + k}\right]_{s=j\omega} \tag{6.43}$$

This result is the same as what we obtained in the previous section as equations (6.33) and (6.37).

It is customary to consider only the magnitude of the complex transmissibility function because the phase angle is not particularly useful in transmissibility studies, where what is key is the relative "strength" of the transmitted signal. This magnitude is termed *transmissibility magnitude* or simply *transmissibility* and is given by

$$T = \left[\frac{\omega^2 b^2 + k^2}{\omega^2 b^2 + \left(k - \omega^2 m^2\right)^2}\right]^{1/2} \tag{6.44}$$

LEARNING OBJECTIVES

1. Determination of the *force transmissibility* of a system using the element TFs and the system structure (series or parallel connection)
2. Determination of the *motion transmissibility* of a system using the element TFs and the system structure (series or parallel connection)
3. Recognizing the equivalence of force transmissibility of a system and the motion transmissibility of the *dual* or the *complementary* or the *reciprocal* system.

Note: Even though the duality of force transmissibility and motion transmissibility is illustrated here using single-degree-of-freedom examples, this characteristic is true for systems with higher degrees of freedom as well. This is illustrated in the end-of-chapter problems.

Summarizing the basics of the presented concepts, we may state the following:

For a structurally complex mechanical system, it is advantageous to consider the analogous electrical-circuit procedures when analyzing the mechanical system (particularly in transmissibility analysis). This is true because generally we are very familiar with the analysis of electrical circuits but not mechanical circuits. Further complexity enters since mechanical impedance is not analogous to electrical impedance. It is the mechanical mobility that is analogous to electrical impedance.

In analyzing a mechanical circuit, a convenient approach is to use the following analogies and then treat the mechanical circuit as an electrical circuit:

1. **Force (or torque) → current**
2. **Velocity → voltage**
3. In **series-connected mechanical (electrical) elements**, velocities (voltages) add → mobilities (electrical impedances) add $(M_e = M_1 + M_2)$; mechanical impedances (electrical admittances) have the inverse relation $\left(Z_e = \dfrac{Z_1 Z_2}{Z_1 + Z_2}\right)$
4. In **parallel-connected mechanical (electrical) elements**, forces (currents) add → mechanical impedances (electrical admittances) add $(Z_e = Z_1 + Z_2)$; mobilities (electrical impedances) have the inverse relation $\left(M_e = \dfrac{M_1 M_2}{M_1 + M_2}\right)$

6.6.3 Vibration Isolation

The concepts of mechanical impedance, mobility, and transmissibility can be extended to other physical domains and to mixed systems (e.g., electromechanical systems, fluid systems, thermal systems, multi-physics systems, or mechatronic systems in general) in a straightforward manner. The procedure follows from the familiar analogies of through-variables (force, current, fluid flow rate, and heat transfer rate) and across-variables (velocity, voltage, pressure, and temperature) in the "unified" approach. Furthermore, the concepts of transmissibility are useful in many practical applications. In particular, transmissibility is applicable in shock and vibration isolation.

Proper operation of engineering systems such as delicate instruments, computer hardware, machine tools, and vehicles can be hampered due to shock and vibration. The purpose of vibration isolation is to "isolate" such devices from vibration and shock disturbances that are coming from its environment (including the supporting structure or road). This is achieved by connecting a *vibration isolator* or *shock mount* or *engine mount* or *suspension system* in between them. This application is discussed now.

6.6.3.1 Force Isolation and Motion Isolation

External disturbance can be force or motion, and depending on that, force isolation (related to force transmissibility) or motion isolation (related to motion transmissibility) would be applicable in the design of the isolator. As we showed in the previous section, the design is quite similar (complementary) for the two situations.

In force isolation, vibration forces that would be ordinarily transmitted directly from a source to a supporting structure (isolated system) are filtered out by an isolator through its flexibility (spring) and dissipation (damping), and part of the force is routed through an inertial path. In motion isolation, vibration motions that are applied to a system (e.g., vehicle, elevator) by a moving platform are absorbed by an isolator through its flexibility and dissipation so that the motion that is transmitted to the system of interest is weakened. The design problem in both cases is to select applicable parameters for the isolator so that the vibrations entering the system of interest are below the specified values within a frequency band of interest (the operating frequency range). This design problem is essentially a situation of "mechanical impedance matching" because impedance parameters (mechanical) of the isolator are chosen depending on the impedance parameters of the system.

Consider again the force-transmissibility model shown in Figure 6.11a and the motion transmissibility model shown in Figure 6.11b. In these circuits, the mechanical impedances of the basic elements are $Z_m = mj\omega$, $Z_b = b$, and $Z_k = \dfrac{k}{j\omega}$, for mass (m), spring (k), and viscous damper (b), respectively. The suspension system (engine mount) is the parallel spring-damper element, with the mechanical impedance $Z_s = Z_b + Z_k$. We know that the expressions for the force transmissibility and the motion transmissibility are identical and given by equation (6.43), and the transmissibility magnitude is given by equations (6.44). Then, with $\dfrac{k}{m} = \omega_n^2$ and $\dfrac{b}{m} = 2\zeta\omega_n$ (or $\omega_n = \sqrt{\dfrac{k}{m}}$ =undamped natural frequency of the system; $\zeta = \dfrac{b}{2\sqrt{km}}$=damping ratio of the system) and dividing (6.44) throughout by ω_n^2, we get the non-dimensional (normalized) result

$$T = \sqrt{\frac{1+4\zeta^2 r^2}{\left(1-r^2\right)^2 + 4\zeta^2 r^2}} \tag{6.45}$$

Here the non-dimensional excitation frequency (or frequency ratio) is defined as $r = \omega/\omega_n$.

Note: The transmissibility function (6.43) has a phase angle as well as a magnitude. In practical applications of vibration isolation, it is the level of attenuation of the vibration excitation that is of primary importance, rather than the phase difference between the vibration excitation and the response. Accordingly, the transmissibility magnitude given by (6.44) and (6.45) is particularly useful.

To determine the peak point of T, differentiate the expression within the square-root sign in (6.45) and equate to zero. The result simplifies to $r\left(2\zeta^2 r^4 + r^2 - 1\right) = 0$. Its roots are $r = 0$ and $r^2 = \dfrac{-1 \pm \sqrt{1+8\zeta^2}}{4\zeta^2}$.

The root $r = 0$ corresponds to the initial stationary state, at zero frequency. It does not represent a peak. Taking only the positive root for r^2 and then its positive square-root, the peak point of the transmissibility magnitude is given by

$$r = \frac{\left[\sqrt{1+8\zeta^2} - 1\right]^{1/2}}{2\zeta} \tag{6.46}$$

For small ζ (i.e., low damping, which is typical in practice), Taylor series expansion gives $\sqrt{1+8\zeta^2} \approx 1 + \dfrac{1}{2} \times 8\zeta^2 = 1 + 4\zeta^2$. With this approximation, (6.46) evaluates to 1. Hence, for low damping, the transmissibility magnitude has a peak at $r=1$, and from Equation (6.45), its value is

$$T \approx \frac{\sqrt{1+4\zeta^2}}{2\zeta} \approx \frac{1+\frac{1}{2}\times 4\zeta^2}{2\zeta} \text{ or}$$

$$T \approx \frac{1}{2\zeta} + \zeta \approx \frac{1}{2\zeta} \tag{6.47}$$

The five curves of T versus r for $\zeta=0$, 0.3, 0.7, 1.0, and 2.0 are shown in Figure 6.13a. These curves use the exact expression (6.45) and were generated using the following MATLAB program:

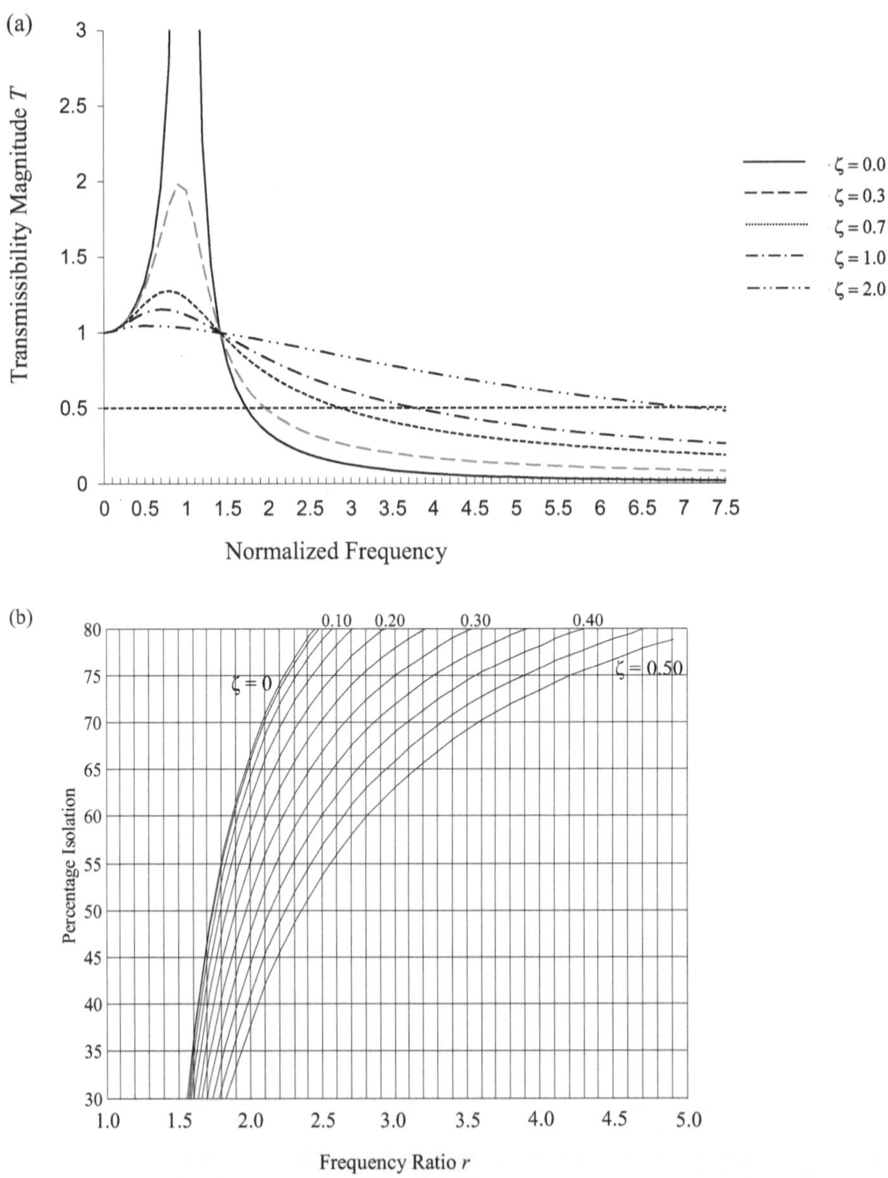

FIGURE 6.13 (a) Transmissibility curves for a damped oscillator model; (b) curves of vibration isolation.

plot(r, T(:, 1), r, T(:, 2), r, T(:, 3), r, T(:, 4), r, T(:, 5));

```
clear;
zeta=[0.0 0.3 0.7 1.0 2.0];
for j=1:5
for i=1:1201
r(i)=(i-1)/200;
T(i, j)=sqrt((1+4*zeta(j)^2*r(i)^2)/((1-r(i)^2)^2+4*zeta(j)^2*r(i)^2));
end
```

From the transmissibility curves in Figure 6.13a, we observe the following:

1. There always exists a non-zero frequency value at which the transmissibility magnitude peaks. This is the resonance (resonant point).
2. For low damping (i.e., small ζ), the peak transmissibility magnitude occurs approximately at $r=1$. As ζ increases, this peak point shifts to the left (i.e., a lower value for peak frequency).
3. The peak magnitude decreases as ζ increases.
4. All transmissibility curves pass through the magnitude value 1.0 at the same frequency $r=\sqrt{2}$.
5. The isolation region (i.e., $T<1$) is given by $r>\sqrt{2}$. In this region, T increases with ζ.
6. In the isolation region, the transmissibility magnitude decreases as r increases.

From the transmissibility curves, we observe the following two specific situations:

For $T<1.0$, $r>\sqrt{2}$ for all ζ
For $T<0.5$, $r>[1.73, 1.964, 2.871, 3.77, 7.075]$, which correspond to $\zeta=[0.0, 0.3, 0.7, 1.0, 2.0]$, respectively.

As a specific example, suppose that the device in Figure 6.11a has a primary, undamped natural frequency of 6 Hz and a damping ratio of 0.2. Suppose that for proper operation, it is required for the system to achieve a force transmissibility magnitude of less than 0.5 with the operating frequency values greater than 12 Hz.

Specifically, we need $\sqrt{\dfrac{1+4\zeta^2r^2}{\left(1-r^2\right)^2+4\zeta^2r^2}}<\dfrac{1}{2}\rightarrow r^4-2r^2-12\zeta^2r^2-3>0$. For $\zeta=0.2$ and $r=12/6=2$, this expression computes to $3.08>0$. Hence, the requirement is met. In fact, for $r=2$, the last expression becomes $2^4-2\times2^2-12\times2^2\zeta^2-3=5-48\zeta^2$. It follows that the requirement would be met for $5-48\zeta^2>0\rightarrow\zeta<0.32$. If the requirement was not met (say, if $\zeta=0.4$), an option would be to reduce damping.

In design problems of vibration isolator, what is normally specified is the *percentage isolation*, which is directly related to the transmissibility, according to

$$I=[1-T]\times100\% \tag{6.48}$$

From Equation (6.45), this corresponds to

$$I=\left[1-\sqrt{\frac{1+4\zeta^2r^2}{\left(r^2-1\right)^2+4\zeta^2r^2}}\right]\times100 \tag{6.49}$$

The isolation curves given by Equation (6.49) are plotted in Figure 6.13b. These curves are useful in the design of vibration isolators.

Note: The models in Figure 6.11 are not limited to sinusoidal vibrations. Any general vibration excitation may be represented by a Fourier spectrum, which is a function of frequency ω. Then, the response vibration spectrum is obtained by multiplying the excitation spectrum by the transmissibility function T. The associated design problem is to select the isolator impedance parameters k and b to meet the specifications of isolation.

Example 6.11

A machine tool, sketched in Figure 6.14a, weighs 1000 kg and normally operates in the speed range of 300–1200 rpm. A set of spring mounts has to be placed beneath the base of the machine so as to achieve a vibration isolation level of at least 70%. A commercially available spring mount has the load-deflection characteristic shown in Figure 6.14b. It is recommended that an appropriate

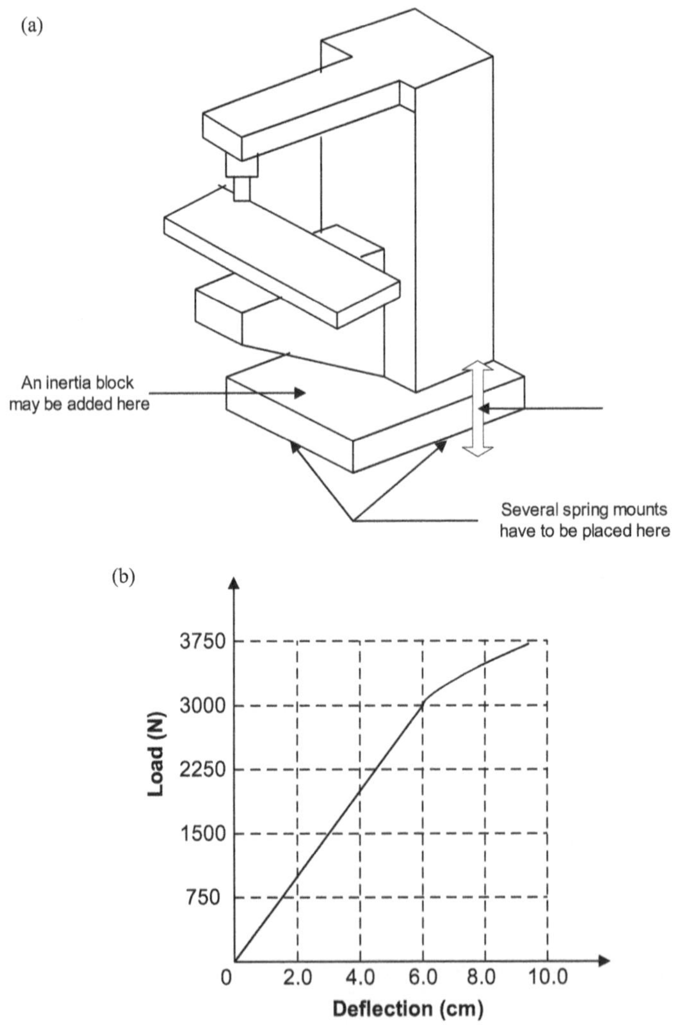

FIGURE 6.14 (a) A machine tool; (b) load-deflection characteristic of a spring mount.

number of these mounts be used, along with an inertia block, if necessary. The damping constant of each mount is 1.56×10^3 $Nm^{-1}s$. Design a vibration isolation system for the machine. Specifically, decide upon the number of spring mounts that are needed and the mass of the inertia block that should be added.

Solution

First we will assume zero damping (since, in practice, the level of damping in a system of this type is small) and design an isolator (spring mount and inertia block) for a level of isolation somewhat greater than the required 70%. Then we will check for the case of damped isolator to see whether the required 70% level of isolation is achieved.

For the undamped case, Equation (6.48) becomes

$$T = \frac{1}{r^2 - 1} \tag{6.50}$$

Note: We have used the case $r>1$ since the isolation region corresponds to $r > \sqrt{2}$.

Assume the conservative value $I=80\% \Rightarrow T=0.2$.

Using equation (6.50), we have $r^2 = \frac{1}{T} + 1 = \frac{1}{0.2} + 1 = 6.0 = \frac{\omega^2}{\omega_n^2} = \frac{m\omega^2}{k}$

The lowest operating speed (frequency) is the most significant one (because it corresponds to the lowest isolation, as clear from Figure 6.13a). Hence,

$$\omega = \frac{300}{60} \times 2\pi = 10\pi \text{ rad/s}$$

From the load-deflection curve of a spring mount (Figure 6.14b),

$$\text{Mount stiffness} = \frac{3000}{6\times10^{-2}} = 50\times10^3 \text{ N/m}$$

We will try 4 mounts. Then $k = 4\times50\times10^3$ N/m

Hence, $\frac{m\times(10\pi)^2}{4\times50\times10^{-3}} = 6.0 \rightarrow m=1.216\times10^3$ kg

It is seen that an inertia block of mass 216 kg has to be added for proper isolation of vibrations.

Now we must check whether the required level of vibration isolation would be achieved in the damped case.

$$\text{Damping ratio } \zeta = \frac{b}{2\sqrt{km}} = \frac{4\times1.56\times10^3}{2\sqrt{4\times50\times10^3\times1.216\times10^3}} = 0.2$$

Substitute in the damped isolator equation (6.45): $T = \sqrt{\frac{1+4\zeta^2r^2}{(r^2-1)^2+4\zeta^2r^2}}$ with $r^2=6$

We get $T = 0.27$. This corresponds to an isolation level of 73%, which is better than the required 70%.

LEARNING OBJECTIVES

1. Application of the concepts of transmissibility in the design of vibration isolators or engine mounts
2. Meanings of transmissibility and isolation
3. The use of the transmissibility curves and isolation curves
4. Practical selection and design of vibration isolators or engine mounts.

6.6.4 Maxwell's Reciprocity Property

For our purposes, Maxwell's property of reciprocity may be stated as follows: In a mechanical dynamic system suppose that, when a force f is applied along the coordinate (degree of freedom) i, the resulting velocity along the coordinate (degree of freedom) j is v. Then, if f is applied along the coordinate j, the resulting velocity along the coordinate i will also be v.

This property is valid for linear, constant-parameter systems in general. Also, for static systems, we can replace "velocity" in the above statement by "displacement." This property can be quite useful in practice, for example, in dynamic testing of complex mechanical systems, to determine a behavior that is difficult to measure. Then, by applying the force at that location and measuring response at the counterpart location, we can obtain the required information.

6.6.4.1 Maxwell's Reciprocity Property in Other Domains

It should be clear that Maxwell's reciprocity property is not limited to the mechanical domain. In view of the fact that force is a through-variable and velocity is an across-variable, the reciprocity property may be extended to any linear dynamic system in terms of a corresponding through-variable and a corresponding across-variable. Then for any physical domain (electrical, fluid, thermal), the reciprocity may be applied using the proper variables and TFs for that domain.

Note: Mechanical impedance is a T-type TF and mobility is an A-type TF.

6.7 SUMMARY SHEET

- **Transfer Function (TF)**: $G(s) = \dfrac{\text{Laplace-transformed output } Y(s)}{\text{Laplace-transformed input } U(s)}$

  For a linear system, this is a numerator polynomial and a denominator polynomial (characteristic polynomial). Roots of numerator=0 → zeros of the system; roots of denominator=0 (i.e., the characteristic equation) → poles (i.e., eigenvalues) of the system

- **Laplace Transform (LT)**: $Y(s) = \int_0^\infty y(t)\exp(-st)\,dt$ or $Y(s) = \mathcal{L}y(t)$
- **Inverse LT**: $y(t) = \dfrac{1}{2\pi j}\int_{\sigma-j\infty}^{\sigma+j\infty} Y(s)\exp(st)\,ds$ or $y(t) = \mathcal{L}^{-1}Y(s)$

  Time domain ↔ Laplace (Complex Frequency) Domain
- **LT of Derivative (General)**: $\mathcal{L}\dfrac{d^n y(t)}{dt^n} = s^nY(s) - s^{n-1}y(0) - s^{n-2}\dot{y}(0) - \cdots - \dfrac{d^{n-1}y}{dt^{n-1}}(0)$
- **LT of Integral**: $\mathcal{L}\int_0^t y(\tau)\,d\tau = \dfrac{1}{s}Y(s)$

  Time derivative ↔ multiplication by Laplace variable s
  Differential equations ↔ algebraic equations (with polynomials in s)
  Time integration ↔ Multiplication by $1/s$

  nth order input–output (I-O) model; time domain ↔ Laplace domain:

$$a_n\frac{d^n y}{dt^n} + a_{n-1}\frac{d^{n-1}y}{dt^{n-1}} + \cdots + a_0 y = b_0 u + b_1\frac{du}{dt} + \cdots + b_m\frac{d^m u}{dt^m} \leftrightarrow \frac{Y(s)}{U(s)} = \frac{b_0 + b_1 s + \cdots + b_m s^m}{a_0 + a_1 s + \cdots + a_n s^n}$$

- **Fourier Transform (FT)**: $Y(j\omega) = \int_{-\infty}^{\infty} y(t)\exp(-j\omega t)\,dt = \mathcal{F}\,y(t)$;

  Note: ω (rad/s) $= 2\pi \times f$ (cycles/s or Hz)

- **Inverse FT**: $y(t) = \dfrac{1}{2\pi}\int_{-\infty}^{\infty} Y(j\omega)\exp(j\omega t)\,d\omega$ or $y(t) = \mathcal{F}^{-1}\,Y(j\omega)$

Laplace result $\underset{\leftarrow j\omega=s}{\overset{s=j\omega\rightarrow}{}}$ Fourier result (one-sided); (e.g., $G(s) \underset{\leftarrow j\omega=s}{\overset{s=j\omega\rightarrow}{}} G(j\omega)$)

- **Magnitude of $G(j\omega)$ (called "Gain")**: $|G(j\omega)| = M$
- **Phase Angle of (called "Phase")** $G(j\omega)$>: $\angle G(j\omega) = \phi$

$$G(j\omega) = M\cos\phi + jM\sin\phi = Me^{j\phi};\ y = u_o Me^{j(\omega t+\phi)}$$

Note: Output is magnified by $M = |G(j\omega)|$; has a *phase lead* wrt input of $\phi = \angle G(j\omega)$ (Typically, for a physically realizable dynamic system provide a *phase lag*—negative phase lead)

- **Bode Plot (curve pair)**: 1. $|G(j\omega)|$ versus ω and 2. $\angle\ G(f)$versus ω
  Log scales are used in: Magnitude axis (e.g., in decibels or $20\log_{10}()$), frequency axis (e.g., in decades, which are multiples of 10; or octaves, which are multiples of 2)

Transfer Function	Definition (in Laplace or Frequency Domain)
Dynamic stiffness	Force/displacement
Receptance (dynamic flexibility or compliance)	Displacement/force
Mechanical impedance (Z)	Force/velocity
Mobility (M)	Velocity/force
Dynamic inertia	Force/acceleration
Accelerance	Acceleration/force
Force transmissibility (T_f)	Transmitted force/applied force
Motion transmissibility (T_m)	Transmitted velocity/applied velocity

- **In Frequency Domain**: Acceleration $=(j\omega)\times$(Velocity);
  Displacement = Velocity/$(j\omega)$ →

$$\text{Dynamic Stiffness} = \text{Force / Displacement} = \text{Mechanical Impedance} \times j\omega$$

$$\text{Receptance} = \text{Displacement / Force} = \text{Mobility} / (j\omega)$$

$$\text{Dynamic Inertia} = \text{Force / Acceleration} = \text{Impedance} / (j\omega)$$

$$\text{Accelerance} = \text{Acceleration / Force} = \text{Mobility} \times j\omega$$

- **Series Connection**: Through-variable is common, across-variables add →

$$\text{Mobility} = \frac{\text{Velocity}}{\text{Force}}:\ M = M_1 + M_2;\ \text{mech impedance: } \frac{1}{Z} = \frac{1}{Z_1} + \frac{1}{Z_2} \text{ or } Z = \frac{Z_1 Z_2}{Z_1 + Z_2}$$

- **Parallel Connection**: Across-variable is common, through-variables add →

$$\text{Mech impedance} = \frac{\text{Force}}{\text{Velocity}}:\ Z = Z_1 + Z_2;\ \text{mobility: } \frac{1}{M} = \frac{1}{M_1} + \frac{1}{M_2} \text{ or } M = \frac{M_1 M_2}{M_1 + M_2}$$

$$\text{Generalized Impedance} = \frac{\text{Across-variable}}{\text{Through-variable}}: \text{an } A\text{-type transfer function}$$

$$\text{Generalized Admittance} = \frac{\text{Through-variable}}{\text{Across-variable}}: \text{a } T\text{-type transfer function}$$

Element	Time-Domain Model	Mechanical Impedance	Mobility (Generalized Impedance)
Mass m	$m\frac{dv}{dt} = f$	$Z_m = ms$	$M_m = \frac{1}{ms}$
Spring k	$\frac{df}{dt} = kv$	$Z_k = \frac{k}{s}$	$M_k = \frac{s}{k}$
Damper b	$f = bv$	$Z_b = b$	$M_b = \frac{1}{b}$

Element	Time-Domain Model	Impedance (Z)	Admittance (W)
Capacitor C	$C\frac{dv}{dt} = i$	$Z_C = \frac{1}{Cs}$	$W_C = Cs$
Inductor L	$L\frac{di}{dt} = v$	$Z_L = Ls$	$W_L = \frac{1}{Ls}$
Resistor R	$Ri = v$	$Z_R = R$	$W_R = \frac{1}{R}$

$$\text{Force Transmissibility } T_f = \frac{\text{Suspension Force } F_s}{\text{Applied Force } F}$$

$$\text{Motion Transmissibility } T_m = \frac{\text{System Motion } V_m}{\text{Support Motion } V}$$

$T_f = T_m$ for complementary (dual, reciprocal) systems
For example, ground-based damped oscillator and oscillator with moving base:

$$T_f = T_m = \left[\frac{bs + k}{ms^2 + bs + k}\right]_{s=j\omega}$$

$$\text{Magnitude } T = \left[\frac{\omega^2 b^2 + k^2}{\omega^2 b^2 + \left(k - \omega^2 m^2\right)^2}\right]^{1/2} = \sqrt{\frac{1 + 4\zeta^2 r^2}{\left(1 - r^2\right)^2 + 4\zeta^2 r^2}}$$

- **Percentage Isolation**: $I = [1 - T] \times 100\% = \left[1 - \sqrt{\frac{1 + 4\zeta^2 r^2}{\left(r^2 - 1\right)^2 + 4\zeta^2 r^2}}\right] \times 100$
- **Maxwell's Reciprocity Property**: Force f is applied along coordinate (dof) i → velocity along j is v → if f is applied along j → velocity along i is also v
  Note: Applicable to other physical domains as well.

PROBLEMS

6.1 State whether true (T) or false (F):

a. The output of a system will depend on the input.
b. The output of a system will depend on the "analytical" transfer function of a linear system.
c. The "analytical" transfer function of a linear system will depend on the input signal.

d. If the Laplace transform of the input signal does not exist (say, infinite), then the transfer function itself does not exist.
e. If the Laplace transform of the output signal does not exist, then the transfer function itself does not exist.

6.2 State whether true (T) or false (F):

a. An analytical transfer function provides an algebraic expression for a system.
b. The Laplace variable s can be interpreted as the time-derivative operator d/dt, assuming zero ICs.
c. The variable $1/s$ may be interpreted as the integration of a signal starting at $t=0$.
d. The numerator of an analytical transfer function is the characteristic polynomial.
e. A SISO, linear, time-invariant (constant-parameter) system has a unique (one and only one) analytical transfer function.

6.3 Consider a mass-spring-damper system with a displacement input $u(t)$ applied to the free end (moving platform) of the spring, as shown in Figure P6.3. *Note*: This will require a "displacement actuator." The "dependent force" of the actuator will react on the ground.

The resulting displacement y of the mass is the output. Also,

m=point mass
b=viscous damping constant
k=stiffness of the spring.

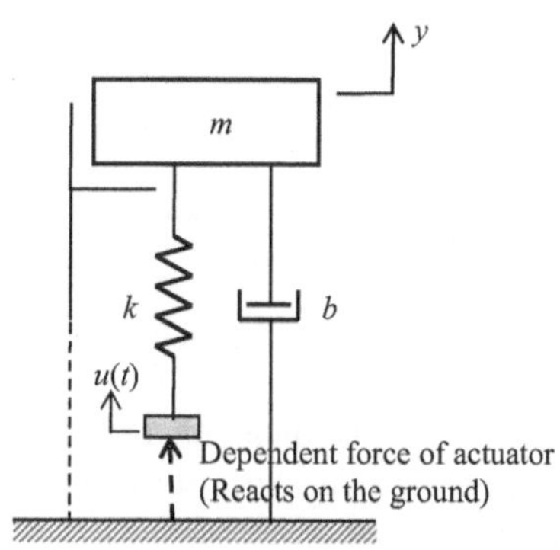

FIGURE P6.3 A mas-spring-damper system with a displacement input.

a. Formulate its input–output differential equation.
b. What is the transfer function of the system?
c. Show that this system is equivalent to a damped oscillator, where both the other ends of the spring and the damper are fixed to the ground, and a force (input) is applied at the mass.
d. How will the system transfer function change if the damper is also attached to the moving platform in Figure P6.3 (with a displacement input $u(t)$ applied to the platform) while the output is still the displacement y of the mass?

6.4 The Fourier transform of a measured position $y(t)$ of an object is $Y(j\omega)$. Select the correct one among the following statements:

i. The Fourier transform of the corresponding velocity signal is
   a. $Y(j\omega)$
   b. $j\omega Y(j\omega)$
   c. $Y(j\omega)/j\omega$
   d. $\omega Y(j\omega)$

ii. The Fourier transform of the acceleration signal is
   a. $Y(j\omega)$
   b. $\omega^2 Y(j\omega)$
   c. $-\omega^2 Y(j\omega)$
   d. $Y(j\omega)/j\omega$

6.5 The movable arm with read/write head of a disk drive unit is modeled as a rotatory damped oscillator (a lumped-parameter model), as shown in Figure P6.5. The unit has an equivalent moment of inertia $J = 1\times 10^{-3}$ g·cm² about its centroid and rotates at an equivalent angle θ radians about the centroid. An equivalent rotation $u(t)$ radians is imparted at the read/write (R/W) head. The bending stiffness at the R/W head with respect to the centroid of the arm is k=10 dyne·cm/rad. Similarly, the bending damping constant at the R/W head with respect to the centroid of the arm is b.

a. Write the input–output differential equation of motion for the read/write arm unit. What are the consistent units for b?
b. What is the undamped natural frequency ω_n of the unit in rad/s?
c. Determine the value of b for 5% of critical damping.
d. Write the FTF of the model.

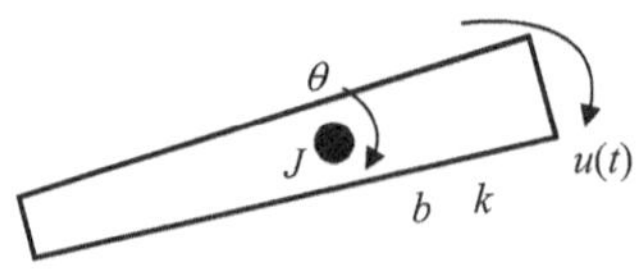

FIGURE P6.5 A damped rotatory oscillator model of a disk-drive arm.

6.6 A system has the transfer function $G(s) = \dfrac{s+1}{\left(s^2+s+4\right)}$. If a sinusoidal input given by $u = 3\cos 2t$ is applied to this system, determine the output (response) y under steady conditions.

6.7 A system is shown in Figure P6.7. It was found to have the following properties:

1. The system transfer function $G(s)$ has two zeros and three poles.
2. The product of the three poles is −4.
3. When the system was excited with a sinusoidal input u (as shown in Figure P6.7) at frequency $\omega = 4$, the output y at steady state was found to be zero (i.e., no response).
4. When the system was excited with a sinusoidal input u (as shown in Figure P6.7) at frequency $\omega = 2$, the output y at steady state was found to have a phase lag of 180° with respect to the input (i.e., the response was in the opposite direction to the input).
5. When the system was excited with a sinusoidal input u (as shown in Figure P6.7) at frequency $\omega = \sqrt{2}$, the output y at steady state was found to have a phase lag of 90° with respect to the input.
6. The DC gain of the system (i.e., the magnitude of the FTF at zero frequency) is 8.

Determine the complete transfer function $G(s)$ of the system (i.e., the numerical values of the five parameters in $G(s)$).

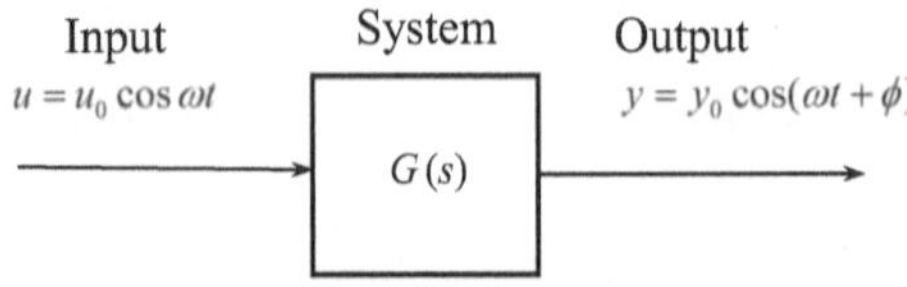

FIGURE P6.7 The system with a harmonic input.

6.8 A system was found to have the following properties:

1. It is a second-order system.
2. It has a zero at $s=-z$ where $z>0$.
3. Its DC gain (i.e., the magnitude of the FTF at zero frequency) is K.
4. When the system is excited at its undamped natural frequency, the magnitude of the FTF is given by rK, where $r>0$, and the phase angle is $-90°$.

In terms of the given parameters z, K, and r, determine the following:

a. Undamped natural frequency of the system.
b. Damping ratio of the system.
c. Complete transfer function of the system.

6.9 The FTF for a damped oscillator is given by

$$G(\omega) = \frac{\omega_n^2}{\left[\omega_n^2 - \omega^2 + 2j\zeta\omega_n\omega\right]}$$

a. If a harmonic excitation $u(t)=a\cos\omega_n t$ is applied to this system, what is the steady-state response?
b. What is the magnitude of the resonant peak?
c. Using your answers to parts (a) and (b) suggest a method to measure damping in a mechanical system.
d. At what excitation frequency does the response amplitude become maximum under steady-state conditions?
e. Giving details, determine an approximate expression for the half-power (3 dB) bandwidth (i.e., the frequency band where the power level is half the peak power, or the TF magnitude drops by $\sqrt{2}$ or 3 dB from the peak value) at low damping, assuming that the resonant frequency $\omega_r \approx \omega_n$, which is true for low damping. Using this result, suggest an alternative method for damping measurement.

6.10 Sketch the Bode magnitude plots (asymptotes only are sufficient when the exact curve cannot be determined without numerical computation) of the following common system elements:

a. Derivative controller: τs
b. Integral controller: $\dfrac{1}{\tau s}$
c. First-order simple lag network: $\dfrac{1}{\tau s+1}$
d. Proportional plus derivative controller: $\tau s+1$.

6.11 The transfer function of a dynamic system is given by

$$G(s) = \frac{(s+3)}{(s^2+4s+16)}$$

Plot the Bode diagram for G and indicate the asymptotes. Comment on the result.

6.12 A rotating machine of mass M is placed on a rigid concrete floor. There is an isolation pad made of elastomeric material between the machine and the floor and is modeled as a viscous damper of damping constant b. In steady operation, there is a predominant harmonic force component $f(t)$, which is acting on the machine in the vertical direction at a frequency equal to the speed of rotation (n rev/s) of the machine. To control the vibrations produced by this force, a dynamic absorber of mass m and stiffness k is mounted on the machine. A model of the system is shown in Figure P6.12.

a. Determine the FTF of the system, with force $f(t)$ as the input and the vertical velocity v of mass M as the output.
b. What is the mass of the dynamic absorber that should be used in order to virtually eliminate the machine vibration (which corresponds to a tuned absorber)?

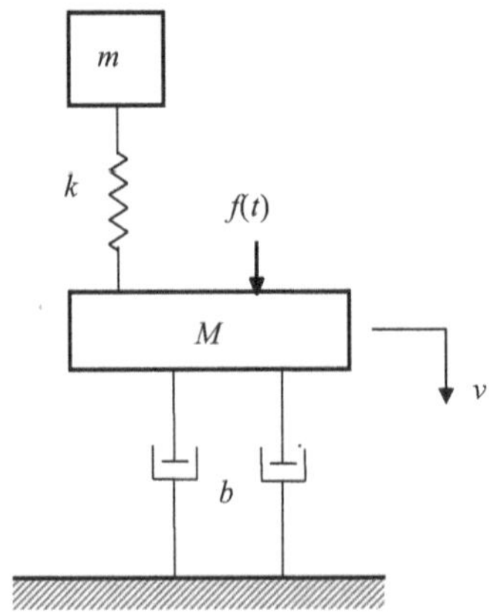

FIGURE P6.12 A mounted machine with a dynamic absorber.

6.13

a. An approximate FTF of a system was determined through Fourier analysis of the measured excitation-response data and fitting the result to an appropriate analytical expression (by curve fitting using the least-squares method). This was found to be $G(f) = \dfrac{5}{10 + j2\pi f}$. What are its magnitude, phase angle, real part, and imaginary part at $f=2\,\text{Hz}$? If the reference frequency is taken as 1 Hz, what is the transfer function magnitude at 2 Hz expressed in dB?
b. A dynamic test on a structure using a portable shaker revealed the following: The *accelerance* between two locations (shaker location and accelerometer location) measured at a frequency ratio of 10 was 35 dB. Determine the corresponding mobility and mechanical impedance at this frequency ratio.

6.14 Answer as true (T) or false (F):

i. Mechanical impedances are additive for two elements connected in parallel.
ii. Mobilities are additive for two elements connected in series.

6.15 Answer as true (T) or false (F):

a. Electrical impedances are additive for two elements connected in parallel.
b. Impedance, both mechanical and electrical, is given by the ratio of effort/flow, in the frequency domain.
c. Impedance, both mechanical and electrical, is given by the ratio of across-variable/through-variable, in the frequency domain.
d. Mechanical impedance is analogous to electrical impedance when determining the equivalent impedance of several interconnected impedances.
e. Mobility is analogous to electrical admittance (current/voltage in the frequency domain) when determining the equivalent value of several interconnected elements.

6.16 A machine of mass m has a rotating device, which generates a harmonic forcing excitation $f(t)$ in the vertical direction. The machine is mounted on the factory floor using a vibration isolator of stiffness k and damping constant b. The harmonic component of the force that is transmitted to the floor, due to the forcing excitation, is $f_s(t)$. A simplified model of the system is shown in Figure P6.16. The corresponding force transmissibility magnitude T

from f to f_s is given by $T = \sqrt{\dfrac{1+4\zeta^2 r^2}{(1-r^2)^2+4\zeta^2 r^2}}$, where $r = \omega/\omega_n$, ζ=damping ratio, ω_n = undamped natural frequency of the system, and ω = excitation frequency (of $f(t)$).

Suppose that $m = 100$ kg and $k = 1.0 \times 10^6$ N/m. Also, the frequency of the excitation force $f(t)$ in the operating range of the machine is known to be 200 rad/s or higher. Determine the damping constant b of the vibration isolator so that the force transmissibility magnitude is not more than 0.5.

Using MATLAB, plot the resulting transmissibility function and verify that the design requirements are met.

Note: 2.0=6 dB; $\sqrt{2}$=3 dB; $1/\sqrt{2} = -3$ dB; 0.5=−6 dB.

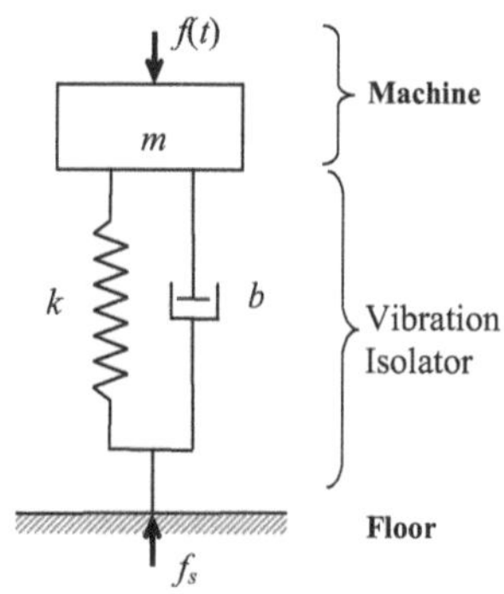

FIGURE P6.16 Model of a machine mounted on a vibration isolator.

6.17 Consider the simplified model of a vehicle shown in Figure P6.17, which can be used to study the heave (vertical up and down) and pitch (front-back rotation) motions due to the road profile and other disturbances. For our purposes, let us assume that the road disturbances that excite the front and back suspensions are independent. The equations of motion for heave (y) and pitch (θ) are written about the static equilibrium configuration of the vehicle model (hence, gravity does not enter into the equations) for small motions:

$$m\ddot{y} = k_1(u_1 - y + l_1\theta) + k_2(u_2 - y + l_2\theta) + b_1(\dot{u}_1 - \dot{y} + l_1\dot{\theta}) + b_2(\dot{u}_2 - \dot{y} + l_2\dot{\theta})$$

$$J\ddot{\theta} = -l_1\left[k_1(u_1 - y + l_1\theta) + b_1(\dot{u}_1 - \dot{y} + l_1\dot{\theta})\right] + l_2\left[k_2(u_2 - y + l_2\theta) + b_2(\dot{u}_2 - \dot{y} + l_2\dot{\theta})\right]$$

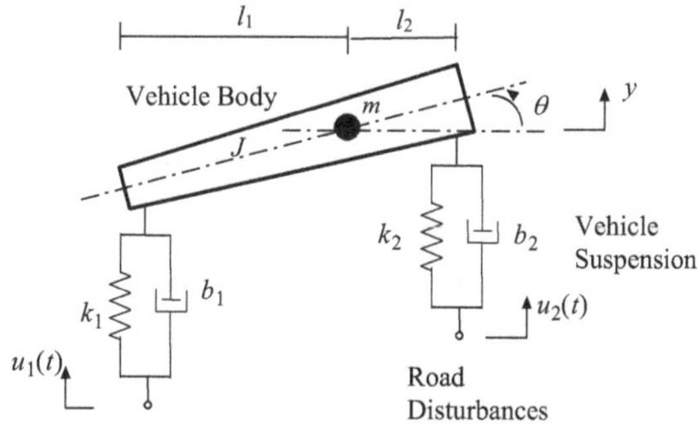

FIGURE P6.17 A model of a vehicle with its suspension system.

The outputs are y and θ. The inputs are the upward movements u_1 and u_2 at the tires. Determine the corresponding four TFs.

6.18 Consider the two-degree-of-freedom systems shown in Figure P6.18. The main system is represented by two masses linked through a spring and a damper. The model shown in Figure P6.18a is used to study force transmissibility. This model may represent a machine tool (e.g., milling machine) with a fixed base. Specifically, we are interested in determining what fraction of the applied force $f(t)$ (this may be the task force of the machine) is transmitted to the floor (i.e., the force transmissibility $\frac{f_s}{f}$). The model shown in Figure P6.18b is used in determining the motion transmissibility. This model may represent a vehicle or an elevator with a mobile base. Here we are interested in determining what fraction of the base/platform velocity $v(t)$ (this may be the upward velocity of the tires due to road irregularities, or the moving velocity of the elevator) is transmitted to a critical mass, such as the vehicle seat or the occupant, of the system (i.e., the motion transmissibility $\frac{v_m}{v}$).

a. Sketch the (mechanical) impedance circuit of the system in Figure P6.18a, and derive the corresponding force transmissibility. Also, sketch the (mechanical) impedance (or, mobility) circuit of the system shown in Figure P6.18b, and derive the corresponding motion transmissibility.
b. Prove that these two transmissibilities are equal.

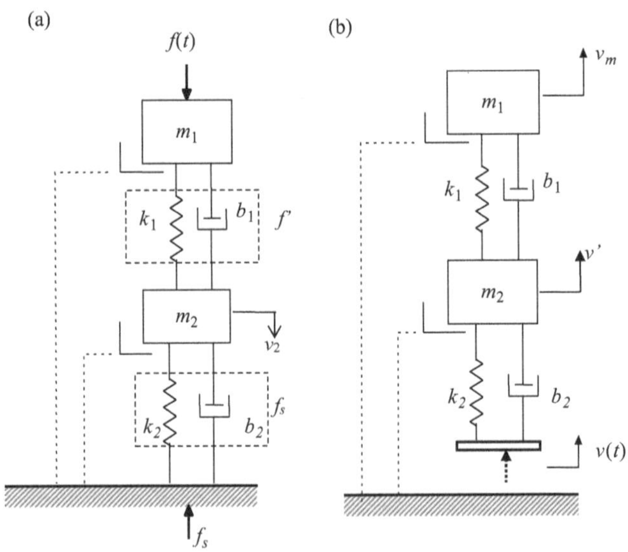

FIGURE P6.18 Systems with two degrees of freedom. (a) Fixed on ground; (b) with support motion.

6.19 A mechanical system is modeled as a linear mass-spring-damper (M-K-B) system together with a vibration absorber modeled as a mass-spring (m-k) system (see Figure P6.19). The input to the system is the vertical force $f(t)$, which is applied on the system mass M. The output of the system is the velocity v of the mass M.

a. Using the mechanical impedance elements:

$$Z_M = Ms;\ Z_K = \frac{K}{s};\ Z_B = B;\ Z_m = ms;\ \text{and } Z_k = \frac{k}{s}$$

and the input element (force source) $F(s)$, sketch the mechanical impedance circuit of the system. Using this mechanical circuit, determine the system transfer function $\dfrac{V(s)}{F(s)}$ in terms of the parameters M, B, K, m, k, and the Laplace variable s.

b. Showing the necessary steps to determine the excitation frequency (expressed in terms of the system parameters) at which the system mass M will remain stationary? *Note:* Neglect the weights of the masses (or assume that the displacements of the masses are measured from their static equilibrium positions).

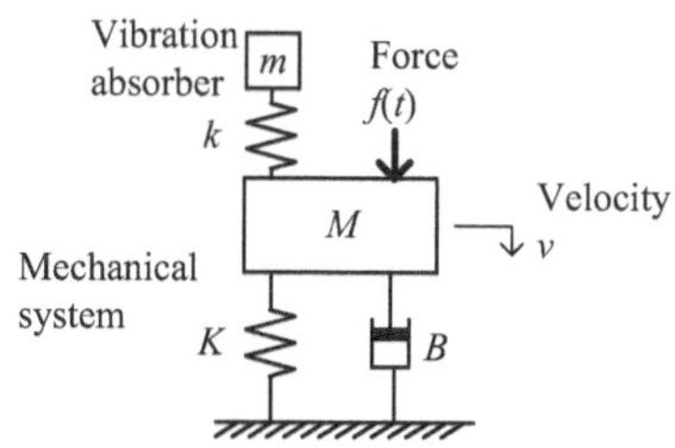

FIGURE P6.19 Model of a mechanical system with a vibration absorber.

6.20 In a music system, the speaker unit will typically have a woofer, which is intended for the lower-frequency sound, and a tweeter, which is intended for the high-frequency sound. Consider the two electrical circuits shown in Figure P6.20a and b. In each circuit, the sound signal reaches the speaker, which is represented as an electrical resistor (with resistance R_1 or R_2), which may be a woofer or a tweeter.

Give expressions of the electrical impedance (A-transfer function) of each circuit branch in Figure P6.20a and b, in terms of the branch parameter ($L_1, L_2, L, C, C_1, C_2, R_1, R_2$) and the frequency ω of the sound signal (the input).

Giving reasons, indicate which circuit corresponds to a woofer and which circuit corresponds to a tweeter.

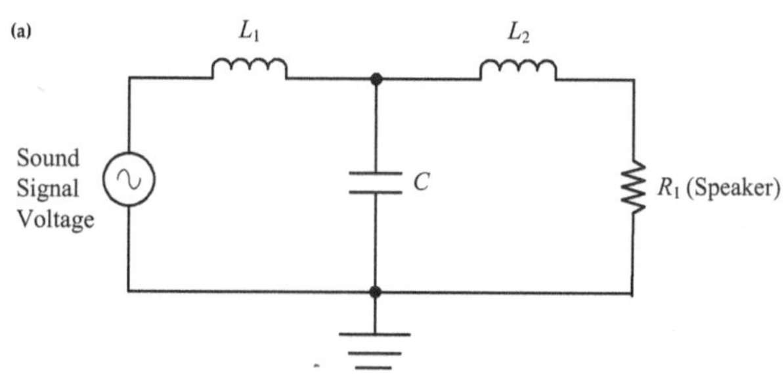

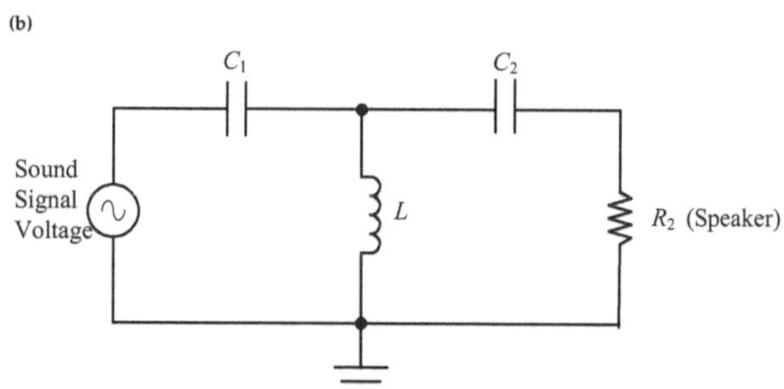

FIGURE P6.20 (a) and (b) Electrical circuits representing a speaker unit of a sound system.

6.21 A hydraulic motion-control system is schematically shown in Figure P6.21. The control input to the system is the volume flow rate $Q(t)$, which is provided by a flow-controlled pump. The hydraulic fluid from the pump enters the cylinder of a positioning ram. The piston (area $= A$) of the ram is light and frictionless and adjusts the motion of the load consisting of a mass m, which moves against a spring of stiffness k (the other end of the spring is fixed, as shown) and a viscous damper of damping constant b.

The motion output of the system is the velocity v of the mass m.

Assume that the fluid in the rigid cylinder is compressible, and the corresponding fluid capacitance is C_f.

a. Sketch a linear graph for the system, orient all its branches, and indicate all the system parameters, and the (through-, across-) variable pairs of the branches.
b. Showing all the steps, systematically, derive a complete state-space model for the system.
c. Determine the input–output models in the time domain (or, TFs in the Laplace domain, where the time derivative $\frac{d}{dt}$ can be interpreted as the Laplace operator s, see Chapter 6): $\frac{v}{Q}$, $\frac{f_3}{Q}$, and $\frac{P_1}{Q}$. Discuss the order of the system based on these and also based on the state-space model.

Note: Express your result only in terms of the given parameters C_f, A, m, k, and b.

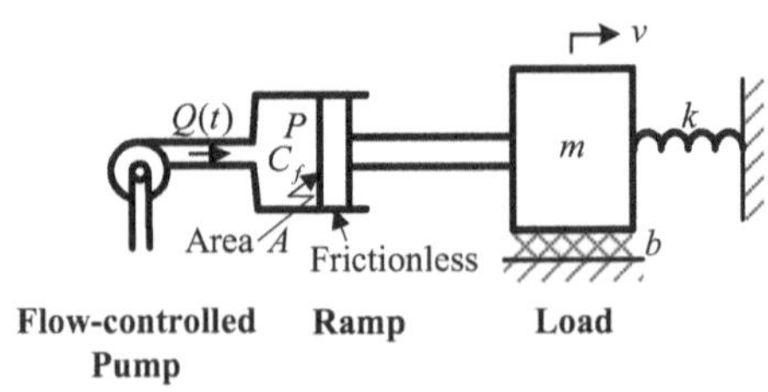

FIGURE P6.21 A hydraulic motion-controlled system.

6.22 A centrifugal pump is used to pump water from a well into an overhead tank, as represented in Figure P6.22. The input pressure from the pump is $P_s(t)$, with respect to the ambient pressure. The output is considered the pressure P_h at the bottom of the water in the tank, with respect to the ambient pressure. The figure shows the three parameters: the equivalent linear fluid resistance of the overall pipe R_{eq}, the equivalent fluid inertance within the overall pipe I_{eq}, and the gravitational fluid capacitance of the overhead tank C_{grv}.

a. Sketch an impedance circuit for the system, and give expressions for the corresponding fluid impedances of the components in the circuit.
b. Obtain the transfer function of the system.

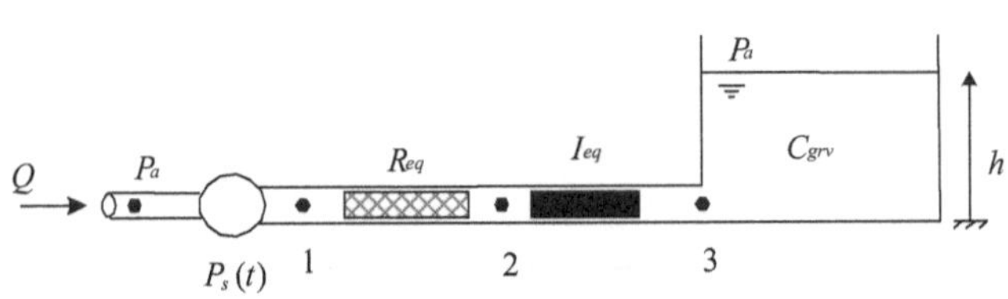

FIGURE P6.22 A lumped-parameter model of the fluid system.

6.23 Consider the two-dof system shown in Figure P6.18(a). Its mechanical impedance circuit was determined in the solution of Problem 6.18. In that problem, the force $f(t)$ is applied on mass m_1. Let the resulting velocity of mass m_2 be v_2. Now reverse this arrangement, and apply the force $f(t)$ on mass m_2. Let the resulting velocity of mass m_1 be v_1, as shown in Figure P6.23. Show that $v_1 = v_2$.

Note: This is an example to illustrate Maxwell's reciprocity property.

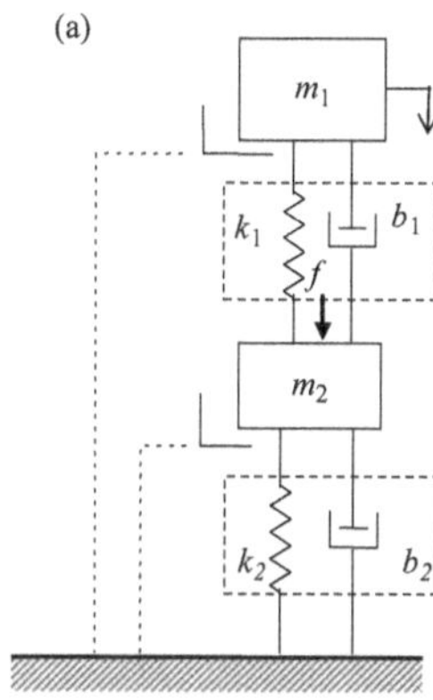

FIGURE P6.23 The reversed system for illustrating Maxwells' reciprocity property.

6.24 Figure P6.24 shows two electrical circuit. The current source $i(t)$ appears in different locations in the two circuits. Furthermore, in circuit (a), the output is the voltage v_2 of the capacitor C_2 while in circuit (b) the output is the voltage v_1 of the capacitor C_1. Show that the voltages v_1 and v_2 in the two circuits are identical.

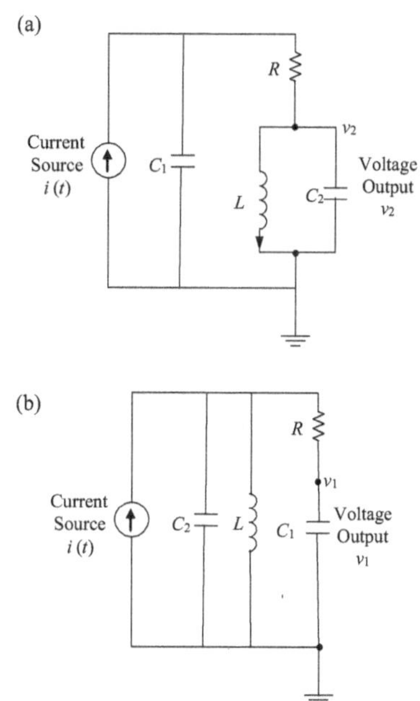

FIGURE P6.24 (a) An electrical circuit; (b) the reciprocal circuit.

Hint: You may apply Maxwell's reciprocity property.

7 Transfer-Function Linear Graphs

HIGHLIGHTS

- Types of Model Equivalence
- Thevenin and Norton Equivalent Circuits
- Extension to Multiple Physical Domains
- Transfer-Function Linear Graphs (TRLGs)
- LG Reduction (Condensation, Simplification)
- Thevenin and Norton Equivalent LGs
- Domain Conversion (Transformation) in Multi-Physics LGs

7.1 INTRODUCTION

In Chapters 2 and 3, we studied the development of time-domain analytical models of engineering dynamic system. In Chapter 4, we studied the linearization of nonlinear systems/models. In Chapter 5, we discussed linear graphs (LGs) and how they facilitate the model formulation in the time domain. The main focus in those chapters has been time-domain state-space models and input–output models. In Chapter 6, we studied the transfer function (TF) methods, particularly the use of frequency-domain TFs in the model formulation and analysis of multi-physics systems. There we saw how the TF methods can simplify the analysis and implementation, particularly because they involve algebraic manipulations rather than calculus and differential equations. With this backdrop, in the present chapter, we integrate the TF approach with the LG approach for the modeling and analysis of systems in various physical domains and of multi-physics systems in general.

Many advantages can be derived through this integration, including those of LGs and TFs; for example, since the component TFs can be combined using the well-established combination rules, unlike in the time-domain situation, there are two main advantages:

1. Not all the variables in a conventional LG need to be indicated in a TF (or frequency-domain) linear graph (TFLG)
2. A TF can be reduced (simplified) by combining the branches according to the rules of TF combination, in a simple and straightforward manner.

Going further, using the TF representation of LGs, other advantages of LGs can be exploited in the model analysis. Notably:

1. Well-established methods of circuit reduction and equivalent circuits in the field of electrical engineering (particularly Thevenin equivalent circuit and Norton equivalent circuit) can be applied to an LG in any physical domain (mechanical, electrical, fluid, thermal)
2. Engineering dynamic systems in multiple physical domains (mixed systems) can be analyzed
3. A multi-domain model can be reduced (transformed) into an equivalent model in a single physical domain in a straightforward manner, thereby simplifying the subsequent analysis and result interpretation.

Added to this list are the specific advantages of LGs, which are highlighted in Chapter 5. We will study and illustrate all these issues in this chapter, through the use of TFLGs.

DOI: 10.1201/9781003124474-7

Recall that this book pays particular attention to *equivalent models*. Notably, *approximate models* concern *approximate equivalence* of models. A model itself is "approximately equivalent" to the actual system. The required level of approximation depends on several factors such as the required accuracy (performance specifications), characteristics of the original system, available resources for modeling (e.g., whether a physical prototype of the system is available), and task timeline. Indeed, any model (be it physical, analytical, computer, experimental, etc.) is only "approximately" equivalent to the actual original system. In the process of modeling, first, an equivalence is established based on such considerations as:

1. Needs of the specific application (purpose of the model: design, analysis, control, etc., and of what aspects are important?)
2. Available resources for modeling: physical information (particularly if a physical model or a prototype is available including its operational history, analytical methodologies, physical system, computer resources, system accessibility and ability to acquire experimental data, ability to develop a physical prototype, past operational and analytical information, etc.)
3. The required accuracy, acceptable model complexity, and allowed timeline for the modeling task.

The decision to develop an *analytical model* can be made based on these considerations.

In the development of an analytical model for a system, we have to first establish some *criteria of equivalence* or "approximate equivalence" of models. For example, we have considered the conversion (analytical) of:

1. A distributed-parameter system (model) into a lumped-parameter model
2. A nonlinear system (model) into a linear model
3. A time-domain system (model) into a TF model
4. A detailed, component-based TF model into a reduced/simplified TF model.

In each case, we (explicitly or implicitly) used some criteria of equivalence; for example, energy equivalence, modal (or, natural-frequency) equivalence, analytical equivalence, computer-simulation (discrete-time or digital, allowed aliasing error) equivalence, and physical equivalence.

In this chapter, we consider another type of equivalence. Specifically, we consider the dynamic equivalence of a complex and extensive subsystem of a system to a simplified model of this complex subsystem, when viewed in relation to the remaining "simple segment" of the system. In other words, we do the following:

1. Identify the simple segment to be separated, or virtually cut (call it subsystem A) of the system. Typically, the output segment is considered as the subsystem A
2. Represent the remaining "complex" subsystem (call it subsystem B) of the system by an equivalent, yet simple model (call it M)
3. Integrated the removed (virtually cut) simple subsystem to this simplified subsystem, and analyze the resulting "simplified" model.

The criteria of equivalence in this case are such that the dynamic response of A within the actual (original) system (i.e., when A and B are together) is identical to the response of A when it is integrated with M. In developing the equivalent model M, we will consider two types of equivalence, which are commonly used in the analysis of electrical circuits:

1. Thevenin equivalence
2. Norton equivalence.

By this approach, different types of linear time-domain models (e.g., complex, single domain, multi-domain) can be represented by equivalent, simpler models. We will integrate the approaches of TFs and LGs for this purpose. Specifically, we will use TFLGs.

At the end, we will use the TFLG methodologies to convert (transform) a multi-physics (multi-domain or mixed) TFLG into an equivalent TFLG that represented entirely in a single physical domain (typically, the output domain). Illustrative examples will be given and the advantages of the presented methodologies of TFLGs will be highlighted.

7.2 CIRCUIT REDUCTION AND EQUIVALENT CIRCUITS

We have observed that TF approaches are more convenient than differential-equation approaches, in dealing with modeling issues of linear systems. This stems primarily from the fact that TF approaches use algebra rather than calculus. Also we have noted that when dealing with circuits (particularly, mechanical impedance and mobility circuits, which are in the mechanical domain, and electrical circuits), TF approaches are quite natural. Since the circuit approaches are extensively used in electrical systems, and as a result, quite mature procedures are available in that context, it is useful to consider extending such approaches to mechanical systems and, hence, to other physical domains such as fluid and thermal and multiple (mixed) domains (e.g., electro-mechanical, fluid-mechanical). Circuit reduction is convenient using Thevenin's equivalence and Norton's equivalence for electrical circuits. LGs, as studied in Chapter 5, can be simplified as well by using TF (frequency-domain) approaches and circuit reduction. Hence, the integration of these two approaches can lead to substantial benefits in model representation and analysis.

In this section, we will address the representation of a complex electrical circuit as a much simpler equivalent circuit (Thevenin form or Norton form). We will justify their equivalence to the original, complex circuit and also to each other. Then we will indicate the generalization of this approach for use in other physical domains.

7.2.1 Thevenin's Theorem for Electrical Circuits

Thevenin's theorem provides a powerful approach to reduce a complex circuit segment into a simpler equivalent representation. Two types of equivalent circuits are generated by this theorem:

1. Thevenin Equivalent Circuit (which consists of an equivalent voltage source and an equivalent electrical impedance Z_e in series)
2. Norton Equivalent Circuit (which consists of an equivalent current source and an equivalent electrical impedance Z_e in parallel).

As implied, the two equivalent impedances are identical in these two equivalent circuits (and they are determined in the same manner from the original complex circuit). The theorem provides means to determine the equivalent source and the equivalent impedance for either of these two equivalent circuits.

7.2.1.1 Circuit Partitioning

Consider a rather complex circuit C. Suppose that for practical purposes, we need to determine the voltage and current in a very small and simple sub-circuit C_1 (which is typically the output segment of the circuit). The remaining (complex) circuit segment is denoted as C_2. This separation of the overall circuit into a much simpler sub-circuit (say, the output segment) and the remaining complex sub-circuit is schematically shown in Figure 7.1.

With Thevenin and Norton equivalent circuits, the complex sub-circuit C_2 can be represented by a simple equivalent circuit with just one source and one impedance. Then, by connecting back the simple sub-circuit C_1 to the "simpler" equivalent sub-circuit, the representation and the analysis

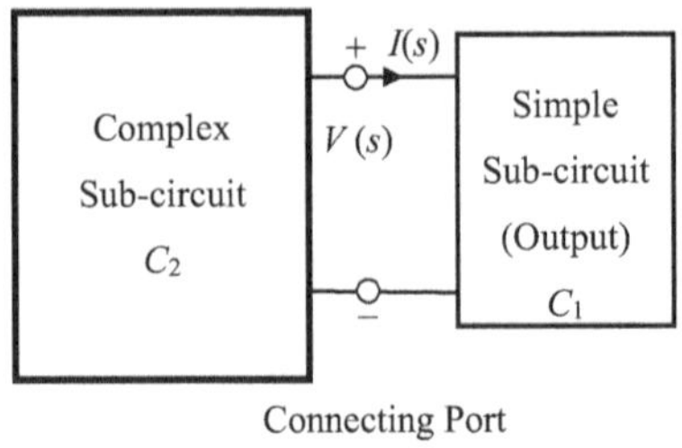

FIGURE 7.1 Isolating a simpler sub-circuit (output) from a complex circuit.

of the overall circuit (system) will be considerably simplified. In fact, much of the analysis goes to determining the equivalent impedance and the equivalent source for the original, complex sub-circuit, which has to be done in the very beginning.

7.2.1.2 Thevenin and Norton Equivalent Circuits

Consider a (rather complex) segment (C_2) of a circuit, consisting of impedances and source elements, as represented in Figure 7.2a, which remains after separating ("virtually" cutting off) the output segment of the original circuit. According to Thevenin's theorem, this sub-circuit can be represented by the Thevenin equivalent circuit, as shown in Figure 7.2b or the Norton equivalent circuit, as shown in Figure 7.2c. For either of the two equivalent circuits, its equivalence to the original complex sub-circuit (C_2) is such that, for any circuit segment connected at the output port (or for the special open-circuit case where no external circuit is connected) the voltage v and the current i at the output port of the original complex sub-circuit (C_2) and those at the output port of the equivalent circuit are identical.

Note: In the circuits, the variables (voltage and current) are indicated by uppercase letters as they denote Laplace-domain variables (or, Fourier spectra, in the frequency domain).

As noted before, the complex circuit segment (Figure 7.2a) is isolated by "virtually" cutting (separating) the original, complex circuit into the complex sub-circuit segment (C_2) and a quite simple (and fully known) segment (C_1), which is typically the output part of the original circuit (which will be connected back to the complex sub-circuit, at the end). Clearly, the "virtual" cut is made at the "port" that links the two sub-circuits. The two terminal ends formed by the virtual cut are the output port of the complex sub-circuit segment (C_2). Normally, this port is not in open-circuit condition under normal conditions because the cut is "virtual," and the simpler and known sub-circuit segment (C_1) is connected to it (and a current flows through the port).

Note: The equivalent source and the equivalent impedance of a "physical" circuit may be determined experimentally, if the analytical details of the circuit are not known.

We define the following variables and parameter:

$V(s)$=voltage across the output port when the entire circuit is complete
$I(s)$=current through the output port when the entire circuit is complete
$V_{oc}(s)$=open-circuit voltage at the output port (i.e., when the terminals open)
$I_{sc}(s)$=short-circuit current at the output port (i.e., when the terminals are shorted)
Z_e=equivalent impedance of the complex circuit segment with its sources killed
(i.e., voltage sources shorted and current sources opened)
= Thevenin impedance.

Note 1: Variables are expressed in the Laplace domain, using the Laplace variable s

Note 2: For a circuit segment with multiple sources, use the principle of superposition, by considering one source at a time and adding the results (because the system is linear).

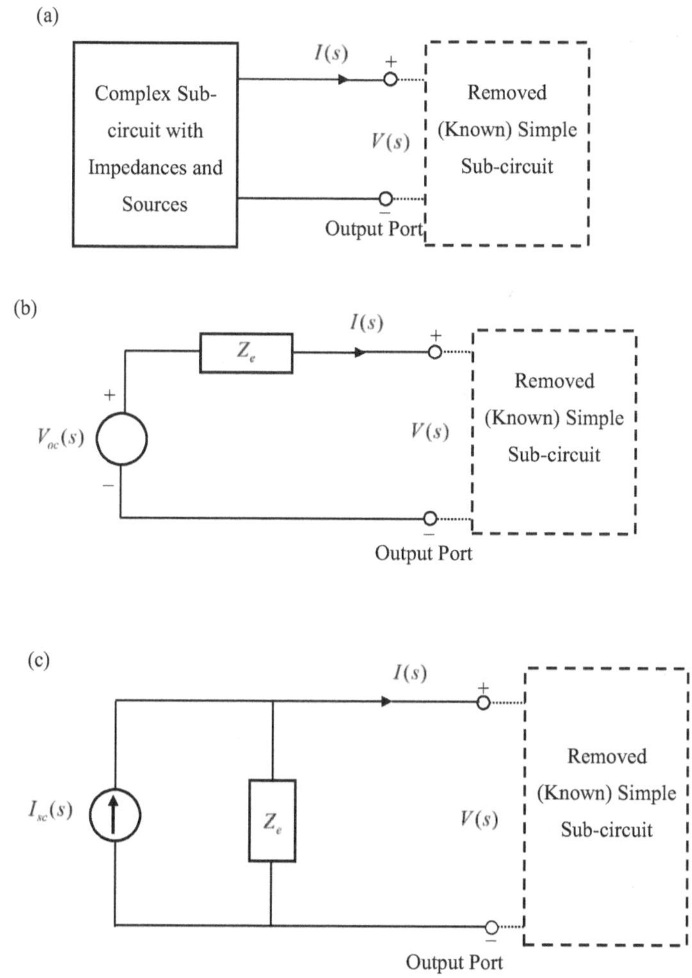

FIGURE 7.2 (a) Complex circuit segment with multiple impedances and sources; (b) Thevenin equivalent circuit; (c) Norton equivalent circuit.

Example 7.1

Illustrative example for Thevenin's theorem

We use an electrical circuit in this illustrative example. Also, electrical impedances are expressed in the Laplace domain.

Note: The corresponding frequency TFs are obtained by setting $s = j\omega$ (see Chapter 6).

Consider the circuit shown in Figure 7.3a. We make a "virtual cut" as indicated by the dotted line and determine the Thevenin and Norton equivalent circuits for the left-side (complex) segment of the circuit.

DETERMINATION OF THE EQUIVALENT IMPEDANCE Z_e

First we kill the two sources (i.e., open the current source and shorten the voltage source so that the source signals are zero). The resulting circuit is shown in Figure 7.3b. Note the series element and two parallel elements. Since the impedances add in series and inversely in parallel, we have

$$Z_e = Z_L + \frac{Z_R Z_C}{Z_R + Z_C} = Ls + \frac{R/(Cs)}{R + 1/(Cs)} = Ls + \frac{R}{RCs + 1} \tag{7.1}$$

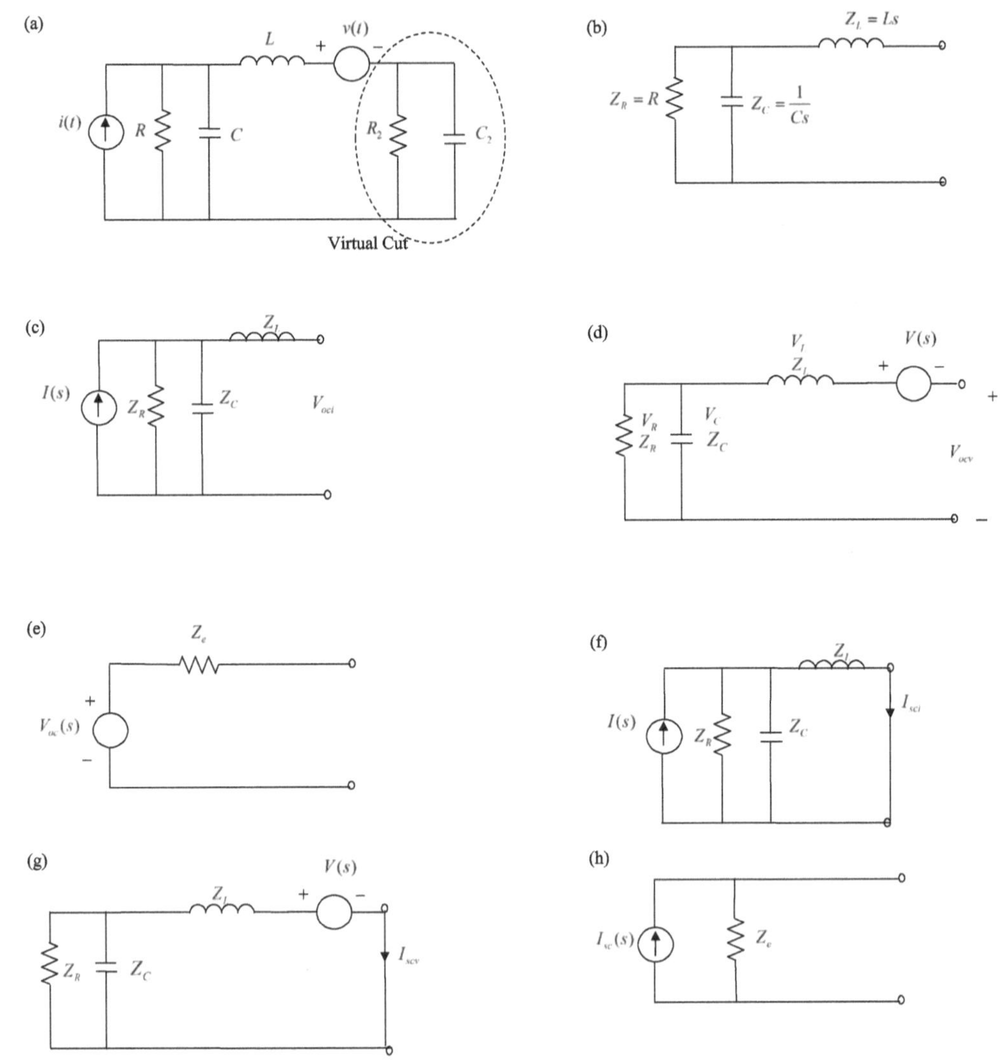

FIGURE 7.3 (a) An electrical impedance circuit; (b) complex sub-circuit with the sources killed; (c) sub-circuit in open circuit with the current source only; (d) sub-circuit in open circuit with the voltage source only; (e) Thevenin equivalent circuit of the sub-circuit; (f) sub-circuit in closed-circuit with current source only; (g) sub-circuit in closed circuit with voltage source only; (h) Norton equivalent circuit of the sub-circuit.

DETERMINATION OF $V_{oc}(s)$ FOR THEVENIN EQUIVALENT CIRCUIT

We find the open-circuit voltage with one source at a time and then use the principle of superposition to determine the overall open-circuit voltage.

a. **With Current Source $I(s)$ Only**

The open-circuit condition with the current source only (i.e., with shorted voltage source) is shown in Figure 7.3c. The source current goes through the two parallel elements only (because the series inductor is in open circuit), whose equivalent impedance is $\frac{Z_R Z_C}{Z_R + Z_C}$. Hence, the voltage across it, which is also the open-circuit voltage (since no current and hence no voltage drop along the inductor), is given by

$$V_{oci} = \frac{Z_R Z_C}{(Z_R + Z_C)} I(s) \tag{7.2a}$$

b. **With Voltage Source *V(s)* Only**
The open-circuit condition with the voltage source only (i.e., with opened current source) is shown in Figure 7.3d. The voltage drop across R should be equal to that across C, and hence, the currents in these two elements must be in the same direction. But, the sum of the currents through these parallel elements must be zero, by the node equation (since the open-circuit current is zero). Hence, each current must be zero and the voltages V_R and V_C must be zero. Furthermore, due to the open circuit, the voltage V_L across the inductor must be zero. Then from the loop equation, we have $V_{ocv} + V(s) = 0$
Or,

$$V_{ocv} = -V(s) \tag{7.2b}$$

Note the positive direction of potential drop for the open-circuit voltage, as needed for the Thevenin equivalent voltage source.
By superposition, the overall open-circuit voltage is

$$V_{oc}(s) = V_{oci} + V_{ocv} = \frac{Z_R Z_C}{(Z_R + Z_C)} I(s) - V(s) \tag{7.2}$$

The resulting Thevenin equivalent circuit is shown in Figure 7.3e.

DETERMINATION OF $I_{sc}(s)$ FOR NORTON EQUIVALENT CIRCUIT

We find the short-circuit current by taking one source at a time and then using the principle of superposition.

a. **With Current Source *I(s)* Only**

The closed-circuit condition with the current source only (shorted voltage source) is shown in Figure 7.3f. The source current goes through the three parallel elements, and the currents are divided inversely with the respective impedances. Hence, the current through the inductor is (note the positive direction as marked, for the Norton equivalent current source),

$$I_{sci} = \frac{1/Z_L}{(1/Z_R + 1/Z_C + 1/Z_L)} I(s) \tag{7.3a}$$

b. **With Voltage Source *V(s)* Only**
The short-circuit condition with the voltage source only (opened current source) is shown in Figure 7.3g. Note from the circuit that the short-circuit current is the current that flows through the overall impedance of the circuit (series inductor and the parallel resistor and capacitor combination). According to the polarity of the voltage source, this current is in the opposite direction to the positive direction marked in Figure 7.3g. We have

$$I_{scv}(s) = -\frac{V(s)}{(Z_L + Z_R Z_C/(Z_R + Z_C))} \tag{7.3b}$$

By superposition, the overall short-circuit current is

$$I_{sc}(s) = I_{sci} + I_{scv} = \frac{1/Z_L}{(1/Z_R + 1/Z_C + 1/Z_L)} I(s) - \frac{V(s)}{(Z_L + Z_R Z_C/(Z_R + Z_C))} \tag{7.3}$$

The resulting Norton equivalent circuit is shown in Figure 7.3h.

LEARNING OBJECTIVES

1. Thevenin's theorem
2. Obtaining the Thevenin equivalent circuit of an electrical circuit
3. Obtaining the Norton equivalent circuit of an electrical circuit
4. Handling of multiple sources, using the principle of superposition.

7.2.2 Justification of Circuit Equivalence

To justify the circuit equivalence, we proceed as follows:

1. Show that the general circuit (Figure 7.2a) is equivalent to the Thevenin equivalent circuit (Figure 7.2b)
2. Show that the general circuit (Figure 7.2a) is equivalent to the Norton equivalent circuit (Figure 7.2c)
3. Show that the Thevenin equivalent circuit (Figure 7.2b) is equivalent to the Norton equivalent circuit (Figure 7.2c). *Note*: Actually, this step is redundant in view of steps 1 and 2.

Step 1: Kill the equivalent source (i.e., short it) in the Thevenin circuit (Figure 7.2b). The resulting overall impedance of the Thevenin circuit is indeed Z_e, which was obtained from the original complex circuit. Next, open the output port of the Thevenin circuit. The open-circuit voltage is indeed $V_{oc}(s)$. By definition, and also in view of how this voltage was determined, this is also the open-circuit voltage of the original complex circuit (Figure 7.2a). Hence, these two circuits must be equivalent.

Step 2: Kill the equivalent source (i.e., open it) in the Norton circuit (Figure 7.2c). The resulting overall impedance of the Norton circuit is Z_e, which was obtained from the original complex circuit. Next, short the output port of the Norton circuit. The short-circuit current is $I_{sc}(s)$. By definition, and also in view of how this current was determined, this is also the short-circuit current of the original complex circuit (Figure 7.2a). Hence, these two circuits must be equivalent.

Step 3: Connect an impedance (load) Z_l at the output port of the Thevenin circuit (Figure 7.2b). The current passing through the load is $\dfrac{V_{oc}}{Z_e + Z_l}$. The voltage across the load (*Note*: Potential divider) is $\dfrac{Z_l}{Z_e + Z_l}V_{oc}$.

Connect an impedance (load) Z_l at the output port of the Norton circuit (Figure 7.2c). The current passing through the load is $\dfrac{Z_e}{Z_e + Z_l}I_{sc}$ (*Note*: Current is inversely proportional to the impedance). The overall impedance of the two parallel impedances is $\dfrac{Z_e Z_l}{Z_e + Z_l}$. Hence, the voltage across the load is $\dfrac{Z_e Z_l}{Z_e + Z_l}I_{sc}$. Now

a. For the currents in these two cases to be equal, we must have $\dfrac{V_{oc}}{Z_e + Z_l} = \dfrac{Z_e}{Z_e + Z_l}I_{sc}$ or

$$V_{oc} = Z_e I_{sc} \tag{7.4}$$

b. For the voltages in these two cases to be equal, we must have $\dfrac{Z_l}{Z_e + Z_l}V_{oc} = \dfrac{Z_e Z_l}{Z_e + Z_l}I_{sc}$ or $V_{oc} = Z_e I_{sc}$

It is seen that the requirements for the current equivalence are the same as the requirement for the voltage equivalence. Hence, the two circuits must be equivalent.

7.2.3 Extension into Other Domains

Circuit reduction and the use of Thevenin and Norton equivalent circuits are not limited to the electrical domain. In view of the fact that current is a through-variable and voltage is an across-variable, the Thevenin and Norton equivalent circuits may be extended to any linear dynamic system in terms of the corresponding through-variable and the corresponding across-variable. Then for any physical domain (mechanical, fluid, thermal), the circuit reduction may be applied using the proper variables and element TFs for the circuits in that domain.

Note: In the mechanical domain, it is the mobility (no the mechanical impedance) that is equivalent to electrical impedance. In fact, this is termed "generalized impedance," which is [across-variable]/[through-variable].

7.3 EQUIVALENT TRANSFER-FUNCTION LINEAR GRAPHS

A TFLG is an LG where only the *A*-type TF (i.e., generalized impedance=[Aaross-variable]/[through-variable]; e.g., electrical impedance, mechanical mobility, fluid impedance, thermal impedance) of the branch is indicated on the branch. This representation of LGs has particular advantages, as presented next. The reduction of a TFLG to obtain a simpler, equivalent LG, by using the concept of equivalent circuits (Thevenin or Norton), is studied subsequently. Multi-domain (multi-physics) TFLGs and their conversion into an equivalent single domain TFLG (typically, the output domain) are presented in Section 7.4.

7.3.1 Transfer-Function Linear Graphs

A TFLG takes a simpler form than a time-domain LG. The main reasons for this relative simplicity are the following:

1. On a TFLG branch, we indicate only the *A*-type TF (generalized impedance; e.g., mobility in the mechanical domain), not the associated through-variable and the across-variable
2. Branches can be combined by combining the corresponding generalized impedances, by using the standard rules for series connection and parallel connection, without manipulating the branch variables themselves. This will produce a simplified (*condensed, reduced*) TFLG
3. Only the variables that are important for the analysis and the end result (e.g., input variables and output variables) are noted on the TFLG
4. The concepts of equivalent circuits (Thevenin and Norton equivalence) are used to further simplify the TFLG.

An illustrative example is given next.

Example 7.2

Consider the electrical circuit shown in Figure 7.4a. In the electrical domain, the generalized impedance (*A*-type TF) is the electrical impedance. The impedance elements in the circuit are as follows:

$$\text{Capacitor}\, C_1 : Z_{C1} = \frac{1}{C_1 s}$$

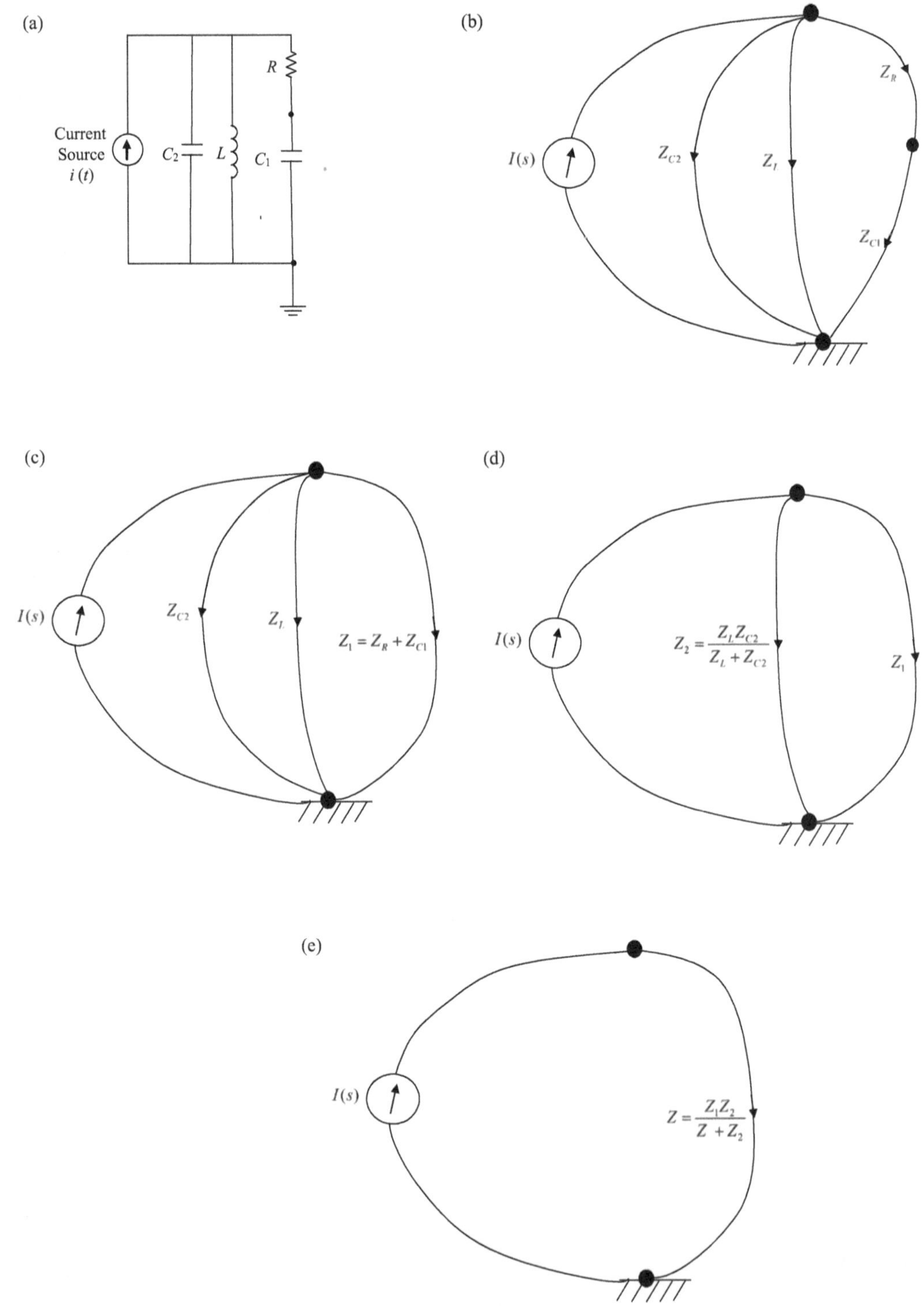

FIGURE 7.4 (a) An electrical circuit; (b) the transfer function linear graph; (c) TFLG simplification, stage 1; (d) TFLG simplification, stage 2; (e) TFLG simplification, stage 3 (final stage).

$$\text{Capacitor}\,C_2 : Z_{C2} = \frac{1}{C_2 s}$$

$$\text{Inductor}\,L : Z_L = Ls$$

$$\text{Resistor}\,R : Z_R = R$$

Note: To convert electrical impedances in the Laplace domain into the corresponding frequency-domain expressions, set $s = j\omega$.

The TFLG of the circuit is shown in Figure 7.4b. We have indicated the electrical impedances of the circuit elements on the corresponding branches of the LG. We have not shown any variables except for the source current (independent variable of the source, which is the input) $I(s)$. System variables may be indicated on the TFLG depending on the requirements of a particular problem (e.g., if we are interested in the voltage of capacitor C_1, we will indicate that voltage (the output variable) on the corresponding branch of the TFLG).

The "preliminary" reduction of the TFLG may be achieved by combining the branches, depending on the problem objective. For instance, if we need to determine the voltage and/or the current of capacitor C_1, we should not combine its branch (impedance Z_{C1}) with any other branch. But, we may combine the branches of the remaining circuit elements where feasible. Specifically, we can combine the branches of Z_L and Z_{C2}. But, we cannot combine the resulting branch with the branch of Z_R (because Z_R is in series with Z_{C1}, and just one of these two elements cannot be combined with the common parallel branch).

A sequence of branch combinations is indicated. Figure 7.4c shows the result when the branches of Z_R and Z_{C1} are combined. Figure 7.4d shows the result when the branches of Z_L and Z_{C2} are combined as well. Figure 7.4e shows the final result of TFLG when the previous two combined branches are combined.

LEARNING OBJECTIVES

1. Development of a TFLG
2. Preliminary simplification (condensation, reduction) of a TFLG by combining the TFLG branches where possible
3. Identification of infeasible combinations of branches.

The concepts of TFLGs can be directly extended from electrical systems to mechanical systems using the familiar force-current analogy. According to this analogy, *electrical impedance* is analogous to mechanical *mobility*, which are *A*-type TFs (or, *generalized impedances*); and electrical admittance is analogous to mechanical impedance, which are *T*-type TFs (or generalized admittances). This analogy is summarized in Table 7.1.

Accordingly, the preliminary reduction (condensation, simplification) of a TFLG is done by the following two steps:

1. On each branch of the LG mark the mobility function (not mechanical impedance), which is the generalized impedance

TABLE 7.1
Analogy of Mechanical and Electrical Transfer-Functions

Mechanical Circuit	Electrical Circuit Analogy
Mobility function	Electrical impedance
Force	Current
Voltage	Velocity

2. Carry out LG analysis and reduction (simplification, condensation) as if we are dealing with an electrical circuit, in view of the analogy given in Table 7.1.

In particular, we do the following:

1. For parallel branches: mobilities are combined by the inverse relation

$$M = \frac{M_1 M_2}{M_1 + M_2} \text{ (product divided by sum)} \tag{7.5}$$

Note: Velocity is common; force is divided inversely to the branch mobility.

2. For series branches: Mobilities add

$$M = M_1 + M_2 \tag{7.6}$$

Note: Force is common; velocity is divided in proportion to the branch mobility.

7.3.2 Equivalent Mechanical Circuit Analysis Using Linear Graphs

The equivalent-circuit analysis (Thevenin equivalence and Norton equivalence) can be extended to mechanical systems as well, by using the force-current analogy. We start by forming the TFLG of the system. Then we simplify (reduce, condense) it as appropriate. Next we perform an appropriate "virtual cut" and determine the Thevenin equivalent circuit or the Norton equivalent circuit, depending on the problem objective.

We have already discussed the formation of a TFLG and the preliminary simplification of a TFLG by combining its branches where feasible. Subsequent steps in the formation of an "equivalent linear graph" are the determination of the equivalent mobility M_e and the equivalent source (an *equivalent velocity source* for a Thevenin circuit and an *equivalent force source* for a Norton circuit). In determining the equivalent mobility, the following operations are applicable:

1. Killing a force source means open-circuiting it (so, transmitted force = 0)
2. Killing a velocity source means short-circuiting it (so, velocity across = 0)

Illustrative examples in the mechanical domain are given next.

Example 7.3

Ground-based mechanical oscillator (revisited)

Consider the ground-based mechanical oscillator shown in Figure 7.5a. Let us use the approach of the LG equivalent circuit to determine the force transmissibility of the system.

Solution

The TFLG of the system is shown in Figure 7.5b. We can simplify/reduce this TFLG by combining the parallel branches in the suspension system (engine mount), as shown in Figure 7.5c, where

Mobility of the mass

$$M_m = \frac{1}{ms} \tag{7.7}$$

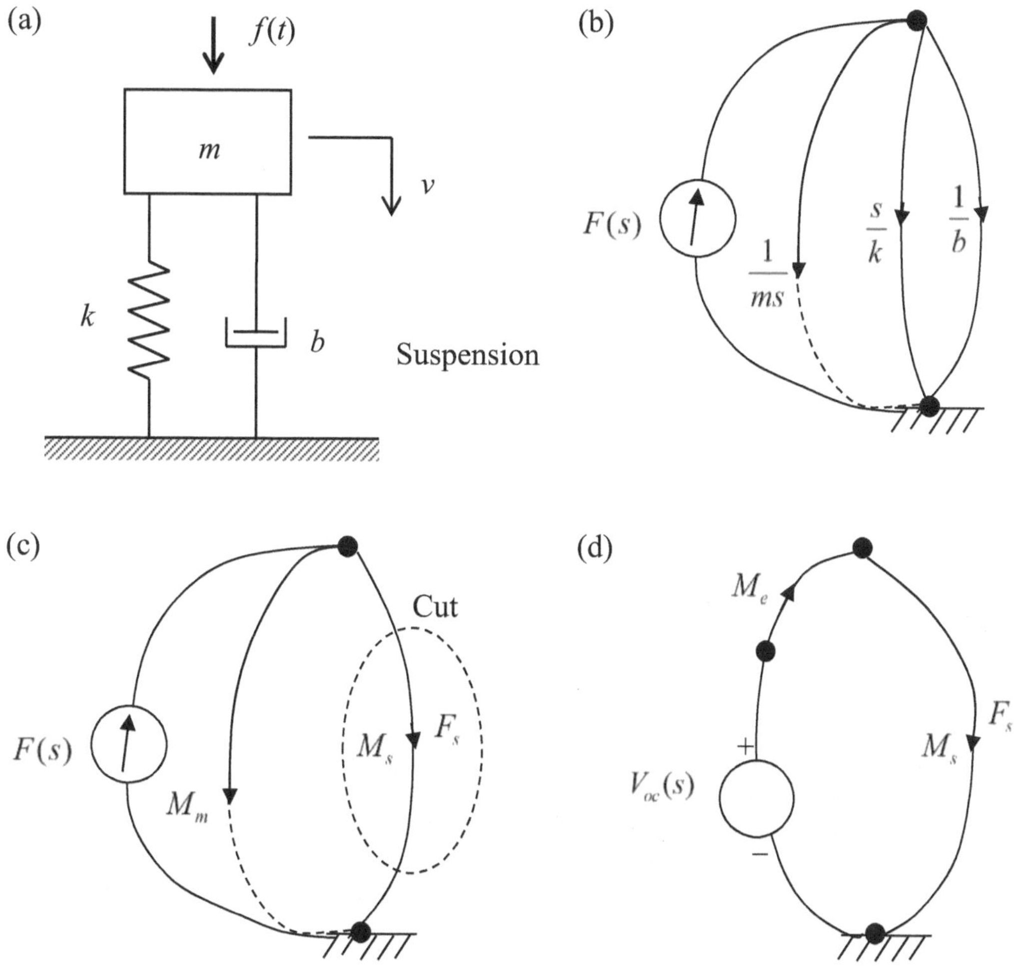

FIGURE 7.5 (a) Ground-based mechanical oscillator; (b) transfer-function linear graph; (c) reduced TFLG; (d) Thevenin equivalent LG.

Mobility of the suspension system

$$M_s = \frac{\frac{s}{k} \times \frac{1}{b}}{\frac{s}{k} + \frac{1}{b}} = \frac{s}{bs + k} \tag{7.8}$$

Note: Equation (7.5) was used to combine the parallel branches.

Now, since we are interested in the force (a through-variable) through the suspension system, we need to do the following:

1. Virtually cut the branch corresponding to the suspension (whose force needs to be determined)
2. Since we need a through-variable, Thevenin equivalent LG is appropriate.

These are carried out, resulting in Figure 7.5d.

To determine the equivalent velocity source, we kill the force source in Figure 7.5c and determine the mobility of the remaining circuit, viewed from the virtually cut port.

Note: For a *T*-source (force source), killing means opening the connection.

Clearly, the resulting equivalent mobility is

$$M_e = M_m \tag{7.9}$$

Also, from Figure 7.5c, the open-circuit velocity is

$$V_{oc}(s) = M_m F(s) \quad 7.10$$

Now from the Thevenin TFLG in Figure 7.5d, the transmitted force is expressed as

$$F_s(s) = \frac{V_{oc}(s)}{M_e + M_s} = \frac{M_m F(s)}{M_m + M_s}$$

Note: We have substituted (7.9) and (7.10).

The resulting force transmissibility is

$$T_f = \frac{M_m}{M_m + M_s} \tag{7.11}$$

This is identical to the result we obtained in Chapter 6.

LEARNING OBJECTIVES

1. Preliminary simplification (condensation) of a TFLG by combining the TFLG branches
2. Identification of the suitable virtual cut and the equivalent TFLG
3. Determination of the Thevenin equivalent TFLG
4. Determination of the force transmissibility using the equivalent IFLG method.

Example 7.4: Oscillator with Support Motion (Revisited)

Consider the mechanical oscillator with support motion, as shown in Figure 7.6a. Let us use the approach of LG equivalent circuits to determine the motion transmissibility of the system.

Solution

The TFLG of the system is shown in Figure 7.6b. We can reduce this TFLG by combining the parallel branches in the suspension system, as shown in Figure 7.6c, where the mobility of the mass, $M_m = \frac{1}{ms}$, and the mobility of the suspension system,

$$M_s = \frac{s}{bs + k} \text{ (See Example 7.3)}$$

Now, since we are interested in the velocity (an across-variable) of the mass, we need to do the following:

1. Virtually cut the branch corresponding to the mass (whose velocity needs to be determined)
2. Since we need an across-variable, Norton equivalent LG is appropriate.

These are carried out, resulting in Figure 7.6d.

To determine the equivalent force source, we kill the velocity source in Figure 7.6c and determine the mobility of the remaining circuit, viewed from the virtually cut port.

Note: For an *A*-source (velocity source), killing means closing the connection.

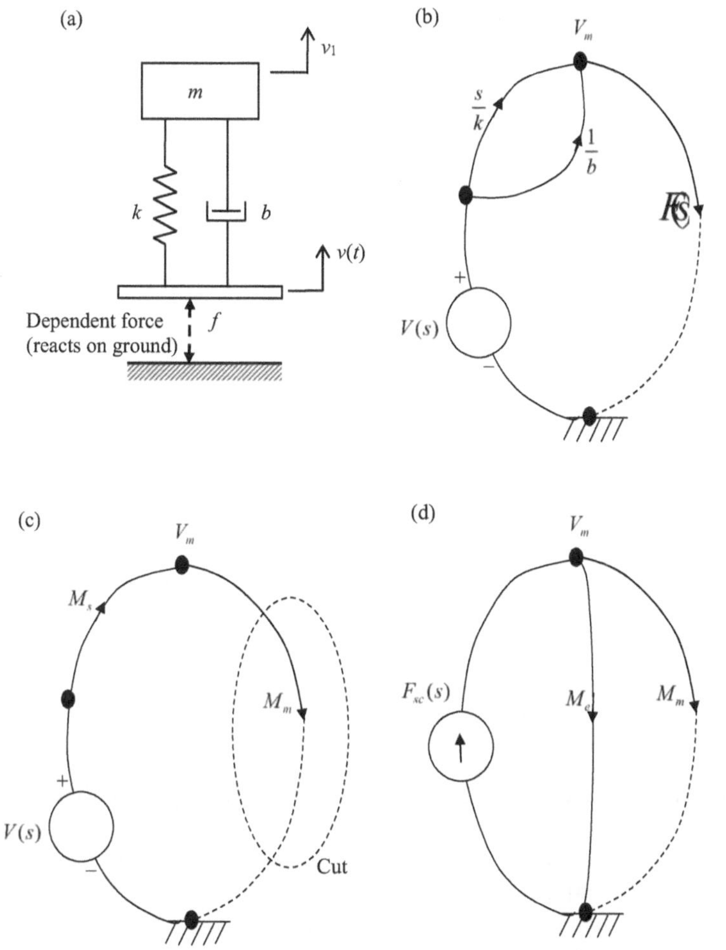

FIGURE 7.6 (a) Mechanical oscillator with support motion; (b) transfer-function linear graph; (c) reduced TFLG; (d) Norton equivalent LG.

Clearly, the resulting equivalent mobility is

$$M_e = M_s \tag{7.12}$$

Also, from Figure 7.6c, the closed-circuit force is

$$F_{sc}(s) = \frac{V(s)}{M_s} \tag{7.13}$$

In the Norton LG of Figure 7.6d, the overall mobility of the two parallel branches is $\frac{M_e M_m}{M_e + M_m}$ (see equation (7.5)). Hence, the velocity at the mass is

$$V_m = \frac{M_e M_m}{M_e + M_m} F_{sc}(s) = \frac{M_s M_m}{M_s + M_m} \frac{V(s)}{M_s} = \frac{M_m V(s)}{M_m + M_s}$$

Note: We have substituted (7.12) and (7.13).

The resulting motion transmissibility is

$$T_m = \frac{M_m}{M_m + M_s} \tag{7.14}$$

This is identical to the result we obtained for the force transmissibility in Example 7.3 and also the results in Chapter 6.

LEARNING OBJECTIVES

1. Preliminary simplification (reduction, condensation) of a TFLG by combining the LG branches
2. Identification of the suitable virtual cut and the needed type of equivalent TFLG
3. Determination of the Norton equivalent TFLG
4. Determination of the motion transmissibility using the equivalent TFLG method
5. Equivalence of the force transmissibility and the motion transmissibility in complementary (dual, reciprocal) systems.

7.3.3 Summary of Thevenin Approach for Mechanical Circuits

We now summarize the general steps in the application of Thevenin's theorem to mechanical circuits that are represented by TFLGs.

7.3.3.1 General Steps

1. Draw the TFLG for the system and mark the mobility functions for all the branches (except the source elements)
2. On the TFLG indicate only those variables that are important for the objective (e.g., the system inputs and outputs)
3. Simplify (reduce, condense) the TFLG by combining branches as appropriate (for series branches: add mobilities; for parallel branches: inverse rule applies for mobilities) and mark the mobilities of the combined branches. Do not combine a branch whose variable needs to be determined (e.g., an output branch), with another branch
4. Depending on the problem objective (e.g., determine a particular force, velocity TF) determine which segment of the circuit (TFLG) should be virtually cut (i.e., the variable or function of interest should be associated with the segment/branch that is removed from the TFLG). The equivalent TFLG of the remaining part needs to be determined
5. Depending on the problem objective, establish whether Thevenin equivalence or Norton equivalence is needed (Specifically, use Thevenin equivalence if a through-variable or T-type TF needs to be determined, because then the equivalent TFLG will have two series branches with a common through-variable; use Norton equivalence if an across-variable or an A-type TF needs to be determined because then the equivalent TFLG will have two parallel branches with a common across-variable)
6. Determine the equivalent source and equivalent mobility of the equivalent TFLG
7. Using the equivalent TFLG determine the variable or function of interest.

7.4 MULTI-DOMAIN TRANSFER-FUNCTION LINEAR GRAPHS

This section studies multi-domain TFLGs. Their analysis is done by first converting the TFLG of the multi-domain system into an equivalent single-domain TFLG. This conversion is done using the constitutive relationships of the two-port elements that connect the different physical domains.

The other key steps in the problem solution are the same as before. In particular, the preliminary steps prior to the domain conversion are the following:

1. Draw the TFLG for the system and mark the A-type TFs (*generalized impedances*) for all the branches (except the source elements)
2. On the TFLG indicate only those variables that are important for the objective (e.g., system inputs and outputs)
3. Simplify (reduce, condense) the TFLG by combining branches as appropriate (for series branches, add mobilities; for parallel branches, inverse rule applies for mobilities) and mark the mobilities of the combined branches. Do not combine a branch whose variable needs to be determined (e.g., an output branch), with another branch.

7.4.1 Conversion into an Equivalent Single Domain

Once the TFLF is pre-processed using the preliminary steps indicated above, the process of converting a two-domain system into an equivalent single domain is done as follows (*Note*: The two domains are coupled either by a *transformer* or a *gyrator*):

1. Decide which physical domain will be converted. This is based on the end objective of the analysis. Typically, the physical domain of the system input is converted into the physical domain of the system output
2. Determine the Thevenin equivalent TFLG of the subsystem that will be converted. This is connected to the input branch of the two-port element (transformer or gyrator), which links to remaining subsystem through the output branch of the two-port element
3. Apply the constitutive equations of the two-port element in the Thevenin equivalent TFLG and determine the equivalent A-source and the equivalent generalized impedance in series, which will replace the output branch of the two-port element.

Note: In step 2, we may use Norton equivalence instead. Also, in step 3, we may determine the equivalent T-source and the equivalent generalized impedance in parallel. The final result will be the same, but the intermediate analysis will be different.

7.4.1.1 Transformer-coupled Systems

Suppose that the two physical domains of the mixed system are coupled through a transformer. First we apply the Thevenin theorem to the subsystem to be converted and determine the equivalent A-source $P_{oc}(s)$ and the equivalent generalized impedance Z_e in series in the Thevenin equivalent TFLG. This result is shown in Figure 7.7a.

Note: If the fluid domain is to be converted, the equivalent Thevenin source $P_{oc}(s)$ is a pressure source, and Z_e is a fluid impedance (pressure/flow rate). If the retaining (equivalent) domain is mechanical, f_1 is a force and v_1 is a velocity. In any other domain, these quantities will take the corresponding meanings.

The governing equations of the TFLG in Figure 7.7a are written now.

$$\text{Equivalent-Impedance Constitutive Equation}: P_e = Z_e Q_e \tag{i}$$

Transformer constitutive equations:

$$v_1 = rP_1 \tag{ii}$$

$$f_1 = -\frac{1}{r}Q_1 \tag{iii}$$

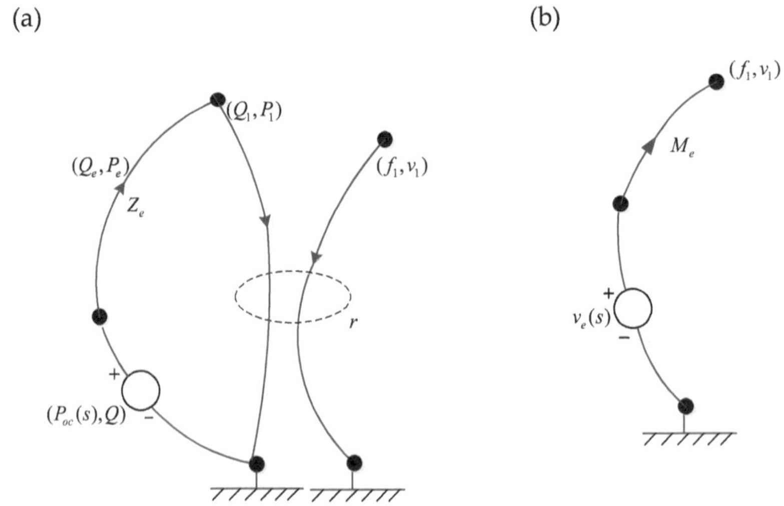

FIGURE 7.7 Transformer-coupled system. (a) Thevenin equivalent TFLG of the subsystem to be converted; (b) equivalent TFLG of the subsystem in the new physical domain.

where r=transformer parameter.

$$\text{Loop Equation}: -P_1 - P_e + P_{\text{oc}}(s) = 0 \tag{iv}$$

$$\text{Node Equation}: Q_e - Q_1 = 0 \tag{v}$$

Now we carry out substitutions as follows:
(ii): $v_1 = rP_1 = r[P_{\text{oc}}(s) - P_e] = rP_{\text{oc}}(s) - rZ_eQ_e = rP_{\text{oc}}(s) - rZ_eQ_1$ (from (iv), (i), and (v))

$$\text{Substitute (iii)}: v_1 = rP_{\text{oc}}(s) + r^2 Z_e f_1 \tag{vi}$$

This may be written as

$$v_1 = V_e(s) + M_e f_1 \tag{7.15}$$

Accordingly, we have the following results:

$$\text{Converted equivalent } A-\text{source } V_e(s) = rP_{\text{oc}}(s) \tag{7.16}$$

$$\text{Converted equivalent generalized impedance (in series)}: M_e = r^2 Z_e \tag{7.17}$$

The equivalent TFLG of the converted subsystem is shown in Figure 7.7b.

The main steps of this conversion are given in Table 7.2.

7.4.1.2 Gyrator-coupled Systems

Now suppose that the two physical domains of the mixed system are coupled through a gyrator (two-domain). Again, we apply the Thevenin theorem to the subsystem to be converted and determine the

TABLE 7.2
Domain Conversion of a Transformer-Coupled System

Formulation Step	Result
Equivalent-impedance constitutive equation	$P_e = Z_e Q_e$
Transformer constitutive equations	$v_1 = rP_1;\ f_1 = -\frac{1}{r}Q_1$; r=transformer parameter
Loop equation	$-P_1 - P_e + P_{oc}(s) = 0$
Node equation	$Q_e - Q_1 = 0$
Result after substitutions	$v_1 = rP_{oc}(s) - rZ_e Q_1$
Final result	$v_1 = rP_{oc}(s) + r^2 Z_e f_1$ or $v_1 = V_e(s) + M_e f_1$
Converted equivalent A-source	$V_e(s) = rP_{oc}(s)$
Converted equivalent generalized impedance (in series)	$M_e = r^2 Z_e$

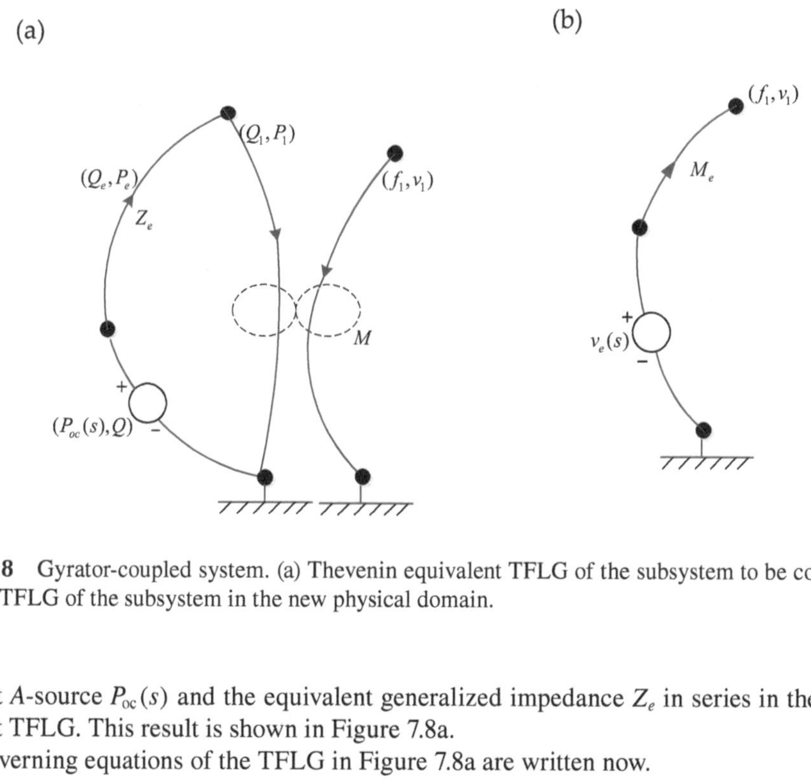

FIGURE 7.8 Gyrator-coupled system. (a) Thevenin equivalent TFLG of the subsystem to be converted; (b) equivalent TFLG of the subsystem in the new physical domain.

equivalent A-source $P_{oc}(s)$ and the equivalent generalized impedance Z_e in series in the Thevenin equivalent TFLG. This result is shown in Figure 7.8a.

The governing equations of the TFLG in Figure 7.8a are written now.

$$\text{Equivalent-Impedance Constitutive Equation}: P_e = Z_e Q_e \tag{i}$$

Gyrator constitutive equations:

$$v_1 = MQ_1 \tag{ii}$$

$$f_1 = -\frac{1}{M}P_1 \tag{iii}$$

TABLE 7.3
Domain Conversion of a Gyrator-Coupled System

Formulation Step	Result
Equivalent-impedance constitutive equation	$P_e = Z_e Q_e$
Gyrator constitutive equations	$v_1 = MQ_1$; $f_1 = -\frac{1}{M}P_1$; M=gyrator parameter
Loop equation	$-P_1 - P_e + P_{oc}(s) = 0$
Node equation	$Q_e - Q_1 = 0$
Result after substitutions	$v_1 = \frac{M}{Z_e}[P_{oc}(s) - P_1]$
Final result	$v_1 = \frac{M}{Z_e}P_{oc}(s) + \frac{M^2}{Z_e}f_1$ or $v_1 = V_e(s) + M_e f_1$
Converted equivalent A-source	$V_e(s) = \frac{M}{Z_e}P_{oc}(s)$
Converted equivalent generalized impedance (in series)	$M_e = \frac{M^2}{Z_e}$

where M=gyrator parameter.

$$\text{Loop Equation}: -P_1 - P_e + P_{oc}(s) = 0 \tag{iv}$$

$$\text{Node Equation}: Q_e - Q_1 = 0 \tag{v}$$

Now we carry out substitutions as follows:

(ii): $v_1 = MQ_1 = MQ_e = \frac{MP_e}{Z_e} = \frac{M}{Z_e}[P_{oc}(s) - P_1] = \frac{M}{Z_e}P_{oc}(s) - \frac{M}{Z_e} \times (-Mf_1)$ (from (v), (i), (iv), and (iii)).

$$\text{We have}: v_1 = \frac{M}{Z_e}P_{oc}(s) + \frac{M^2}{Z_e}f_1 \tag{vi}$$

This may be written as

$$v_1 = V_e(s) + M_e f_1 \tag{7.18}$$

Accordingly, we have the following results:

$$\text{Converted equivalent } A - \text{source } V_e(s) = \frac{M}{Z_e}P_{oc}(ss) \tag{7.19}$$

$$\text{Converted equivalent generalized impedance (in series)}: M_e = \frac{M^2}{Z_e} \tag{7.20}$$

The equivalent TFLG of the converted subsystem is shown in Figure 7.8b.

The main steps of this conversion are given in Table 7.3.

7.4.2 Illustrative Examples

Once the equivalent TFLG of the domain-converted subsystem is obtained, the remaining subsystem (of the same domain) is directly connected to it (the same way the remaining subsystem had been connected to the output branch of the two-port element). This gives the overall equivalent TFLG, expressed in a single domain. Then this equivalent TFLG, expressed in a single domain, can be analyzed in the same way as done in the previous section. Illustrative examples are given now.

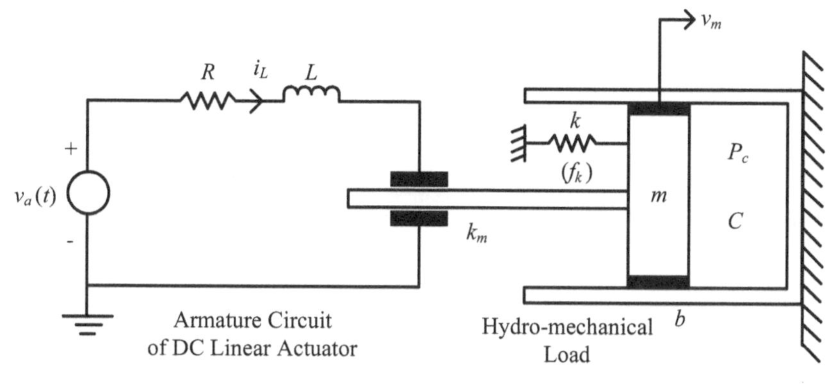

FIGURE 7.9 A hydro-mechanical load driven by a DC linear motor.

Example 7.5

Consider a positioning mechanism, whose lumped-parameter linear model is shown in Figure 7.9, which has an armature-controlled linear dc motor that drives a hydro-mechanical (mixed-domain) load of mass *m*, viscous damping of damping constant *b*, resisting spring of stiffness *k*, and a hydraulic capacitor of capacitance *C*. The hydraulic piston and cylinder (capacitor) unit serves as a shock absorber for the load and has hydraulic pressure P_c.

The following model parameters are given:

R=resistance of the armature coil
L=leakage inductance of the armature circuit
k_m="force constant" of the linear dc motor (i.e., the conversion factor for armature current to actuator force)
m=overall mass of the load (including the armature)
k=resisting stiffness of the load
b=viscous damping constant of the load
C=fluid capacitance of the shock absorber
A=area of the shock-absorber piston

a. Sketch a complete LG for the system model.
   Note: Orient the branches and indicate all the system parameters and branch variables (as through- and across-pairs).
b. Using the LG, systematically develop a complete, linear, and proper state-space model for the system. Initially, use the following state variables, which correspond to the energy storage elements in the system:
   i_L=current through the leakage inductance of the armature
   v_m=velocity of the mechanical load (and also of the actuator rod)
   f_k=spring force of the mechanical load (attached to the actuator piston)
   P_c=hydraulic pressure in the shock-absorber cylinder
   System input=$v_a(t)$=voltage applied to the armature.
   System output=v_m=velocity of the load.
   You must give all the vectors and matrices of the final state-space model.
c. From the state-space model, determine the input–output differential equation (input=$v_a(t)$, output=v_m). From that equation, write the system TF.
d. Draw the TFLG corresponding to the state-space model obtained in Part (b). Convert it into a TFLG that is entirely in the mechanical domain. From that, systematically obtain the system TF. (*Note*: This result should be identical to the result obtained in Part (c).

Solution

(a)

The LG of the system is shown in Figure 7.10a.

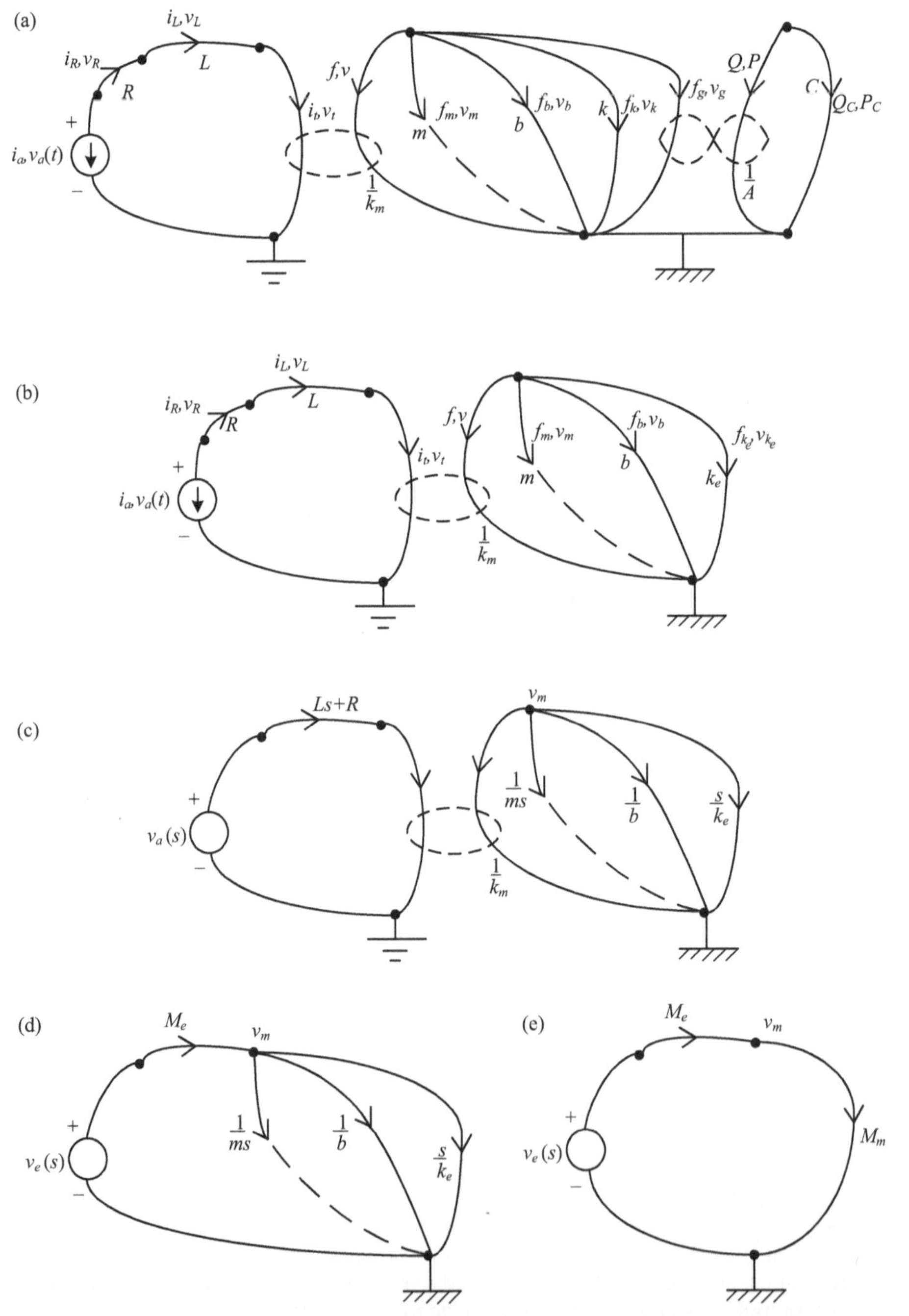

FIGURE 7.10 (a) Linear graph of the system; (b) Linear graph of the equivalent minimal system with independent energy storage elements; (c) The transfer-function linear graph of the multi-domain (electro-mechanical) system; (d) The equivalent TFLG entirely in the mechanical domain; (e) The reduced TFLG in the mechanical domain.

(b)

State-Space Shell:

$$L\dot{i}_L = v_L$$

$$m\dot{v}_m = f_m$$

$$\dot{f}_k = kv_k$$

$$C\dot{P}_C = Q_C$$

Other Constitutive Equations:

$$v_R = Ri_R$$

$$f_b = bv_b$$

$$\left.\begin{aligned} v &= \frac{v_t}{k_m} \\ f &= -k_m i_t \end{aligned}\right\}\text{DC motor (electro-mechanical transformer)}$$

$$\left.\begin{aligned} P &= \frac{f_g}{A} \\ Q &= -Av_g \end{aligned}\right\}\text{Shock-absorber (fluid gyrator)}$$

Node Equations:

$$i_a - i_R = 0 \text{ (useless)};\ i_R - i_L = 0;\ i_L - i_t = 0;\ -f - f_m - f_b - f_k - f_g = 0;\ -Q - Q_C = 0$$

Loop Equations:

$$-v_t - v_L - v_R + v_a(t) = 0;\ -v_m + v = 0;\ -v_b + v_m = 0;\ -v_k + v_m = 0;\ -v_g + v_m = 0;\ -P_C + P = 0$$

Eliminate Auxiliary Variables:

$$v_L = -v_t - v_R + v_a(t) = -k_m v - Ri_R + v_a(t) = -k_m v_m - Ri_L + v_a(t)$$

$$f_m = -f - f_b - f_k - f_g = k_m i_t - bv_b - f_k - AP = k_m i_L - bv_m - f_k - AP_C$$

$$v_k = v_m$$

$$Q_C = -Q = Av_g = Av_m$$

$$\text{Output} = v_m$$

Hydro-Mechanical Load:

State vector $\boldsymbol{x} = \begin{bmatrix} x_1 & x_2 & x_3 & x_4 \end{bmatrix}^T = \begin{bmatrix} i_L, & v_m, & f_k, & P_C \end{bmatrix}^T$

Input vector $\boldsymbol{u} = [u] = [v_a(t)]$

Output vector $\boldsymbol{y} = [y] = [v_m]$

State-Space Model:

$$\dot{\boldsymbol{x}} = \boldsymbol{A}\boldsymbol{x} + \boldsymbol{B}\boldsymbol{u}$$
$$\boldsymbol{y} = \boldsymbol{C}\boldsymbol{x} + \boldsymbol{D}\boldsymbol{u}$$

The corresponding model matrices are

$$\boldsymbol{A}=\begin{bmatrix} -\frac{R}{L} & -\frac{k_m}{L} & 0 & 0 \\ \frac{k_m}{m} & -\frac{b}{m} & -\frac{1}{m} & -\frac{A}{m} \\ 0 & k & 0 & 0 \\ 0 & \frac{A}{C} & 0 & 0 \end{bmatrix}; \boldsymbol{B}=\begin{bmatrix} \frac{1}{L} \\ 0 \\ 0 \\ 0 \end{bmatrix}; \boldsymbol{C}=\begin{bmatrix} 0 & 1 & 0 & 0 \end{bmatrix}; \boldsymbol{D}=[0]$$

This is not the final result of Part (a), as discussed next.

Important Observation

It is clear that in the system matrix (A), the fourth row is directly proportional to the third row (through a constant multiplier). This tells us that the state variables x_3 and x_4 are not independent. Since the state-space model must contain the least number of state variables that completely represent the dynamics of the system, we have to eliminate one of these two state variables. There are many ways to perform this elimination, but we will use the most direct and simplest way. Before doing that, we should also recall that the system order (which is equal to the order of the state vector) is equal to the number of "independent" energy storage elements in the system. Through physical examination of the model (Figure 7.9), it should be clear that the fluid capacitor and the spring (of stiffness k) are not independent. Since these two energy storage elements are not independent, we cannot use two separate state variables to represent them in the final state-space model. All these tell us that the state-space model (fourth order) that was obtained before is not the correct final result, and it has to be reduced to a third-order model. This is accomplished now.

From the proportionality of the third and the fourth state equations, it is clear that

$$P_C=\frac{A}{Ck}f_k$$

Substitute this in the second-state equation

$$m\dot{v}_m=k_m i_L-bv_m-f_k-AP_C=k_m i_L-bv_m-f_k-\frac{A^2}{Ck}f_k$$

$$=k_m i_L-bv_m-f_k-\frac{A^2}{Ck}f_k=k_m i_L-bv_m-\left(1+\frac{A^2}{Ck}\right)f_k=k_m i_L-bv_m-f_{ke}$$

where

$$f_{ke}=\left(1+\frac{A^2}{Ck}f_k\right) \tag{i}$$

Then, the third state equation may be modified as

$$\dot{f}_{ke}=k_e v_m$$

where

$$k_e=k\left(1+\frac{A^2}{Ck}\right)=k+\frac{A^2}{C} \tag{ii}$$

Note: It is seen that the equivalent spring stiffness of the hydraulic capacitor is $\frac{A^2}{C}$

In this manner, we have the following "proper," minimal (third order) state-space model, with the state vector $\boldsymbol{x} = \begin{bmatrix} x_1 & x_2 & x_{3e} \end{bmatrix}^T = \begin{bmatrix} i_L, & v_m, & f_{ke} \end{bmatrix}^T$ and the model matrices:

$$\boldsymbol{A} = \begin{bmatrix} -\dfrac{R}{L} & -\dfrac{k_m}{L} & 0 \\ \dfrac{k_m}{m} & -\dfrac{b}{m} & -\dfrac{1}{m} \\ 0 & k_e & 0 \end{bmatrix}; \boldsymbol{B} = \begin{bmatrix} \dfrac{1}{L} \\ 0 \\ 0 \end{bmatrix}; \boldsymbol{C} = \begin{bmatrix} 0 & 1 & 0 \end{bmatrix}; \boldsymbol{D} = [0]$$

where k_e is given by (ii).

The corresponding LG is shown in Figure 7.10b.

(c)

The state-space model may be written as

$$L\dot{x}_1 = -Rx_1 - k_m y + u \tag{iii}$$

$$m\dot{y} = k_m x_1 - by - x_{3e} \tag{iv}$$

$$\dot{x}_{3e} = k_e y \tag{v}$$

We have to express these three equations as a single equation in input u and output y only. This is done as follows:

$$\frac{d(\text{iv})}{dt} \text{ and substitite (iii) and (v): } m\ddot{y} = \frac{k_m}{L}(-Rx_1 - k_m y + u) - b\dot{y} - k_e y$$

This gives

$$m\ddot{y} + b\dot{y} + \left(\frac{k_m^2}{L} + k_e\right)y - \frac{k_m}{L}u = -\frac{k_m R}{L}x_1$$

Or

$$x_1 = -\frac{1}{a_2}(m\ddot{y} + b\dot{y} + a_1 y - a_3 u) \tag{vi}$$

where

$$a_1 = \left(\frac{k_m^2}{L} + k_e\right);\ a_2 = \frac{k_m R}{L};\ a_3 = \frac{k_m}{L} \tag{vii}$$

Substitute (vi) in (iii):

$$-\frac{L}{a_2}(m\dddot{y} + b\ddot{y} + a_1\dot{y} - a_3\dot{u}) = \frac{R}{a_2}(m\ddot{y} + b\dot{y} + a_1 y - a_3 u) - k_m y + u$$

Rearrange:

$$\frac{Lm}{a_2}\dddot{y} + \frac{1}{a_2}(Lb + Rm)\ddot{y} + \frac{1}{a_2}(La_1 + Rb)\dot{y} + \left(\frac{Ra_1}{a_2} - k_m\right)y = \frac{La_3}{a_2}\dot{u} + \left(\frac{Ra_3}{a_2} - 1\right)u$$

Or

$$Lm\dddot{y}+(Lb+Rm)\ddot{y}+(La_1+Rb)\dot{y}+(Ra_1-k_m a_2)y=La_3\dot{u}+(Ra_3-a_2)u \qquad \text{(viii)}$$

Substitute (vii) in the coefficients of (viii):

$$La_1+Rb=L\left(\frac{k_m^2}{L}+k_e\right)+Rb=k_m^2+Lk_e+Rb$$

$$Ra_1-k_m a_2=R\left(\frac{k_m^2}{L}+k_e\right)-k_m\frac{k_m R}{L}=Rk_e$$

$$La_3=L\frac{k_m}{L}=k_m$$

$$Ra_3-a_2=R\frac{k_m}{L}-\frac{k_m R}{L}=0$$

Substitute in (viii). We have the input–output differential equation:

$$Lm\dddot{y}+(Lb+Rm)\ddot{y}+\left(k_m^2+Lk_e+Rb\right)\dot{y}+Rk_e y=k_m\dot{u}$$

Note: At this stage, it is a good idea to check the physical units of each of the coefficients and verify that they are consistent. If not, that means, you have made errors.

As expected, we obtained a third-order input–output differential equation model for this third-order system (having three independent energy storage elements).

The corresponding system TF:

$$\frac{V_m(s)}{V_a(s)}=\frac{k_m s}{Lms^3+(Lb+Rm)s^2+\left(k_m^2+Lk_e+Rb\right)s+Rk_e}$$

(d)

The TFLG of the system in Figure 7.10b is shown in Figure 7.10c.

Now, we convert the electrical domain (the LHS segment of the electro-mechanical transformer in Figure 7.10c) into an equivalent mechanical domain.

The domain conversion of the transformer-coupled two-domain system is carried out using the standard result of converting a Thevenin segment in one domain into an equivalent Thevenin segment in the other domain.

In the converted segment, the equivalent velocity source is $V_e(s)=\frac{1}{k_m}V_a(s)$ and the series mobility in that segment is $M_e=\frac{(Ls+R)}{k_m^2}$

Note: In the present problem, the parameter of the electro-mechanical transformer is $r=\frac{1}{k_m}$.

We get the equivalent TFLG shown in Figure 7.10d. This equivalent system is entirely in the mechanical domain.

In Figure 7.10d, the three parallel branches of *m*, *b*, and *k* can be combined to give the mobility M_m, as follows: $\frac{1}{M_m}=ms+b+\frac{k_e}{s}$

Hence, $M_m(s)=\frac{s}{ms^2+bs+k_e}$

The corresponding equivalent TFLG is shown in Figure 7.10e.

Applying potential division, we get

$$V_m=\frac{M_m}{M_e+M_m}\times V_e(s)=\frac{V_e(s)}{M_e/M_m+1}=\frac{1}{\left[\frac{(Ls+R)}{k_m^2}\times\frac{\left(ms^2+bs+k_e\right)}{s}+1\right]}\frac{1}{k_m}V_a(s)$$

$$= \frac{k_m s}{\left[(Ls+R)\left(ms^2+bs+k_e\right)+k_m^2 s\right]} V_a(s)$$

Hence, the system TF is

$$\frac{V_m(s)}{V_a(s)} = \frac{k_m s}{\left[Lms^3+(Lb+Rm)s^2+\left(Rb+k_e L+k_m^2\right)s+Rk_e\right]}$$

This TF is identical to what was obtained previously, by the time-domain approach.

Example 7.6

Consider the multi-domain (multi-physics or mixed) system consisting of both mechanical components and fluid components, as shown in Figure 7.11. In this system, a pump of pressure $P_s(t)$, which is a pressure source (the system input), pumps water through a uniform horizontal thin pipe, into a horizontal cylinder of area of cross-section A and a leak-proof piston, which serve as the hydraulic actuator that drives a mechanical load. The combined mass of the actuator piston and the mechanical load is m, the resisting stiffness of the mechanical load is k, and the combined viscous damping constant of the actuator piston and the mechanical load is b. The fluid inertance in the pipe is represented by I and the fluid resistance in the pipe is represented by R. The pressure ripples in the water flow of the pipe are suppressed before the flow enters the actuator, by means of an energy absorber consisting of an open tank of fluid capacitance C_f (assumed to be constant). The water in this buffer tank goes through a valve of fluid resistance R_v before entering the actuator cylinder.

Note: Even though the fluid resistance R_v of the valve is adjustable (by operating the valve), assume that it is a constant in the present example.

The velocity v_m of the load m (also, of the actuator piston) is the system output.

i. Draw a complete LG for the system. Orient (with arrows) all the branches. Mark all the variables and parameters of the LG branches. Giving proper justification, indicate the order of the system.
ii. Using the LG, and giving all the details, systematically develop a complete (and linear) state-space model for the system. Use the following state variables:
   Q_I = volume flow rate of the water in the pipe before reaching the buffer tank
   P_f = gauge pressure of the water at the bottom of the buffer tank
   v_m = velocity of the mechanical load (and also of the actuator piston)
   f_k = spring force of the mechanical load (attached to the actuator piston)
   Express the state-space model in the vector-matrix form.

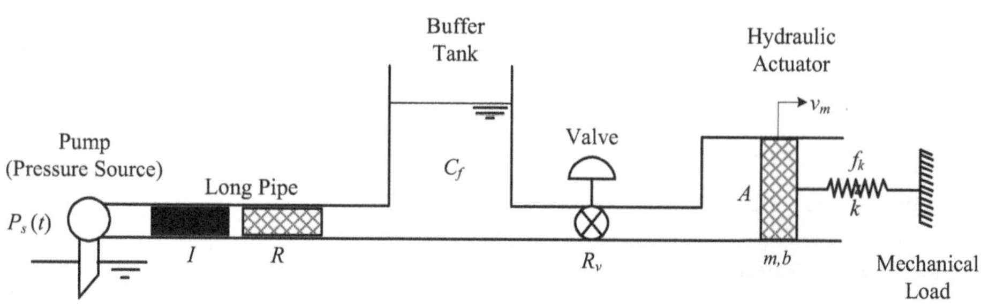

FIGURE 7.11 A hydraulic actuator system that operates a mechanical load.

iii. From the state-space model, determine the input–output differential equation (input= $P_s(t)$, output= v_m), working entirely in the time domain. From that equation, write the system TF.
iv. Give the TFLG corresponding to the LG obtained in Part (i). Appropriately reduce it, by combining branches. Convert it into a TFLG that is entirely in the mechanical domain. From that, systematically obtain the system TF.

Solution

(i)

The LG of the system is shown in Figure 7.12a.

The particular choice of the primary loops (five) and the primary nodes (6 – 1 = 5) are indicated on the LG.

The system has four independent energy storage elements (I, C_f, m, k). Hence, the system is fourth order, with four state variables (which will be chosen according to our systematic approach).

(ii)

State-Space Shell:

$$I\frac{dQ_I}{dt} = P_I$$
$$C_f\frac{dP_f}{dt} = Q_f$$
$$m\frac{dv_m}{dt} = f_m$$
$$\frac{df_k}{dt} = kv_k$$

Other Constitutive Equations:

$$P_R = RQ_R$$
$$P_v = R_vQ_v$$
$$f_b = bv_b$$

$$\left.\begin{aligned} v_2 &= \frac{Q_1}{A} \\ f_2 &= -AP_1 \end{aligned}\right\}\text{Fluid-mechanical gyrator}$$

Node Equations:

$$\text{Node 1: } Q_s - Q_I = 0 \text{ (useless); Node 2: } Q_I - Q_R = 0; \text{ Node 3: } Q_R - Q_f - Q_v = 0;$$
$$\text{Node 4: } Q_v - Q_1 = 0; \text{ Node 5: } -f_2 - f_m - f_k - f_b = 0$$

Loop Equations:

$$\text{Loop 1: } -P_f - P_R - P_I + P_s(t) = 0; \text{ Loop 2: } -P_1 - P_v + P_f = 0; \text{ Loop 3: } -v_m + v_2 = 0;$$
$$\text{Loop 4: } -v_k + v_m = 0; \text{ Loop 5: } -v_b + v_m = 0$$

Eliminate Auxiliary Variables:

$P_I = -P_f - P_R + P_s(t) = -P_f - RQ_R + P_s(t) = -P_f - RQ_I + P_s(t)$ {From Loop 1, R-constitutive, and Node 2 equations}

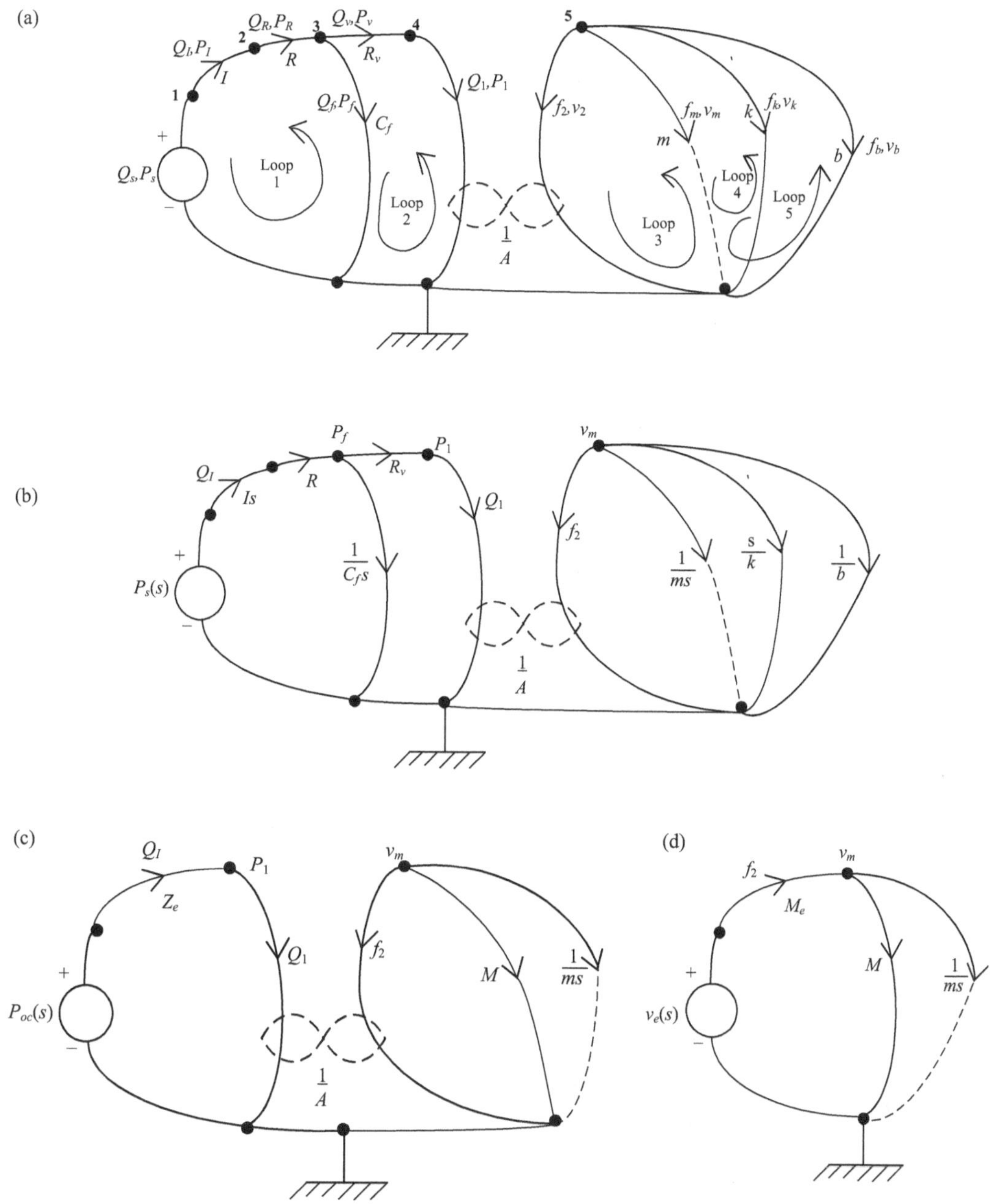

FIGURE 7.12 (a) Linear graph of the system; (b) The TFLG of the multi-domain system (fluid-mechanical); (c) The TFLG with the fluid domain expressed in the Thevenin form; (d) The equivalent TFLG entirely in the mechanical domain.

$Q_f = Q_R - Q_v = Q_I - Q_1 = Q_I - Av_2 = Q_I - Av_m$ {From Node 3, Node 2, Node 4, gyrator-constitutive, and Loop 3 equations}

$$
\begin{aligned}
f_m &= -f_2 - f_k - f_b = AP_1 - f_k - bv_b = A\left(-P_v + P_f\right) - f_k - bv_m \\
&= A\left(-R_vQ_v + P_f\right) - f_k - bv_m = -AR_vQ_1 + AP_f - f_k - bv_m \\
&= -AR_vAv_2 + AP_f - f_k - bv_m = -A^2R_vv_m + AP_f - f_k - bv_m = -\left(A^2R_v + b\right)v_m + AP_f - f_k
\end{aligned}
$$

{From Node 5, gyrator-constitutive, Loop 2, Loop 5, R_v-constitutive, Node 4, gyrator-constitutive, and Loop 3 equations}

$v_k = v_m$ {From Loop 4 equation}

State Equations:

$$I\frac{dQ_I}{dt} = -RQ_I - P_f + P_s(t) \tag{i}$$

$$C_f\frac{dP_f}{dt} = Q_I - Av_m \tag{ii}$$

$$m\frac{dv_m}{dt} = AP_f - \left(A^2R_v + b\right)v_m - f_k \tag{iii}$$

$$\frac{df_k}{dt} = kv_m \tag{iv}$$

$$\text{Output} = v_m$$

Vector-Matrix Formulation:

State vector $\boldsymbol{x} = \begin{bmatrix} Q_I & P_f & v_m & f_k \end{bmatrix}^T$

Input vector $\boldsymbol{u} = [u] = [P_s(t)]$

Output vector $\boldsymbol{y} = [y] = [v_m]$

State-Space Model:

$$\dot{\boldsymbol{x}} = \boldsymbol{A}\boldsymbol{x} + \boldsymbol{B}\boldsymbol{u}$$
$$\boldsymbol{y} = \boldsymbol{C}\boldsymbol{x} + \boldsymbol{D}\boldsymbol{u}$$

with

$$\boldsymbol{A} = \begin{bmatrix} -\frac{R}{I} & -\frac{1}{I} & 0 & 0 \\ \frac{1}{C_f} & 0 & -\frac{A}{C_f} & 0 \\ 0 & \frac{A}{m} & -\frac{\left(A^2R_v + b\right)}{m} & -\frac{1}{m} \\ 0 & 0 & k & 0 \end{bmatrix}; \boldsymbol{B} = \begin{bmatrix} \frac{1}{I} \\ 0 \\ 0 \\ 0 \end{bmatrix}; \boldsymbol{C} = \begin{bmatrix} 0 & 0 & 1 & 0 \end{bmatrix}; \boldsymbol{D} = [0]$$

(iii)

The state equations may be written as

$$\dot{x}_1 = -a_{11}x_1 - a_{12}x_2 + b_1u \tag{i}$$

$$\dot{x}_2 = a_{21}x_1 - a_{23}y \tag{ii}$$

$$\dot{y} = a_{32}x_2 - a_{33}y - a_{34}x_4 \tag{iii}$$

$$\dot{x}_4 = a_{43}y \tag{iv}$$

with

$$a_{11} = \frac{R}{I}; a_{12} = \frac{1}{I}; a_{21} = \frac{1}{C_f}; a_{23} = \frac{A}{C_f}; a_{32} = \frac{A}{m}; a_{33} = \frac{\left(A^2R_v + b\right)}{m}; a_{34} = \frac{1}{m}; a_{43} = k; b_1 = \frac{1}{I}$$

We have to express these four state equations as a single differential equation in *u* and *y* only. This is done as follows.

Substitute (iii) into (iv) to eliminate x_4:

$$m\frac{d}{dt}\left[-\dot{y}+a_{32}x_2-a_{33}y\right]=a_{43}y \rightarrow m\left[\ddot{y}+a_{33}\dot{y}\right]+ky=A\dot{x}_2 \quad \text{(v)}$$

Substitute (ii) into (i) to eliminate x_1:

$$C_f\frac{d}{dt}\left[\dot{x}_2+a_{23}y\right]=-a_{11}C_f\left[\dot{x}_2+a_{23}y\right]-a_{12}x_2+b_1u \rightarrow IC_f\ddot{x}_2+RC_f\dot{x}_2+x_2=-AI\dot{y}-ARy+u \quad \text{(vi)}$$

Substitute (v) into differentiated (vi) to eliminate x_2:

$$\frac{IC_f}{A}\left[m\frac{d^4y}{dt^4}+\left(A^2R_v+b\right)\frac{d^3y}{dt^3}+k\frac{d^2y}{dt^2}\right]+\frac{RC_f}{A}\left[m\frac{d^3y}{dt^3}+\left(A^2R_v+b\right)\frac{d^2y}{dt^2}+k\frac{dy}{dt}\right]$$
$$+\frac{1}{A}\left[m\frac{d^2y}{dt^2}+\left(A^2R_v+b\right)\frac{dy}{dt}+ky\right]=-AI\frac{d^2y}{dt^2}-AR\frac{dy}{dt}+\frac{du}{dt}$$

$$\frac{IC_f}{A}\left[m\frac{d^4y}{dt^4}+\left(A^2R_v+b\right)\frac{d^3y}{dt^3}+k\frac{d^2y}{dt^2}\right]+\frac{RC_f}{A}\left[m\frac{d^3y}{dt^3}+\left(A^2R_v+b\right)\frac{d^2y}{dt^2}+k\frac{dy}{dt}\right]$$
$$+\frac{1}{A}\left[m\frac{d^2y}{dt^2}+\left(A^2R_v+b\right)\frac{dy}{dt}+ky\right]=-AI\frac{d^2y}{dt^2}-AR\frac{dy}{dt}+\frac{du}{dt}$$

By substituting for a_{33} and rearranging, we have the I-O differential equation:

$$\frac{IC_f}{A}\left[m\frac{d^4y}{dt^4}+\left(A^2R_v+b\right)\frac{d^3y}{dt^3}+k\frac{d^2y}{dt^2}\right]+\frac{RC_f}{A}\left[m\frac{d^3y}{dt^3}+\left(A^2R_v+b\right)\frac{d^2y}{dt^2}+k\frac{dy}{dt}\right]$$
$$+\frac{1}{A}\left[m\frac{d^2y}{dt^2}+\left(A^2R_v+b\right)\frac{dy}{dt}+ky\right]=-AI\frac{d^2y}{dt^2}-AR\frac{dy}{dt}+\frac{du}{dt}$$

$$\frac{IC_fm}{A}\frac{d^4y}{dt^4}+\frac{IC_f\left(A^2R_v+b\right)}{A}\frac{d^3y}{dt^3}+\frac{IC_fk}{A}\frac{d^2y}{dt^2}+\frac{RC_fm}{A}\frac{d^3y}{dt^3}+\frac{RC_f\left(A^2R_v+b\right)}{A}\frac{d^2y}{dt^2}$$
$$+\frac{RC_fk}{A}\frac{dy}{dt}+\frac{m}{A}\frac{d^2y}{dt^2}+\frac{\left(A^2R_v+b\right)}{A}\frac{dy}{dt}+\frac{k}{A}y=-AI\frac{d^2y}{dt^2}-AR\frac{dy}{dt}+\frac{du}{dt}$$

Now, group the like terms, and multiply throughout by *A*:

$$IC_fm\frac{d^4v_m}{dt^4}+\left[IC_f\left(A^2R_v+b\right)+RC_fm\right]\frac{d^3v_m}{dt^3}+\left[IC_fk+RC_f\left(A^2R_v+b\right)+m+A^2I\right]\frac{d^2v_m}{dt^2}$$
$$+\left[RC_fk+A^2\left(R_v+R\right)+b\right]\frac{dv_m}{dt}+kv_m=A\frac{dP_s(s)}{dt}$$

Or

$$IC_fm\frac{d^4v_m}{dt^4}+\left[IC_f\left(A^2R_v+b\right)+RC_fm\right]\frac{d^3v_m}{dt^3}+\left[IC_fk+RC_f\left(A^2R_v+b\right)+m+A^2I\right]\frac{d^2v_m}{dt^2}$$
$$+\left[RC_fk+A^2\left(R_v+R\right)+b\right]\frac{dv_m}{dt}+kv_m=A\frac{dP_s(s)}{dt}$$

As expected, we obtained a fourth-order input–output differential equation model for this fourth-order system (which has four independent energy storage elements).

The corresponding system TF (by changing $\frac{d}{dt}$ to s):

$$\frac{v_m(s)}{P_s(s)} = \frac{As}{IC_f m s^4 + \left[IC_f(A^2R_v + b) + RC_f m\right]s^3 + \left[IC_f k + RC_f(A^2R_v + b) + m + A^2 I\right]s^2 + \left[RC_f k + A^2(R_v + R) + b\right]s + k}$$

(iv)

The TFLG of the LG in Figure 7.12a is shown in Figure 7.12b.

The fluid domain is now converted into the Thevenin form, as shown in Figure 7.12c.

Further reduction of the TFLG has been done as well, by combining the two parallel branches: k-branch and the b-branch in the mechanical domain, into a single branch with mobility M. Here,

$$M = \frac{\frac{s}{k} \times \frac{1}{b}}{\left(\frac{s}{k} + \frac{1}{b}\right)} = \frac{s}{bs + k} \tag{vi}$$

By following the usual procedure for Thevenin circuit development, we have:

Equivalent (open-circuit) pressure source

$$P_{oc}(s) = \frac{1}{\left(Is + R + \frac{1}{C_f s}\right)} \times \frac{1}{C_f s} P_s(s) = \frac{1}{C_f s(Is + R) + 1} P_s(s) \tag{vii}$$

Note: The potential divider (pressure divider, in the fluid domain) method is used in writing this equation.

$$\text{Equivalent fluid impedance } Z_e = \frac{(Is + R) \times \frac{1}{C_f s}}{(Is + R) + \frac{1}{C_f s}} + R_v = \frac{Is + R}{C_f s(Is + R) + 1} + R_v \tag{viii}$$

Note: Here, after killing (i.e., shorting) the fluid source, we combined the resulting two parallel branches and then added the series branch.

Next, we convert the fluid domain into an equivalent mechanical domain, through the gyrator (fluid-mechanical energy converter). We get the equivalent TFLG shown in Figure 7.12d. This equivalent system is entirely in the mechanical domain.

The domain conversion of the original gyrator-coupled two-domain system is carried out as follows:

Constitutive Equations for Gyrator: $v_m = \frac{1}{A} Q_1;\ f_2 = -AP_1$

Loop Equation (with constitution equation for Z_e substituted): $-P_1 - Z_e Q_l + P_{oc}(s) = 0$

Substitute the node equation, $Q_l - Q_1 = 0: -P_1 - Z_e Q_1 + P_{oc}(s) = 0$

Substitute the gyrator equations, to eliminate the fluid-domain variables:

$$\frac{f_2}{A} - Z_e A v_m + P_{oc}(s) = 0 \text{ or } v_m = \frac{1}{AZ_e} P_{oc}(s) + \frac{1}{A^2 Z_e} f_2$$

This result has only the mechanical side variables of the output branch of the gyrator (i.e., v_m and f_2).

From this result, we have

$$\text{Equivalent velocity source, } v_e(s) = \frac{1}{AZ_e} P_{oc}(s) \tag{ix}$$

$$\text{Equivalent series mobility, } M_e = \frac{1}{A^2 Z_e} \tag{x}$$

In Figure 7.12d, the equation of the first branch in the equivalent system is

$$v_m = v_e(s) - M_e f_2 \tag{xi}$$

Note: The direction of f_2 in the first branch of the TFLG that is entirely in the mechanical (see Figure 7.12d) is opposite of what is in the output branch of the gyrator in the original TFLG (see Figure 7.12c). This is the reason for the sign reversal of the f_2 term in (xi).

The combined mobility of the two parallel branches in Figure 7.12d is

$$M_m = \frac{M \times \dfrac{1}{ms}}{\left(M + \dfrac{1}{ms}\right)} = \frac{M}{Mms+1} \tag{xii}$$

To get the system TF, we need an expression for v_m (the output). For this, we use the familiar, "potential divider" approach (in the present "mechanical" situation, we use the analogous, "velocity divider" approach) in Figure 7.12d:

$$v_m = \frac{M_m}{(M_e + M_m)} \times v_e(s)$$

Into this equation, substitute (ix), then (vii):

$$v_m = \frac{M_m}{(M_e + M_m)} \times \frac{1}{AZ_e} P_{oc}(s) = \frac{1}{AZ_e\left(M_e/M_m + 1\right)} \times \frac{1}{\left[C_f s(Is+R)+1\right]} P_s(s)$$

The system TF is

$$\text{TF} = \frac{v_m}{P_s(s)} = \frac{1}{AZ_e\left(M_e / M_m + 1\right)\left[C_f s(Is+R)+1\right]}$$

Now substitute (xii), (x), and (vi):

$$\text{TF} = \frac{1}{AZ_e\left(\dfrac{1}{A^2 Z_e} \times \dfrac{Mms+1}{M} + 1\right)\left[C_f s(Is+R)+1\right]} = \frac{A}{\left(ms + \dfrac{1}{M} + A^2 Z_e\right)\left[C_f s(Is+R)+1\right]}$$

$$= \frac{A}{\left(ms + \dfrac{bs+k}{s} + A^2 Z_e\right)\left[C_f s(Is+R)+1\right]}$$

Now substitute (viii):

$$\text{TF} = \frac{A}{\left[ms + \dfrac{bs+k}{s} + A^2\left\{\dfrac{Is+R}{C_f s(Is+R)+1} + R_v\right\}\right]\left[C_f s(Is+R)+1\right]}$$

$$= \frac{As}{\left[ms^2 + bs + k + A^2 s\left\{\dfrac{Is+R}{C_f s(Is+R)+1} + R_v\right\}\right]\left[C_f s(Is+R)+1\right]}$$

$$= \frac{As}{\left[\left(ms^2+bs+k\right)\left\{C_f s(Is+R)+1\right\} + A^2 s(Is+R) + A^2 R_v s\left\{C_f s(Is+R)+1\right\}\right]}$$

Simplify:

$$\mathrm{TF} = \frac{As}{\left[\left(ms^2+bs+k\right)\left\{C_f s(Is+R)+1\right\}+A^2Is^2+A^2Rs+A^2R_vC_fIs^3+A^2R_vC_fRs^2+A^2R_vs\right]}$$

Or

$$\mathrm{TF} = \frac{As}{\left[ms^2C_f s(Is+R)+bsC_f s(Is+R)+kC_f s(Is+R)+ms^2+bs+k+\cdots\right.\\ \left.A^2Is^2+A^2Rs+A^2R_vC_fIs^3+A^2R_vC_fRs^2+A^2R_vs\right]}$$

Or

$$\mathrm{TF} = \frac{As}{\left[mC_fIs^4+mC_fRs^3+bC_fIs^3+RbC_fs^2+kC_fIs^2+RkC_fs+ms^2+bs+k+\cdots\right.\\ \left.A^2Is^2+A^2Rs+A^2R_vC_fIs^3+A^2R_vC_fRs^2+A^2R_vs\right]}$$

Or

$$\mathrm{TF} = \frac{As}{mC_fIs^4+\left[A^2R_vC_fI+bC_fI+mC_fR\right]s^3+\left[kC_fI+A^2R_vC_fR+RbC_f+m+A^2I\right]s^2+\\ \cdots+\left[RkC_f+A^2R+A^2R_v+b\right]s+k}$$

This result is identical to what we obtained from the time-domain approach, in Part (iii).

7.5 SUMMARY SHEET

- **Model Conversion/Approximation**: 1. Distributed parameter to lumped parameter; 2. nonlinear to linear; 3. time domain to TF (Laplace domain, frequency domain); 4. detailed component based to a reduced TF; 5. complex structure to simple equivalent (Thevenin or Norton) structure
- **Thevenin Equivalence**: Equivalent *A*-type source with equivalent *A*-type TF (generalized impedance) in series
- **Thevenin Source**: Open-circuit across-variable → $V_{oc}(s)$
- **Thevenin Impedance**: Overall impedance with sources killed $= Z_e$
- Killing a force source → open-circuiting it → transmitted force $= 0$
- Killing a velocity source → short-circuiting it → velocity across $= 0$
- **Norton Equivalence**: Equivalent *T*-type source with equivalent *A*-type TF (generalized impedance) in parallel
- **Norton Source**: Short-circuit through-variable → $I_{sc}(s)$; *Note*: $V_{oc} = Z_e I_{sc}$
- **Norton Impedance**: Overall impedance with sources killed = Thevenin impedance $= Z_e$
- **Series Branches:** Combined mobility (or, generalized impedance) $M = M_1 + M_2$
- **Parallel Branches:** Combined mobility (or, generalized impedance) $M = \dfrac{M_1M_2}{M_1 + M_2}$

- **Steps of Equivalent TFLG Approach**
  1. Draw TFLG; mark *A*-type TFs (generalized impedances) for all the branches (except sources)
  2. On TFLG indicate important variables only (e.g., system inputs and outputs)
  3. Simplify (reduce, condense) TFLG by combining branches. Do not combine a branch whose variable needs to be determined (e.g., output branch)
  4. Depending on problem objective, determine which TFLG segment should be cut. The equivalent TFLG of the remaining part needs to be determined
  5. Depending on problem objective, establish whether Thevenin equivalence or Norton equivalence is needed. Use Thevenin equivalence if a through-variable needs to be determined; use Norton equivalence if an across-variable needs to be determined
  6. Determine the equivalent source and generalized impedance of the equivalent LG
  7. Using the equivalent LG, determine the variable or function of interest.

- **Domain Transformation (Conversion into Equivalent Single Domain)**: 1. Determine the Thevenin equivalent LG (associated source $P_{oc}(s)$ and generalized impedance Z_e) of the subsystem to be converted; 2. determine the converted equivalent *A*-source and series *A*-TF (generalized impedance) depending on the two-port coupling element
- **Transformer-Coupled Systems:** Converted equivalent *A*-source $V_e(s) = rP_{oc}(s)$; converted equivalent generalized impedance (in series): $M_e = r^2 Z_e$

  Note: r = transformer parameter
- **Gyrator-Coupled Systems:** Converted equivalent *A*-source $V_e(s) = \frac{M}{Z_e} P_{oc}(s)$; converted equivalent generalized impedance (in series): $M_e = \frac{M^2}{Z_e}$
- *Note*: M = gyrator parameter

PROBLEMS

7.1 Consider the ground-based mechanical oscillator shown in Figure P7.1.

a. Sketch the TFLG of the system. Indicate the source element and the mobility functions of the branches.

b. Combine the spring and the damper branches.

c. Virtually cut the mass branch and find the Thevenin equivalent TFLG of the remaining TFLG. Then, determine the force transmissibility.

d. Determine the force (inertial force) through the mass and the velocity of the mass using 1. simplified TFLG; 2. Thevenin equivalent TFLG. Show that the results from the two methods are identical.

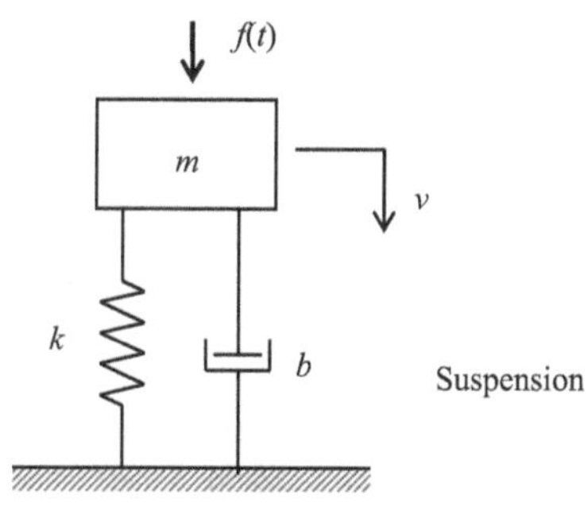

FIGURE P7.1 Ground-based mechanical oscillator.

7.2 Consider the mechanical oscillator with support motion, as shown in Figure P7.2.

a. Sketch the TFLG of the system. Indicate the source element and the mobility functions of the branches.
b. Combine the spring and damper branches.
c. Virtually cut the mass branch and find the Norton equivalent TFLG of the remaining TFLG.
d. Determine the force (inertial force) through the mass and the velocity of the mass using 1. simplified TFLG; 2. Norton equivalent TFLG. Show that the results from the two methods are identical.

7.3 Consider the mass-spring-damper system that has two inputs: force source $f(t)$ acting on the mass; velocity $v(t)$ source at the base, as shown in Figure P7.3. The resulting velocity of mass m is v_m. The force associated with the velocity source $v(t)$ is f_s, which is also the force transmitted through the suspension unit consisting of the spring and the damper (*Note*: $v(t)$, not f_s, is the input. f_s is the "dependent variable" of the velocity source).

a. Sketch the TFLG for the system and mark the mobility parameters of the branches and the source variables $F(s)$ and $V(s)$, in the Laplace domain. Also mark the mass velocity V_m and the support force F_s on the TFLG.
b. Redraw the TFLG so that the two parallel branches corresponding to k and b are combined into a single branch with mobility M_s. Express M_s in terms of k and b.
c. Make a virtual cut of the branch corresponding to the mass element in the TFLG of Part (b) and determine the corresponding Norton equivalent TFLG.

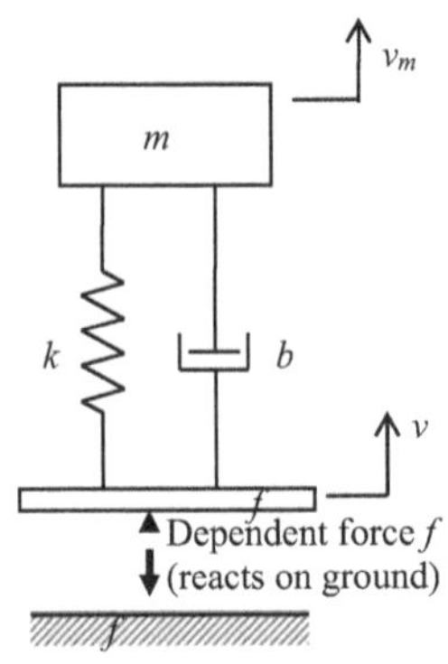

FIGURE P7.2 Oscillator with support motion.

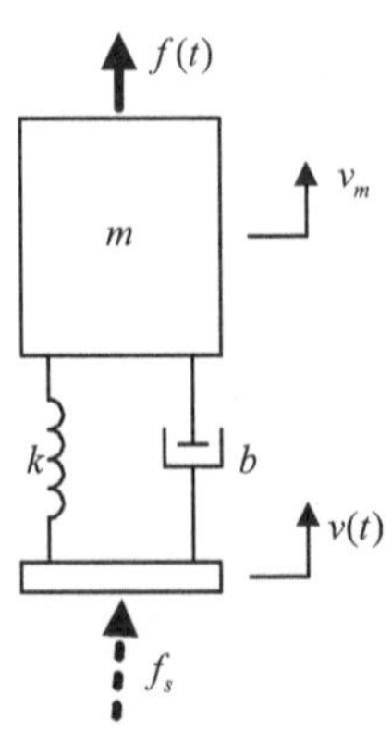

FIGURE P7.3 Mechanical oscillator with a force source and a velocity source.

Using the Norton TFLG express V_m in terms of the inputs $F(s)$ and $V(s)$. From this result, determine the motion transmissibility $V_m/V(s)$ for the special case: $F(s) = 0$.

d. Make a virtual cut of the branch corresponding to the suspension element (M_s) in the TFLG of Part (b) and determine the corresponding Thevenin equivalent TFLG.

Using the Thevenin TFLG, express F_s in terms of the inputs $F(s)$ and $V(s)$.

From this result, determine the force transmissibility $F_s/F(s)$ for the special case of $V(s) = 0$ (i.e., when the base is fixed).

7.4 An incompressible liquid is pumped into a tank, as schematically shown in Figure P7.4. The pump is modeled as a pressure source of pressure $P(t)$, which is measured with respect to the ambient (atmospheric) pressure. The overall resistance of the liquid flow in the piping is R_f and the overall inertance of the liquid flow in the piping is I_f. The liquid capacitance (due to gravity head) of the tank is C_f.

a. Sketch an LG for the system (*Note*: You may introduce auxiliary variables as necessary).

b. Taking $\boldsymbol{x} = [P_H \;\; Q]^T$ as the state vector, $\boldsymbol{u} = [P(t)]$ as the input, and $\boldsymbol{y} = [H]$ as the output, determine a complete state-space model for the system (i.e., determine the model matrices $\boldsymbol{A}$, $\boldsymbol{B}$, $\boldsymbol{C}$, and $\boldsymbol{D}$) systematically, using the LG. Nomenclature:

P_H = liquid pressure at the bottom of the tank with respect to the ambient pressure
Q = volume flow rate of the liquid in the piping
H = height of liquid in the tank
ρ = mass density of the liquid
g = acceleration due to gravity

c. From the state-space model of Part (b), determine the input–output differential equation of the system.

d. Represent the LG of Part (a) in the Laplace domain, as a TFLG and indicate the impedance functions (i.e., [pressure]/[flow rate] ratio in the Laplace domain) of each branch (except the source). Cut the branch corresponding to the liquid capacitance. Obtain the Norton equivalent circuit of the remaining circuit (i.e., determine an equivalent liquid flow source $Q_{sc}(s)$ and an equivalent liquid impedance Z_e in parallel, to which the cut capacitance branch will be connected in parallel. From this circuit, determine the TF between the capacitance pressure $P_H(s)$ and the input pressure $P(s)$.

Determine the corresponding input–output differential equation (in the time domain; i.e., change s to $\frac{d}{dt}$) relating H and $P(t)$. Verify that this result is identical to what was obtained in Part (c).

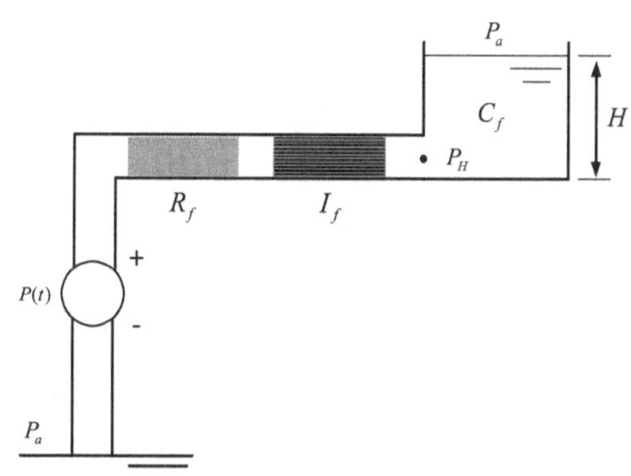

FIGURE P7.4 A liquid pumping system.

7.5 Figure P7.5 shows two systems (a) and (b), which may be used to study force transmissibility and motion transmissibility, respectively. Clearly discuss whether the force transmissibility F_s/F (in the Laplace domain) in System (a) is equal to the motion transmissibility V_m/V (in the Laplace domain) in System (b) (i.e., check the "duality," or "complementarity" of the two systems), by carrying out the following steps:

1. Draw the LGs for the two systems and mark the mobility functions for all the branches (except the source elements); that is, sketch the TFLGs
2. Simplify the two TFLGs by combining branches as appropriate (series branches: mobilities add; parallel branches; inverse rule applies for mobilities) and mark the mobilities of the combined branches
3. Based on the objectives of the problem (i.e., determination of the force transmissibility of System (a), and motion transmissibility of System (b)), for applying Thevenin's theorem, determine which segment of the TFLG should be cut (*Note*: The variable of interest in the particular transmissibility function should be associated with the segment of the circuit that is cut)
4. Based on the objectives of the problem, establish whether Thevenin equivalence or Norton equivalence is appropriate (Specifically: Use Thevenin equivalence if a through-variable needs to be determined, because the Thevenin LG gives two series elements with a common through-variable; use Norton equivalence if an across-variable needs to be determined, because the Norton LG gives two parallel elements with a common across-variable)
5. Determine the equivalent sources and mobilities of the equivalent TFLGs of the two systems
6. Using the two equivalent TFLGs, determine the transmissibility functions of interest
7. By analysis, examine whether the two mobility functions obtained in this manner are equivalent.

Note: Neglect the effects of gravity (i.e., assume that the systems are horizontal, supported on frictionless rollers; or the static deflections of the springs compensate for the gravitational forces).

Bonus: Extend your results to an n-degree-of-freedom system (i.e., one with n mass elements), structured as in Figure P7.5a and b

7.6 Consider the rotatory load driven by a turbine through a long shaft and a step-down gear, as shown in Figure P7.6.

Its state-space model is given by

$$\dot{T}_k = k\left(\omega_s(t) - p\omega_l\right)$$

$$J_l\dot{\omega}_l = pT_k - b_l\omega_l$$

Input$=\omega_s(t)$; output$=\omega_l$

The I/O differential equation model is obtained by eliminating T_k as follows:

The second-state equation gives $T_k = \dfrac{1}{p}\left(J_l\dot{\omega}_l + b_l\omega_l\right)$

Substitute in the first-state equation after differentiating

$$\frac{1}{p}\left(J_l\ddot{\omega}_l + b_l\dot{\omega}_l\right) = k\left(\omega_s(t) - p\omega_l\right) \Rightarrow J_l\ddot{\omega}_l + b_l\dot{\omega}_l + kp^2\omega_l = kp\omega_s(s)$$

The corresponding TF is $\dfrac{\omega_l}{\omega_s} = \dfrac{kp}{J_l s^2 + b_l s + kp^2}$

Obtain this result by using the approach of TFLG (Laplace domain) and equivalent TFLG (Thevenin or Norton).

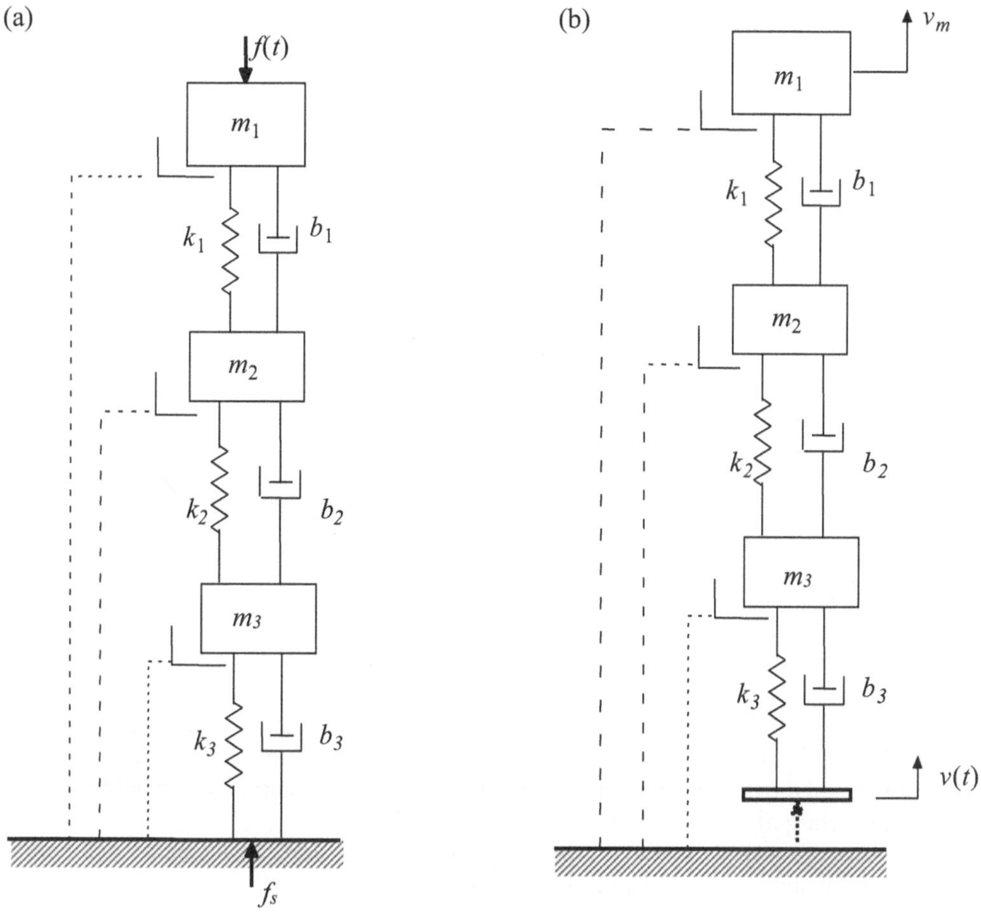

FIGURE P7.5 Two mechanical systems. (a) For determining force transmissibility; (b) for determining motion transmissibility.

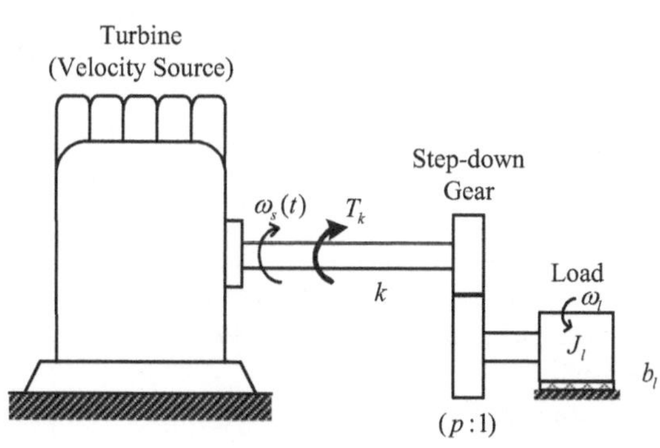

FIGURE P7.6 A rotatory load driven by a turbine through a long shaft and a step-down gear.

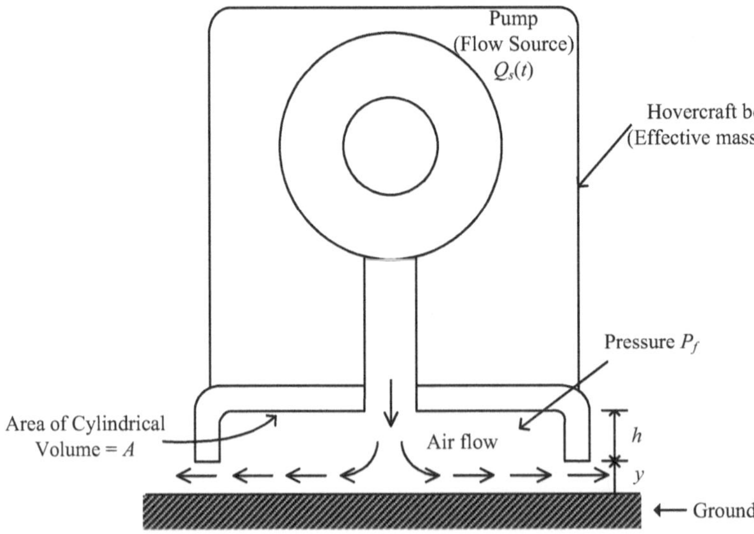

FIGURE P7.7 A simplified model for vertical dynamics of a hovercraft.

7.7 Consider the simplified model for the vertical dynamics of a hovercraft, as sketched in Figure P7.7.

Input $\boldsymbol{u} = [Q_s(t)]$; output $\boldsymbol{y} = [v]$. Also, $Q_s(t)$=volume flow rate of the pump; P_f=gage pressure of air in the cylindrical space underneath the hovercraft; A=area of cross-section of the cylindrical space; C_f=fluid capacitance due to air compressibility; R_f=fluid flow resistance of the exiting air from the edges of the cylindrical space; v=vertical velocity of the hovercraft; M_h=effective mass of the hovercraft.

a. Sketch a conventional LG (in the time domain) and using it, systematically determine a linear state-space model for the system
b. Using the state-space model obtain the input–output differential equation and the corresponding TF of this model
c. Sketch a TFLG, in the Laplace domain, for the system
d. Convert the fluid domain into an equivalent mechanical domain and sketch the corresponding TFLG, which is entirely in the mechanical domain
e. Make a virtual cut off the branch that represents the mass M. Based on that, obtain the Norton equivalent TFLG
f. From the Norton TFLG, determine the system TF $\left(v/Q_s\right)$ and show that this result is identical to what was obtained in Part (b).

7.8 A mechanical load of mass m is driven by a force source $f(t)$ and is buffered using a gas shock absorber, as shown in Figure P7.8. The gas cylinder is properly sealed (i.e., gas does not leak out of the cylinder compartment). The piston moves against the cylinder without experiencing appreciable friction (or, that friction component may be incorporated into the damping in the motion of the mass, assuming that the shaft is rigid). The following parameters are given:

m=mass of the moving load
b=equivalent viscous damping constant of the load motion
C_f=capacitance of the gas in the shock absorber
A=area of the piston

a. Using Laplace domain variables and A-type TFs (e.g., mobility functions for mechanical elements), sketch a TFLG for the system.
b. Suppose that we are interested in the force transmissibility function from the applied force $F(s)$ to the damping force $F_b(s)$ at the mechanical damper b. Make a suitable cut

of a branch and determine a suitable equivalent TFLG (you must decide whether to use Thevenin equivalence or Norton equivalence, based on what needs to be determined). Using that equivalent TFLG, determine the transmissibility function $F_b(s)/F(s)$, where $F(s)$ = Laplace-transformed $f(t)$.

7.9 An armature-controlled linear dc actuator (similar in principle to a rotatory dc motor, but for rectilinear motions rather than rotatory motions) is used to adjust the linear motion of a load (see Figure P7.9). Nomenclature

L_a = armature leakage inductance
R_a = armature resistance
$v_a(t)$ = drive voltage in the armature circuit (input)
k_m = force constant (the constant of proportionality between the generated magnetic force and the armature current); analogous to torque constant
m_a = armature mass
m_l = load mass
b_l = damping constant of the resisting force on the load mass
k = stiffness of the link between m_a and m_l
v_l = speed of load mass (output)

Note 1: In ideal energy conversion, k_m is equal to the constant of proportionality between the back emf and the speed of the motor.

Note 2: In armature control, the field (stator) conditions are assumed constant.

a. Draw a TFLG in the Laplace domain, for the system, indicating the *A*-type TFs (e.g., electrical impedance and mechanical mobility) of the branches
b. Obtain an equivalent TFLG of the given electromechanical system, entirely in the mechanical domain
c. Make a cut at the branch corresponding to the load mass and obtain the Norton equivalent TFLG of the converted system
d. Using the Norton equivalent TFLG, determine the TF of the system.

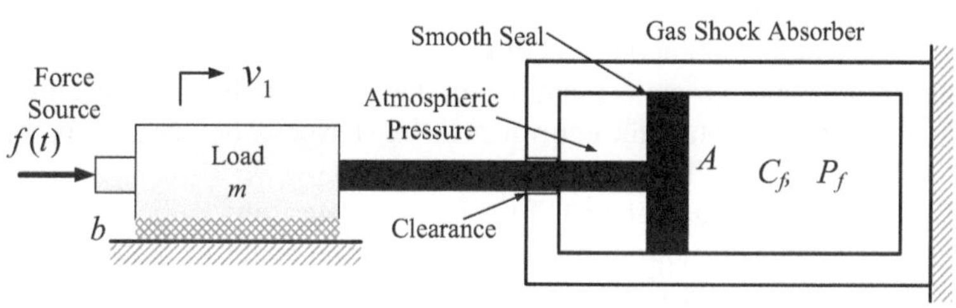

FIGURE P7.8 Mechanical load with a shock absorber.

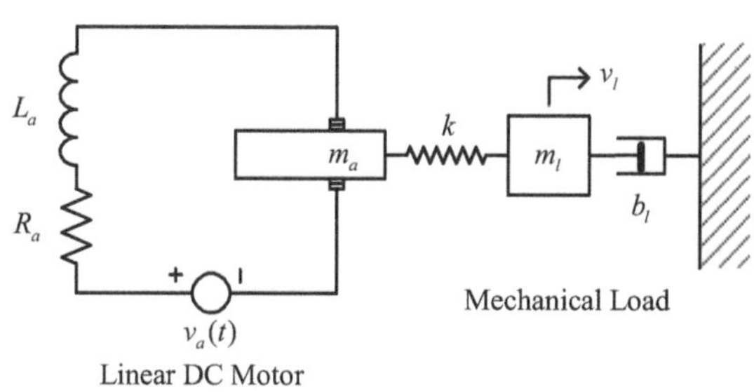

FIGURE P7.9 DC linear actuator that moves a linear mechanical load.

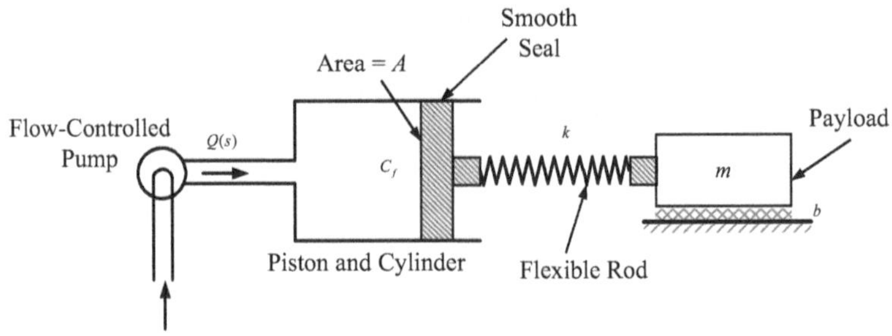

FIGURE P7.10 Motion control of a payload using a pump-cylinder device.

7.10 A flow-controlled liquid pump drives a piston within a cylinder. The piston rod, which is flexible with stiffness k, moves a payload of mass m against some mechanical resistance that is approximated by a linear viscous damper of damping constant b (see Figure P7.10). Also,

C_f = fluid capacitance of the piston-cylinder unit
A = area of the piston on the liquid side.

a. Determine the TF between the input $Q(s)$ and the output V_m, where
$Q(s)$ = volume flow rate of the pump
V_m = speed of the mass m.
Note: All the variables are expressed in the Laplace domain.

b. From the result of Part (a), determine the input–output differential equation of the system.
In solving this problem, you must use the following steps:
1. Draw a TFLG of the system, in the Laplace domain. Indicate the pertinent A-type TFs of the TFLG branches (in terms of the parameters C_f, A, k, b, and m)
2. Convert the fluid domain into an equivalent mechanical domain. Give the corresponding, equivalent TFLG in the mechanical domain. Clearly indicate the input (source) and the output on the TFLG
3. Perform an appropriate cut and obtain the pertinent TFLG (Thevenin or Norton, as appropriate). Obtain the equivalent source (velocity or force, as appropriate) and the equivalent mobility for the equivalent TFLG
4. By analyzing the equivalent TFLG of step 3, determine V_m in terms of $Q(s)$.

7.11 An armature-controlled dc motor is used to operate a fluid pump, as schematically shown in Figure P7.11a. The equivalent armature (rotor) circuit of the motor is shown in Figure P7.11b, where the field windings (stationary) are assumed to provide a steady magnetic field.

$v_a(t)$ = armature voltage (input)
R_a = resistance in the armature windings
L_a = leakage inductance of the armature
i_L = current through the inductance.

The mechanical subsystem is modeled as in Figure P7.11c.

J = overall moment of inertia (motor rotor and pump impeller combined; this may also include the "added mass" of the fluid in the pump
b_l = linear viscous damping constant representing the pump fluid load (corresponding to resisting torque at constant speed, which excludes the added mass effect), and mechanical friction in the motor and the pump
ω = speed of the inertia (= motor speed ω_m = pump speed ω_l).
Note: The shaft connecting the motor to the pump is rigid.

Also, k_m = torque constant of the motor.

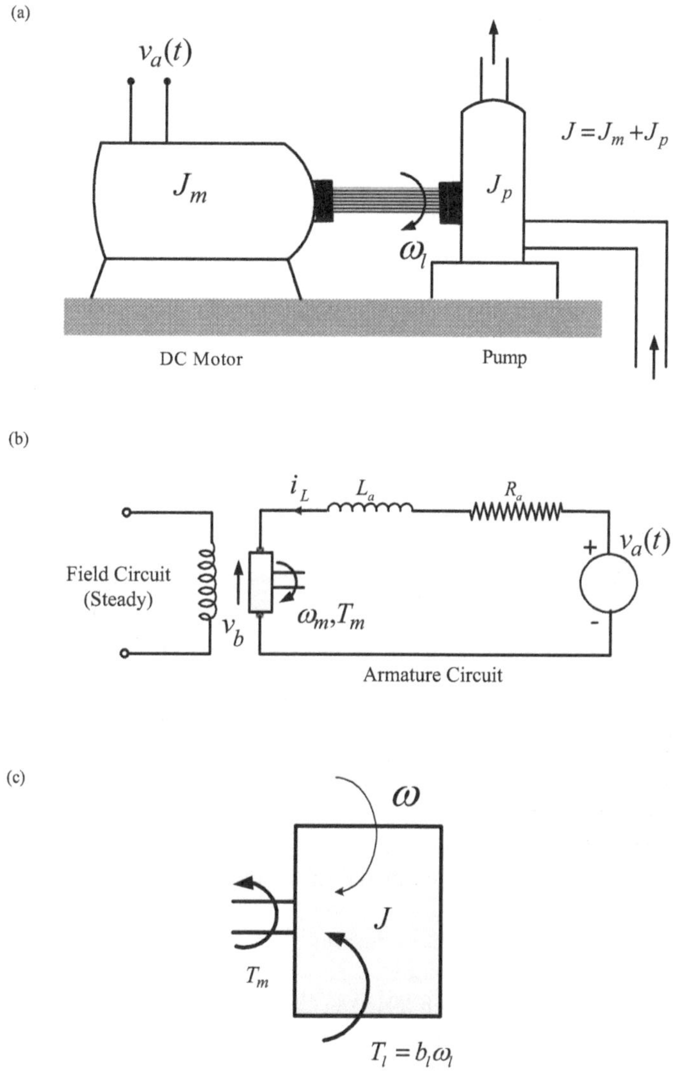

FIGURE P7.11 (a) An armature-controlled dc motor operating a fluid pump; (b) Armature circuit (field circuit is steady); (c) Mechanical model.

a. Sketch a conventional (time-domain) LG for the system (*Note*: You may introduce auxiliary variables as necessary)
b. Taking $\boldsymbol{x} = \begin{bmatrix} \omega & i_L \end{bmatrix}^T$ as the state vector, $\boldsymbol{u} = [v_a(t)]$ as the input, and $\boldsymbol{y} = [\omega_l]$ as the output, determine a complete state-space model for the system, systematically, using the LG
c. From the state-space model, determine the input–output differential equation relating the output ω_l to the input $v_a(t)$.
d. Represent the LG of Part (a) in the Laplace domain as a TFLG, and indicate the electrical impedances and mechanical mobilities of all the branches (except the source and the transformer). Convert the input side of the TFLG (electrical domain) into an equivalent mechanical domain. Cut the branch corresponding to the load b_l and obtain the corresponding Norton equivalent TFLG (i.e., an equivalent torque source in parallel with an equivalent mechanical mobility) to which the branch b_l will be connected in parallel.

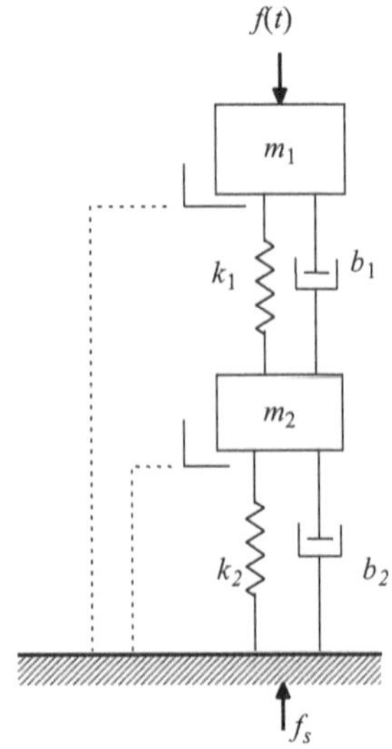

FIGURE P7.12 Ground-based two-degree-of-freedom oscillator.

From this Norton equivalent TFLG, determine the TF between the output ω_l and the input v_a. Determine the corresponding input–output differential equation. Verify that this result is identical to what was obtained in Part (c).

7.12 Consider the ground-based two-degree-of freedom oscillator shown in Figure P7.12.
Using Thevenin's theorem and LGs, determine an expression for the force transmissibility of the system.

7.13 Consider the two-degree-of freedom mechanical oscillator with support motion, shown in Figure P7.13. Using Thevenin's theorem and LGs, determine an expression for the motion transmissibility of the system.

7.14 The model of a robotic hand that is turning a door knob may be represented as in Figure P7.14. The actuator is a dc motor.
The flowing parameters are given (also see Chapter 5):
k_m = torque constant of the dc motor
L_a = leakage inductance of the motor armature
R_a = resistance of the motor armature
J_d = equivalent moment of inertia of the motor rotor and the mechanical load (door knob)
k_d = torsional stiffness of the door knob
b_d = torsional damping constant of the motor bearings and the load.
The following variables are defined: input = armature voltage = $v_a(t)$; output = speed of the mechanical load (door knob) = ω_d.

a. Draw an LG for the system, and using it, systematically, obtain a complete state-space model.
b. By mathematically manipulating the state-space model, determine the input–output model (differential equation in ω_d and v_a) of the system (in the time domain).
c. Sketch the TFLG of the system. Indicate the *A*-type TFs of the branches (except the source branch).
d. Convert the electrical domain segment (dc motor, which is the "input segment") of the TFLG into an equivalent mechanical-domain representation, and sketch the corresponding overall TFLG entirely in the mechanical domain.
e. With the objective of determining the Laplace TF ω_d/v_a, do the following: 1. Reduce the equivalent mechanical TFLG by combining appropriate branches; 2. virtually cut

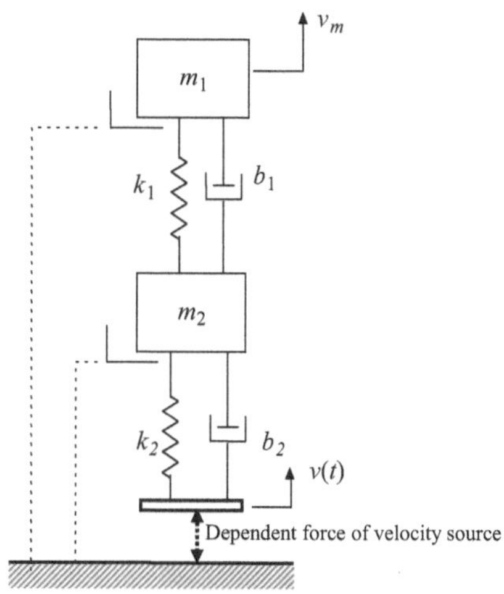

FIGURE P7.13 Two-degree-of-freedom oscillator with support motion.

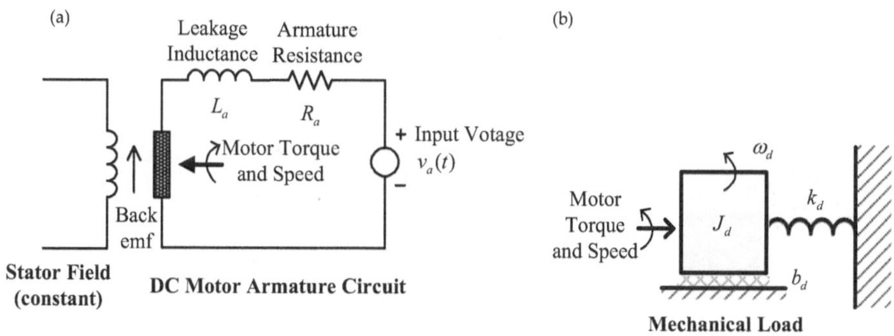

FIGURE P7.14 (a) Motor circuit; (b) mechanical load.

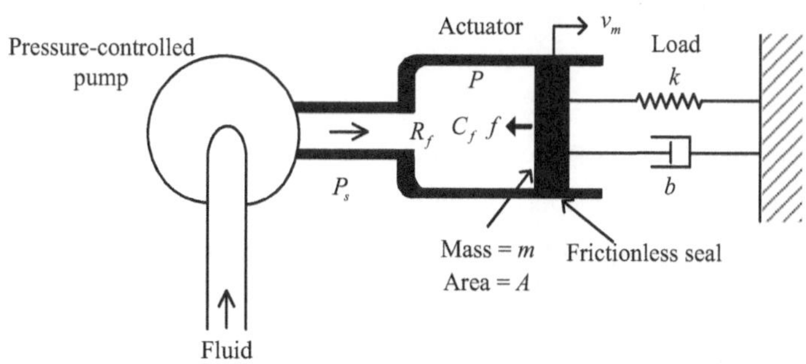

FIGURE P7.15 A fluid actuator moving a mechanical load.

an appropriate branch of the reduced TFLG; 3. determine an appropriate (Norton or Thevenin) equivalent TFLG for the system (in the mechanical domain).

f. Using the equivalent TFLG that was obtained in Part (e), determine the system TF ω_d/v_a (Laplace) and show that it is identical to the result obtained in Part (b).

7.15 A pressure-controlled fluid actuator is used to move a mechanical load (see Figure P7.15). The pump provides hydraulic fluid at pressure P_S, which enters the actuator chamber through a valve of fluid flow resistance R_f. The actuator chamber has a fluid capacitance C_f due to the compressibility of the fluid, whose pressure is P. This pressure moves the load whose mass is m and the area of cross-section is A, against a mechanical resistance of stiffness k and damping constant b. The mass is moved inside the actuator chamber at speed v_m.

a. Draw an LG for the system. Using $\boldsymbol{x} = \left[P, f_k, v_m\right]^T$ as the state vector, $\boldsymbol{u} = \left[P_s(t)\right]$ as the input, and $y = [x]$ as the output, give the state equations of the system. What is order of the system. Justify the answer. *Note*: f_k = spring force; x = position of the load (with velocity v_m).

b. By direct manipulation of the state equations in the Laplace domain, determine the TF $\frac{v_m}{P_s(s)}$ of the system.

c. Determine an equivalent TFLG of the system entirely in the mechanical domain. Using it, determine the TF $\frac{v_m}{P_s(s)}$ of the system. Show that this result is identical to what was obtained in Part (b).

7.16 A hydraulic motion-controlled system is schematically shown in Figure P7.16a. The control input to the system is the volume flow rate $Q(t)$, which is provided by a flow-controlled pump. The hydraulic fluid from the pump enters the cylinder of a positioning ram. The piston (area = A) of the ram is light and frictionless and adjusts the motion of the load consisting of a mass m, which moves at speed v against a spring of stiffness k (the other end of the spring is fixed, as shown) and compressive force f_3 and a viscous damper of damping constant b. The motion output of the system is velocity m of the mass m. Assume that the fluid in the rigid cylinder is compressible, and the corresponding fluid capacitance is C_f and the pressure is P. An LG for the system is shown in Figure 7.16b.

By the systematic, unified, and integrated approach, it can be shown that the state equation of the system is

$$C_f \dot{P}_1 = Q(t) - \frac{v}{M}$$

$$m\dot{v} = \frac{1}{M} P_1 - bv - f_3$$

$$\dot{f}_3 = kv$$

$$\text{with } M = \frac{1}{A}$$

i. By direct manipulation of the state equations in the Laplace domain, determine the TF models (input–output models in the Laplace domain) $\frac{v}{Q(s)}, \frac{f_3}{Q(s)}$, and $\frac{P_1}{Q(s)}$. Discuss the order of the system based on these models and also based on the state-space model.

ii. Determine an equivalent TFLG of the system entirely in the mechanical domain. Using it, determine the TF $\frac{v}{Q(s)}$ of the system. Show that this result is identical to what was obtained in Part (i).

Note: Express your result only in terms of the given parameters C_f, A, m, k, and b.

7.17 Consider the multi-domain (mixed) system consisting of both mechanical components and fluid components, as shown in Figure P7.17. A pump of pressure $P_s(t)$, which is a pressure source, pumps water into a uniform horizontal cylinder of area of cross-section A, which serves as the hydraulic actuator that drives a mechanical load. The combined mass of the actuator piston and the mechanical load is m, the resisting stiffness of the mechanical load is k, and the combined viscous damping constant of the actuator piston and the mechanical load is b. The water is pumped through a short pipe of circular cross-section.

Note: Assume that the water is incompressible.

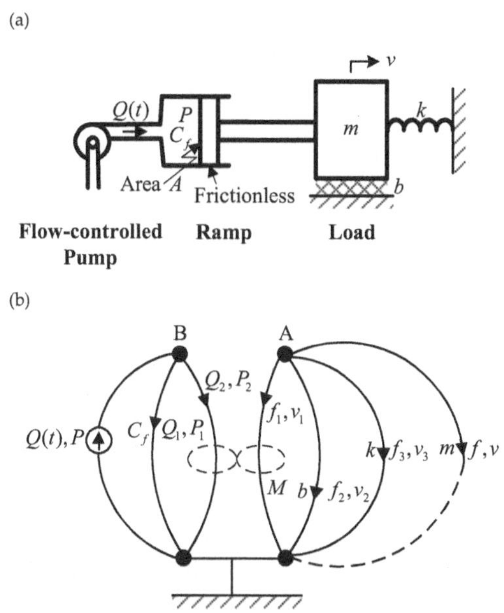

FIGURE P7.16 (a) A hydraulic motion-controlled system; (b) linear graph of the system.

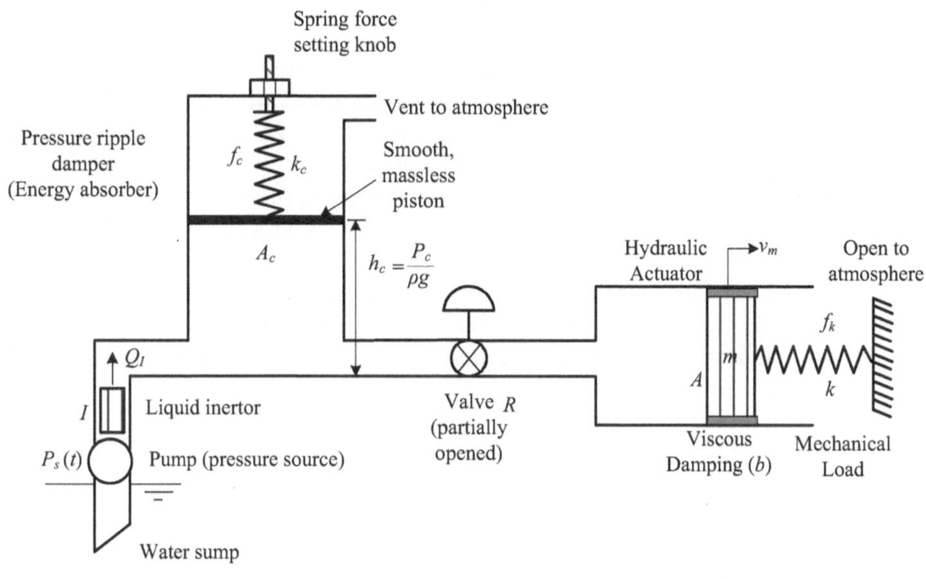

FIGURE P7.17 Water pumping system for a hydraulic actuator.

The pressure ripples in the water flow of the pipe are reduced before entering the actuator, by means of an energy absorber (hydraulic capacitor) consisting of a small fluid tank of area of cross-section A_c and a spring-loaded, light (massless) and smooth (no energy dissipation) piston with the resisting stiffness k_c. *Note*: This stiffness is adjustable using a nut, as shown, but in this question, assume it to be a constant.

The water flow into the actuator cylinder can be adjusted by means of a valve, as shown. It offers a fluid resistance R. Even though this is also adjustable, assume that it is a constant in the present question. This is the only notable fluid resistance that is present in the entire system (i.e., neglect any other hydraulic resistances).

Also given,

I = fluid inertance in the pipe from the pump up to the energy absorber (passive pressure controller). Neglect any other fluid inertances.

ρ = mass density of the water

g = acceleration due to gravity.

i. Draw a complete LG for the system. Orient all the branches. Mark all the variables and parameters of the LG branches.
ii. Using the LG, systematically develop a complete (and linear) state-space model for the system. Use the following state variables:

   v_m = velocity of the mechanical load (and also of the actuator piston)

   f_k = spring force of the mechanical load (attached to the actuator piston)

   f_c = compressive force of the spring of the energy absorber

   Q_I = volume flow rate of the water in the pipe before reaching the energy absorber

   P_c = pressure difference of the water column (of height h_c) in the energy absorber tank (not the pressure at the bottom of this water column). *Note*: $P_c = \rho g h_c$

   System output = v_m = velocity of the load (also, of the actuator piston).

   Note: Neglect the bulk modulus of water and the flexibility of the pipes, actuator cylinder, and the absorber tank.

   You must give all the vectors and matrices of the state-space model.
iii. From the state-space model, determine the input–output differential equation (input = $P_s(t)$, output = v_m). From that equation, write the system TF.
iv. Give the TFLG corresponding to the LG obtained in Part i. Appropriately reduce it, by combining branches. Convert it into a TF LG that is entirely in the mechanical domain. From that, systematically obtain the system TF. (*Note*: This result should be the same as the result obtained in Part iii. Even if your two answers are not identical, you will receive appropriate credit if the procedures are correct).

7.18 A mechanical load consists of mass m and a restraining spring of stiffness k. It is moved by a pressure-controlled hydraulic pump of pressure $P_s(t)$ through a long pipe and an end cylinder inside which the load is located (see Figure P7.18). The load m serves as the piston of the hydraulic cylinder. Neglect the friction between the piston and the cylinder. In addition to the load parameters m and k, the following system parameters are given:

R_f = hydraulic resistance in the pipe that connects the drive pump to the load cylinder

I_f = hydraulic inertance in the pipe

C_f = hydraulic capacitance in the cylinder of the load

A = piston/cylinder sectional area

Also,

System input = drive pressure of the pump = $P_s(t)$

System output = velocity of the load mass = v_m

a. Sketch a complete LG for the system. In particular, orient the LG (i.e., show the arrows), indicate the system parameters, and the through-across-variable pairs of the LG branches (*Note*: You will have to introduce auxiliary variables).
b. Systematically, using the unified and integrated approach (see Chapter 5), derive a state-space model for the system, using the LG. Give the corresponding matrices **A**, **B**, **C**, and **D**.
c. By mathematically manipulating the state equations, obtain the input–output differential equation relating the output v_m to the input $P_s(t)$. What is the corresponding Laplace TF?

d. Sketch the "TFLG" of the system. On the LG, indicate the A-type TFs of the branches and some key variables (which may be useful later in the LG simplification/reduction). Convert the hydraulic (fluid)-domain segment of the LG into an equivalent mechanical-domain segment. Sketch the resulting equivalent mechanical-domain LG. Reduce this LG into a Norton-equivalent LG, by performing a suitable virtual cut (for determining the system output). Using this equivalent LG, determine the Laplace TF v_m/P_s of the system. Show that this result is identical to what you obtained in Part(c).

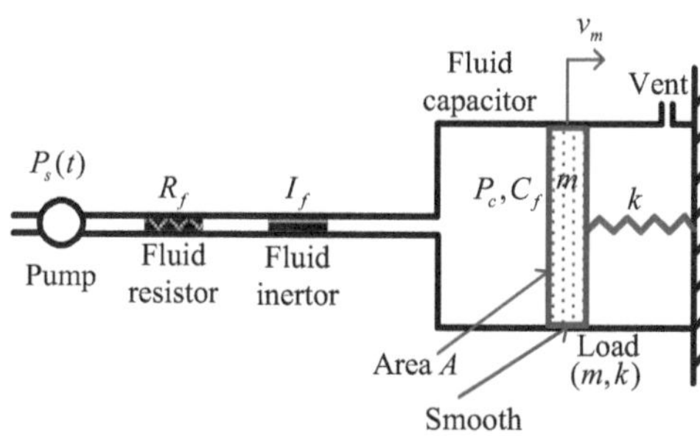

FIGURE P7.18 A mechanical load moved by a hydraulic device.

Appendix A
Graph-Tree Concepts for Linear Graphs

A graph tree, particularly a normal tree, provides a systematic procedure in model formulation linear graphs (LGs). The concepts and generation of a normal tree for a system are outlined in this appendix, and examples are given to illustrate them.

A.1 CONTINUITY EQUATIONS

A continuity equation may be written for through-variables passing through any closed contour drawn on the LG. This contour may enclose more than one node. Then, the resulting continuity equation = sum of the continuity equations of the nodes enclosed in the contour.

Proof: The internal through-variables have opposite signs with respect to the adjoining nodes and hence cancel out. Consider the example in Figure A.1.

Continuity Equation for Node A:

$$f_1 + f_2 - f_3 - f_4 + f_5 = 0 \tag{i}$$

Continuity Equation for Node B:

$$-f_5 + f_4 + f_6 - f_7 = 0 \tag{ii}$$

Continuity Equation for the Closed Contour:

$$f_1 + f_2 - f_3 + f_6 - f_7 = 0 \tag{iii}$$

It is seen that (iii) = (i) + (ii)

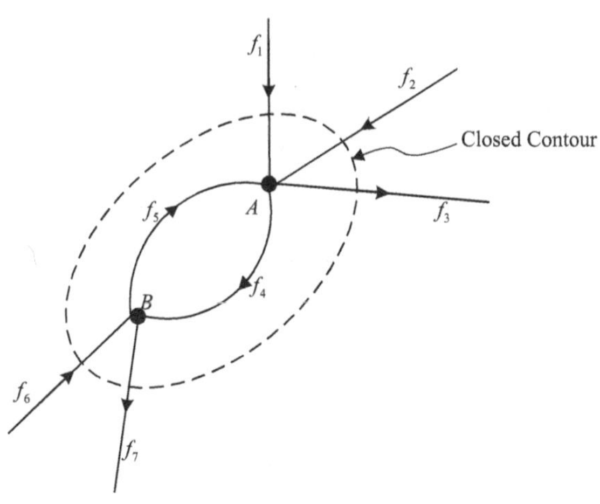

FIGURE A.1 Continuity equation through a closed contour.

A.2 GRAPH TREE

A graph tree of an LG is formed by all the nodes and the maximum number of branches of the LG such that no closed loops are created.

Note 1: For an LG with n nodes, there will be $n-1$ branches in its graph tree (proved later).

Note 2: Graph tree of an LG is not unique. More than one graph tree is possible for a given LG. As an example, consider the LG with four branches as shown in Figure A.2a.

It can generate four graph trees as shown in Figure A.2b.

Property: There are $n-1$ branches in a graph tree (n = number of nodes in the LG).

Proof: Consider any node in the LG. Connect it to a second node $\Rightarrow$ one branch with two nodes. The next branch in the graph tree will connect one of these nodes to just one new node (because it will not form loop—by the definition of graph tree) and so on until the graph tree is complete, connecting all n nodes.

A.2.1 LINK

A link is a branch of an LG that is not included in its graph tree.

Property: By definition, each link will form a new loop in the LG. Hence,

Number of primary loops (l) in an LG = number of links corresponding to any one of its graph trees. Then, by definition,

Number of links = $b-(n-1)$, where b = number of branches in the LG

Note: $n-1$ = number of branches in its graph tree. Hence,

$$l = b - (n-1)$$

(a)

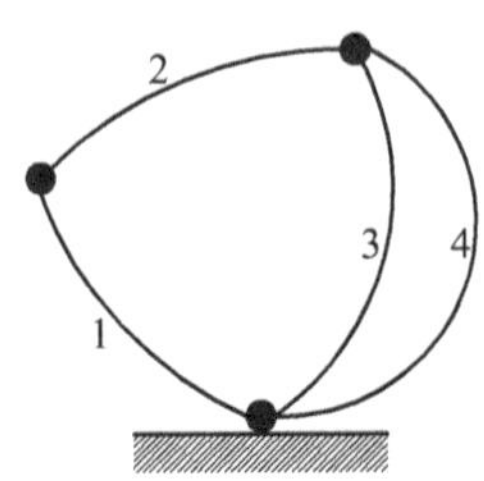

(b)

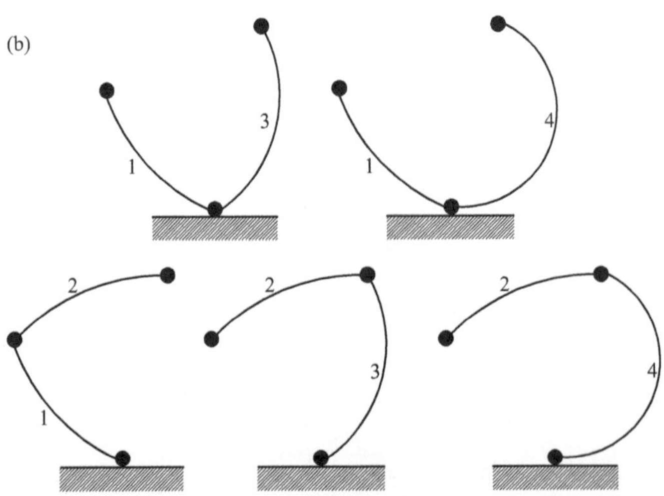

FIGURE A.2 (a) A linear graph example; (b) Possible graph trees of the LG in (a).

A.3 CONFLICT IDENTIFICATION IN AN LG USING ITS GRAPH TREE

There are several types of conflicts that can arise in an LG, which need to be resolved. Some situations are indicated now.

1. **Two or more independent *A*-sources forming a loop**
   See the example in Figure A.3a. The corresponding loop equation is $-v_2(t)+v_1(t)=0$. This violates the assumption that the *A*-sources are "independent," and hence, that $v_1(t)$ and $v_2(t)$ are independent.

2. **Two or more independent *T*-sources joining at a node**
   See the example in Figure A.3b. The node equation at N is $f_1(t)+f_2(t)=0$
   This violates the assumption that the *T*-sources are "independent" and hence, that $f_1(t)$ and $f_2(t)$ are independent.

3. **Two or more *A*-sources in a loop with some passive elements**
   See the example in Figure A.3c. The loop equation is $-v_2(t)-v+v_1(t)=0$. Hence, one of the two sources is redundant; the two sources act as a single *A*-source $v_1(t)-v_2(t)$.

4. **Two or more *T*-sources joining at a node with some passive elements**
   See the example in Figure A.3d. The node equation at N is $f_1(t)+f_2(t)-f=0$. Hence, one of the two sources is redundant; the two sources act as a single *T*-source $f_1(t)+f_2(t)$.

5. **An *A*-element closing a loop with other elements whose primary variables are *A*-variables**
   See the example in Figure A.4a. It has two *A*-type elements (C_1 and C_2) and a *D*-type element (R) in the loop. Loop equation $-v_1+v_2+v_3=0$.

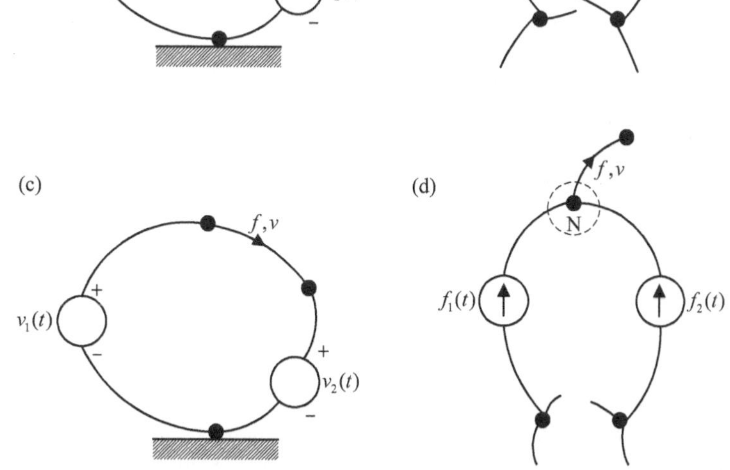

FIGURE A.3 (a) Two *A*-sources in a loop—a conflict; (b) Two *T*-sources at a node—a conflict; (c) Two *A*-sources and a passive element in a loop—a redundant case; (d) Two *T*-sources and a passive element at a node—a redundant case.

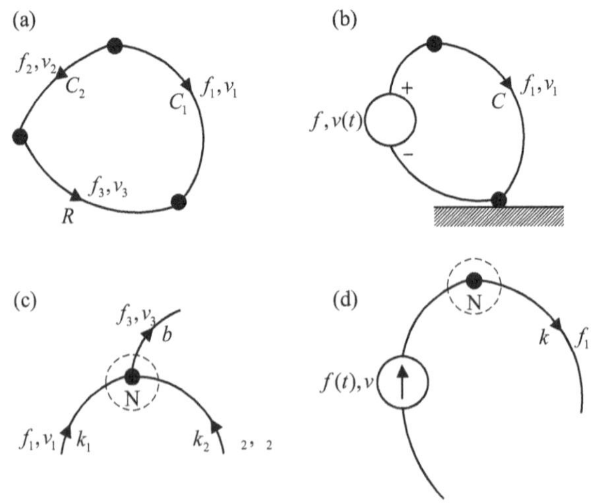

FIGURE A.4 (a) Two *A*-elements in a loop with a *D*-element—a case of dependent elements; (b) An *A*-source and an *A*-element in a loop—a case of decoupling; (c) Two *T*-elements in a node with a *D*-element—a case of dependent elements; (d) A *T*-source and an *A*-element in a loop—a case of decoupling.

In the *D*-element, the two variables f_3 and v_3 are completely determined by other state variables in the system (because *D*-type elements do not introduce new state variables.)

Hence, v_1 and v_2 cannot both act as state variables; one of them will be completely determined by other state variables (thereby violating a requirement for a state variable).

It follows that C_1 and C_2 are not independent *A*-type elements.

Now see the second example, shown in Figure A.4b. It has the *A*-type element *C* forming a loop with an *A*-type source.

Loop equation: $-v_1 + v(t) = 0$

Hence, v_1 is completely determined by the input $v(t)$. It follows that v_1 cannot serve as a state variable because it acts as an input variable (independent). In fact, element *C* has no effect on any other components that may be connected to this system (because it is in parallel with the *A*-source) and is effectively decoupled from the rest of the system (as discussed before concerning parallel connections with *A*-sources).

6. **A *T*-element forming a node with other elements whose primary variables are *T*-variables**

See the example in Figure A.4c. Here, two *T*-type elements K_1, K_2 and a *D*-type element *b* form a node. Node equation $f_1 + f_2 - f_3 = 0$

Since *b* is a *D*-type element, its variable f_3 is completely determined by other state variables in the system. Hence, f_1 and f_2 cannot both function as state variables; one of them will be completely determined by other state variables.

Hence, K_1 and K_2 are not independent *T*-type elements.

Now see the second example as shown in Figure A.4d. It has a *T*-type element *k* connected in series with a *T*-type source, forming a node.

Node equation $f(t) - f_1 = 0$ ® $f_1 = f(t)$

Hence, f_1 is completely determined by the input $f(t)$. It follows that f_1 cannot serve as a state variable because it acts as an input variable (independent). The element *k* has no effect on any other components that may be connected to this system (because it is in series with the *T*-source) and is effectively decoupled from the rest of the system (as discussed before concerning series connections with *T*-sources).

A.4 DEPENDENT ENERGY STORAGE ELEMENTS

There are two configurations of dependent energy storage elements which should be represented by an equivalent single energy storage element before addressing any other possibilities of dependence and conflicts in an LG. These two configurations are discussed now.

a. Several *A*-type Elements Connected in Series

Consider the case of two capacitor elements C_1 and C_2 connected in series. This is equivalent to a single capacitance C_{eq}, as indicated in Figure A.5.

The equivalent capacitor is derived now.

$$\text{We have}: f = f_1 = f_2$$

$$v = v_1 + v_2$$

with $C_1\dot{v}_1 = f_1$ and $C_2\dot{v}_2 = f_2$

Hence, $\dot{v} = \dot{v}_1 + \dot{v}_2 = \frac{1}{C_1}f_1 + \frac{1}{C_2}f_2 = \left(\frac{1}{C_1} + \frac{1}{C_2}\right)f$

Compare with $\dot{v} = \frac{1}{C_{eq}}f$

We have $\frac{1}{C_{eq}} = \frac{1}{C_1} + \frac{1}{C_2}$

This result may be generalized for the case of more than two series-connected *A*-type elements.

b. Several *T*-type Elements in Parallel

Consider the case of two springs k_1 and k_2 connected in parallel. This is equivalent to a single spring k_{eq}, as indicated in Figure A.6.

We have $v = v_1 = v_2$ and $f = f_1 + f_2$ with $\dot{f}_1 = k_1 v_1$ and $\dot{f}_2 = k_2 v_2$. Hence, $\dot{f} = \dot{f}_1 + \dot{f}_2 = k_1 v_1 + k_2 v_2 = (k_1 + k_2)v$. Compare with $\dot{f} = k_{eq} v$. We have $k_{eq} = k_1 + k_2$. This result may be generalized for the case of more than two parallel-connected *T*-type elements.

FIGURE A.5 Two capacitors in series and the equivalent capacitor.

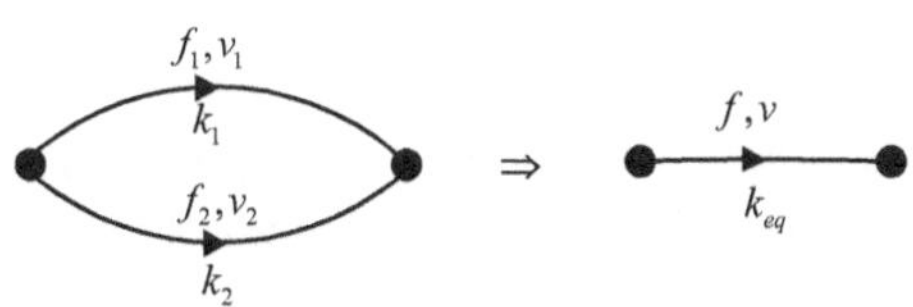

FIGURE A.6 Two springs in series and the equivalent spring.

A.5 NORMAL TREE OF A LINEAR GRAPH

We have defined "graph tree" which is a particular type of sub-structure of an LG. We found that more than one graph tree is possible for a given LG. A particularly useful graph tree is a *normal tree*. By obtaining a normal tree of an LG, it is possible to find

a. Any dependent energy storage elements
b. Conflicting source elements (i.e., source elements connected in unacceptable/redundant/ infeasible manner)
c. System order
d. A systematic way to obtain state-equations

A.5.1 Steps of Obtaining a Normal Tree

Step 1: Draw the LG of the system using *dotted lines* to represent the branches
Step 2: Convert any series-connected *A*-type elements into a single equivalent *A*-type element. Convert any parallel-connected *T*-type elements into a single equivalent *T*-type element
Step 3: Convert the dotted lines of any *A*-type sources into solid lines as long as they do not create loops. If two or more *A*-type sources form a loop, there is a conflict, which must be resolved before proceeding further (e.g., remove one source to open the loop and combine the remaining sources into a single source)
Step 4: If there is a transformer, convert one of its branches into a solid line.
Step 5: Convert the dotted lines of any *A*-type energy storage elements into solid lines (use half solid lines for inertia elements) as long as they do not create loops. If *A*-type energy storage elements form a loop, keep one of these elements aside. It is a dependent energy storage element (which does not generate a new state variable)
Step 6: If there is a gyrator, convert both of its branches into solid lines provided they do not create loops.
Step 7: Convert the dotted lines of any *D*-type elements into solid lines as long as they do not create loops.
Step 8: Convert the dotted lines of any *T*-type energy storage elements into solid lines as long as they do not create loops. The *T*-type elements included in the tree in this manner are dependent energy storage elements (they do not result in new state variables).
Step 9: If one or more *T*-type sources need to be included to complete the tree, they are conflicting source elements (they violate continuity). They have to be removed from the system and replaced by new elements, to complete the tree.
The resulting graph tree is called a normal tree. What information it provides and why are explained now.

A.5.2 Main Result from Normal Tree

Independent energy storage elements = *A*-type storage elements in normal tree + *T*-type storage elements to be added at links of the tree in forming the LG of the system.

Note: System order = the number of independent storage elements

Proof:

a. Since a graph tree (normal tree in the present case) does not have loops, no compatibility equations can be written. Hence, the across-variables of its *A*-type elements are linearly independent and provide a subset of state variables.
b. If a loop-forming *A*-type storage element exists, the corresponding loop equation provides an equation for its across-variable as a linear combination of the variables of the other

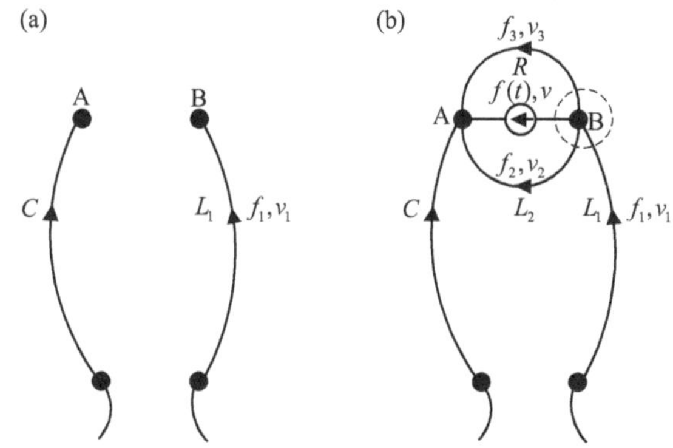

FIGURE A.7 (a) A segment of normal tree showing link AB; (b) Completed link AB and node B.

A-type elements in the loop. Hence, this *A*-type storage element is not independent, and its across-variable does not provide an additional state variable.

c. If a *T*-type storage element is present in the normal tree, a contour can be drawn through its branch and other branches of the LG that are links. Hence, a continuity equation can be written to express its through-variable in terms of the through-variables of the links (which can only be *T*-type storage elements, *T*-type sources, and/or *D*-type elements). Hence, this through-variable can be expressed as a linear combination of other state variables and input variables. It follows that this *T*-type storage element is not independent and its through-variable does not provide an additional state variable.

   As an illustrative example, suppose that a normal tree has two open nodes A and B which form the link AB. Also suppose that the *T*-type storage element L_1 (an inductor) is in the normal tree and B is a terminal of its branch, as shown in Figure A.7a. When completing the LG, suppose that link AB takes a *T*-type storage element L_2, a *T*-type source, and a *D*-type element *R* in parallel, as shown in Figure A.7b.
   The continuity equation for node B is $f_1 - f_2 - f(t) - f_3 = 0$. Now f_2 is a state variable (through-variable of the *T*-type independent storage element L_2) and $f(t)$ is an input variable (through variable of the *T*-type source). The through-variable f_3 of the *D*-type element *R* is expressible in terms of state variables and input variables (because a *D*-type element does not introduce any new variables).

   It follows that f_1 is expressed as a linear combination of state variables and input variables. Hence, it cannot be a new state variable, and the *T*-type element L_1 is not independent.

d. In view of step 2, no more than one *T*-type storage element can be present at a link of a normal tree. Then from the counter-argument of that given in item(c) above, its through-variable cannot be expressed as a linear combination of other state variables (and input variables). Hence, it is an independent storage element.

Justification of Step 5

Inclusion of one branch of the transformer is needed in order to recognize the possible presence of an *A*-type storage element that is dependent on another *A*-type element (dependence arising from the constitutive relation $v_o = rv_i$).

Both branches of the transformer should not be included in the normal tree because that can form a loop with an independent *A*-type storage element, which needs to be retained in the normal tree.

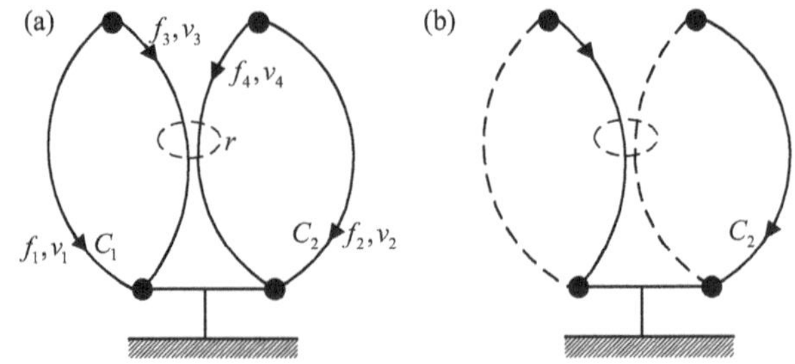

FIGURE A.8 (a) Both branches of the transformer should be in the normal tree; (b) Only one branch of the transformer should be in the normal tree.

As an illustrative example, consider the two *A*-type storage elements C_1 and C_2 linked by a transformer with parameter r, as shown in Figure A.8a.

Constitutive Equations: $C_1\dot{v}_1 = f_1$; $C_2\dot{v}_2 = f_2$; $v_4 = rv_3$; $f_4 = -\frac{1}{r}f_3$

Loop Equations: $-v_3 + v_1 = 0$; $-v_2 + v_4 = 0$

Node Equations: $-f_1 - f_3 = 0$; $-f_4 - f_2 = 0$

We note that $v_2 = v_4 = rv_3 = rv_1$

Hence, v_1 and v_2 cannot constitute two independent state variables. The elements C_1 and C_2 are not independent, and only one of them can remain in the normal tree. This is possible if and only if one of the two branches of the transformer is included in the normal tree. One possibility of a normal tree of the LG is shown in Figure A.8b.

Justification of Step 6

The constitutive relations of a gyrator are:

$$v_o = M f_i;\ f_o = -\frac{1}{M}v_i$$

Hence, unlike a transformer (which couples an across variable with an across variable), a gyrator couples an across-variable with a through-variable. Hence, it can help identify dependence between an *A*-type element and a *T*-type element.

In particular, on removing both branches of the gyrator from the normal tree, a need may arise to include a *T*-type element in the tree (either a dependent or conflicting element). Alternatively, by including both branches in the normal tree, a need may arise to remove an *A*-type element from the tree that forms a loop with a branch of the gyrator (again indicating a dependent element or a conflicting element).

As an illustrative example, consider an *A*-type storage element C_1 and a *T*-type storage element L linked through a gyrator (M) along with another *A*-type storage element C_2, as shown by the LG in Figure A.9a.

The equations of the LG are as follows.

Constitutive Equations:

$$C_1\dot{v}_1 = f_1$$

$$C_2\dot{v}_2 = f_2$$

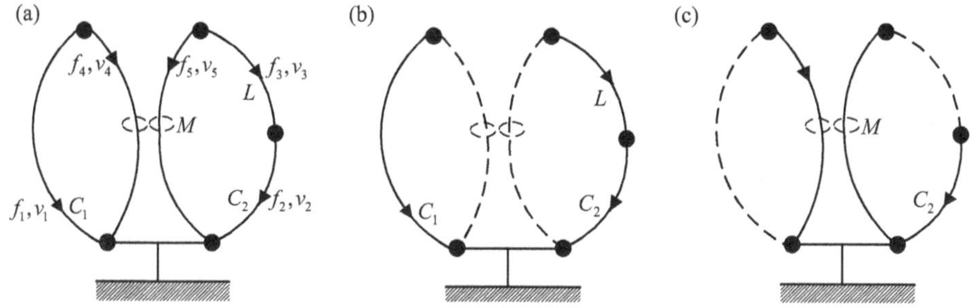

FIGURE A.9 (a) Example concerning possible inclusion of a gyrator in a normal tree; (b) A possible graph tree; (c) Another possible graph tree.

$$L\dot{f}_3 = v_3$$

$$v_5 = M f_4$$

$$f_5 = -\frac{1}{M} v_4$$

Loop Equations:

$$-v_4 + v_1 = 0$$

$$-v_2 - v_3 + v_5 = 0$$

Node Equations:

$$-f_1 - f_4 = 0$$

$$-f_3 - f_5 = 0$$

$$-f_2 + f_3 = 0$$

State Equations:

$$C_2\dot{v}_2 = f_2 = f_3 = -f_5 = \frac{1}{M}v_4 = \frac{1}{M}v_1 \quad \text{(i)}$$

Now consider the constitutive equation of the first capacitor:

$$C_1\dot{v}_1 = f_1 = -f_4 = -\frac{1}{M}v_5 = -\frac{1}{M}(v_2 + v_3) = -\frac{1}{M}\left(v_2 + L\dot{f}_3\right)$$

or

$$C_1\dot{v}_1 + \frac{L}{M}\dot{f}_2 = -\frac{1}{M}v_2 \tag{ii}$$

Equation (ii) shows the dependence of the A-type storage element C_1 and the T-type storage element L.

In fact, define the state variable

$$x = C_1 v_1 + \frac{L}{M} f_2$$

$$\text{(ii) becomes : } \dot{x} = -\frac{1}{M}v_2 \tag{ii'}$$

Equations (i) and (ii)' form a complete set of state equations ⇒ The model is second order for this system with three energy storage elements. This further confirms the dependence of the elements C_1 and L.

The possible graph trees for the system are shown in Figures A.9b and A.9c.

A.6 FORMULATION OF A STATE-SPACE MODEL USING A NORMAL TREE

The steps of formulating a state-space model, while systematically identifying conflicts in source elements and dependence in energy storage elements, are summarized now.

Step 1: Sketch the LG of the system, using dotted lines

Step 2: Superimpose a normal tree on the LG, using solid lines (half-solid line for an inertia element) using the usual steps. Resolve any conflicts among source elements. Identify the independent energy storage elements (and system order). Identify the state variables. Write constitutive equations for the elements in the normal tree

Step 3: Add missing branches (links of the normal tree) one by one into the normal tree. Write the resulting loop equations and constitutive equations

Step 4: Once the LG is complete, write the node equations. *Note*: In doing this, it may be convenient to draw a contour that intersects only one branch in the normal tree at a time. Write the continuity equation for the contour. Repeat this process until all the links of the normal tree (except those corresponding to A-type sources) are covered

Step 5: Eliminate the auxiliary variables in the state-space shell (constitutive equations of the independent energy storage elements)

Step 6: Formulate the algebraic output equation in terms of the state variables.

Example: Geared Load Driven by Velocity Source through Flexible Coupling

Consider the system having an A-source $u(t)$, A-type storage elements m_1 and m_2, T-type storage element k, D-type element b, and a transformer r, as shown in Figure A.10a. The output of the system is v_1, which is the velocity of the inertia m_1.

The LG of the system is sketched using dotted lines in Figure A.10b.

Note that this LG represents the same system, which we solved before.

A possible normal tree for the LG is shown in Figure A.10c.

We note from the normal tree that

Independent energy storage elements are m_1 and k

State vector $x = \begin{bmatrix} v_1 & f_2 \end{bmatrix}^T$

System order = 2

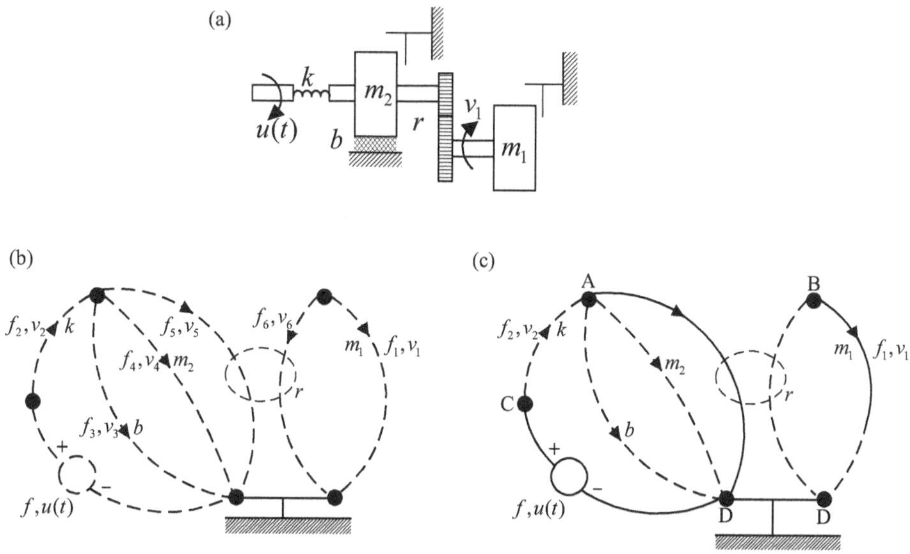

FIGURE A.10 (a) A dynamic system; (b) Linear graph of the system; (c) A normal tree of the LG.

State-Space Shell:

$m_1 : m_1\dot{v}_1 = f_1$ (i)

$k : \dot{f}_2 = kv_2$ (ii)

Links of the normal tree are AC, AD for b and m_2 and BD.
Insert link AC (k). Corresponding loop equation $-v_5 + v_2 + u(t) = 0$.
Insert link AD (b). The corresponding constitutive equation and loop equation are $b : f_3 = bv_3$

Loop equation: $-v_5 + v_3 = 0$

Insert link AD (m_2). The corresponding constitutive equation and loop equation are $m_2 : m_2\dot{v}_4 = f_4$

Loop equation: $-v_5 + v_4 = 0$

Insert link BD (r). The corresponding constitutive equations and loop equation are

$$v_6 = rv_5$$

$$f_6 = -\frac{1}{r} f_5$$

$$-v_1 + v_6 = 0$$

Node Equations:

Three independent node equations can be written for nodes A, B, and C. However, the equation for C involves the through-variable f of the A-source. Hence, it is not needed.

Node A: $f_2 - f_3 - f_4 - f_5 = 0$
Node B: $-f_1 - f_6 = 0$

Eliminate the auxiliary variable f_1 in (i):

$$f_1 = -f_6 = \frac{1}{r} f_5 = \frac{1}{r}(f_2 - f_3 - f_4) = \frac{1}{r}(f_2 - bv_3 - m_2\dot{v}_4) = \frac{1}{r}(f_2 - bv_5 - m_2\dot{v}_5)$$

$$= \frac{1}{r}\left(f_2 - \frac{b}{r}v_6 - \frac{m_2}{r}\dot{v}_6\right) = \frac{1}{r}\left(f_2 - \frac{b}{r}v_1 - \frac{m_2}{r}\dot{v}_1\right) = -\frac{m_2}{r^2}\dot{v}_1 - \frac{b}{r^2}v_1 + \frac{f_2}{r}$$

Hence,

$$m_1\dot{v}_1 = -\frac{1}{r^2}m_2\dot{v}_1 - \frac{b}{r^2}v_1 + \frac{f_2}{r}$$

or

$$\left(m_1 + \frac{1}{r^2}m_2\right)\dot{v}_1 = -\frac{b}{r^2}v_1 + \frac{f_2}{r}$$

or

$$m\dot{v}_1 = -bv_1 + rf_2 \quad \text{(i)}'$$

where $m = r^2 m_1 + m_2$

Eliminate v_2 in (ii):

$$v_2 = -v_5 + u(t) = -\frac{v_6}{r} + u(t) = -\frac{v_1}{r} + u(t)$$

Hence,

$$\dot{f}_2 = -\frac{K}{r}v_1 + Ku(t) \qquad \text{(ii)}'$$

This is identical to the result we obtained previously.

The algebraic output equation is $v_1 = x_1$. This corresponds to $\boldsymbol{C} = \begin{bmatrix} 1 & 0 \end{bmatrix}$

Example: Mechanical Load Moved by a Flow-Controlled Pump

A flow-controlled pump with volume flow rate $Q(t)$ operates a ramp (piston-cylinder device) which moves a mechanical load consisting of a spring (k) and mass (m) against damping (b), as shown in Figure A.11a. The fluid in the cylinder is compressible, and this effect is represented by the fluid capacitance C_f. Area of the piston is A. Velocity v of the mass is the output variable.

We will derive a state-space model for the system using the systematic approach, starting with a normal tree of the system LG.

The LG of the system is sketched in Figure A.11b. Note the hydraulic-mechanical gyrator (mixed-domain).

A normal tree of the LG is shown in Figure A.11c.

Note from the normal tree that the *T*-type storage element k is dependent.

Independent storage elements: C and m

State vector $\underline{x} = \begin{bmatrix} P_2 & v \end{bmatrix}^T$.

System order = 2

State-Space Shell:

$$m : m\dot{v} = f \qquad \text{(i)}$$

$$C_f : C_f\dot{P}_2 = Q_2 \qquad \text{(ii)}$$

Dependent Source:

$$k : \dot{f}_2 = kv_2$$

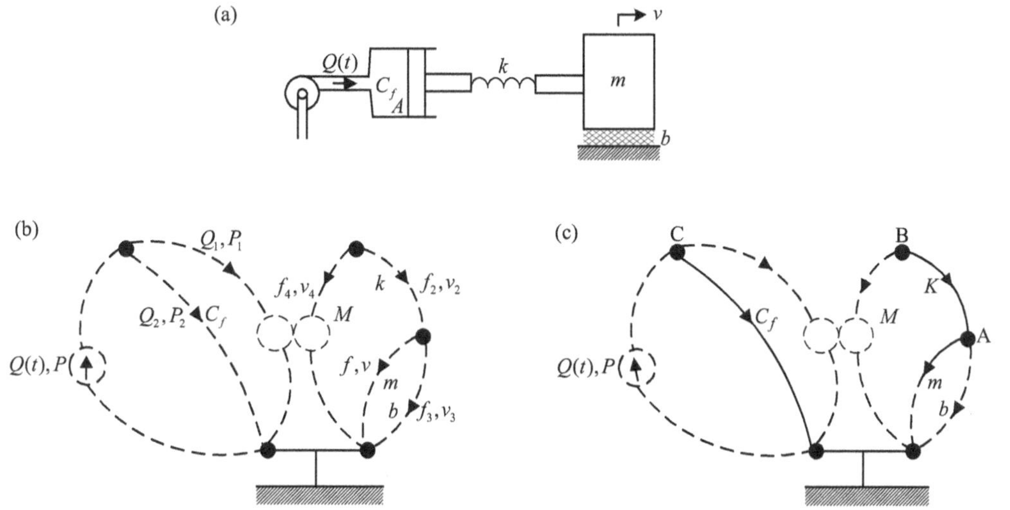

FIGURE A.11 (a) Flow-controlled ramp with a mechanical load; (b) Linear graph of the system; (c) A normal tree of the LG.

Insert Link *b*:

$$b: f_3 = b\, v_3$$

$$\text{Loop}: -v_3 + v = 0$$

Insert Links for Gyrator *M*:

$$\text{Constitutive Equations}: Q_1 = Av_4 \text{ (flow continuity)}$$

$$AP_1 + f_4 = 0 \text{ (force balance)}$$

Hence, with $M = \dfrac{1}{A}$

$$v_4 = MQ_1$$

$$f_4 = -\frac{1}{M} P_1$$

Loops: $-v - v_2 + v_4 = 0$

$$-P_1 + P_2 = 0$$

Note: Ignore the loop equation that includes the flow source because we do not need to determine its across variable P.

Node Equations:

Node A: $-f_3 - f + f_2 = 0$

Node B: $-f_4 - f_2 = 0$

Node C: $Q(t) - Q_2 - Q_1 = 0$

Eliminate f in (i):

$$f = f_2 - f_3 = -f_4 - bv_3 = \frac{1}{M}P_1 - bv = \frac{1}{M}P_2 - bv$$

Eliminate Q_2 in (ii):

$$Q_2 = Q(t) - Q_1 = Q(t) - \frac{1}{M}v_4 = Q(t) - \frac{1}{M}(v + v_2)$$

$$= Q(t) - \frac{1}{M}\left(v - \frac{1}{k}\dot{f}_2\right) = Q(t) - \frac{1}{M}\left(v + \frac{1}{k}\dot{f}_4\right)$$

$$= Q(t) - \frac{1}{M}\left[v + \frac{1}{k}\left(-\frac{1}{M}\dot{P}_1\right)\right] = Q(t) - \frac{1}{M}v + \frac{1}{M^2k}\dot{P}_2$$

State Equations:

$$m\dot{v} = -bv + \frac{1}{M}P_2 \quad \text{(i)}'$$

$$\left(C_f - \frac{1}{M^2k}\right)\dot{P}_2 = -\frac{1}{M}v + Q(t) \qquad \text{(ii)}'$$

We have

$$\boldsymbol{A} = \begin{bmatrix} -\dfrac{b}{m} & \dfrac{1}{mM} \\ -\dfrac{1}{M\left(C_f - \dfrac{1}{M^2k}\right)} & 0 \end{bmatrix}; \boldsymbol{B} = \begin{bmatrix} 0 \\ \dfrac{1}{C_f - M^2k} \end{bmatrix}; \boldsymbol{C} = \begin{bmatrix} 1 & 0 \end{bmatrix}; \boldsymbol{D} = [0]$$

Appendix B
MATLAB® Toolbox for Linear Graphs

Haoxiang Lang, Eric McCormick, and Clarence W. de Silva

This appendix presents a custom MATLAB® toolbox, called LGtheory, which provides a robust and automated method for generating linear graph (LG) models of multi-domain (multi-physics or mixed) engineering systems, within the MATLAB® programming environment. The need for such a toolbox is required because, while the LG approach is easy to perform manually for low-order systems, it is beneficial to automate this process to evaluate larger, more complex multi-domain systems. The toolbox can be downloaded from the following link:

https://github.com/GRASP-ONTechU/Linear_Graph

B.1 INPUTTING A LINEAR GRAPH INTO THE LGTHEROY TOOBOX

B.1.1 LGtheory Tooblbox Inputs

In order to properly use the LGtheory MATLAB toolbox, users must understand the requirements for converting LG models into the acceptable inputs of the toolbox. The main function of the toolbox, which converts the input LG model into the corresponding state-space representation, is

```
[Model] = LGtheory(LG);
```

where the input to the function (LG) is a structure array containing the following fields:

`S`—Source Vector:

A vector, which defines the starting (reference) nodes of each system element (represented using directed branches) in a column-wise manner.

`T`—Target Vector:

A vector, which defines the end nodes (points of action) of each system element (represented using directed branches) in a column-wise manner.

`Type`—Type Vector:

A vector, which defines the element type of each branch of the LG model in a column-wise manner. The element types are specified using indexing values as given in Table B.1.

`Domain`—Domain Vector:

A vector, which defines the energy domain of each system element in a column-wise manner. These values ensure that the parameters of the system elements are correctly accounted for (specifically in the case of spring elements in the mechanical domain) and ensure that the appropriate variable types are applied to each element. The energy domains are specified using the indexing values given in Table B.1.

`Var _ Names`—Variable Names Vector:

A vector, which defines the variable names of each system element in a column-wise manner. The variable name inputs must be symbolic variables or functions defined using either the sym or syms commands in MATLAB. These variable names will correspond to the system parameters used to symbolically represent the matrices of the state-space model.

Table B.1
Index Values of Element Types and Energy Domains for Inputs of LGtheory Toolbox

Index	Element Type	Index	Energy Domain
1	Across-variable source	0	Generalized
2	*A*-type element	1	Electrical
3	Transformer	2	Mechanical Translational
4	Gyrator	3	Mechanical rotational
5	*D*-type element	4	Hydraulic/Fluid
6	*T*-type element	5	Thermal
7	Through-variable source		

Table B.2
Format for Defining State-Space Outputs to LGtheory for Each Energy Domain and Variable Type

Energy Domain	Across-Variable	Through-Variable
Generalized	f_<name>(t)	v_<name>(t)
Electrical	V_<name>(t)	i_<name>(t)
Mechanical Translational	v_<name>(t)	F_<name>(t)
Mechanical rotational	Omega_<name>(t)	Tau_<name>(t)
Hydraulic/fluid	P_<name>(t)	Qf_<name>(t)
Thermal	T_<name>(t)	Q_<name>(t)

<name> is replaced with the symbolic variable name for the element of interest

`y`—Output Vector:

A vector, which defines the desired output variables of the state-space model. The output variable names must be symbolic variables defined using either the sym or syms commands in MATLAB. The symbolic variables must be created based on the desired (across- or through-) variable(s) and the energy domain(s) of the system element(s) using the formats given in Table B.2.

B.1.2 LGtheory Toolbox Outputs

Similar to the inputs, the output of the function, to the user (model), is a structure array containing various information used to construct the state-space model through the LG approach, as well as, the vectors and the matrices of the state-space model itself. The following fields are contained within the structure array:

`In`—Incidence Matrix:

A sparse matrix representation of the topology of an LG model using "1" and "−1" to represent the directionality of the system elements and the connection to the system nodes and zeros to represent the absence of connection between the elements and the nodes.

`Tree`—Normal Tree Matrix:

A matrix that represents the normal tree of the LG model in the form of an incidence matrix.

`Branches`—Branches Vector:

A vector that defines the column-wise indexes and the element types of the normal tree branches with non-zero values and the column-wise indexes of the co-tree links with zeros.

`CoTree`—Co-Tree Matrix:

A matrix that represents the co-tree of the LG model in the form of an incidence matrix.

`Links`—Links Vector:

A vector that defines the column-wise indices and the element types of the co-tree links with non-zero values and the column-wise indices of the normal tree branches with zeros.

`Across _ Vars`—Across-Variables Vector:

A vector that defines the across-variables of all the system elements in a column-wise manner using the naming convention that is defined in Table B.2.

`Through _ Vars`—Through-Variables Vector:

A vector that defines the through-variables of all the system elements in a column-wise manner using the naming convention that is defined in Table B.2.

`Prime`—Primary-Variables Vector:

A vector that defines the primary-variables of the system as the across-variables of normal tree branches and the through-variables of co-tree links. It is important to identify the primary-variables because they are the variables that determine the energy stored by each independent energy storage element within the system.

`Secon`—Secondary-Variables Vector:

A vector that defines the secondary-variables of the system as the through-variables of normal tree branches and the across-variables of co-tree links. These variables have no impact on the energy storage of the independent system elements.

`Params`—State-Space Parameters Vector:

A vector that defines the parameters of each system element in a column-wise manner based on the variable name defined by the user. While similar to the Var_Names vector, this vector accounts for the requirement to inverse the spring constants of the mechanical domains, and the resistance parameters, in some cases.

`x`—State Vector:

A column vector containing the state-variables of the state-space model, defined as the across-variables of the *A*-type elements in the normal tree and the through-variables of the *T*-type elements in the co-tree.

`u`—Input Vector

A column vector containing the input-variables of the state-space model, as defined by the source elements of the LG model.

`elem _ eqns`—Elemental/Constitutive Equations Vector:

A column vector containing the constitutive (elemental) equations of each passive element, as defined in Table B.2.

`cont _ eqns`—Continuity Equations Vector:

A column vector containing the continuity (node) equations for each passive branch of the normal tree, isolated for the secondary-variable of that element.

`comp _ eqns`—Compatibility Equations Vector:

A column vector containing the compatibility (loop) equations for each passive link of the normal tree, isolated for the secondary-variable of that element.

`A`—State Matrix

An $n \times n$ matrix (where n is the order of the system), which defines how the current state-variables affect the rate of change of the future system states.

`B`—Input Matrix

An $n \times r$ matrix (where r is the number of system inputs), which defines how the current system input-variables affect the rate of change of the future system states.

`C`—Output Matrix

An $m \times n$ matrix (where m is the number of outputs), which defines how the current state-variables affect the outputs of the system.

D—Feedforward Matrix

An $m \times r$ matrix, which defines how the current system input-variables directly affect the outputs of the system.

E—Input Derivative Matrix

An $n \times r$ matrix, which defines how the derivatives of the input-variables affect the rate of change of the future system states. This matrix only occurs for some cases where the LG model contains dependent energy storage elements.

F—Output Derivative Matrix

An $m \times r$ matrix, which defines how the derivative of the input-variables affects the system outputs. This matrix only occurs for some cases where the LG model contains dependent energy storage elements.

B.1.3 Suggested Best Practices

While the LGtheory toolbox is robust in relation to how it handles inputs, there are several best practices that should be employed by the user to ensure that they are providing the inputs to the LG model properly.

The user should specify the ground node in the source and the target vector input fields as a "1." This is required as MATLAB index arrays and matrices start from 1 instead of 0, making the work done by the toolbox much simpler. Each node added to the system will be incremented by 1. If a value for a node is skipped, or the ground node is not specified as "1," it will likely generate an error message or an incorrect result from the toolbox.

For systems that contain transfer elements (transformers or gyrators), the first (input) and the second (output) ports of these two-port elements should be placed in successive order within the toolbox input fields without any other transfer port elements in between. This is because the toolbox couples and evaluates the successive transfer ports in the order in which they are listed in the inputs. Because of this, it is considered best practice to list the two ports of the same transfer element successively (immediately next to each other) in the LG inputs in order to ensure that they will be coupled and evaluated together and correctly. While the successive ordering of the transfer port elements is important, the order of the nodes to which they are attached is not required to be successive (i.e., port 1 can be connected to node 5 and ground, while port 2 is connected to node 8 and the ground).

Similarly, the variable names of the two ports of a transfer element should be differentiated from each other in the Var_Names input field. While this is not required for the transfer elements spanning multiple energy domains because the variable types on the two sides of a domain-transfer element will be different from one another, it is a requirement for single-domain transfer elements. Typically, this is done using an alphabetical differentiation of each port (i.e., TF1*a* and TF1*b*), as transfer element pairs are typically differentiated from one another numerically in systems that contain multiple two-port elements (transfer elements).

When specifying the domain indices of the LG model, it is important that the user defines these index values correctly for each element contained in the system, as the across- and the through-variables associated with each element will be determined based on these values. Likewise, the parameters of specific system element types will be changed based on the requirements of each domain. If the user wishes to leave the domains in their generalized terms in order to evaluate systems outside of the primary domains supported by the toolbox (the primary domains are electrical, mechanical translational and translational, hydraulic/fluid, and thermal), they must keep in mind that the toolbox will be unable to automatically adjust these parameters, and that this must be done manually when the parameter values are being substituted into the state-space model.

Before the user converts their model into inputs for the toolbox, it must be ensured that the LG model is sufficiently simplified to ensure that there are no redundant state-variables (in addition to the number governed by the system order). This is done by combining *A*-type elements that are directly in series and *T*-type elements that are directly in parallel. Similarly, the user must ensure that their model is complete with no loose elements (no nodes with only a single element attached). In either of these cases, the LGtheory toolbox will return an error message.

The equations for converting direct series *A*-type elements and direct parallel *T*-type elements are

$$C_e = \frac{1}{\sum_i \frac{1}{C_i}} \tag{B.1}$$

$$L_e = \frac{1}{\sum_i \frac{1}{L_i}} \tag{B.2}$$

B.1.4 Example of Inputting a Linear Graph Model into the LGtheory Toolbox

A DC motor with an inertial load is used as an example now, to demonstrate the processes associated with the LGtheory toolbox for generating the state-space representation of an LG model.

Based on the LG model of the system presented in Figure B.1, the following inputs should be defined for the LGtheory MATLAB toolbox:

```
LG.S = [2 2 3 4 5 5 5];
LG.T = [1 3 4 1 1 1 1];
LG.Type = [1 5 6 3 3 5 2];
LG.Domain = [1 1 1 1 3 3 3];
syms s R L TFa TFb B J
LG.Var_Names = [s R L TFa TFb B J];
syms i_TFa(t) Tau_TFb(t) Omega_J(t)
LG.y= [i_TFa(t) Tau_TFb(t) Omega_J(t)];
[Model] =LGtheory(LG);
```

For this example, the outputs of the state-space model were specified as the current of the electrical port of the DC motor (i_TFa(t)), the torque output of the motor on the mechanical side (Tau_TFb(t)), and the rotational velocity (angular velocity) of the inertial load (Omega_J(t)).

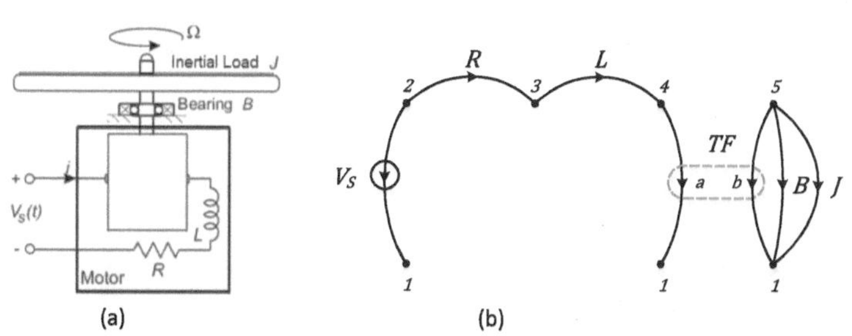

FIGURE B.1 (a) Schematic diagram and (b) LG model of DC motor with inertial load.

B.2 LGTHEORY TOOLBOX FUNCTIONS

This section describes the various processes and algorithms used by the sub-functions of the LG toolbox to convert the LG model representation that is inputted to the LGtheory function into the corresponding state-space model of the system.

B.2.1 Check Model Inputs

```
CheckModel(LG);
```

Before the LG model can be evaluated, the inputs must be checked to ensure that they meet the requirements of the LGtheory toolbox.

First, the toolbox will check to ensure that the model is complete (closed) and that no loose elements (open loops) exist within the model inputs. A loose element can be detected by examining the source and the target vectors for any node values that only occur once between the two vectors. If such a node is found, then it is known that the node has only a single element attached to it, meaning that the model contains an open loop.

Next, the toolbox will check for and the presence of excess (redundant) state-variables. This is done by checking the inputs for A-type elements that are in direct series or for T-type elements that are in direct parallel. If either case is found, an error message is displayed, informing the user of the presence of excess state-variables.

If neither of these cases is detected, the evaluation of the model inputs shall continue.

B.2.2 Conversion to Incidence Matrix Representation

```
[Model]=IncidenceMatrix(LG);
```

In order to mathematically represent the topology and directionality of an LG model in MATLAB, an incidence matrix representation is utilized. Incidence matrices, commonly used in graph theory, are sparse matrices used for representing relationships between two sets of objects. In the case of LG models, an incidence matrix is used to represent the relationship between the system elements (as columns) and the system nodes (as rows). Similarly, the directionality of the system elements is captured in this representation by a "−1" in the row corresponding to the node that the element is leaving (source node) and a "1" in the row corresponding to the node that the element is entering (target node).

The toolbox constructs the incidence matrix of the LG model by initializing a zero matrix with as many rows there are as nodes (determined by the highest number in the source and target vectors) and as many columns there are as elements (determined by the length of the source and the target vectors). The program then indexes through the columns of both vectors and the incidence matrix and assigns a "−1" or "1" in the incidence matrix row indices corresponding to the source and the target vectors, respectively.

From the LG model presented in Figure B.1, and the source and target vectors specified in the example input to MATLAB, the following incidence matrix is produced for the DC motor with the inertial load

$$
\begin{array}{cc}
 & \begin{array}{ccccccc} V_s & R & L & \mathrm{TF}_a & \mathrm{TF}_b & B & J \end{array} \\
\text{In} = \begin{array}{c} 1 \\ 2 \\ 3 \\ 4 \\ 5 \end{array} & \begin{bmatrix} 1 & 0 & 0 & 1 & 1 & 1 & 1 \\ -1 & -1 & 0 & 0 & 0 & 0 & 0 \\ 0 & 1 & -1 & 0 & 0 & 0 & 0 \\ 0 & 0 & 1 & -1 & 0 & 0 & 0 \\ 0 & 0 & 0 & 0 & -1 & -1 & -1 \end{bmatrix}
\end{array}
\tag{B.3}
$$

B.2.3 Building the Normal Tree

```
[Model]=BuildNormalTree(LG,Model);
```

The normal tree is a sub-graph of the LG model, which connects all nodes of the LG while forming no loops (see Appendix A). The normal tree is important in the LG approach as it allows for the classification of the primary and the secondary-variables and also for providing a systematic process of identifying independent and dependent energy storage elements.

The process for constructing the normal tree starts by creating an empty three-dimensional incidence matrix for the tree (and one for the co-tree), which has the same size as the incidence matrix of the LG model but with a depth of 2^T, where T is the number of transfer elements contained within the LG model. This depth is required for evaluating all possible normal tree configurations that can result due to the inclusion of transfer elements, where each page (or layer) of the three-dimensional matrix represents a different permutation of the normal tree (see Figure B.2).

Next, the program adds all A-type (across-type) source elements to each of the normal tree configurations. This is done for each element, by adding the corresponding column of the LG model incidence matrix to the same column of the normal tree matrix.

Transfer elements are then added to the normal tree incidence matrices with only one transfer port and either both or neither gyrator ports being included in the normal tree. This is done in a looping process, which ensures that all possible combinations of the transfer element branches are created throughout the 2^T layers.

The algorithm then cycles through the passive elements in the order: A-type elements, D-type elements, and T-type elements. For each column-wise element that is added to a layer of the normal tree matrix, a depth-first-search loop detecting algorithm is used to determine if the inclusion of the most recent element resulted in the creation of a loop in the normal tree. If no loop is detected, the element remains in the normal tree. If a loop is detected, the last element added to the normal tree is removed from that layer and is instead added to the corresponding layer of the co-tree matrix.

Finally, this process is attempted again for the T-type (through-type) source elements in every layer of the normal tree matrix. Since it is required that no T-type source elements are included in the normal tree, the layer index of any permutation of the normal tree matrix that requires the addition of a T-type source in order to be completed shall be flagged in a logical vector as an invalid normal tree configuration.

Once this process is completed for all elements of the LG model, each normal tree configuration is revaluated for any potential loops that may have been created as the result of the required inclusion of A-type source elements and transfer port elements. Similarly, each configuration is evaluated

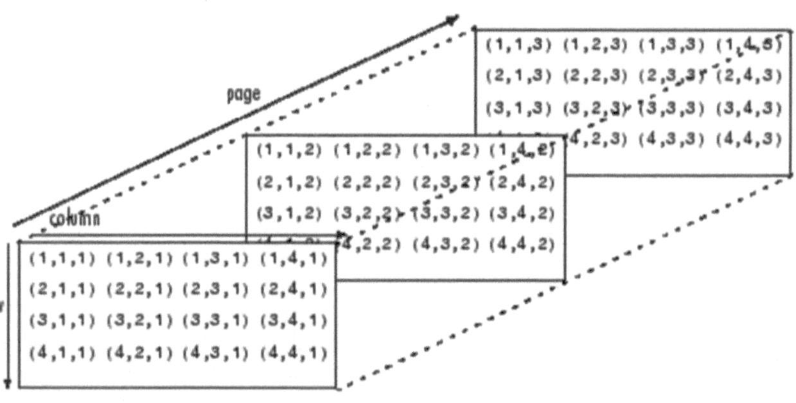

FIGURE B.2 An example of indexing values of a multidimensional array in MATLAB.

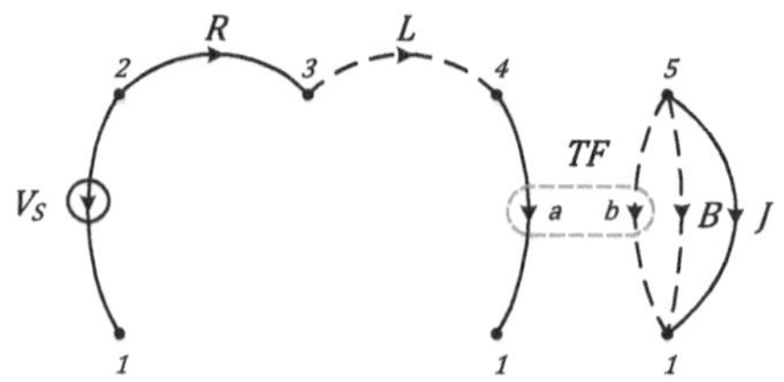

FIGURE B.3 Normal tree of the DC motor with inertial load LG model.

to determine whether all nodes are contained within the normal tree matrix. Any configuration that is determined to violate either of these requirements is assigned a logical value for its layer index, denoting it as being invalid.

In order to make the final decision on which configuration will be selected as the normal tree, the algorithm finds the valid permutations that contain the most *A*-type elements. If there are multiple layers that contain the most *A*-type elements, the algorithm will then select the layer that contains the least *T*-type elements, as this combination of most *A*-type and least *T*-type elements ensures that the program identifies the most independent energy storage elements in the system.

The normal tree resulting from this process for the example system is shown in Figure B.3, where the solid lines represent branches and the dashed lines represent links.

The resulting incidence matrix representation of the normal tree is

$$\text{Tree} = \begin{array}{c} \\ 1 \\ 2 \\ 3 \\ 4 \\ 5 \end{array} \begin{array}{c} \begin{array}{ccccccc} V_s & R & L & \text{TF}_a & \text{TF}_b & B & J \end{array} \\ \begin{bmatrix} 1 & 0 & 0 & 1 & 0 & 0 & 1 \\ -1 & -1 & 0 & 0 & 0 & 0 & 0 \\ 0 & 1 & 0 & 0 & 0 & 0 & 0 \\ 0 & 0 & 0 & -1 & 0 & 0 & 0 \\ 0 & 0 & 0 & 0 & 0 & 0 & -1 \end{bmatrix} \end{array} \quad \text{(B.4)}$$

The corresponding incidence matrix representation of the co-tree is

$$\text{CoTree} = \begin{array}{c} \\ 1 \\ 2 \\ 3 \\ 4 \\ 5 \end{array} \begin{array}{c} \begin{array}{ccccccc} V_s & R & L & \text{TF}_a & \text{TF}_b & B & J \end{array} \\ \begin{bmatrix} 0 & 0 & 0 & 0 & 1 & 1 & 0 \\ 0 & 0 & 0 & 0 & 0 & 0 & 0 \\ 0 & 0 & -1 & 0 & 0 & 0 & 0 \\ 0 & 0 & 1 & 0 & 0 & 0 & 0 \\ 0 & 0 & 0 & 0 & -1 & -1 & 0 \end{bmatrix} \end{array} \quad \text{(B.5)}$$

B.2.4 Variable Classification

```
[Model]=ClassifyVariables(LG,Model);
```

Once the normal tree of the LG model has been identified, it can be utilized to assist in the process of variable classification. First, the toolbox creates two arrays that contain the across- and through-variables of all the system elements in symbolic form, based on the formats specified in Table B.2. For the present example, these two arrays would be as follows:

$$\text{Across\_Vars} = \begin{bmatrix} V_s & V_R & V_L & V_{\text{TF}_a} & \omega_{\text{TF}_b} & \omega_B & \omega_J \end{bmatrix} \tag{B.6}$$

$$\text{Through\_Vars} = \begin{bmatrix} i_s & i_R & i_L & i_{\text{TF}_a} & T_{\text{TF}_b} & T_B & T_J \end{bmatrix} \tag{B.7}$$

Similarly, the two vectors for the primary- and secondary-variables, where the primary-variables are the across-variables of the branches and the through-variables of the links, and the secondary-variables are the through-variables of the branches and the across-variables of the links:

$$\text{Prime} = \begin{bmatrix} V_s & V_R & i_L & V_{\text{TF}_a} & T_{\text{TF}_b} & T_B & \omega_J \end{bmatrix} \tag{B.8}$$

$$\text{Secon} = \begin{bmatrix} i_s & i_R & V_L & i_{\text{TF}_a} & \omega_{\text{TF}_b} & \omega_B & T_J \end{bmatrix} \tag{B.9}$$

Finally, the program identifies the state-variables as a vector containing the across-variables of the *A*-type branches, and the through-variables of the *T*-type links, and the input-variables as a vector containing the source variable type of each source element:

$$\boldsymbol{x} = \begin{bmatrix} \omega_J & i_L \end{bmatrix}^T \tag{B.10}$$

$$\boldsymbol{u} = \begin{bmatrix} V_s \end{bmatrix}^T \tag{B.11}$$

B.2.5 Constitutive Equations

```
[Model]=ElementalEquations(LG,Model);
```

The constitutive equations of the system are created for all passive elements. Once formed, the program rearranges each equation to isolate for the primary-variable (or its derivative) associated with that element.

For the example of the DC motor with inertial load, the constitutive equations are found to be

$$\begin{aligned}
\frac{d\omega_J}{dt} &= \frac{1}{J} T_J \\
\frac{di_L}{dt} &= \frac{1}{L} V_L \\
V_R &= R \cdot i_R \\
T_B &= B \cdot \omega_B \\
V_{\text{TF}_a} &= \text{TF} \cdot \omega_{\text{TF}_b} \\
T_{\text{TF}_b} &= -\text{TF} \cdot i_{\text{TF}_a}
\end{aligned} \tag{B.12}$$

B.2.6 Network Equations

```
[Model]=NetworkEquations(Model);
```

The continuity (node) equations of an LG model are formed using the contouring method. This method involves "virtual cutting" around a node or a set of nodes (as shown in Figure B.4 for the contour of the *K* element) in such a way that only a single branch is intersected by the contour.

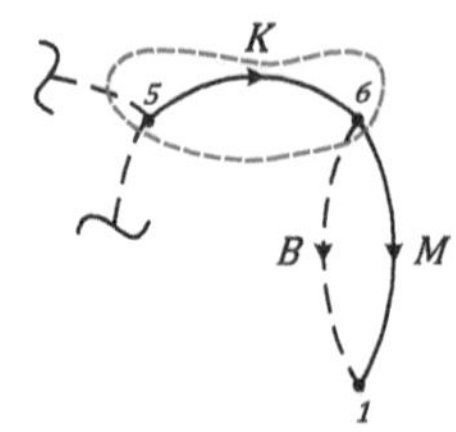

FIGURE B.4 Example of contour enveloping multiple nodes.

This contour can then be treated in a similar manner as a junction in the Kirchhoff's current law, where the sum of all the through-variables entering and exiting the contour is equal to zero. A continuity equation is constructed for each passive branch of the normal tree, where each equation is rearranged to isolate the secondary-variable of the passive branch.

Similarly, the compatibility (loop) equations of an LG model are constructed by temporarily including each passive link into the normal tree and writing the equation of the resulting loop formed from that element's inclusion. This method is treated in a similar manner as a loop in the Kirchhoff's voltage law, where the sum of all across-variables in the loop is equal to zero. A compatibility equation is constructed for each passive link that is not contained in the normal tree, where each equation is rearranged to isolate the passive link's secondary-variable.

Figure B.5 illustrates the contours of each passive normal tree branch (shown by red broken lines), and the re-inclusion of the passive co-tree links (shown by light blue lines) to the LG model.

While the manual process of identify and evaluating the loops and contours of this model is simple, the computer-automation of the process, using a program, for recognizing these patterns can be much more difficult. So, in order to accomplish this using a computer program, the concept of the fundamental cut set is employed.

This concept involves partitioning the LG model's incidence matrix into two sub-matrices of the normal tree and the co-tree. The incidence matrices found when building the normal tree can be used here, but the columns representing the elements that are not contained in either tree must be removed first. Likewise, row one, representing the ground node of the model, must be removed from the matrices. This results in the following sub-matrices

$$A_{\mathrm{tr}} = \begin{array}{c} \\ 2 \\ 3 \\ 4 \\ 5 \end{array} \begin{array}{c} \begin{array}{cccc} V_s & R & \mathrm{TF}_a & J \end{array} \\ \begin{bmatrix} -1 & -1 & 0 & 0 \\ 0 & 1 & 0 & 0 \\ 0 & 0 & -1 & 0 \\ 0 & 0 & 0 & -1 \end{bmatrix} \end{array} \tag{B.13}$$

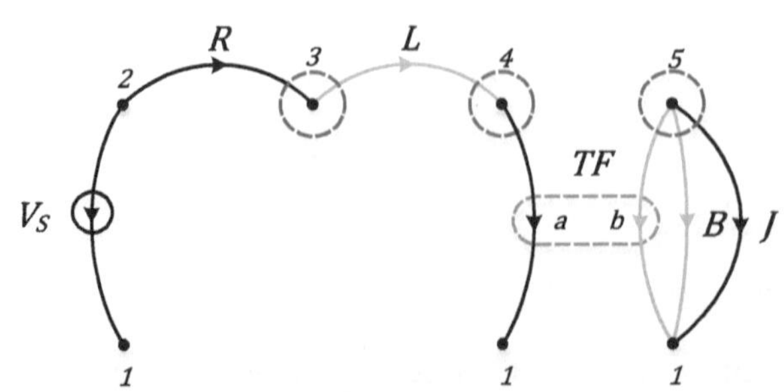

FIGURE B.5 LG model with contour and temporary link inclusion overlay.

$$A_{co} = \begin{array}{c} \\ 2 \\ 3 \\ 4 \\ 5 \end{array} \begin{array}{c} \begin{array}{ccc} L & TF_b & B \end{array} \\ \left[\begin{array}{ccc} 0 & 0 & 0 \\ -1 & 0 & 0 \\ 1 & 0 & 0 \\ 0 & -1 & -1 \end{array}\right] \end{array} \tag{B.14}$$

Next, the fundamental cut set matrix is found as

$$\boldsymbol{F}_c = \begin{bmatrix} \boldsymbol{H} & \boldsymbol{I} \end{bmatrix} \tag{B.15}$$

where

$$\boldsymbol{H} = \boldsymbol{A}_{tr}^{-1}\boldsymbol{A}_{co} \tag{B.16}$$

Here, $\boldsymbol{I}$ is an identity matrix of corresponding size to $\boldsymbol{H}$. For the present example, the fundamental cut set matrix will be

$$\boldsymbol{F}_c = \begin{bmatrix} 1 & 0 & 0 & 1 & 0 & 0 & 0 \\ -1 & 0 & 0 & 0 & 1 & 0 & 0 \\ -1 & 0 & 0 & 0 & 0 & 1 & 0 \\ 0 & 1 & 1 & 0 & 0 & 0 & 1 \end{bmatrix} \tag{B.17}$$

Next, the across- and through-variables of the LG model must be separated based on whether their corresponding elements exist in the normal tree or the co-tree. The following is the resulting vectors for the present example:

$$\boldsymbol{v}_{tr} = \begin{bmatrix} V_s \\ V_R \\ V_{TF_a} \\ \omega_J \end{bmatrix}, \boldsymbol{v}_{co} = \begin{bmatrix} V_L \\ \omega_{TF_b} \\ \omega_B \end{bmatrix} \tag{B.18}$$

$$\boldsymbol{f}_{tr} = \begin{bmatrix} i_s \\ i_R \\ i_{TF_a} \\ T_J \end{bmatrix}, \boldsymbol{f}_{co} = \begin{bmatrix} i_L \\ T_{TF_b} \\ T_B \end{bmatrix} \tag{B.19}$$

The continuity equations for each passive branch (and, in this case, the across-variable source element) can subsequently be found as

$$\boldsymbol{f}_{tr} = -\boldsymbol{H}\boldsymbol{f}_{co} \tag{B.20}$$

For the present example, this would result in:

$$\begin{bmatrix} i_s \\ i_R \\ i_{TF_a} \\ T_J \end{bmatrix} = -\begin{bmatrix} 1 & 0 & 0 \\ -1 & 0 & 0 \\ -1 & 0 & 0 \\ 0 & 1 & 1 \end{bmatrix}\begin{bmatrix} i_L \\ T_{TF_b} \\ T_B \end{bmatrix} \tag{B.21}$$

$$i_s = -i_L$$

$$i_R = i_L$$

$$i_{\mathrm{TF}_a} = i_L$$

$$T_J = -T_{\mathrm{TF}_b} - T_B \tag{B.22}$$

Subsequently, the compatibility equation for each passive link can be found as

$$\boldsymbol{v}_{\mathrm{co}} = \boldsymbol{H}^T \boldsymbol{v}_{\mathrm{tr}} \tag{B.23}$$

For the present example, this would result in

$$\begin{bmatrix} V_L \\ \omega_{\mathrm{TF}_b} \\ \omega_B \end{bmatrix} = \begin{bmatrix} 1 & -1 & -1 & 0 \\ 0 & 0 & 0 & 1 \\ 0 & 0 & 0 & 1 \end{bmatrix} \begin{bmatrix} V_s \\ V_R \\ V_{\mathrm{TF}_a} \\ \omega_J \end{bmatrix} \tag{B.24}$$

$$V_L = V_s - V_R - V_{\mathrm{TF}_a}$$

$$\omega_{\mathrm{TF}_b} = \omega_J$$

$$\omega_B = \omega_J \tag{B.25}$$

B.2.7 Creating the State-Space Matrices

```
[Model]=StateSpaceMatrices(LG,Model);
```

On construction of the constitutive, continuity, and compatibility equations, a symbolic substitution of the continuity and compatibility equations into the constitutive equations is performed in order to reduce the set of equations and eliminate all the secondary (redundant, auxiliary) variables.

$$\frac{d\omega_J}{dt} = \frac{-T_B - T_{\mathrm{TF}_b}}{J} \tag{B.26}$$

$$\frac{di_L}{dt} = \frac{-V_R - V_{\mathrm{TF}_a} + V_s}{L} \tag{B.27}$$

$$V_R = R \cdot i_L \tag{B.28}$$

$$T_B = B \cdot \omega_J \tag{B.29}$$

$$V_{\mathrm{TF}_a} = \mathrm{TF}_a \cdot \omega_J \tag{B.30}$$

$$T_{\mathrm{TF}_b} = -\mathrm{TF}_a \cdot i_L \tag{B.31}$$

The newly created equations are then classified into one of three column vectors depending on the isolated primary-variable associated with the element: vector $\boldsymbol{x}$ for primary-variables of independent storage elements (state-variables); vector $\boldsymbol{d}$ for primary-variables of dependent storage elements; and vector $\boldsymbol{p}$ for primary-variables of non-energy storage elements. For the present example, the column vector $\boldsymbol{x}$ will consist of equations (3.26) and (3.27), the column vector $\boldsymbol{d}$ will be empty as there are no dependent storage elements (as determined by the normal tree) and the column vector $\boldsymbol{p}$ will consist of equations (B.27)–(B.31). These vectors can be written as the following matrix equations:

$$\dot{\boldsymbol{x}} = \boldsymbol{Px} + \boldsymbol{Qp} + \boldsymbol{Rd} + \boldsymbol{Su} \tag{B.32}$$

$$\boldsymbol{d} = \boldsymbol{M}\dot{\boldsymbol{x}} + \boldsymbol{N}\dot{\boldsymbol{u}} \tag{B.33}$$

$$\boldsymbol{p} = \boldsymbol{Hx} + \boldsymbol{Jp} + \boldsymbol{Kd} + \boldsymbol{Lu} \tag{B.34}$$

Therefore,

$$\dot{\boldsymbol{x}} = [0]_{2\times 2}\,\boldsymbol{x} + \begin{bmatrix} 0 & -\frac{1}{J} & 0 & -\frac{1}{J} \\ -\frac{1}{L} & 0 & -\frac{1}{L} & 0 \end{bmatrix} \boldsymbol{p} + [0]\boldsymbol{d} + \begin{bmatrix} 0 \\ \frac{1}{L} \end{bmatrix} \boldsymbol{u} \tag{B.35}$$

$$\boldsymbol{d} = [0]\dot{\boldsymbol{x}} + [0]\dot{\boldsymbol{u}} \tag{B.36}$$

$$\boldsymbol{p} = \begin{bmatrix} 0 & R \\ B & 0 \\ \mathrm{TF}_a & 0 \\ 0 & -\mathrm{TF}_a \end{bmatrix} \boldsymbol{x} + [0]_{4\times 4}\,\boldsymbol{p} + [0]\boldsymbol{d} + [0]_{4\times 1}\,\boldsymbol{u} \tag{B.37}$$

The general solution to the state-space equation is formed by isolating $\boldsymbol{d}$ in (3.33) and $\boldsymbol{p}$ in (3.34) and substituting the results into (3.32). Once simplified, this process results in the following general formulation of the state-space model:

$$\dot{\boldsymbol{x}} = \boldsymbol{Ax} + \boldsymbol{Bu} + \boldsymbol{E}\dot{\boldsymbol{u}} \tag{B.38}$$

where

$$\boldsymbol{A} = \left[\boldsymbol{I} - (\boldsymbol{QK}' + \boldsymbol{R})\boldsymbol{M}\right]^{-1}(\boldsymbol{P} + \boldsymbol{QH}') \tag{B.39}$$

$$\boldsymbol{B} = \left[\boldsymbol{I} - (\boldsymbol{QK}' + \boldsymbol{R})\boldsymbol{M}\right]^{-1}(\boldsymbol{S} + \boldsymbol{QL}') \tag{B.40}$$

$$\boldsymbol{E} = \left[\boldsymbol{I} - (\boldsymbol{QK}' + \boldsymbol{R})\boldsymbol{M}\right]^{-1}(\boldsymbol{R} + \boldsymbol{QK}')\boldsymbol{N} \tag{B.41}$$

And

$$\boldsymbol{K}' = [\boldsymbol{I} - \boldsymbol{J}]^{-1}\boldsymbol{K} \tag{B.42}$$

$$H' = [I - J]^{-1} H \tag{B.43}$$

$$L' = [I - J]^{-1} L \tag{B.44}$$

Depending on the system that is being considered, this general solution can be simplified in the following two scenarios:

1. If the system contains no dependent energy storage elements (i.e., $\boldsymbol{d} = 0$), the general solution can be simplified by eliminating $\boldsymbol{R}$, $\boldsymbol{M}$, $\boldsymbol{N}$, and $\boldsymbol{K}$. This results in the following state-space model matrices:

$$A = P + QH' \tag{B.45}$$

$$B = S + QL' \tag{B.46}$$

2. If the system contains dependent energy storage elements (i.e., $\boldsymbol{d} \neq 0$) but contains no input derivatives (i.e., $\dot{\boldsymbol{u}} = 0$), the general solution can be simplified by eliminating $\boldsymbol{N}$. This results in the elimination of the $\boldsymbol{E}$ matrix, while $\boldsymbol{A}$ and $\boldsymbol{B}$ are still calculated using (3.39) and (3.40), respectively.

As previously stated, for the present example, the LGtheory toolbox program determines from the normal tree that it contains no dependent energy storage elements ($\boldsymbol{d} = 0$), meaning that this system falls into scenario 1, as described above. The MATLAB program subsequently extracts the necessary matrices and performs calculations for the state-space matrices using (B.45) and (B.46), to obtain:

$$A = [0]_{2\times 2} + \begin{bmatrix} 0 & -\frac{1}{J} & 0 & -\frac{1}{J} \\ -\frac{1}{L} & 0 & -\frac{1}{L} & 0 \end{bmatrix} \left[I_{4\times 4} - [0]_{4\times 4} \right]^{-1} \begin{bmatrix} 0 & R \\ B & 0 \\ \mathrm{TF}_a & 0 \\ 0 & -\mathrm{TF}_a \end{bmatrix} \tag{B.47}$$

$$A = \begin{bmatrix} -\frac{B}{J} & \frac{\mathrm{TF}}{J} \\ -\frac{\mathrm{TF}}{L} & -\frac{R}{L} \end{bmatrix} \tag{B.48}$$

and

$$B = \begin{bmatrix} 0 \\ \frac{1}{L} \end{bmatrix} + \begin{bmatrix} 0 & -\frac{1}{J} & 0 & -\frac{1}{J} \\ -\frac{1}{L} & 0 & -\frac{1}{L} & 0 \end{bmatrix} \left[I_{4\times 4} - [0]_{4\times 4} \right]^{-1} [0]_{4\times 1} \tag{B.49}$$

$$B = \begin{bmatrix} 0 \\ \frac{1}{L} \end{bmatrix} \tag{B.50}$$

The output equations are then constructed as an algebraic relationship between the variables of interest, as defined by the user in the output array, and the state- and input-variables. This is achieved in LGtheory by examining the continuity and compatibility equations and also the substituted constitutive equations from this section and selecting the equations that can be isolated for the desired output variables. Once these equations are identified, substitution and manipulation operations are conducted in order to express the output variables exclusively in terms of the state- and input-variables; the $\boldsymbol{C}$ and $\boldsymbol{D}$, and potentially $\boldsymbol{F}$, matrices are extracted from these equations.

For the present example, the variables of interest have been specified as the current supplied to the motor, the torque output by the motor, and the rotational (angular) velocity of the inertial load. For these output variables, the following $\boldsymbol{C}$ and $\boldsymbol{D}$ matrices are produced:

$$\boldsymbol{C} = \begin{bmatrix} 0 & 1 \\ 0 & -\mathrm{TF} \\ 1 & 0 \end{bmatrix} \quad \boldsymbol{D} = 0 \tag{B.51}$$

B.2.8 Standard State-Space Form Conversion

```
[Model] =StandardForm(Model);
```

In some LG models, the state-space matrices produced by the toolbox will result in a non-standard form of state-space model:

$$\dot{x} = \boldsymbol{A}x + \boldsymbol{B}u + \boldsymbol{E}\dot{u}$$

$$y = \boldsymbol{C}x + \boldsymbol{D}u + \boldsymbol{F}\dot{u} \tag{B.52}$$

The additional matrices, $\boldsymbol{E}$ and $\boldsymbol{F}$, represent the effects that the derivatives of the input variables have on the rate of change of the state-variables and on the output-variables.

While the main LGtheory function of the toolbox will return all state matrices in a non-standard form (assuming the input LG model results in the additional matrices), the StandardForm function is included in the LGtheory toolbox library in order to automate the conversion of a non-standard state-space model into the standard form.

This function works by transforming the state-variables of the system, and also the output ($\boldsymbol{B}$) and feedforward matrices ($\boldsymbol{D}$), to a modified set that accounts for the derivative(s) of the input variable(s) and eliminates the $\boldsymbol{E}$ matrix:

$$\boldsymbol{x}' = \boldsymbol{x} - \boldsymbol{E}\boldsymbol{u} \tag{B.53}$$

$$\boldsymbol{B}' = \boldsymbol{A}\boldsymbol{E} + \boldsymbol{B} \tag{B.54}$$

$$\boldsymbol{D}' = \boldsymbol{C}\boldsymbol{E} + \boldsymbol{D} \tag{B.55}$$

The state-space model can then be expressed as follows:

$$\dot{x} = \boldsymbol{A}x' + \boldsymbol{B}'u$$

$$y = \boldsymbol{C}x' + \boldsymbol{D}'u + \boldsymbol{F}\dot{u} \tag{B.56}$$

B.3 ADDITIONAL EXAMPLES USING LGTHEORY TOOLBOX

B.3.1 Example 1: Translational Mechanical System

The following system consists of two mass elements attached to each other through a spring, and each mass element is attached to ground through a damper element (Figure B.6).

The following are the inputs to the LGtheory toolbox for this system:

```
LG.S = [1 2 2 2 3 3 1]; %Source vector
LG.T = [2 1 1 3 1 1 3]; %Target vector
LG.Type = [7 2 5 6 2 5 7]; %Type vector
LG.Domain = [2 2 2 2 2 2 2]; %Domain vector
syms m m_m b_m K m_c b_c c
LG.Var_Names = [m m_m b_m K m_c b_c c];
syms v_m_m(t) v_m_c(t)
LG.y = [v_m_m(t) v_m_c(t)];
[Model] = LGtheory(LG);
```

The following equations represent the state-space model produced by the toolbox with outputs specified as the velocities of the mass elements:

$$\begin{bmatrix} \dot{v}_{m_m} \\ \dot{v}_{m_c} \\ \dot{F}_K \end{bmatrix} = \begin{bmatrix} -\frac{b_m}{m_m} & 0 & -\frac{1}{m_m} \\ 0 & -\frac{b_c}{m_c} & \frac{1}{m_c} \\ K & -K & 0 \end{bmatrix} \begin{bmatrix} v_{m_m} \\ v_{m_c} \\ F_K \end{bmatrix} + \begin{bmatrix} \frac{1}{m_m} & 0 \\ 0 & \frac{1}{m_c} \\ 0 & 0 \end{bmatrix} \begin{bmatrix} F_m \\ F_c \end{bmatrix} \quad \text{(B.57)}$$

$$\begin{bmatrix} v_{m_m} \\ v_{m_c} \end{bmatrix} = \begin{bmatrix} 1 & 0 & 0 \\ 0 & 1 & 0 \end{bmatrix} \begin{bmatrix} v_{m_m} \\ v_{m_c} \\ F_K \end{bmatrix} + \begin{bmatrix} 0 & 0 \\ 0 & 0 \end{bmatrix} \begin{bmatrix} F_m \\ F_c \end{bmatrix} \quad \text{(B.58)}$$

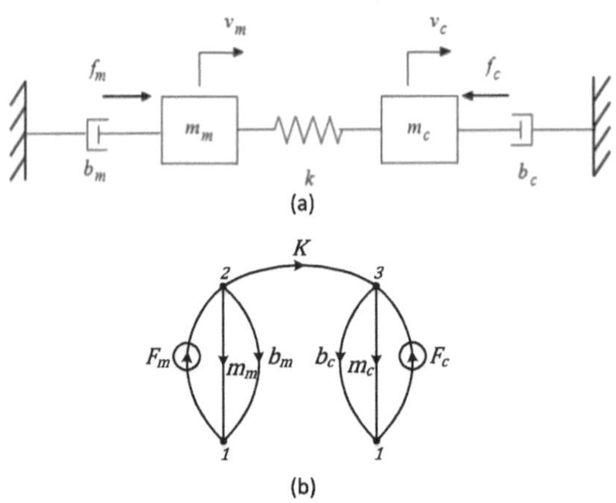

FIGURE B.6 (a) System model and (b) LG model of a translational mechanical system.

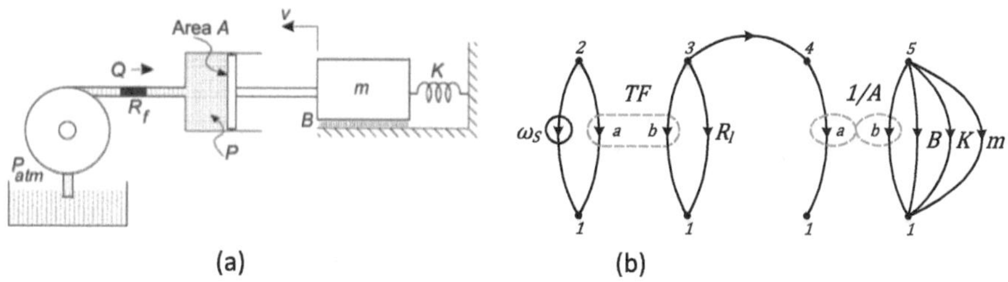

FIGURE B.7 (a) System schematic model and (b) LG model of a hydro-mechanical system.

B.3.2 Example 2: Hydro-Mechanical System

The following multi-physics (multi-domain or mixed) system consists of a rotational torque source, representing the power from an electric motor, powering a positive-displacement pump and a piston, which actuates a mass element attached to ground through a spring. This is translational (rectilinear) system (Figure B.7).

The following are the inputs to the LGtheory toolbox for this system:

```
LG.S = [2 2 3 3 3 4 5 5 5 5];
LG.T = [1 1 1 1 4 1 1 1 1 1];
LG.Type = [1 3 3 5 5 4 4 5 6 2];
LG.Domain = [3 3 4 4 4 4 2 2 2 2];
syms s TF R_l R_f A B K m
LG.Var_Names = [s TF TF R_l R_f 1/A 1/A B K m];
syms P_R_f(t) v_m(t)
LG.y = [P_R_f(t) v_m(t)];
[Model] = LGtheory(LG);
```

The following equations represent the state-space model that is produced by the toolbox, with the outputs specified as the pressure of the fluid between the pump and the piston, and the velocity of the mass element:

$$\begin{bmatrix} \dot{v}_m \\ \dot{F}_k \end{bmatrix} = \begin{bmatrix} \dfrac{A^2\left(R_f + R_l - R_f R_l TF\right)}{m(R_l TF - 1)} - \dfrac{B}{m} & -\dfrac{1}{m} \\ K & 0 \end{bmatrix} \begin{bmatrix} v_m \\ F_k \end{bmatrix} + \begin{bmatrix} 0 \\ 0 \end{bmatrix} [\omega_s] \tag{B.59}$$

$$\begin{bmatrix} P_{R_f} \\ v_m \end{bmatrix} = \begin{bmatrix} -AR_f & 0 \\ 1 & 0 \end{bmatrix} \begin{bmatrix} v_m \\ F_k \end{bmatrix} + \begin{bmatrix} 0 \\ 0 \end{bmatrix} [\omega_s] \tag{B.60}$$

Index